# Renewable Energy Resources

*Renewable Energy Resources* is a numerate and quantitative text covering subjects of proven technical and economic importance worldwide. Energy supplies from renewables (such as solar, thermal, photovoltaic, wind, hydro, biofuels, wave, tidal, ocean and geothermal sources) are essential components of every nation's energy strategy, not least because of concerns for the environment and for sustainability. In the years between the first and this second edition, renewable energy has come of age: it makes good sense, good government and good business.

This second edition maintains the book's basis on fundamentals, whilst including experience gained from the rapid growth of renewable energy technologies as secure national resources and for climate change mitigation, more extensively illustrated with case studies and worked problems. The presentation has been improved throughout, along with a new chapter on economics and institutional factors. Each chapter begins with fundamental theory from a scientific perspective, then considers applied engineering examples and developments, and includes a set of problems and solutions and a bibliography of printed and web-based material for further study. Common symbols and cross referencing apply throughout, essential data are tabulated in appendices. Sections on social and environmental aspects have been added to each technology chapter.

*Renewable Energy Resources* supports multi-disciplinary master degrees in science and engineering, and specialist modules in first degrees. Practising scientists and engineers who have not had a comprehensive training in renewable energy will find this book a useful introductory text and a reference book.

**John Twidell** has considerable experience in renewable energy as an academic professor, a board member of wind and solar professional associations, a journal editor and contractor with the European Commission. As well as holding posts in the UK, he has worked in Sudan and Fiji.

**Tony Weir** is a policy adviser to the Australian government, specialising in the interface between technology and policy, covering subjects such as energy supply and demand, climate change and innovation in business. He was formerly Senior Energy Officer at the South Pacific Forum Secretariat in Fiji, and has lectured and researched in physics and policy studies at universities of the UK, Australia and the Pacific.

# Renewable Energy Resources

Second edition

John Twidell and Tony Weir

LONDON AND NEW YORK

First published 1986
by E&FN Spon Ltd
Second edition published 2006
by Taylor & Francis
2 Park Square, Milton Park, Abingdon, Oxon OX14 4RN

Simultaneously published in the USA and Canada
by Taylor & Francis
270 Madison Ave, New York, NY 10016, USA

*Taylor & Francis is an imprint of the Taylor & Francis Group*



Typeset in Sabon by
Integra Software Services Pvt. Ltd, Pondicherry, India
Printed and bound in Great Britain by
MPG Books Ltd, Bodmin



*British Library Cataloguing in Publication Data*
A catalogue record for this book is available
from the British Library

*Library of Congress Cataloging in Publication Data*
Twidell, John.
Renewable energy resources / John Twidell and
Anthony Weir. — 2nd ed.
p. cm.
Includes bibliographical references and index.
ISBN 0–419–25320–3 (hardback) — ISBN 0–419–25330–0 (pbk.)
1. Renewable energy sources. I. Weir, Anthony D. II. Title.

TJ808.T95 2005
621.042—dc22

2005015300

ISBN10: 0–419–25320–3 ISBN13: 978–0–419–25320–4 Hardback
ISBN10: 0–419–25330–0 ISBN13: 978–0–419–25330–3 Paperback

# Contents

# Preface

## Our aim

*Renewable Energy Resources* is a numerate and quantitative text covering subjects of proven technical and economic importance worldwide. Energy supply from renewables is an essential component of every nation's strategy, especially when there is responsibility for the environment and for sustainability.

This book considers the timeless principles of renewable energy technologies, yet seeks to demonstrate modern application and case studies. *Renewable Energy Resources* supports multi-disciplinary master degrees in science and engineering, and also specialist modules in science and engineering first degrees. Moreover, since many practising scientists and engineers will not have had a general training in renewable energy, the book has wider use beyond colleges and universities. Each chapter begins with fundamental theory from a physical science perspective, then considers applied examples and developments, and finally concludes with a set of problems and solutions. The whole book is structured to share common material and to relate aspects together. After each chapter, reading and web-based material is indicated for further study. Therefore the book is intended both for basic study and for application. Throughout the book and in the appendices, we include essential and useful reference material.

## The subject

Renewable energy supplies are of ever increasing environmental and economic importance in all countries. A wide range of renewable energy technologies are established commercially and recognised as growth industries by most governments. World agencies, such as the United Nations, have large programmes to encourage the technology. In this book we stress the scientific understanding and analysis of renewable energy, since we believe these are distinctive and require specialist attention. The subject is not easy, mainly because of the spread of disciplines involved, which is why we aim to unify the approach within one book.

This book bridges the gap between descriptive reviews and specialised engineering treatises on particular aspects. It centres on demonstrating how fundamental physical processes govern renewable energy resources and their application. Although the applications are being updated continually, the fundamental principles remain the same and we are confident that this new edition will continue to provide a useful platform for those advancing the subject and its industries. We have been encouraged in this approach by the ever increasing commercial importance of renewable energy technologies.

## Why a second edition?

In the relatively few years between the first edition, with five reprinted revisions, and this second edition, renewable energy has come of age; its use makes good sense, good government and good business. From being (apart from hydro-power) small-scale 'curiosities' promoted by idealists, renewables have become *mainstream technologies*, produced and operated by companies competing in an increasingly open market where consumers and politicians are very conscious of sustainability issues.

In recognition of the social, political and institutional factors which continue to drive this change, this new edition includes a new final chapter on institutional and economic factors. The new chapter also discusses and demonstrates some tools for evaluating the increasingly favourable economics of renewable energy systems. There is also a substantial new section in Chapter 1 showing how renewable energy is a key component of sustainable development, an ideal which has become much more explicit since the first edition. Each technology chapter now includes a brief concluding section on its social and environmental impacts.

The book maintains the same general format as the first edition, but many improvements and updates have been made. In particular we wish to relate to the vibrant developments in the individual renewable energy technologies, and to the related commercial growth. We have improved the presentation of the fundamentals throughout, in the light of our teaching experience. Although the book continues to focus on fundamental physical principles, which have not changed, we have updated the technological applications and their relative emphases to reflect market experience. For electricity generation, wind-power and photovoltaics have had dramatic growth over the last two decades, both in terms of installed capacity and in sophistication of the industries. In all aspects of renewable energy, composite materials and microelectronic control have transformed traditional technologies, including hydro-power and the use of biomass.

Extra problems have been added at the end of each chapter, with hints and guidance for all solutions as an appendix. We continue to emphasise simplified, order-of-magnitude, calculations of the potential outputs of the various technologies. Such calculations are especially useful in indicating the potential applicability of a technology for a particular site. However we appreciate that specialists increasingly use computer modelling of whole, complex systems; in our view such modelling is essential but only after initial calculation as presented here.

## Readership

We expect our readers to have a basic understanding of science and technology, especially of physical science and mathematics. It is not necessary to read or refer to chapters consecutively, as each aspect of the subject is treated, in the main, as independent of the other aspects. However, some common elements, especially heat transfer, will have to be studied seriously if the reader is to progress to any depth of understanding in solar energy. The disciplines behind a proper understanding and application of renewable energy also include environmental science, chemistry and engineering, with social science vital for dissemination. We are aware that readers with a physical science background will usually be unfamiliar with life science and agricultural science, but we stress the importance of these subjects with obvious application for biofuels and for developments akin to photosynthesis. We ourselves see renewable energy as within human-inclusive ecology, both now and for a sustainable future.

## Ourselves

We would like our readers to enjoy the subject of renewable energy, as we do, and to be stimulated to apply the energy sources for the benefit of their societies. Our own interest and commitment has evolved from the work in both hemispheres and in a range of countries. We first taught, and therefore learnt, renewable energy at the University of Strathclyde in Glasgow (JWT) and the University of the South Pacific in Fiji (ADW and JWT). So teaching, together with research and application in Scotland and the South Pacific, has been a strong influence for this book. Since the first edition we have made separate careers in universities and in government service, whilst experiencing the remarkable, but predicable, growth in relevance of renewable energy. One of us (JWT) became Director of the Energy Studies Unit, in the Faculty of Engineering at the University of Strathclyde in Glasgow, Scotland, and then accepted the Chair in Renewable Energy at the AMSET Centre, De Montfort University, Leicester, England. He is editor of the academic journal *Wind Engineering*, has been a Council and Board member of the British Wind Energy Association and the UK Solar Energy Society, and has supervised many postgraduates for their dissertations. The AMSET Centre is now a private company, for research, education and training in renewables; support is given to MSc courses at Reading University, Oxford University and City University, and there are European

Union–funded research programmes. TW was for several years the Senior Energy Officer of the South Pacific Forum Secretariat, where he managed a substantial program of renewable energy pilot projects. He then worked for the Australian Government as an adviser on climate change, and later on new economy issues.

We do not see the world as divided sharply between developed industrialised countries and developing countries of the Third World. Renewables are essential for both, and indeed provide one way for the separating concepts to become irrelevant. This is meaningful to us personally, since we wish our own energies to be directed for a just and sustainable society, increasingly free of poverty and the threat of cataclysmic war. We sincerely believe the development and application of renewable energy technology will favour these aspirations. Our readers may not share these views, and this fortunately does not affect the content of the book. One thing they will have to share, however, is contact with the outdoors. Renewable energy is drawn from the environment, and practitioners must put on their rubber boots or their sun hat and move from the closed environment of buildings to the outside. This is no great hardship however; the natural environment is the joy and fulfilment of renewables.

## Suggestions for using the book in teaching

How a book is used in teaching depends mainly on how much time is devoted to its subject. For example, the book originated from short and one-semester courses to senior undergraduates in Physics at the University of the South Pacific and the University of Strathclyde, namely 'Energy Resources and Distribution', 'Renewable Energy' and 'Physics and Ecology'. When completed and with regular revisions, the book has been mostly used worldwide for MSc degrees in engineering and science, including those on 'renewable energy' and on 'energy and the environment'. We have also taught other lecture and laboratory courses, and have found many of the subjects and technologies in renewable energy can be incorporated with great benefit into conventional teaching.

This book deliberately contains more material than could be covered in one specialist course. This enables the instructor and readers to concentrate on those particular energy technologies appropriate in their situation. To assist in this selection, each chapter starts with a preliminary outline and estimate of each technology's resource and geographical variation, and ends with a discussion of its social and environmental aspects.

The chapters are broadly grouped into similar areas. Chapter 1 (Principles of Renewable Energy) introduces renewable energy supplies in general, and in particular the characteristics that distinguish their application from that for fossil or nuclear fuels. Chapter 2 (Fluid Mechanics) and Chapter 3 (Heat Transfer) are background material for later chapters. They contain nothing that a senior student in mechanical engineering will not already know. Chapters 4–7 deal with various aspects of direct solar energy. Readers interested in this area are advised to start with the early sections of Chapter 5 (Solar Water Heating) or Chapter 7 (Photovoltaics), and review Chapters 3 and 4 as required. Chapters 8 (Hydro), 9 (Wind), 12 (Waves) and 13 (Tides) present applications of fluid mechanics. Again the reader is advised to start with an applications chapter, and review the elements from Chapter 2 as required. Chapters 10 and 11 deal with biomass as an energy source and how the energy is stored and may be used. Chapters 14 (OTEC) and 15 (Geothermal) treat sources that are, like those in Chapters 12 (wave) and 13 (tidal), important only in fairly limited geographical areas. Chapter 16, like Chapter 1, treats matters of importance to all renewable energy sources, namely the storage and distribution of energy and the integration of energy sources into energy systems. Chapter 17, on institutional and economic factors bearing on renewable energy, recognises that science and engineering are not the only factors for implementing technologies and developments. Appendices A (units), B (data) and C (heat transfer formulas) are referred to either implicitly or explicitly throughout the book. We keep to a common set of symbols throughout, as listed in the front. Bibliographies include both specific and general references of conventional publications and of websites; the internet is particularly valuable for seeking applications. Suggestions for further reading and problems (mostly numerical in nature) are included with most chapters. Answer guidance is provided at the end of the book for most of the problems.

## Acknowledgements

As authors we bear responsibility for all interpretations, opinions and errors in this work. However, many have helped us, and we express our gratitude to them. The first edition acknowledged the many students, colleagues and contacts that had helped and encouraged us at that stage. For this second edition, enormously more information and experience has been available, especially from major international and national R&D and from commercial experience, with significant information available on the internet. We acknowledge the help and information we have gained from many such sources, with specific acknowledgement indicated by conventional referencing and listing in the bibliographies. We welcome communications from our readers, especially when they point out mistakes and possible improvement.

Much of TW's work on this second edition was done while he was on leave at the International Global Change Institute of the University of Waikato, New Zealand, in 2004. He gratefully acknowledges the academic hospitality of Neil Ericksen and colleagues, and the continuing support of the [Australian Government] Department of Industry Tourism and

Resources. JWT is especially grateful for the comments and ideas from students of his courses.

And last, but not least, we have to thank a succession of editors at Spon Press and Taylor & Francis and our families for their patience and encouragement. Our children were young at the first edition, but had nearly all left home at the second; the third edition will be for their future generations.

*John Twidell MA DPhil*

AMSET Centre, Horninghold
Leicestershire, LE16 8DH, UK

and

Visiting Professor in Renewable Energy
University of Reading, UK

email <info.renewable@tandf.co.uk>
see <www.amset.com>

*A.D. (Tony) Weir BSc PhD*

Canberra
Australia

# List of symbols

| Symbol | Main use | Other use or comment |
|---|---|---|
| *Capitals* | | |
| $A$ | Area ($m^2$) | Acceptor; ideality factor |
| AM | Air-mass-ratio | |
| $C$ | Thermal capacitance ($J\,K^{-1}$) | Electrical capacitance ($F$); constant |
| $C_P$ | Power coefficient | |
| $C_r$ | Concentration ratio | |
| $C_\Gamma$ | Torque coefficient | |
| $D$ | Distance (m) | Diameter (of pipe or blade) |
| $E$ | Energy (J) | |
| $E_F$ | Fermi level | |
| $E_g$ | Band gap (eV) | |
| $E_K$ | Kinetic energy (J) | |
| EMF | Electromotive force (V) | |
| $F$ | Force (N) | Faraday constant ($C\,mol^{-1}$) |
| $F'_{ij}$ | | Radiation exchange factor ($i$ to $j$) |
| $G$ | Solar irradiance ($W\,m^{-2}$) | Gravitational constant ($N\,m^2\,kg^{-2}$); Temperature gradient ($K\,m^{-1}$); Gibbs energy |
| $G_b, G_d, G_h$ | Irradiance (beam, diffuse, on horizontal) | |
| $H$ | Enthalpy (J) | Head (pressure height) of fluid (m); wave crest height (m); insolation ($J\,m^{-2}\,day^{-1}$); heat of reaction ($\Delta H$) |
| $I$ | Electric current (A) | Moment of inertia ($kg\,m^2$) |
| $J$ | Current density ($A\,m^{-2}$) | |
| $K$ | Extinction coefficient ($m^{-1}$) | Clearness index ($K_T$); constant |
| $L$ | Distance, length (m) | Diffusion length (m); litre ($10^{-3}\,m^3$) |
| $M$ | Mass (kg) | Molecular weight |
| $N$ | Concentration ($m^{-3}$) | Hours of daylight |
| $N_0$ | Avogadro number | |
| $P$ | Power (W) | |
| $P'$ | Power per unit length ($W\,m^{-1}$) | |
| PS | Photosystem | |

(Continued)

| *Symbol* | *Main use* | *Other use or comment* |
|---|---|---|
| $Q$ | Volume flow rate ($m^3\,s^{-1}$) | |
| $R$ | Thermal resistance ($K\,W^{-1}$) | Radius (m); electrical resistance ($\Omega$); reduction level; tidal range (m); gas constant ($R_0$); |
| $R_m$ | Thermal resistance (*m*ass transfer) | |
| $R_n$ | Thermal resistance (co*n*duction) | |
| $R_r$ | Thermal resistance (*r*adiation) | |
| $R_v$ | Thermal resistance (con*v*ection) | |
| RFD | Radiant flux density ($W\,m^{-2}$) | |
| $S$ | Surface area ($m^2$) | entropy |
| $S_v$ | Surface recombination velocity ($m\,s^{-1}$) | |
| STP | Standard temperature and pressure | |
| $T$ | Temperature (K) | Period ($s^{-1}$) |
| $U$ | Potential energy (J) | Heat loss coefficient ($W\,m^{-2}\,K^{-1}$) |
| $V$ | Volume ($m^3$) | Electrical potential (V) |
| $W$ | Width (m) | Energy density ($J\,m^{-3}$) |
| $X$ | Characteristic dimension (m) | Concentration ratio |
| *Script capitals* | (Non-dimensional numbers characterising fluid flow) | |
| $\mathscr{A}$ | Rayleigh number | |
| $\mathscr{G}$ | Grashof number | |
| $\mathscr{N}$ | Nusselt number | |
| $\mathscr{P}$ | Prandtl number | |
| $\mathscr{R}$ | Reynolds number | |
| $\mathscr{S}$ | Shape number (of turbine) | |
| *Lower case* | | |
| $a$ | Amplitude (m) | Wind interference factor; radius (m) |
| $b$ | Wind profile exponent | Width (m) |
| $c$ | Specific heat capacity ($J\,kg^{-1}\,K^{-1}$) | Speed of light ($m\,s^{-1}$); phase velocity of wave ($m\,s^{-1}$); chord length (m); Weibull speed factor ($m\,s^{-1}$) |
| $d$ | Distance (m) | Diameter (m); depth (m); zero plane displacement (wind) (m) |
| e | Electron charge (C) | Base of natural logarithms (2.718) |
| $f$ | Frequency of cycles ($Hz = s^{-1}$) | Pipe friction coefficient; fraction; force per unit length ($N\,m^{-1}$) |
| $g$ | Acceleration due to gravity ($m\,s^{-2}$) | |
| $h$ | Heat transfer coefficient ($W\,m^{-2}\,K^{-1}$) | Vertical displacement (m); Planck constant (Js) |

| Symbol | Main use | Other use or comment |
|---|---|---|
| $i$ | $\sqrt{-1}$ | |
| $k$ | Thermal conductivity ($W m^{-1} K^{-1}$) | Wave vector ($=2\pi/\lambda$); Boltzmann constant ($=1.38 \times 10^{-23} J K^{-1}$) |
| $l$ | Distance (m) | |
| $m$ | Mass (kg) | Air-mass-ratio |
| $n$ | Number | Number of nozzles, of hours of bright sunshine, of wind-turbine blades; electron concentration ($m^{-3}$) |
| $p$ | Pressure ($N m^{-2} = Pa$) | Hole concentration ($m^{-3}$) |
| $q$ | Power per unit area ($W m^{-2}$) | |
| $r$ | Thermal resistivity of unit area ('R-value' = RA) ($m^2 K W^{-1}$) | Radius (m); distance (m) |
| $s$ | Angle of slope (degrees) | |
| $t$ | Time (s) | Thickness (m) |
| $u$ | Velocity along stream ($m s^{-1}$) | Group velocity ($m s^{-1}$) |
| $v$ | Velocity (not along stream) ($m s^{-1}$) | |
| $w$ | Distance (m) | Moisture content (dry basis, %); moisture content (wet basis, %) ($w'$) |
| $x$ | Co-ordinate (along stream) (m) | |
| $y$ | Co-ordinate (across stream) (m) | |
| $z$ | Co-ordinate (vertical) (m) | |
| *Greek capitals* | | |
| $\Gamma$ (gamma) | Torque (N m) | Gamma function |
| $\Delta$ (delta) | Increment of . . . (other symbol) | |
| $\Lambda$ (lambda) | Latent heat ($J kg^{-1}$) | |
| $\Sigma$ (sigma) | Summation sign | |
| $\Phi$ (phi) | Radiant flux (W) | Probability function |
| $\Phi_u$ | Probability distribution of wind speed ($(m s^{-1})^{-1}$) | |
| $\Omega$ (omega) | Solid angle (steradian) | Phonon frequency ($s^{-1}$); angular velocity of blade ($rad s^{-1}$) |
| *Greek lower case* | | |
| $\alpha$ (alpha) | Absorptance | Angle of attack (deg) |
| $\alpha_\lambda$ | Monochromatic absorptance | |
| $\beta$ (beta) | Angle (deg) | Volumetric Expansion coefficient ($K^{-1}$) |
| $\gamma$ (gamma) | Angle (deg) | Blade setting angle (deg) |
| $\delta$ (delta) | Boundary layer thickness (m) | Angle of declination (deg) |
| $\varepsilon$ epsilon | Emittance | Wave 'spectral width'; permittivity; dielectric constant |
| $\varepsilon_\lambda$ | Monochromatic emittance | |
| $\eta$ (eta) | Efficiency | |

(Continued)

| Symbol | Main use | Other use or comment |
|---|---|---|
| $\theta$ (theta) | Angle of incidence (deg) | Temperature difference (°C) |
| $\kappa$ (kappa) | Thermal diffusivity ($m^2 s^{-1}$) | |
| $\lambda$ (lambda) | Wavelength (m) | Tip speed ratio of wind-turbine |
| $\mu$ (mu) | Dynamic viscosity ($N m^{-2} s$) | |
| $\nu$ (nu) | Kinematic viscosity ($m^2 s^{-1}$) | |
| $\xi$ (xi) | Electrode potential (V) | Roughness height (m) |
| $\pi$ (pi) | 3.1416 | |
| $\rho$ (rho) | Density ($kg m^{-3}$) | Reflectance; electrical resistivity ($\Omega$ m) |
| $\rho_\lambda$ | Monochromatic reflectance | |
| $\sigma$ (sigma) | Stefan–Boltzmann constant | |
| $\tau$ (tau) | Transmittance | Relaxation time (s); duration (s); shear stress ($N m^{-2}$) |
| $\tau_\lambda$ | Monochromatic transmittance | |
| $\phi$ (phi) | Radiant flux density (RFD) ($W m^{-2}$) | Wind-blade angle (deg); potential difference ($V$); latitude (deg) |
| $\phi_\lambda$ | Spectral distribution of RFD ($W m^{-3}$) | |
| $\chi$ (chi) | Absolute humidity ($kg m^{-3}$) | |
| $\psi$ (psi) | Longitude (deg) | Angle (deg) |
| $\omega$ (omega) | Angular frequency ($= 2\pi f$) ($rad s^{-1}$) | Hour angle (deg); solid angle (steradian) |
| *Subscripts* | | |
| B | Black body | Band |
| D | Drag | Dark |
| E | Earth | |
| F | Force | |
| G | Generator | |
| L | Lift | |
| M | Moon | |
| P | Power | |
| R | Rated | |
| S | Sun | |
| T | Tangential | Turbine |
| a | Ambient | Aperture; available (head); aquifer |
| abs | Absorbed | |
| b | Beam | Blade; bottom; base; biogas |
| c | Collector | Cold |
| ci | Cut-in | |
| co | Cut-out | |
| cov | Cover | |
| d | Diffuse | Dopant; digester |
| e | Electrical | Equilibrium; energy |
| f | Fluid | Forced; friction; flow |
| g | Glass | Generation current; band gap |
| h | Horizontal | Hot |

| *Symbol* | *Main use* | *Other use or comment* |
|---|---|---|
| i | Integer | Intrinsic |
| in | Incident (incoming) | |
| int | Internal | |
| j | Integer | |
| m | mass transfer | Mean (average); methane |
| max | Maximum | |
| n | conduction | |
| net | Heat flow across surface | |
| o | (read as numeral zero) | |
| oc | Open circuit | |
| p | Plate | Peak; positive charge carriers (holes) |
| r | radiation | Relative; recombination; room; resonant; rock |
| rad | Radiated | |
| refl | Reflected | |
| rms | Root mean square | |
| s | Surface | Significant; saturated; Sun |
| sc | Short circuit | |
| t | Tip | Total |
| th | Thermal | |
| trans | Transmitted | |
| u | Useful | |
| v | convection | Vapour |
| w | Wind | Water |
| z | Zenith | |
| $\lambda$ | Monochromatic, e.g. $\alpha_\lambda$ | |
| 0 | Distant approach | Ambient; extra-terrestrial; dry matter; saturated; ground-level |
| 1 | Entry to device | First |
| 2 | Exit from device | Second |
| 3 | Output | Third |
| *Superscript* | | |
| *m* or *max* | Maximum | |
| * | Measured perpendicular to direction of propagation (e.g. $G_b^*$) | |
| · (dot) | Rate of, e.g. $\dot{m}$ | |
| *Other symbols* | | |
| Bold face | Vector, e.g. $\boldsymbol{F}$ | |
| $=$ | Mathematical equality | |
| $\approx$ | Approximate equality (within a few %) | |
| $\sim$ | Equality in order of magnitude (within a factor of 2–10) | |
| $\equiv$ | Mathematical identity (or definition), equivalent | |

Chapter 1

# Principles of renewable energy

## 1.1 Introduction

The aim of this text is to analyse the full range of renewable energy supplies available for modern economies. Such renewables are recognised as vital inputs for sustainability and so encouraging their growth is significant. Subjects will include power from wind, water, biomass, sunshine and other such continuing sources, including wastes. Although the scale of local application ranges from tens to many millions of watts, and the totality is a global resource, four questions are asked for practical application:

1. How much energy is available in the immediate environment – what is the resource?
2. For what purposes can this energy be used – what is the end-use?
3. What is the environmental impact of the technology – is it sustainable?
4. What is the cost of the energy – is it cost-effective?

The first two are technical questions considered in the central chapters by the type of renewables technology. The third question relates to broad issues of planning, social responsibility and sustainable development; these are considered in this chapter and in Chapter 17. The environmental impacts of specific renewable energy technologies are summarised in the last section of each technology chapter. The fourth question, considered with other institutional factors in the last chapter, may dominate for consumers and usually becomes the major criterion for commercial installations. However, cost-effectiveness depends significantly on:

a. Appreciating the *distinctive scientific principles* of renewable energy (Section 1.4).
b. Making each stage of the energy supply process *efficient* in terms of both minimising losses and maximising economic, social and environmental benefits.
c. Like-for-like *comparisons*, including externalities, with fossil fuel and nuclear power.

When these conditions have been met, it is possible to calculate the costs and benefits of a particular scheme and compare these with alternatives for an economic and environmental assessment.

Failure to understand the distinctive scientific principles for harnessing renewable energy will almost certainly lead to poor engineering and uneconomic operation. Frequently there will be a marked contrast between the methods developed for renewable supplies and those used for the non-renewable fossil fuel and nuclear supplies.

## 1.2 Energy and sustainable development

### 1.2.1 Principles and major issues

*Sustainable development* can be broadly defined as living, producing and consuming in a manner that meets the needs of the present without compromising the ability of future generations to meet their own needs. It has become a key guiding principle for policy in the 21st century. Worldwide, politicians, industrialists, environmentalists, economists and theologians affirm that the principle must be applied at international, national and local level. Actually applying it in practice and in detail is of course much harder!

In the international context, the word 'development' refers to improvement in quality of life, and, especially, standard of living in the less developed countries of the world. The aim of sustainable development is for the improvement to be achieved whilst maintaining the ecological processes on which life depends. At a local level, progressive businesses aim to report a positive *triple bottom line*, i.e. a positive contribution to the *economic, social and environmental* well-being of the community in which they operate.

The concept of sustainable development became widely accepted following the seminal report of the World Commission on Environment and Development (1987). The commission was set up by the United Nations because the scale and unevenness of economic development and population growth were, and still are, placing unprecedented pressures on our planet's lands, waters and other natural resources. Some of these pressures are severe enough to threaten the very survival of some regional populations and, in the longer term, to lead to global catastrophes. Changes in lifestyle, especially regarding production and consumption, will eventually be forced on populations by ecological and economic pressures. Nevertheless, the economic and social pain of such changes can be eased by foresight, planning and political (i.e. community) will.

Energy resources exemplify these issues. Reliable energy supply is essential in all economies for lighting, heating, communications, computers, industrial equipment, transport, etc. Purchases of energy account for 5–10% of gross national product in developed economies. However, in some developing countries, energy imports may have cost over half the value of total exports; such economies are unsustainable and an economic challenge for sustainable development. World energy use increased more than tenfold over the 20th century, predominantly from fossil fuels (i.e. coal, oil and gas) and with the addition of electricity from nuclear power. In the 21st century, further increases in world energy consumption can be expected, much for rising industrialisation and demand in previously less developed countries, aggravated by gross inefficiencies in all countries. Whatever the energy source, there is an overriding need for efficient generation and use of energy.

Fossil fuels are not being newly formed at any significant rate, and thus present stocks are ultimately finite. The location and the amount of such stocks depend on the latest surveys. Clearly the dominant fossil fuel type by mass is coal, with oil and gas much less. The reserve lifetime of a resource may be defined as the known accessible amount divided by the rate of present use. By this definition, the lifetime of oil and gas resources is usually only a few decades; whereas lifetime for coal is a few centuries. Economics predicts that as the lifetime of a fuel reserve shortens, so the fuel price increases; consequently demand for that fuel reduces and previously more expensive sources and alternatives enter the market. This process tends to make the original source last longer than an immediate calculation indicates. In practice, many other factors are involved, especially governmental policy and international relations. Nevertheless, the basic geological fact remains: fossil fuel reserves are limited and so the present patterns of energy consumption and growth are not sustainable in the longer term.

Moreover, it is the *emissions* from fossil fuel use (and indeed nuclear power) that increasingly determine the fundamental limitations. Increasing concentration of $CO_2$ in the Atmosphere is such an example. Indeed, from an ecological understanding of our Earth's long-term history over billions of years, carbon was in excess in the Atmosphere originally and needed to be sequestered below ground to provide our present oxygen-rich atmosphere. Therefore from arguments of: (i) the finite nature of fossil and nuclear fuel materials, (ii) the harm of emissions and (iii) ecological sustainability, it is essential to expand renewable energy supplies and to use energy more efficiently. Such conclusions are supported in economics if the full external costs of both obtaining the fuels and paying for the damage from emissions are internalised in the price. Such fundamental analyses may conclude that renewable energy and the efficient use of energy are cheaper for society than the traditional use of fossil and nuclear fuels.

The detrimental environmental effects of burning the fossil fuels likewise imply that current patterns of use are unsustainable in the longer term. In particular, $CO_2$ emissions from the combustion of fossil fuels have significantly raised the concentration of $CO_2$ in the Atmosphere. The balance of scientific opinion is that if this continues, it will enhance the *greenhouse effect*[1] and lead to significant *climate change* within a century or less, which could have major adverse impact on food production, water supply and human, e.g. through floods and cyclones (IPCC). Recognising that this is a global problem, which no single country can avert on its own, over 150 national governments signed the UN Framework Convention on Climate Change, which set up a framework for concerted action on the issue. Sadly, concrete action is slow, not least because of the reluctance of governments in industrialised countries to disturb the lifestyle of their voters. However, potential climate change, and related sustainability issues, is now established as one of the major drivers of energy policy.

In short, renewable energy supplies are much more compatible with sustainable development than are fossil and nuclear fuels, in regard to both resource limitations and environmental impacts (see Table 1.1).

Consequently almost all national energy plans include four vital factors for improving or maintaining social benefit from energy:

1 increased harnessing of renewable supplies
2 increased efficiency of supply and end-use
3 reduction in pollution
4 consideration of lifestyle.

### *1.2.2 A simple numerical model*

Consider the following simple model describing the need for commercial and non-commercial energy resources:

$$R = EN \tag{1.1}$$

Here $R$ is the total yearly energy requirement for a population of $N$ people. $E$ is the per capita energy-use averaged over one year, related closely to provision of food and manufactured goods. The unit of $E$ is energy per unit time, i.e. power. On a world scale, the dominant supply of energy is from commercial sources, especially fossil fuels; however, significant use of non-commercial energy may occur (e.g. fuel wood, passive solar heating), which is often absent from most official and company statistics. In terms of total commercial energy use, the average per capita value of $E$ worldwide is about 2 kW; however, regional average values range widely, with North America 9 kW, Europe as a whole 4 kW, and several regions of Central Africa as small as 0.1 kW. The inclusion of non-commercial energy increases

1 As described in Chapter 4, the presence of $CO_2$ (and certain other gases) in the atmosphere keeps the Earth some 30 degrees warmer than it would otherwise be. By analogy with horticultural greenhouses, this is called the 'greenhouse effect'.

*Table 1.1* Comparison of renewable and conventional energy systems

| | *Renewable energy supplies (green)* | *Conventional energy supplies (brown)* |
|---|---|---|
| Examples | Wind, solar, biomass, tidal | Coal, oil, gas, radioactive ore |
| Source | Natural local environment | Concentrated stock |
| Normal state | A current or flow of energy. An income | Static store of energy. Capital |
| Initial average intensity | Low intensity, dispersed: $\leq 300\,W\,m^{-2}$ | Released at $\geq 100\,kW\,m^{-2}$ |
| Lifetime of supply | Infinite | Finite |
| Cost at source | Free | Increasingly expensive. |
| Equipment capital cost per kW capacity | Expensive, commonly $\approx$US\$1000 $kW^{-1}$ | Moderate, perhaps \$500 $kW^{-1}$ without emissions control; yet expensive >US\$1000 $kW^{-1}$ with emissions reduction |
| Variation and control | Fluctuating; best controlled by change of load using positive feedforward control | Steady, best controlled by adjusting source with negative feedback control |
| Location for use | Site- and society-specific | General and invariant use |
| Scale | Small and moderate scale often economic, large scale may present difficulties | Increased scale often improves supply costs, large scale frequently favoured |
| Skills | Interdisciplinary and varied. Wide range of skills. Importance of bioscience and agriculture | Strong links with electrical and mechanical engineering. Narrow range of personal skills |
| Context | Bias to rural, decentralised industry | Bias to urban, centralised industry |
| Dependence | Self-sufficient and 'islanded' systems supported | Systems dependent on outside inputs |
| Safety | Local hazards possible in operation: usually safe when out of action | May be shielded and enclosed to lessen great potential dangers; most dangerous when faulty |
| Pollution and environmental damage | Usually little environmental harm, especially at moderate scale<br>Hazards from excess biomass burning<br>Soil erosion from excessive biofuel use<br>Large hydro reservoirs disruptive<br>Compatible with natural ecology | Environmental pollution intrinsic and common, especially of air and water<br>Permanent damage common from mining and radioactive elements entering water table. Deforestation and ecological sterilisation from excessive air pollution<br>Climate change emissions |
| Aesthetics, visual impact | Local perturbations may be unpopular, but usually acceptable if local need perceived | Usually utilitarian, with centralisation and economy of large scale |

all these figures and has the major proportional benefit in countries where the value of $E$ is small.

Standard of living relates in a complex and an ill-defined way to $E$. Thus per capita gross national product $S$ (a crude measure of standard of living) may be related to $E$ by:

$$S = fE \tag{1.2}$$

Here $f$ is a complex and non-linear coefficient that is itself a function of many factors. It may be considered an efficiency for transforming energy into wealth and, by traditional economics, is expected to be as large as possible. However, $S$ does not increase uniformly as $E$ increases. Indeed $S$ may even decrease for large $E$ (e.g. because of pollution or technical inefficiency). Obviously unnecessary waste of energy leads to a lower value of $f$ than would otherwise be possible. Substituting for $E$ in (1.1), the national requirement for energy becomes:

$$R = \frac{(SN)}{f} \tag{1.3}$$

so

$$\frac{\Delta R}{R} = \frac{\Delta S}{S} + \frac{\Delta N}{N} - \frac{\Delta f}{f} \tag{1.4}$$

Now consider substituting global values for the parameters in (1.4). In 50 years the world population $N$ increased from 2500 million in 1950 to over 6000 million in 2000. It is now increasing at approximately 2–3% per year so as to double every 20–30 years. Tragically, high infant mortality and low life expectancy tend to hide the intrinsic pressures of population growth in many countries. Conventional economists seek exponential growth of $S$ at 2–5% per year. Thus in (1.4), at constant efficiency $f$, the growth of total world energy supply is effectively the sum of population and economic growth, i.e. 4–8% per year. Without new supplies such growth cannot be maintained. Yet at the same time as more energy is required, fossil and nuclear fuels are being depleted and debilitating pollution and climate change increase; so an obvious conclusion to overcome such constraints is to increase renewable energy supplies. Moreover, from (1.3) and (1.4), it is most beneficial to increase the parameter $f$, i.e. to have a positive value of $\dot{f}$. Consequently there is a growth rate in energy efficiency, so that $S$ can increase, while $R$ decreases.

### *1.2.3 Global resources*

Considering these aims, and with the most energy-efficient modern equipment, buildings and transportation, a justifiable target for energy use in a modern society with an appropriate lifestyle is $E = 2\,\mathrm{kW}$ per person. Such a target is consistent with an energy policy of 'contract and converge' for global equity, since worldwide energy supply would total approximately the present global average usage, but would be consumed for a far higher standard of living. Is this possible, even in principle, from renewable energy? Each square metre of the earth's habitable surface is crossed by, or accessible to, an average energy flux from all renewable sources of about 500 W (see Problem 1.1). This includes solar, wind or other renewable energy forms in an overall estimate. If this flux is harnessed at just 4% efficiency, 2 kW of power can be drawn from an area of $10\,\mathrm{m} \times 10\,\mathrm{m}$, assuming suitable methods. Suburban areas of residential towns have population densities of about 500 people per square kilometre. At 2 kW per person, the total energy demand of $1000\,\mathrm{kW\,km^{-2}}$ could be obtained in principle by using just 5% of the local land area for energy production. Thus renewable energy supplies can provide a satisfactory standard of living, but only if the technical methods and institutional frameworks exist to extract, use and store the energy in an appropriate form at realistic costs. This book considers both the technical background of a great variety of possible methods and a summary of the institutional factors involved. Implementation is then everyone's responsibility.

## 1.3 Fundamentals

### 1.3.1 Definitions

For all practical purposes energy supplies can be divided into two classes:

1. *Renewable energy.* 'Energy obtained from natural and persistent flows of energy occurring in the immediate environment'. An obvious example is solar (sunshine) energy, where 'repetitive' refers to the 24-hour major period. Note that the energy is already passing through the environment as a *current* or *flow*, irrespective of there being a device to intercept and harness this power. Such energy may also be called *Green Energy* or *Sustainable Energy*.
2. *Non-renewable energy.* 'Energy obtained from static stores of energy that remain underground unless released by human interaction'. Examples are nuclear fuels and fossil fuels of coal, oil and natural gas. Note that the energy is initially an isolated energy *potential*, and external action is required to initiate the supply of energy for practical purposes. To avoid using the ungainly word 'non-renewable', such energy supplies are called *finite supplies* or *Brown Energy*.

These two definitions are portrayed in Figure 1.1. Table 1.1 provides a comparison of renewable and conventional energy systems.

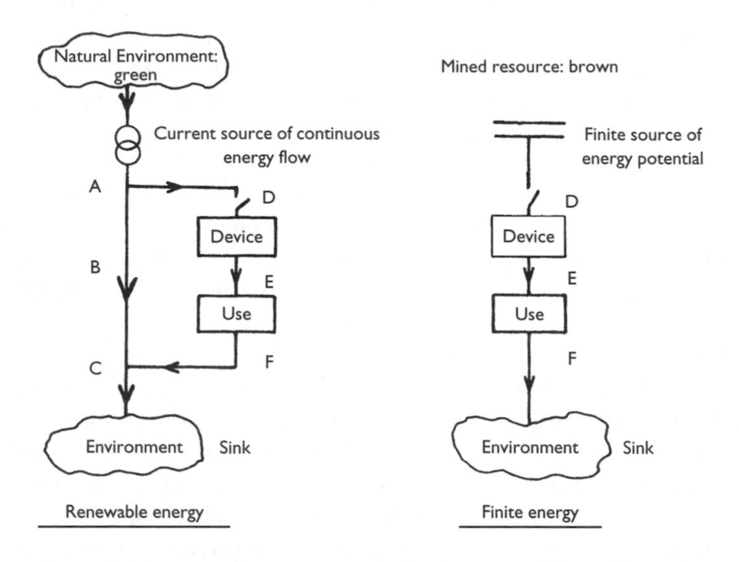


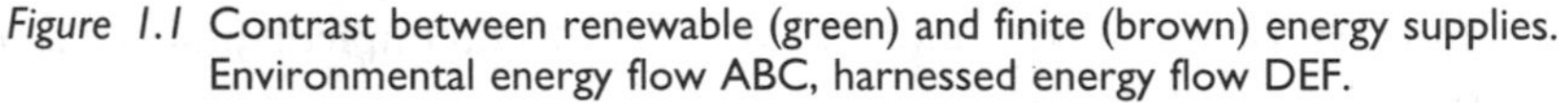
*Figure 1.1* Contrast between renewable (green) and finite (brown) energy supplies. Environmental energy flow ABC, harnessed energy flow DEF.

### *1.3.2 Energy sources*

There are five ultimate primary sources of useful energy:

1. The Sun.
2. The motion and gravitational potential of the Sun, Moon and Earth.
3. Geothermal energy from cooling, chemical reactions and radioactive decay in the Earth.
4. Human-induced nuclear reactions.
5. Chemical reactions from mineral sources.

Renewable energy derives continuously from sources 1, 2 and 3 (aquifers). Finite energy derives from sources 1 (fossil fuels), 3 (hot rocks), 4 and 5. The sources of most significance for global energy supplies are 1 and 4. The fifth category is relatively minor, but useful for primary batteries, e.g. dry cells.

### *1.3.3 Environmental energy*

The flows of energy passing continuously as renewable energy through the Earth are shown in Figure 1.2. For instance, total solar flux absorbed at sea level is about $1.2 \times 10^{17}$ W. Thus the solar flux reaching the Earth's surface is ~20 MW per person; 20 MW is the power of ten very large

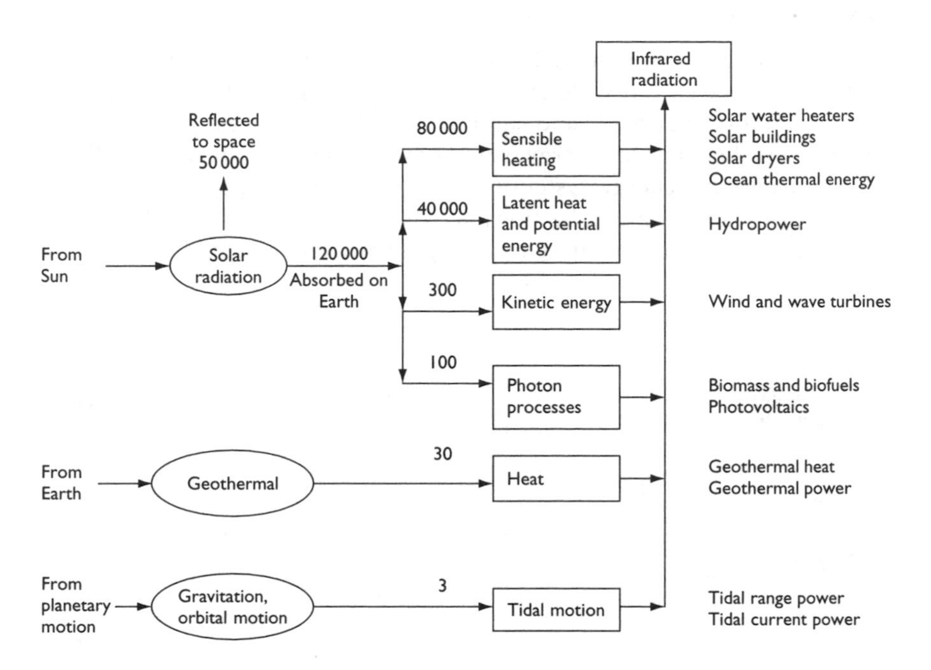


*Figure 1.2* Natural energy currents on earth, showing renewable energy system. Note the great range of energy flux ($1:10^5$) and the dominance of solar radiation and heat. Units terawatts ($10^{12}$ W).

diesel electric generators, enough to supply all the energy needs of a town of about 50 000 people. The maximum solar flux density (irradiance) perpendicular to the solar beam is about $1\,\mathrm{kW\,m^{-2}}$; a very useful and easy number to remember. In general terms, a human being is able to intercept such an energy flux without harm, but any increase begins to cause stress and difficulty. Interestingly, power flux densities of $\sim 1\,\mathrm{kW\,m^{-2}}$ begin to cause physical difficulty to an adult in wind, water currents or waves.

However, the global data of Figure 1.2 are of little value for practical engineering applications, since particular sites can have remarkably different environments and possibilities for harnessing renewable energy. Obviously flat regions, such as Denmark, have little opportunity for hydro-power but may have wind power. Yet neighbouring regions, for example Norway, may have vast hydro potential. Tropical rain forests may have biomass energy sources, but deserts at the same latitude have none (moreover, forests must not be destroyed so making more deserts). Thus practical renewable energy systems have to be matched to particular local environmental energy flows occurring in a particular region.

### *1.3.4 Primary supply to end-use*

All energy systems can be visualised as a series of pipes or circuits through which the energy currents are channelled and transformed to become useful in domestic, industrial and agricultural circumstances. Figure 1.3(a) is a Sankey diagram of energy supply, which shows the energy flows through a national energy system (sometimes called a 'spaghetti diagram' because of its appearance). Sections across such a diagram can be drawn as pie charts showing primary energy supply and energy supply to end-use

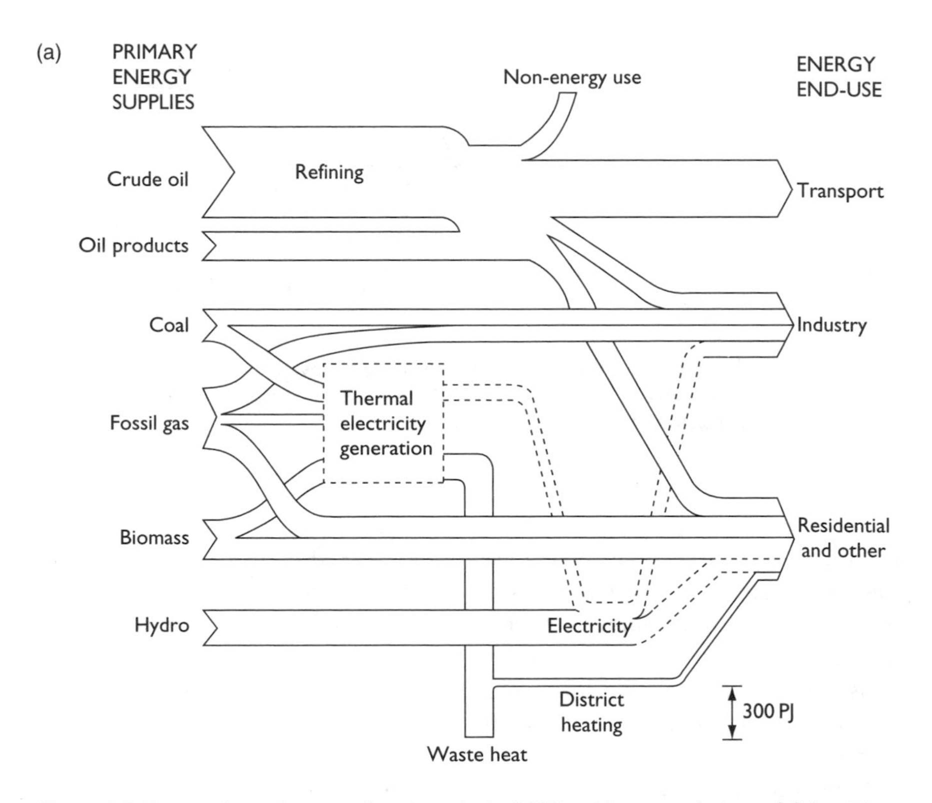


*Figure 1.3* Energy flow diagrams for Austria in 2000, with a population of 8.1 million. (a) Sankey ('spaghetti') diagram, with flows involving thermal electricity shown dashed. (b)–(c) Pie diagrams. The contribution of hydropower and biomass (wood and waste) is greater than in most industrialised countries, as is the use of heat produced from thermal generation of electricity ('combined heat and power'). Energy use for transport is substantial and very dependent on (imported) oil and oil products, therefore the Austrian government encourages increased use of biofuels. Austria's energy use has grown by over 50% since 1970, although the population has grown by less than 10%, indicating the need for greater efficiency of energy use. [Data source: simplified from International Energy Agency, *Energy Balances of OECD countries 2000–2001*.]

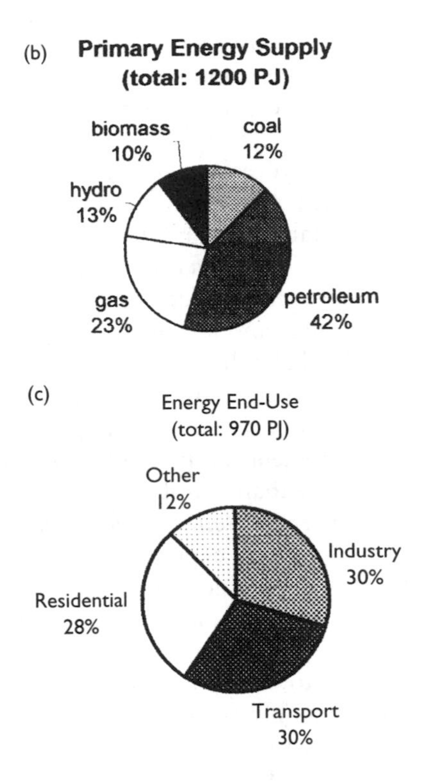


*Figure 1.3* (Continued).

(Figure 1.3(b)). Note how the total energy end-use is less than the primary supply because of losses in the transformation processes, notably the generation of electricity from fossil fuels.

### *1.3.5 Energy planning*

1 *Complete energy systems* must be analysed, and supply should not be considered separately from end-use. Unfortunately precise *needs* for energy are too frequently forgotten, and supplies are not well matched to end-use. Energy losses and uneconomic operation therefore frequently result. For instance, if a dominant domestic energy requirement is heat for warmth and hot water, it is irresponsible to generate grid quality electricity from a fuel, waste the majority of the energy as thermal emission from the boiler and turbine, distribute the electricity in lossy cables and then dissipate this electricity as heat. Sadly

such inefficiency and disregard for resources often occurs. Heating would be more efficient and cost-effective from direct heat production with local distribution. Even better is to combine electricity generation with the heat production using CHP – combined heat and power (electricity).

2 *System efficiency* calculations can be most revealing and can pinpoint unnecessary losses. Here we define 'efficiency' as the ratio of the useful energy output from a process to the total energy input to that process. Consider electric lighting produced from 'conventional' thermally generated electricity and lamps. Successive energy efficiencies are: electricity generation ~30%, distribution ~90% and incandescent lighting (energy in visible radiation, usually with a light-shade) 4–5%. The total efficiency is 1–1.5%. Contrast this with cogeneration of useful heat and electricity (efficiency ~85%), distribution (~90%) and lighting in modern low consumption compact fluorescent lamps (CFL) (~22%). The total efficiency is now 14–18%; a more than tenfold improvement! The total life cycle cost of the more efficient system will be much less than for the conventional, despite higher per unit capital costs, because (i) less generating capacity and fuel are needed, (ii) less per unit emission costs are charged, and (iii) equipment (especially lamps) lasts longer (see Problems 1.2 and 1.3).

3 *Energy management* is always important to improve overall efficiency and reduce economic losses. No energy supply is free, and renewable supplies are usually more expensive in practice than might be assumed. Thus there is no excuse for wasting energy of any form unnecessarily. Efficiency with finite fuels reduces pollution; efficiency with renewables reduces capital costs.

## 1.4 Scientific principles of renewable energy

The definitions of renewable (green) and finite (brown) energy supplies (Section 1.3.1) indicate the fundamental differences between the two forms of supply. As a consequence the efficient use of renewable energy requires the correct application of certain principles.

### *1.4.1 Energy currents*

It is essential that a sufficient renewable current is *already present* in the local environment. It is not good practice to try to create this energy current especially for a particular system. Renewable energy was once ridiculed by calculating the number of pigs required to produce dung for sufficient methane generation to power a whole city. It is obvious, however, that biogas (methane) production should only be contemplated as a *by-product* of an animal industry already established, and not vice versa. Likewise for a biomass energy station, the biomass resource must exist locally to avoid large inefficiencies in transportation. The practical implication of this principle is that the local environment has to be monitored and analysed over a long period to establish precisely what energy flows are present. In Figure 1.1 the energy current ABC must be assessed before the diverted flow through DEF is established.

### 1.4.2 Dynamic characteristics

End-use requirements for energy vary with time. For example, electricity demand on a power network often peaks in the morning and evening, and reaches a minimum through the night. If power is provided from a finite source, such as oil, the input can be adjusted in response to demand. Unused energy is not wasted, but remains with the source fuel. However, with renewable energy systems, not only does end-use vary uncontrollably with time but so too does the natural supply in the environment. Thus a renewable energy device must be matched dynamically at both D and E of Figure 1.1; the characteristics will probably be quite different at both interfaces. Examples of these dynamic effects will appear in most of the following chapters.

The major periodic variations of renewable sources are listed in Table 1.2, but precise dynamic behaviour may well be greatly affected by irregularities. Systems range from the very variable (e.g. wind power) to the accurately predictable (e.g. tidal power). Solar energy may be very predicable in some regions (e.g. Khartoum) but somewhat random in others (e.g. Glasgow).

### 1.4.3 Quality of supply

The quality of an energy supply or store is often discussed, but usually remains undefined. We define *quality* as the proportion of an energy source that can be converted to mechanical work. Thus electricity has high quality because when consumed in an electric motor >95% of the input energy may be converted to mechanical work, say to lift a weight; the heat losses are correspondingly small, <5%. The quality of nuclear, fossil or biomass fuel in a single stage thermal power station is moderately low, because only about 33% of the calorific value of the fuel can be made to appear as mechanical work and about 67% is lost as heat to the environment. If the fuel is used in a combined cycle power station (e.g. methane gas turbine stage followed by steam turbine), then the quality is increased to ~50%. It is possible to analyse such factors in terms of the thermodynamic variable energy, defined here as 'the theoretical maximum amount of work obtainable, at a particular environmental temperature, from an energy source'.

*Table 1.2* Intensity and periodical properties of renewable sources

| *System* | *Major periods* | *Major variables* | *Power relationship* | *Comment* | *Text reference (equation)* |
|---|---|---|---|---|---|
| Direct sunshine | 24 h, 1 y | Solar beam irradiance $G_b^*(\mathrm{W\,m^{-2}})$; Angle of beam from vertical $\theta_z$ | $P \propto G_b^* \cos\theta_z$ $P_{max} = 1\,\mathrm{kW\,m^{-2}}$ | Daytime only | (4.2) |
| Diffuse sunshine | 24 h, 1 y | Cloud cover, perhaps air pollution | $P \ll G$; $P \leq 300\,\mathrm{W\,m^{-2}}$ | Significant energy over time | (4.3) |
| Biofuels | 1 y | Soil condition, insolation, water, plant species, wastes | Stored energy $\sim 10\,\mathrm{MJ\,kg^{-1}}$ | Very many chemical types and sources. Linked to agriculture and forestry. Stored energy | Table 11.1 Table 11.4 |
| Wind | 1 y | Wind speed $u_0$ Height nacelle above ground $z$; height of anemometer mast $h$ | $P \propto u_0^3$ $u_z/u_h = (z/h)^b$ | Most variable $b \sim 0.15$ | (9.2) (9.54) |
| Wave | 1 y | 'Significant wave height' $H_s$ wave period $T$ | $P \propto H_s^2 T$ | Relatively large power density $\sim 50\,\mathrm{kW\,m^{-1}}$ across wave front | (12.47) |
| Hydro | 1 y | Reservoir height $H$ water volume flow rate $Q$ | $P \propto HQ$ | Established resource | (8.1) |
| Tidal | 12 h 25 min | Tidal range $R$; contained area $A$; estuary length $L$, depth $h$ Tidal stream/current; peak current $u_0$, density sea water $\rho(\sim 1000 \times \mathrm{air})$ | $P \propto R^2 A$ $P \propto \rho u_0^3$ | Enhanced tidal range if $L/\sqrt{h} = 36\,000\,\mathrm{m^{1/2}}$ Enhanced tidal currents between certain islands | (13.35) and (13.28) (13.30) |
| Ocean thermal energy conversion | Constant | Temperature difference between sea surface and deep water, $\Delta T$ | $P \propto (\Delta T)^2$ | Some tropical locations have $\Delta T \sim 20\,°\mathrm{C}$, so potentially harnessable, but small efficiency | (14.5) |

Renewable energy supply systems divide into three broad divisions:

1 *Mechanical supplies*, such as hydro, wind, wave and tidal power. The mechanical source of power is usually transformed into electricity at high efficiency. The proportion of power in the environment extracted by the devices is determined by the mechanics of the process, linked to the variability of the source, as explained in later chapters. The proportions are, commonly, wind 35%, hydro 70–90%, wave 50% and tidal 75%.
2 *Heat supplies*, such as biomass combustion and solar collectors. These sources provide heat at high efficiency. However, the maximum proportion of heat energy extractable as mechanical work, and hence electricity, is given by the second law of thermodynamics and the Carnot Theorem, which assumes reversible, infinitely long transformations. In practice, maximum mechanical power produced in a dynamic process is about half that predicted by the Carnot criteria. For thermal boiler heat engines, maximum realisable quality is about 35%.
3 *Photon processes*, such as photosynthesis and photochemistry (Chapter 10) and photovoltaic conversion (Chapter 7). For example, solar photons of a single frequency may be transformed into mechanical work via electricity with high efficiency using a matched solar cell. In practice, the broad band of frequencies in the solar spectrum makes matching difficult and photon conversion efficiencies of 20–30% are considered good.

### *1.4.4 Dispersed versus centralised energy*

A pronounced difference between renewable and finite energy supplies is the energy flux density at the initial transformation. Renewable energy commonly arrives at about $1\,\text{kW}\,\text{m}^{-2}$ (e.g. solar beam irradiance, energy in the wind at $10\,\text{m}\,\text{s}^{-1}$), whereas finite centralised sources have energy flux densities that are orders of magnitude greater. For instance, boiler tubes in gas furnaces easily transfer $100\,\text{kW}\,\text{m}^{-2}$, and in a nuclear reactor the first wall heat exchanger must transmit several $\text{MW}\,\text{m}^{-2}$. At end-use after distribution, however, supplies from finite sources must be greatly reduced in flux density. Thus apart from major exceptions such as metal refining, end-use loads for both renewable and finite supplies are similar. In summary *finite energy is most easily produced centrally and is expensive to distribute. Renewable energy is most easily produced in dispersed locations and is expensive to concentrate.* With an electrical grid, the renewable generators are said to be 'embedded' within the (dispersed) system.

A practical consequence of renewable energy application is development and increased cash flow in the *rural* economy. Thus the use of renewable energy favours rural development and not urbanisation.

### 1.4.5 Complex systems

Renewable energy supplies are intimately linked to the natural environment, which is not the preserve of just one academic discipline such as physics or electrical engineering. Frequently it is necessary to cross disciplinary boundaries from as far apart as, say, plant physiology to electronic control engineering. An example is the energy planning of integrated farming (Section 11.8.1). Animal and plant wastes may be used to generate methane, liquid and solid fuels, and the whole system integrated with fertilizer production and nutrient cycling for optimum agricultural yields.

### 1.4.6 Situation dependence

No single renewable energy system is universally applicable, since the ability of the local environment to supply the energy and the suitability of society to accept the energy vary greatly. It is as necessary to 'prospect' the environment for renewable energy as it is to prospect geological formations for oil. It is also necessary to conduct energy surveys of the domestic, agricultural and industrial needs of the local community. Particular end-use needs and local renewable energy supplies can then be matched, subject to economic and environmental constraints. In this respect renewable energy is similar to agriculture. Particular environments and soils are suitable for some crops and not others, and the market pull for selling the produce will depend on particular needs. The main consequence of this 'situation dependence' of renewable energy is the impossibility of making simplistic international or national energy plans. Solar energy systems in southern Italy should be quite different from those in Belgium or indeed in northern Italy. Corn alcohol fuels might be suitable for farmers in Missouri but not in New England. A suitable scale for renewable energy planning might be 250 km, but certainly not 2500 km. Unfortunately present-day large urban and industrialised societies are not well suited for such flexibility and variation.

## 1.5 Technical implications

### 1.5.1 Prospecting the environment

Normally, monitoring is needed for several years at the site in question. Ongoing analysis must insure that useful data are being recorded, particularly with respect to dynamic characteristics of the energy systems planned. Meteorological data are always important, but unfortunately the sites of official stations are often different from the energy generating sites, and the methods of recording and analysis are not ideal for energy prospecting. However, an important use of the long-term data from official monitoring stations is as a base for comparison with local site variations. Thus wind velocity may be monitored for several months at a prospective generating site and compared with data from the nearest official base station. Extrapolation using many years of base station data may then be possible. Data unrelated to normal meteorological measurements may be difficult to obtain. In particular, flows of biomass and waste materials will often not have been previously assessed, and will not have been considered for energy generation. In general, prospecting for supplies of renewable energy requires specialised methods and equipment that demand significant resources of finance and manpower. Fortunately the links with meteorology, agriculture and marine science give rise to much basic information.

### 1.5.2 End-use requirements and efficiency

As explained in Section 1.3.5, energy generation should always follow quantitative and comprehensive assessment of energy end-use requirements. Since no energy supply is cheap or occurs without some form of environmental disruption, it is also important to use the energy efficiently with good methods of energy conservation. With electrical systems, the end-use requirement is called the *load*, and the size and dynamic characteristics of the load will greatly affect the type of generating supply. Money spent on energy conservation and improvements in end-use efficiency usually gives better long-term benefit than money spent on increased generation and supply capacity. The largest energy requirements are usually for heat and transport. Both uses are associated with energy storage capacity in thermal mass, batteries or fuel tanks, and the inclusion of these uses in energy systems can greatly improve overall efficiency.

### 1.5.3 Matching supply and demand

After quantification and analysis of the separate dynamic characteristics of end-use demands and environmental supply options, the total demand and supply have to be brought together. This may be explained as follows:

1. The maximum amount of environmental energy must be utilised within the capability of the renewable energy devices and systems. In Figure 1.4(a), the resistance to energy flow at D, E and F should be small. The main benefit of this is to reduce the size and amount of generating equipment.
2. *Negative feedback* control from demand to supply is *not* beneficial since the result is to waste or spill harnessable energy (Figure 1.4(b)); in effect the capital value of the equipment is not fully utilised. Such control should only be used at times of emergency or when all conceivable end-uses have been satisfied. Note that the disadvantage of negative feedback control is a consequence of *renewable* energy being flow or

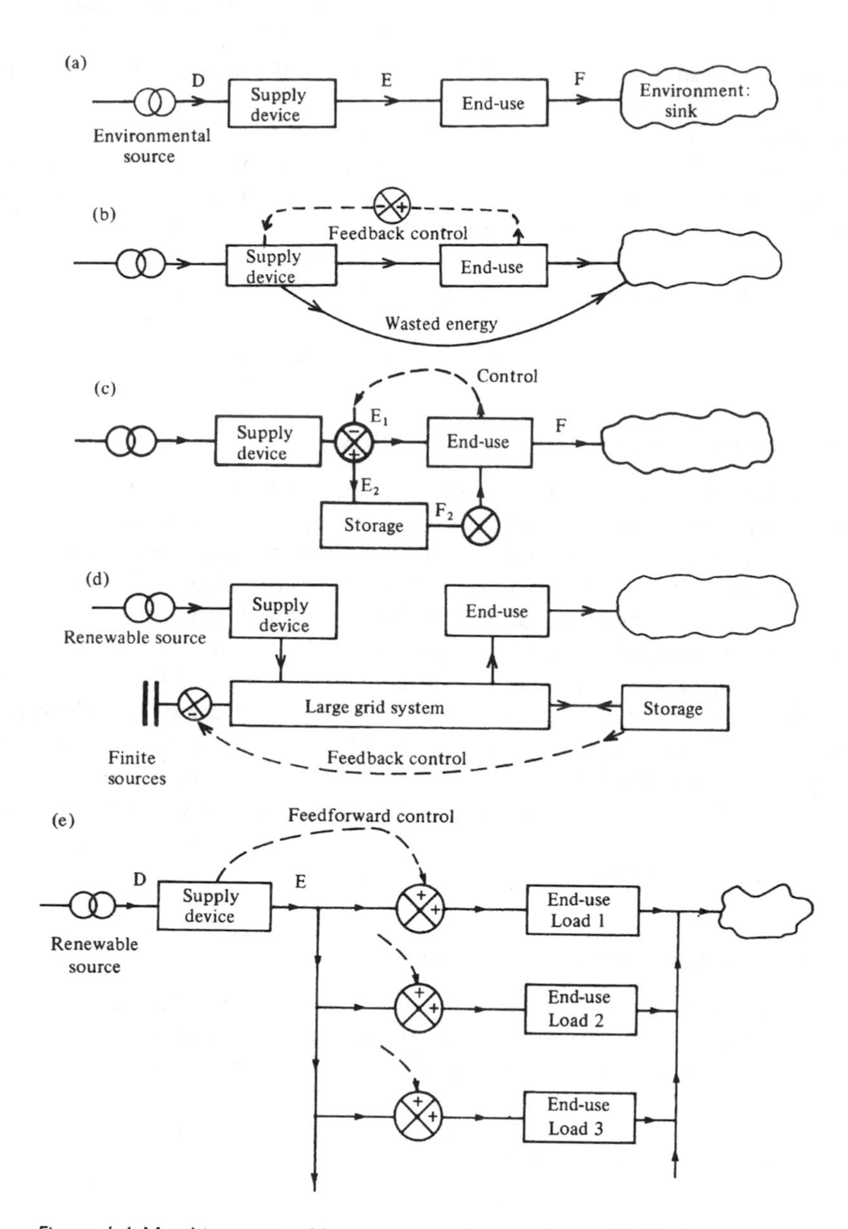


*Figure 1.4* Matching renewable energy supply to end-use. (a) Maximum energy flow for minimum size of device or system requires low resistance to flow at D, E and F. (b) Negative feedback control wastes energy opportunity and capital value. (c) Energy storage allows the dynamic characteristics of end-use to be decoupled from the supply characteristics. (d) Decoupling with a large grid system. (e) Feedforward load management control of the supply; arguably the most efficient way to use renewable energy. Total load at E may be matched to the available supply at D at all times and so control the supply device.

current sources that can never be stopped. With *finite* energy sources, negative feedback control to the energy source is beneficial, since less fuel is used.

3 The natural periods and dynamic properties of end-use are most unlikely to be the same as those of the renewable supply, as discussed in Section 1.4.2. The only way to match supply and demand that have different dynamic characteristics, and yet not to waste harnessable energy, is to incorporate storage (Figure 1.4(c)). Satisfactory energy storage is expensive (see Chapter 16), especially if not incorporated at the earliest stages of planning.
4 The difficulties of matching renewable energy supplies to end-use in stand-alone systems are so great that one common approach is to decouple supply from local demand by connection to an energy network or grid (Figure 1.4(d)). Here the renewable supply is embedded in an energy grid network having input from finite sources having feedback control. Such systems imply relatively large scale operation and include electricity grids for transmission and distribution. As in (3) the addition of substantial energy storage in the system, say pumped hydro or thermal capacity for heating, can improve efficiency and allow the proportion of renewable supply to increase. By using the grid for both the export and the import of energy, the grid becomes a 'virtual store'.
5 The most efficient way to use renewable energy is shown in Figure 1.4(e). Here a range of end-uses is available and can be switched or adjusted so that the total load equals the supply at any one time. Some of the end-use blocks could themselves be adjustable (e.g. variable voltage water heating, pumped water storage). Such systems require *feedforward control* (see Section 1.5.4). Since the end-use load increases with increase in the renewable energy supply, this is *positive feedforward control.*

### *1.5.4 Control options*

Good matching of renewable energy supply to end-use demand is accomplished by control of machines, devices and systems. The discussion in Section 1.5.3 shows that there are three possible categories of control: (1) spill the excess energy, (2) incorporate storage and (3) operate load management control. These categories may be applied in different ways, separately or together, to all renewable energy systems, and will be illustrated here with a few examples (Figure 1.5).

1 *Spill excess energy*. Since renewable energy derives from energy flow sources, energy not used is energy wasted. Nevertheless spilling excess

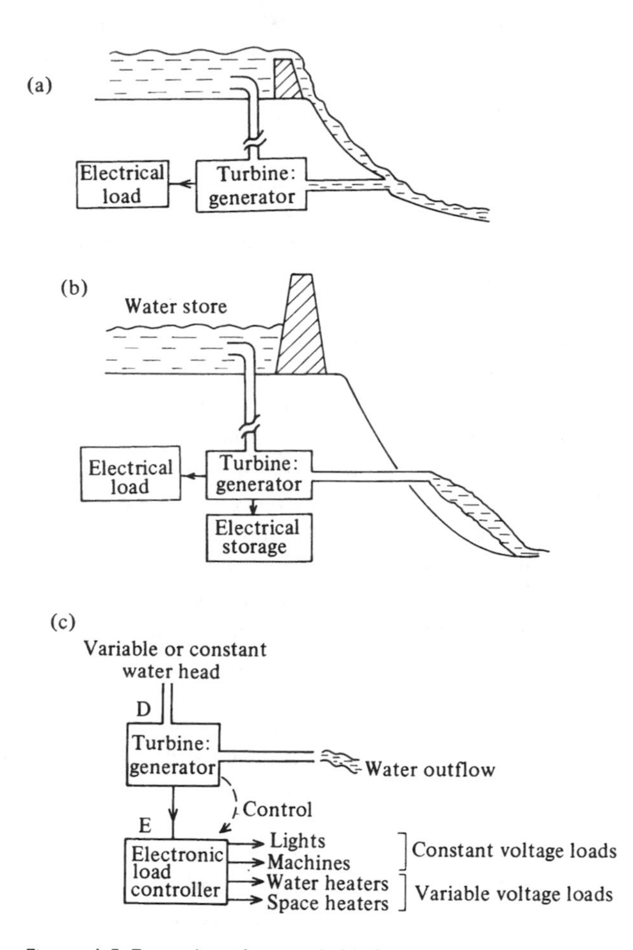


*Figure 1.5* Examples of control. (a) Control by spilling excess energy: constant pressure maintained for the turbine. (b) Control incorporating storage in hydroelectric catchment dam. (c) Control by load variation: feedforward control. Load controller automatically shunts power between end-uses, maintaining constant generator load at E. Turbine also has constant load and hence constant frequency: only rudimentary mechanical control of turbine is needed.

energy provides easy control and may be the cheapest option. Examples occur with run-of-the-river hydroelectric systems (Figure 1.5(a)), shades and blinds with passive solar heating of buildings, and wind turbines with adjustable blade pitch.

2 *Incorporate storage.* Storage before transformation allows a maximum amount of energy to be trapped from the environment and eventually

harnessed or used. Control methods are then similar to conventional methods with finite sources, with the store equivalent to fuel. The main disadvantages are the large relative capital costs of storage, and the difficulty of reducing conventional control methods to small-scale and remote operation. In the example of Figure 1.5(b), hydro storage is usually only contemplated for generation at more than ~10 MW. The mechanical flow control devices become unwieldy and expensive at a microhydro scale of ~10 kW. A disadvantage of hydro storage may be the environmental damage caused by reservoirs.

Storage after energy transformation, e.g. battery charging or hydrogen production, is also possible and may become increasingly important especially in small systems. Thermal storage is already common.

3 *Load control.* Parallel arrangements of end-uses may be switched and controlled so as to present optimum total load to the supply. An example of a microhydro load controller for household power is shown in Figure 1.5(c) (see also Section 8.6). The principle may be applied on a small or large scale, but is perhaps most advantageous when many varied end-uses are available locally. There are considerable advantages if load control is applied to renewable energy systems:

   a No environmental energy need be wasted if parallel outputs are opened and closed to take whatever input energy flow is available. Likewise, the capital-intensive equipment is well used.
   b Priorities and requirements for different types of end-use can be incorporated in many varied control modes (e.g. low priority uses can receive energy at low cost, provided that they can be switched off by feedforward control; electrical resistive heaters may receive variable voltage and hence variable power).
   c End-uses having storage capability (e.g. thermal capacity of water heating and building space conditioning) can be switched to give the benefits of storage in the system at no extra cost.
   d Electronic and microprocessor-based control may be used with benefits of low cost, reliability, and extremely fast and accurate operation.

Feedforward load control may be particularly advantageous for autonomous wind energy systems (see Chapter 9, especially Section 9.8.2). Wind fluctuates greatly in speed and the wind turbine should change rotational frequency to maintain optimum output. Rapid accurate control is necessary without adding greatly to the cost or mechanical complexity, and so electronically based feedforward control into several parallel electrical loads is most useful. An example is shown in Figure 1.6.

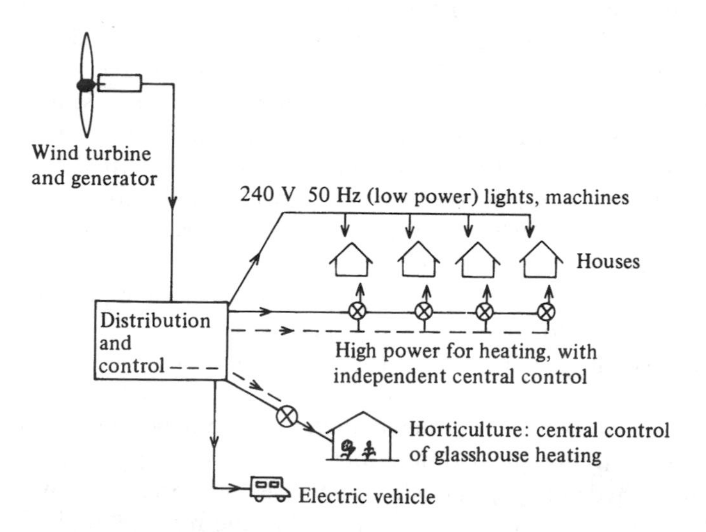


*Figure 1.6* Wind energy conversion system for Fair Isle, Scotland. Electrical loads are switched by small changes in the supply frequency, so presenting a matched load to the generator over a wide range of wind speeds.

## 1.6 Social implications

The Industrial Revolution in Europe and North America and industrial development in all countries have profoundly affected social structures and patterns of living. The influence of changing and new energy sources has been the driving function for much of this change. Thus there is a historic relationship between coal mining and the development of industrialised countries, which will continue for several hundred years. In the non-industrialised countries, relatively cheap oil supplies became available in the 1950s at the same time as many countries obtained independence from colonialism. Thus in all countries the use of fossil fuels has led to profound changes in lifestyle.

### *1.6.1 Dispersed living*

In Sections 1.1 and 1.4.4 the dispersed and small energy flux density of renewable sources was discussed. Renewable energy arrives dispersed in the environment and is difficult and expensive to concentrate. By contrast finite energy sources are energy stores that are easily concentrated at source and expensive to disperse. Thus electrical distribution grids from fossil fuel and nuclear sources tended to radiate from central, intensive distribution points, typically with ~1000 $MW_e$ capacity. Industry has developed on these grids, with heavy industry closest to the points of intensive supply. Domestic populations have grown in response to the employment opportunities of industry and commerce. Similar effects have occurred with the relationships between coal mining and steel production, oil refining and chemical engineering and the availability of gas supplies and urban complexes.

This physical review of the effect of the primary flux density of energy sources suggests that widespread application of renewable energy will favour dispersed, rather than concentrated, communities. Electricity grids in such situations are powered by smaller-scale, embedded, generation, with power flows moving intermittently in both directions according to local generation and local demand. In Section 1.2.2 an approximate estimate of 500 people per square kilometre was made of maximum population density for communities relying on renewable sources. This is considerably greater than for rural communities ($\sim$100 people per square kilometre) and corresponds with the population densities of the main administration and commercial towns of rural regions. Thus the gradual acceptance of significant supplies of renewable energy could allow relief from the concentrated metropolises of excessive urbanisation, yet would not require unacceptably low population densities. A further advantage is the increased security for a nation having its energy supplies from such indigenous and dispersed sources.

### *1.6.2 Pollution and environmental impact*

Harmful emissions can be classified as chemical (as from fossil fuel and nuclear power plant), physical (including acoustic noise and radioactivity) and biological (including pathogens); such pollution from energy generation is overwhelmingly a result of using 'brown' fuels, fossil and nuclear. In contrast, renewable energy is always extracted from flows of energy already compatible with the environment (Figure 1.1). The energy is then returned to the environment, so no thermal pollution can occur on anything but a small scale. Likewise material and chemical pollution in air, water and refuse tend to be minimal. An exception is air pollution from incomplete combustion of biomass or refuses (see Chapter 11). Environmental pollution does occur if brown energy is used for the materials and manufacture of renewable energy devices, but this is small over the lifetime of the equipment.

The environmental impact of renewables depends on the particular technology and circumstances. We consider these in the last section of each technology chapter that follows. General institutional factors, often related to the abatement of pollution, are considered in the last chapter.

### *1.6.3 The future*

In short, we see that many changes in social patterns are related to energy supplies. We can expect further changes to occur as renewable energy systems become widespread. The influence of modern science and technology ensures that there are considerable improvements to older technologies, and subsequently standards of living can be expected to rise, especially in rural and previously less developed sectors. It is impossible to predict exactly the long-term effect of such changes in energy supply, but the sustainable nature of renewable energy should produce greater socio-economic stability than has been the case with fossil fuels and nuclear power. In particular we expect the great diversity of renewable energy supplies to be associated with a similar diversity in local economic and social characteristics.

## Problems

1.1 a Show that the average solar irradiance absorbed during 24 h over the whole Earth's surface is about 230 W (see Figure 1.2)

b Using devices, the average local power accessible can be increased, e.g. by tilting solar devices towards the Sun, by intercepting winds. Is it reasonable to state that 'each square metre of the Earth's *habitable* surface is crossed or accessible to an average flux of about 500 W'?

1.2 a Compare the direct costs to the consumer of using:

i a succession of ten 100 W incandescent light bulbs with an efficiency for electricity to visible light of 5%, life of 1 000 h, price €0.5;

ii one compact fluorescent lamp (CFL) giving the same illumination at 22% efficiency, life of 10 000 h, price €3.0. Use a fixed electricity price of €0.10 $kW\,h^{-1}$;

b what is the approximate payback time in lighting-hours of (b) against (a). [See also Problem 17.1 that allows for the more sophisticated discounted costs.]

1.3 Repeat the calculation of Problem 1.2, with tariff prices of your local lamps and electricity. Both the price of CFL's in local shops and of electricity vary markedly, so your answers may differ significantly. Nevertheless it is highly likely the significant lifetime savings will still occur.

1.4 Economists argue that as oil reserves become smaller, the price will increase, so demand will decrease and previously uneconomic supplies will come into production. This tends to make the resource last longer than would be suggested by a simple calculation (based on 'today's reserves' divided by 'today's use') . On the other hand, demand increases driven by increased economic development in developing countries tend to shorten the life of the reserve. Discuss.

## Bibliography

Refer to the bibliographies at the end of each chapter for particular subjects and technologies.

### *Surveys of renewable energy technology and resources*

Boyle, G. (ed.) (2004, 2nd edn) *Renewable Energy*, Oxford University Press. {Excellent introduction for both scientific and non-scientific readers.}

Jackson, T. (ed.) (1993) *Renewable Energy: Prospects for implementation*, Butterworth-Heinemann, Oxford. {Collection of a series of articles from the journal *Energy Policy*, with focus on implementation rather than technical detail.}

Johansson, T.J., Kelly, H., Reddy, A.K.N., Williams, R.H. (eds) (1993) *Renewable Energy: Sources for fuels and electricity*, produced for the UN Solar Energy Group for Environment and Development, Earthscan, London, and Island Press, Washington DC, 1000pp. {An authoritative study; but does not attempt to include the built environment.}

Sørensen, B. (2004, 3rd edn) *Renewable Energy*, Academic Press, London. {Outstandingly the best theoretical text at postgraduate level, considering energy from the environment to final use.}

US Department of Energy (1997) *Renewable Energy Characterisations*, US-DOE Topical Report TR-109496. [Available on website www.eere.gov] {Emphasis on prospects for electricity generation and R&D requirements.}

### *Energy, society and the environment (including 'sustainable development')*

[see also the bibliography for chapter 17]

Boyle, G., Everett, R. and Ramage, J. (eds) (2003) *Energy Systems and Sustainability: Power for a Sustainable Future*, Oxford UP in association with the Open University. {Good non-technical account for 'science and society' courses.}

Cassedy, E.S. and Grossman, P.G. (2002, 2nd edn) *Introduction to Energy: Resources, Technology and Society*, Cambridge UP. {Good non-technical account for 'science and society' courses.}

Elliot, D. (2003, 2nd edn) *Energy, Society and the Environment*, Routledge. {Brief survey of technologies, but more extensive discussion of institutional and societal aspects.}

Goldemberg, J. (1996) *Energy, Environment and Development*, Earthscan (with James and James), London. {Wide-ranging and readable exposition of the links between energy and social and economic development and sustainability, with consideration of equity within and between countries by a Brazilian expert.}

Houghton, J.T. (1997, 2nd edn) *Global Warming: The Complete Briefing*, Cambridge UP. {Less technical and more committed than the official IPCC report (Sir John Houghton was co-chair of IPCC).}

Intergovernmental Panel on Climate Change (IPCC) (2001) *Third Assessment Report* – 3 vols – see especially 'Summary for Policy Makers: synthesis report' (see IPCC website listed below). {The full report is 3 large volumes.}

International Energy Agency (IEA) (2001) *Toward a Sustainable Energy Future*, Paris. {Emphasises the 'economic dimension' of sustainable development.}

McNeill, J. R. (2000) *Something New Under the Sun: An Environmental History of the Twentieth Century*, Penguin, London. {The growth of fossil-fuel-fired cities and their impacts on water, air and the biosphere.}

Ruedisili, L.C. and Firebaugh, M.W. (eds) (1978, 2nd edn) *Perspectives on Energy*, Oxford University Press. {Well-chosen collection of reprints, often of contrasting views. Illustrates that many of the issues (including some of the funding issues concerning renewable energy) have not changed from the 1970s, when the policy motivation for 'alternative energy sources' was oil shortages rather than greenhouse gas emissions.}

Twidell, J., Hounam, I. and Lewis, C. (1986) *Energy for Highlands and Islands*, IV proceedings of fourth annual conference on this subject, Pergamon, Oxford. {Descriptions of some of the early circumstances outlined in Chapter 1.}

Von Weizsacker, E., Lovins, A.B. and Lovins, H. (2000) *Factor Four: Doubling Wealth, Halving Resource Use*, Penguin, London. {Explores the wider social and political issues of energy supply, especially those associated with renewable and nuclear supplies.}

World Commission on Environment and Development (1987) *Our Common Future*, Oxford University Press (the 'Bruntland report'). {A seminal work, warning about the key issues in plain language for politicians.}

### *Official publications (including energy statistics and projections)*

See also below under journals and websites, as many official publications, especially those of a statistical nature, are updated every year or two. United Nations agencies produce a wide range of essential publications regarding energy. These are especially important for data. For instance we recommend:

United Nations *World Energy Supplies*, UN document no. ST/ESA/STAT/SER.J/19, annual. Gives statistics of energy consumption around the world, classified by source, country, continent, etc., but counts only 'commercial energy' (i.e. excludes firewood, etc.).

Government publications are always important. For instance, UK Department of Energy Series of Energy Papers. Such publications are usually clearly written and include economic factors at the time of writing. Basic principles are covered, but usually without the details required for serious study. Annual updates of many of government and UN publications are also available through the corresponding websites.

World Energy Council (2001) *Survey of world energy resources*. {Compiled every 5 years or so by the WEC, which comprises mainly energy utility companies from around the world; covers both renewable and non-renewable resources.}

International Energy Agency, *World Energy Outlook* (annual), Paris. {Focus is on fossil fuel resources and use, based on detailed projections for each member country, and for those non-member countries which are significant in world energy markets, e.g. OPEC and China.}

### *Do-it-yourself publications*

There are many publications for the general public and for enthusiasts. Do not despise these, but take care if the tasks are made to look easy. Many of these publications give stimulating ideas and are attractive to read, e.g. Merrill, R. and Gage, T. (eds) *Energy Primer*, Dell, New York (several editions).

Berrill, T. *et al.* (eds) (2003, 4th edn) *Introduction to Renewable Energy Technologies: Resource book*, Australian Centre for Renewable Energy, Perth. {Written with tradespeople in mind; clearly describes available equipment with basic account of principles; emphasis on practicalities of installation and operation.}

### *Journals, trade indexes and websites*

Renewable energy and more generally energy technology and policy are continually advancing. Use a web search engine for general information and for technical explanations and surveys. For serious study it is necessary to refer to the periodical literature (journals and magazines), which is increasingly available on the web, with much more available by payment. Websites of key organisations, such as those listed below and including governmental sites with statistical surveys, also carry updates of our information.

We urge readers to scan the serious scientific and engineering journals, e.g. *New Scientist*, *Annual Review of Energy and the Environment*, and magazines, e.g. *Electrical Review*, *Modern Power Systems*. These publications regularly cover renewable energy projects among the general articles. The magazine *Refocus*, published for the International Solar Energy Society, carries numerous well-illustrated articles on all aspects of renewable energy. The series *Advances in Solar Energy*, published by the American Solar Energy Society, comprises annual volumes of high level reviews, including all solar technologies and some solar-derived technologies (e.g. wind power and biomass). There are also many specialist journals, such as *Renewable and Sustainable Energy Reviews*, *Solar Energy*, *Wind Engineering*, and *Biomass and Bioenergy*, referred to in the relevant chapters.

As renewable energy has developed commercially, many indexes of companies and products have been produced; most are updated annually, e.g. *European Directory of Renewable Energy Supplies and Services*, annual, ed. B. Cross, James and James, London.

www.iea.org
The International Energy Agency (IEA) comprises the governments of about 20 industrialised countries; its publications cover policies, energy statistics and trends, and to a lesser extent technologies; it also co-ordinates and publishes much collaborative international R&D, including clearly written appraisals of the state of the art of numerous renewable energy technologies. Its publications draw on detailed inputs from member countries.

www.wec.org
The World Energy Council comprises mainly energy utility companies for around the world, who cooperate to produce surveys and projections of resources, technologies and prices.

www.ipcc.ch
The Intergovernmental Panel on Climate Change (IPCC) is a panel of some 2000 scientists convened by the United Nations to report on the science, economics and mitigation of greenhouse gases and climate change; their reports, issued every five years or so, are regarded as authoritative. Summaries are available on the website.

www.itdg.org
ITDG (the Intermediate Technology Development Group) develops and promotes simple and cheap but effective technology – including renewable energy technologies – for use in rural areas of developing countries. They have an extensive publication list plus on-line 'technical briefs'.

www.caddet.org
The acronym CADDET stands for Centre for the Analysis and Dissemination of *Demonstrated* Energy Technologies. The site gives information and contact details about renewable energy and energy efficiency projects from many countries.

www.ewea.org
European Wind Energy Association is one of many renewable energy associations, all of which have useful websites. Most such associations are 'trade associations', as funded by members in the named renewable energy industry. However, they are aware of the public and educational interest, so will have information and give connections for specialist information.

Chapter 2

# Essentials of fluid dynamics

## 2.1 Introduction

Several renewable energy resources derive from the natural movement of air and water. Therefore the transfer of energy to and from a moving fluid is the basis of meteorology and of hydro, wind, wave and some solar power systems. Examples of such applications include hydropower turbines (Figures 8.3, 8.5 and 8.6), wind turbines (picture on front cover and Figure 9.4), solar air heaters (Figure 6.1) and wave energy systems (Figure 12.14).

To understand such systems, we must start with the basic laws of mechanics as they apply to fluids, notably the laws of conservation of mass, energy and momentum. The term *fluid* includes both liquids and gases, which, unlike solids, do not remain in equilibrium when subjected to shearing forces. The hydrodynamic distinction between liquids and gases is that gases are easily compressed, whereas liquids have volumes varying only slightly with temperature and pressure. Gaseous volumes vary directly with temperature and inversely with pressure, approximately as the perfect-gas law ($pV = nRT$). Nevertheless, for air, flowing at speeds $<100\,\text{m}\,\text{s}^{-1}$ and not subject to large imposed variations in pressure or temperature, density change is negligible; this is the situation for the renewable energy systems analysed quantitatively in this book. It does not apply to the analysis of gas turbines, for which specialist texts should be consulted. Therefore, throughout this text, moving air is considered to have the fluid dynamics of an *incompressible* fluid. This considerably simplifies the analysis of most renewable energy systems.

Many important fluid flows are also *steady*, i.e. the particular type of flow *pattern* at a location does not vary with time. So it is useful to picture a set of lines, called *streamlines*, parallel with the velocity vectors at each point. A further distinction is between laminar and turbulent flow (Section 2.5). For example, watch the smoke rising from a smouldering taper in still air. Near the taper, the smoke rises in an orderly, *laminar*, stream, with the paths of neighbouring smoke particles parallel. Further from the taper, the flow becomes chaotic, *turbulent*, with individual smoke particles intermingling in three dimensions. Turbulent flow approximates to a steady mean flow, subject to internal friction caused by the velocity fluctuations. However, even in turbulence, the airflow remains within well-defined (though imaginary) *streamtubes*, as bounded by streamlines.

## 2.2 Conservation of energy: Bernoulli's equation

Consider the most important case of steady, incompressible flow. At first, we assume no work is done by the moving fluid on, say, a hydro turbine.

Figure 2.1 shows a section of a streamtube between heights $z_1$ and $z_2$. The tube is narrow in comparison with other dimensions, so $z$ is considered constant over each cross-section of the tube. A mass $m = \rho A_1 u_1 \Delta t$ enters the control volume at 1, and an equal mass $m = \rho A_2 u_2 \Delta t$ leaves at 2 (where $\rho$ is the density of the fluid, treated as constant). Then the energy balance on the fluid within the control volume is

potential energy lost + work done by pressure forces

= gain in kinetic energy + heat losses due to friction

and may be written as

$$mg(z_1 - z_2) + [(p_1 A_1)(u_1 \Delta t) - (p_2 A_2)(u_2 \Delta t)] = \frac{1}{2} m(u_2^2 - u_1^2) + E_f \qquad (2.1)$$

where the pressure force $p_1 A_1$ acts through a distance $u_1 \Delta t$, and similarly for $p_2 A_2$, and $E_f$ is the heat generated by friction.

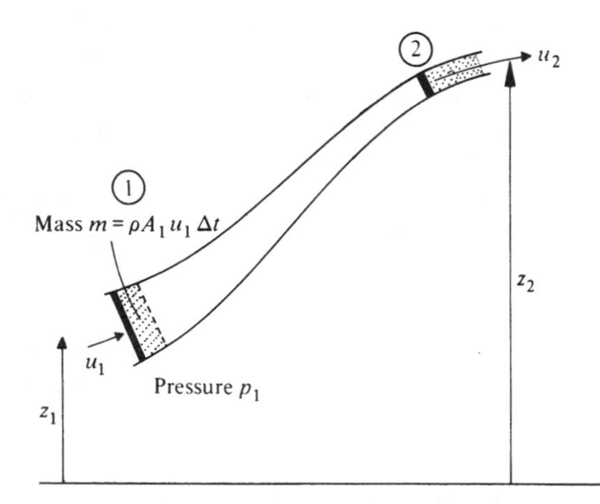


*Figure 2.1* Illustrating conservation of energy: a streamtube rises from height $z_1$ to $z_2$.

We neglect fluid friction, $E_f$, now, but we will examine some of its effects in Section 2.5. In this ideal, frictionless case, (2.1) reduces to

$$\frac{p_1}{\rho} + gz_1 + \frac{1}{2}u_1^2 = \frac{p_2}{\rho} + gz_2 + \frac{1}{2}u_2^2 \tag{2.2}$$

or, equivalently, so each term has the dimension of height (m)

$$\frac{p}{\rho g} + z + \frac{u^2}{2g} = \text{constant along a streamline, with no loss of energy} \tag{2.3}$$

Either of these forms of the equation is called *Bernoulli's equation*. Equations (8.10) and (9.19) are examples of its application in hydro and wind power respectively.

The sum of the terms on the left of (2.3) is called the total *head* of fluid ($H$). It relates to the total energy of a unit mass of fluid, however the constant in (2.3) may vary from streamline to streamline. Moreover, for many situations, the friction losses, $E_f$, have to be included. Head has the dimensions of length. For hydropower, head is the effective height of the moving water column incident on the turbine – see Section 8.3.

The main limitation of (2.2) and (2.3) is that they apply only to fluids treated as ideal, i.e. with zero viscosity, zero compressibility and zero thermal conductivity. However, this is applicable to wind and hydro turbines with their relatively low-speed movement of air and water, and with no internal heat sources. The energy equation can however be modified to include non-ideal characteristics (see Bibliography), as for combustion engines and many other thermal devices, e.g. high temperature solar collectors.

In solar heating systems and heat exchangers, power $P_{th}$ is added to the fluid from heat sources (Figure 2.2). Heat $E = P_{th}\Delta t$ is added to the energy inputs on the left hand side of (2.1). The mass $m$ coming into the control volume at temperature $T_1$ has heat content $mcT_1$ (where $c$ is the specific heat capacity of the fluid), and that going out has heat content $mcT_2$. Thus we add to the right hand side of (2.1) the net heat carried out of the control volume in time $\Delta t$, namely $mc(T_2 - T_1)$. This gives an equation corresponding to (2.2), namely

$$\frac{p_1}{\rho} + gz_1 + \frac{1}{2}u_1^2 + cT_1 + \frac{P_{th}}{\rho Q} = \left(\frac{p_2}{\rho}\right) + gz_2 + \frac{1}{2}u_2^2 + cT_2 \tag{2.4}$$

where the volume flow rate is

$$Q = Au \tag{2.5}$$

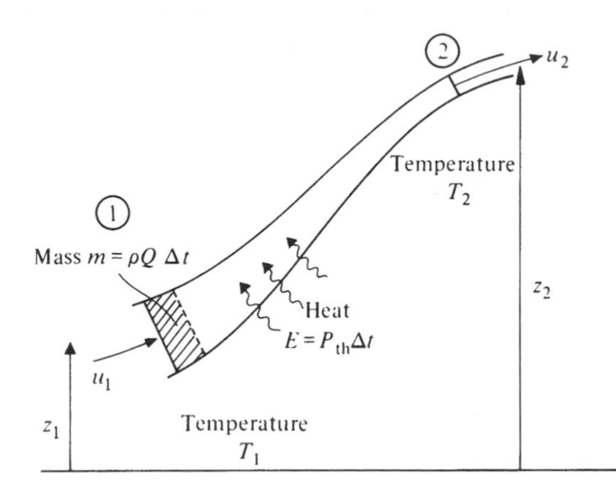


*Figure 2.2* As for Figure 2.1, but with heat sources present. Thermal power $P_{th}$ is added to the flow.

In most heating systems, including active solar heating (see Equation 5.2 onwards), thermal contributions dominate the energy balance. So (2.4) reduces to

$$P_{th} = \rho c Q(T_2 - T_1) \tag{2.6}$$

## 2.3 Conservation of momentum

Newton's second law of motion may be defined for fluids as: 'At any instant in steady flow, the resultant force acting on the moving fluid within a fixed volume of space equals the net rate of outflow of momentum from the closed surface bounding that volume.' This is known as the *momentum theorem*.

As an example, consider a fluid passing across a turbine in a pipe. In Figure 2.3, fluid flowing at speed $u_1$ into the left of the control surface carries momentum $\rho u_1$ *per unit volume* in the direction of flow. In time $\Delta t$, the volume entering the surface is $A_1 u_1 \Delta t$. Therefore the *rate* at which the momentum is entering the control surface along the $x$ direction is

$$\frac{(A_1 u_1 \Delta t)(\rho u_1)}{\Delta t} = \rho A_2 u_1^2 \tag{2.7}$$

Similarly the rate at which momentum is leaving the control volume is $\rho A_1 u_2^2$. The momentum theorem tells us that the rate of change of momentum equals the force, $F$ on the fluid and the reaction, $-F$ is the force exerted on the turbine and pipe by the fluid. So

$$F = \rho(A_2 u_2^2 - A_1 u_1^2) = \dot{m} u_2 - \dot{m} u_1 \tag{2.8}$$

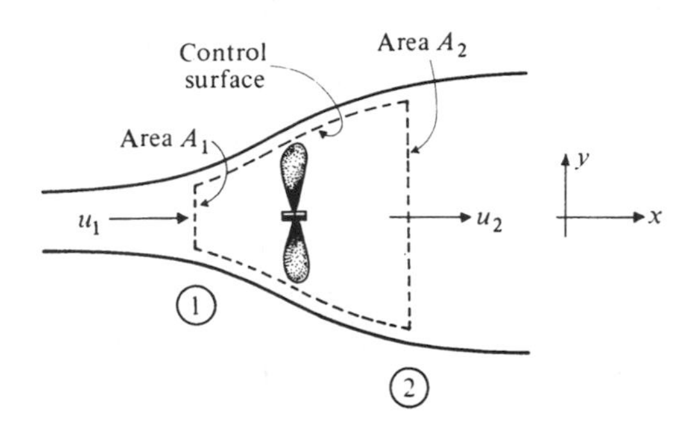


*Figure 2.3* A turbine in a pipe. The dotted line shows the control surface over which the momentum theorem is applied.

where $\dot{m} = \rho A_1 u_1 = \rho A_2 u_2$ is the mass flow and the signs indicate directions, which are obvious in this case. However, in more complex cases, such as inside a turbine, it is important to remember that momentum and force are vectors (i.e. their direction matters!).

## 2.4 Viscosity

Suppose we have two parallel plates with fluid filling the gap between them, and the top plate moving at a velocity $u_1$ relative to the bottom one, as shown in Figure 2.4. We choose axes as shown, with $x$ in the direction of motion, and $y$ across the gap between the plates. It is found experimentally at a macroscopic scale that *fluid does not slip at a solid surface*. Thus the bulk fluid immediately adjacent to each plate is considered to have the same speed and direction of movement as the plate.

However, at much smaller scale in the fluid, the molecules have the velocity of the fluid plus additional random (thermal) motion. This random molecular motion has the effect of transferring larger momentum (acquired from the top plate) downwards and smaller momentum (acquired from the bottom plate) upwards. This net *diffusion of momentum* limits the velocity

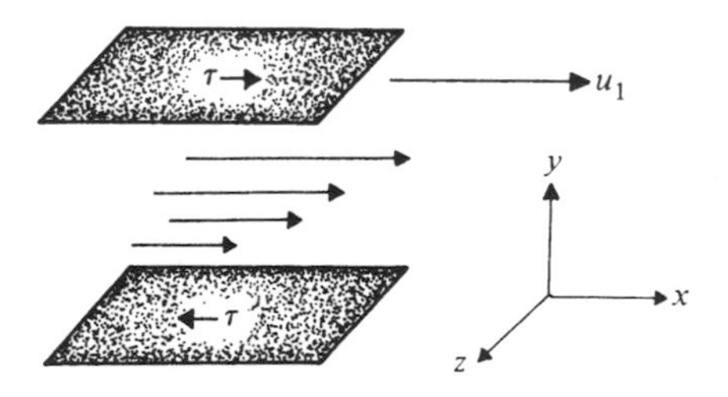


*Figure 2.4* Flow between two parallel plates.

gradient that the fluid can sustain, producing an internal friction opposing the horizontal slip in the flow. It is found that the shear stress (i.e. the force per unit area, in the direction indicated in Figure 2.4) is

$$\tau = \mu\,(\partial u/\partial y) \tag{2.9}$$

where $\mu$ is the *dynamic viscosity* (unit $\mathrm{N\,s\,m^{-2}}$). This viscosity is independent of $\tau$ and $\partial u/\partial y$, and depends only on the composition and temperature of the fluid.

A closely related fluid parameter is *kinematic viscosity*, defined as

$$v = \mu/\rho \tag{2.10}$$

In incompressible fluids, the flow pattern often depends more directly on $\nu$ than on $\mu$. By combining (2.9) and (2.10), we find that the units of $\nu$ are

$$\frac{(\mathrm{kg\,m\,s^{-2}})\mathrm{m^{-2}}}{\mathrm{kg\,m^{-3}}}\frac{\mathrm{m}}{\mathrm{m\,s^{-1}}} = \mathrm{m^2\,s^{-1}}$$

Thus $\nu$ has the character of a diffusivity. That is to say, the time taken for a change in momentum to diffuse a distance $x$ is $x^2/\nu$ (cf. thermal diffusivity $\kappa$ defined in (3.15)). Typical values of $\nu$ are given in Appendix B.

## 2.5 Turbulence

Turbulent flow occurs because, in general, rapid fluid motion is unstable. Suppose fluid is initially flowing through a pipe in the orderly manner, illustrated by the pathlines in Figure 2.5(a). Eventually, something will disturb the motion, e.g. an obstruction in or a knock on the pipe. Consequently, elemental volumes, described here as 'elements', will alter course and, if they are moving rapidly enough, fluid friction will not be able to restore them to their original paths. Moreover the disturbed elements disturb other elements of fluid from their original path, and soon the entire flow is in the semi-chaotic state called *turbulence*, illustrated in Figure 2.5(b).

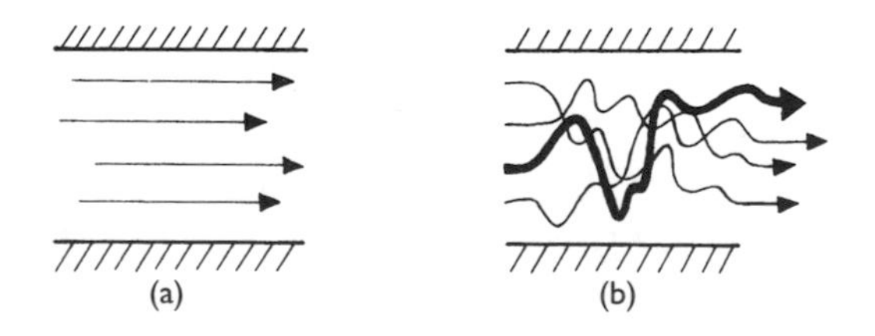


*Figure 2.5* Pathlines of flow in a pipe (a) laminar (b) turbulent.

It is the ratio of fluid momentum (arising from 'inertia forces') to viscous friction which determines whether the flow becomes smooth (i.e. laminar) or turbulent. This ratio is usually characterised by the non-dimensional *Reynolds number*

$$\mathcal{R} = \frac{uX}{\nu} \tag{2.11}$$

Here $u$ is the mean speed of the flow, $X$ is a nominated *characteristic length* of the system (in this case the diameter of the pipe) and $\nu$ $(= \mu/\rho$; as in (2.10)) is the kinematic viscosity of the fluid. The value of $\mathcal{R}$ is important for characterising types of fluid flow, e.g. turbulence. For instance in pipes, flow will be usually turbulent if $\mathcal{R}$ is larger than about 2300 ($\mathcal{R} \gtrsim 2300$). Criteria for laminar or turbulent flow in heat transfer are discussed in Chapter 3.

In turbulent flow, random local fluctuations of velocity in three dimensions are imposed on the mean flow. Thus small elements of fluid moving along the pipe also move rapidly inwards and outwards across the pipe, as illustrated by Figure 2.5(b). Since fluid does not slip at the pipe surface (Section 2.4), the mean speed near the surface is smaller than the average and the mean speed near the centre of the pipe is correspondingly larger. Therefore the effect of the sideways motions of the fluid elements is to carry fluid of larger velocity outwards, and fluid of smaller velocity inwards. This transfer of momentum by elements of fluid is much larger than the corresponding transfer by molecular motions described in Section 2.4 because an element of fluid may move significantly across the pipe in a single jump. In this case with water as the fluid, the mean free path of a molecule in the liquid is of the order of nanometres.

This transfer of momentum from the fluid to the walls constitutes a sizeable friction force opposing the motion of the fluid. Thus the presence of turbulence *increases* friction as compared with laminar flow; it is important to appreciate this characteristic of turbulent flow as, for instance, in the aerodynamics of wind turbines.

If the walls of the pipe are hotter than the incoming fluid, then these rapid inward and outward motions transfer heat rapidly to the bulk of the fluid. An element of cold fluid can jump from the centre of the pipe, pick up heat by conduction from the hot wall, and then carry it much more rapidly back into the centre of the pipe than could molecular conduction. Thus *turbulence* increases heat transfer, as discussed in more detail in Section 3.4. and as applicable in the design of active solar systems.

## 2.6 Friction in pipe flow

Due to friction, useful energy and pressure are 'lost' or 'dissipated' when a fluid flows through pipes. Such factors may cause significant inefficiency in hydropower (Chapter 8), in ocean thermal energy conversion (OTEC,

Chapter 14) and in all applications where heat is transferred by mass flow (e.g. Section 3.7 and, for solar energy, Section 5.5).

Let $\Delta p$ be the pressure overcoming friction, as fluid moves at average speed $u$, through the pipe of length $L$ and diameter $D$. Since the flow is statistically uniform along the pipe, each meter of pipe is considered to contribute the same friction. Therefore $\Delta p$ increases with $L$. Since much of the resistance to the flow originates from the no-slip condition at the walls (Section 2.4), moving the walls closer to the bulk movement of the fluid increases the friction. Therefore $\Delta p$ increases as $D$ decreases. Equation (2.9) implies that fluid friction increases with flow speed, so that $\Delta p$ increases with $u$. Bernoulli's equation (2.3) shows that the quantity $\rho u^2/2$ has the same dimensions as $p$ (i.e. $\mathrm{kg\,(m\,s^2)^{-1}}$). All these characteristics can be expressed in the single equations

$$\Delta p = 2f(L/D)(\rho u^2) = 4f(L/D)(\rho u^2/2) \tag{2.12}$$

or

$$\Delta p = f'(L/D)(\rho u^2/2) \tag{2.13}$$

Here $f$ and $f'(=4f)$ are dimensionless pipe *friction coefficients* that change value with experimental conditions. Two equations are given because there are (unfortunately) alternative conventions for the definition of friction coefficient. As with many non-dimensional factors in engineering, the magnitudes of $f$ and $f'$ characterise the physical conditions independently of the scale, depending only on the *pattern* of flow, i.e. the *shape* of the streamlines.

This is because the factor $\rho u^2/2$ in (2.12) and (2.13) represents a natural unit of pressure drop in the pipe. The friction coefficient is the *proportion* of the kinetic energy $(\rho u^2/2)$ entering unit area of the pipe that has to be applied as external work $(\Delta p)$ to overcome frictional forces. This will depend on the time that a typical fluid element spends in contact with the pipe wall, expressed as a *proportion* of the time the element takes to move a unit length along the pipe. This proportion is much larger for the turbulent paths (Figure 2.5(b)) than for the laminar paths (Figure 2.5(a)). Fluid flow depends mainly on the dimensionless Reynolds number $\mathcal{R}$ of (2.11). A plot of $f$ or $f'$ against $\mathcal{R}$ should give a single curve applying to pipes of any length and diameter, carrying any fluid at any speed. There is no particular reason why this curve should be a straight line, or even continuous. Indeed we might expect a discontinuity at $\mathcal{R} \geq 2000$, where the flow pattern changes from laminar to turbulent. This curve, shown in Figure 2.6, does indeed have a discontinuity at $\mathcal{R} \approx 2000$. If we consider real pipes with rough walls, it is reasonable to suppose that the flow pattern depends on the ratio of the height, $\xi$, of the surface bumps to the diameter

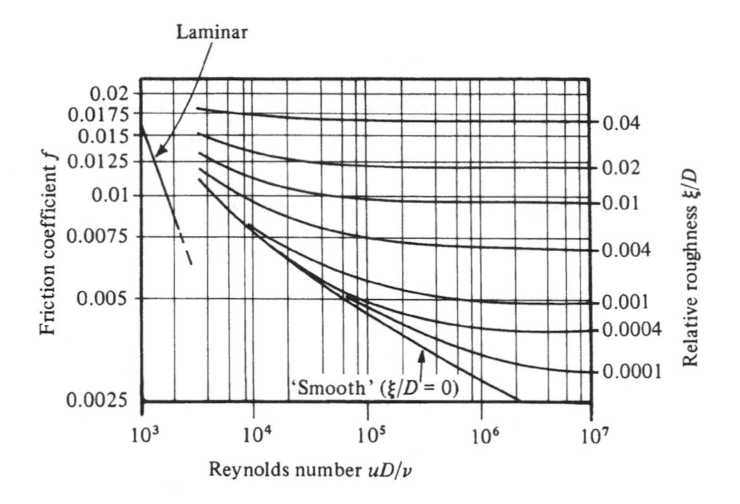


*Figure 2.6* Friction coefficient $f$ for pipe flow (see (2.12)).

*Table 2.1* Approximate pipe roughness $\xi$

| *Material* | $\xi$ (mm) |
|---|---|
| PVC | 0 ('smooth') |
| Asbestos cement | 0.012 |
| New steel | 0.1 |
| 'Smooth' concrete | 0.4 |

of the pipe, $D$. Plotting the experimental data on this basis, Figure 2.6, we obtain a series of curves, with one curve for each roughness ratio $\xi/D$.

Provided the appropriate value of $\xi$ is used, these curves give a reasonable estimate of pipe friction. Typical values of $\xi$ are given in Table 2.1, but it should be realised that the roughness of a pipe tends to increase with age and, very noticeably, with accretion of sediments and encrustations. For example, this presents a significant maintenance factor for hydropower installations.

*Example 2.1*

What is the pressure head required to force $0.10\,\mathrm{m^3\,s^{-1}}$ of water at $20\,^\circ\mathrm{C}$, through a concrete pipe of length 200 m and diameter 0.30 m?

*Solution*

The mean speed is

$$u = \frac{Q}{A} = \frac{0.1\,\mathrm{m^3\,s^{-1}}}{\pi(0.15\,\mathrm{m})^2} = 1.4\,\mathrm{m\,s^{-1}}$$

From (2.11), the Reynolds number

$$\mathcal{R} = \frac{uD}{v} = \frac{(1.4\,\text{m s}^{-1})(0.3\,\text{m})}{1.0 \times 10^{-6}\,\text{m}^2\,\text{s}^{-1}} = 0.4 \times 10^6 \geq 2000$$

Therefore flow is turbulent.
For concrete (from Table 2.1), $\xi = 0.4\,\text{mm}$. Thus the ratio

$$\frac{\xi}{D} = \frac{0.4\,\text{mm}}{300\,\text{mm}} = 0.0013$$

For this $\mathcal{R}$ and $\xi/D$, Figure 2.6 gives

$$f = 0.0050$$

and

$$f' = 0.020$$

Expressing (2.12) in terms of the head loss due to friction,

$$H_\text{f} = \frac{\Delta p}{\rho g} = \frac{2fLu^2}{Dg} \tag{2.14}$$

Hence

$$\begin{aligned} H_\text{f} &= \frac{(2)(5.0 \times 10^{-3})(200\,\text{m})(1.4\,\text{m s}^{-1})^2}{(0.3\,\text{m})(9.8\,\text{m s}^{-2})} \\ &= 1.3\,\text{m} \end{aligned}$$

Figure 2.6 shows only one curve for $\mathcal{R} < 2000$, indicating that the friction coefficient is independent of pipe roughness $\xi$ in this range. This is because the flow is laminar, and the bumps hardly disturb the smoothness of the flow. In this laminar case, the pressure drop $\Delta p$ can be explicitly calculated from (2.9) for viscous shear stress, as indicated in Problem 2.4. The corresponding expression for the friction coefficient is

$$f = \frac{16v}{uD} \quad \text{(laminar)} \tag{2.15}$$

or

$$f' = 4f = \frac{64v}{uD} \quad \text{(laminar)} \tag{2.15a}$$

## 2.7 Lift and drag forces: fluid and turbine machinery

Here we introduce the forces of lift and drag which are as fundamental to turbine motion as they are to sailing yachts and airplanes, for they apply to any solid object immersed in a fluid flow. Obtaining rotary motion on a shaft from a flow of water or air is the basis of every turbine, relating, in this book, to hydro turbines (Section 8.4 onwards for conventional hydropower, Sections 13.4 and 13.5 for tidal power), wind turbines (Section 9.2 onwards), and wave power turbines, (included in Section 12.5).

In Figure 2.7(a) a solid object is immersed asymmetrically in a fluid so there is a relative fluid velocity flowing from left to right. However, because of intricacies of the flow pattern passing the object, the resulting force on the object is unlikely to be parallel to the upstream flow. If the total (vector)

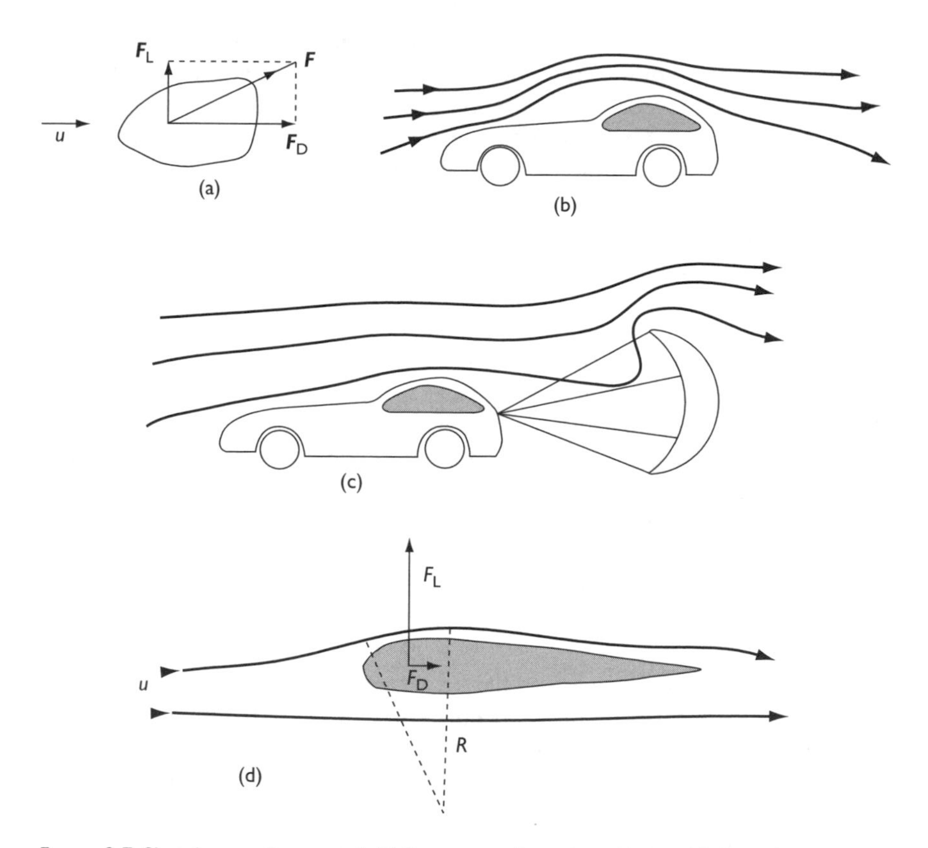


*Figure 2.7* Sketches to illustrate fluid flow around moving objects. (a) Any object moving at relative velocity $u$ will experience both lift and drag forces. (b) Smooth streamlines to reduce drag. (c) Contorted streamlines to increase drag. (d) Section of airfoil wing, shaped to increase lift and reduce drag.

force exerted on the body is $\boldsymbol{F}$; the *drag* force $\boldsymbol{F}_D$ is the component of that force in the direction of the upstream flow and the *lift* force $\boldsymbol{F}_L$ is the component normal to the flow. It is the lift force that twists and turns the object. To illustrate how these forces arise, consider vehicles driving down a long straight open road in otherwise still air. The drag force from air resistance is reduced if the streamlines flow smoothly around the vehicle, as in the 'streamlined' car shown in Figure 2.7(b). However, even with perfect streamlining, the drag force will not be zero because of the viscous friction between the fluid and the surface of the solid, as described in Section 2.4. Since turbulence greatly increases the effect of such friction, as explained in Section 2.5, a vehicle designer seeking minimum fuel consumption and good performance reduces drag with fluid flow around the vehicle as smooth as possible, with few sharp corners and projecting parts.

Figure 2.7(c) illustrates a vehicle being slowed down by a parachute, as in some cars seeking speed records. Increased drag, the opposite of streamlining, is now required. Some typical fluid pathlines are indicated from the *relative* motion of the body and the fluid that determines the flow pattern, and as measured by an observer moving beside at the velocity of the car. In the observer's frame of reference, the fluid air hits the parachute and loses most of its forward momentum. By the momentum theorem, a force has been exerted on the fluid by the body (i.e. by the car and its parachute). Consequently, the fluid has exerted an equal and opposite force on the body. This force is in the direction of the flow and is a classic example of a drag force. This set of forces is analogous to that in an impulse turbine (compare Figure 8.3 for hydro power, a cup anemometer (Figure 9.4b) and a wind 'drag machine' (Section 9.3.4)).

In Figure 2.7(d), the 'vehicle' is an airplane wing; such an *aerofoil* is thin, with a sharp trailing edge and more curved on the top than on the bottom. This deflects and changes the flow so a positive upward force (lift) occurs, so enabling the aircraft to take off and fly. Lift force can be experienced by holding one's arm out of a moving car rear window and shaping the hand as an aerofoil; drag, of course, will also be experienced. Aerofoils are used in many other applications besides airplanes, in particular for the blades of wind turbines, where the mechanics are dominated by lift forces (Section 9.5).

Basically, an aerofoil generates lift because the flow near the top surface follows a curved path. If the local curvature of the streamline above the top surface is $R$ (as indicated in Figure 2.7(d)) then the fluid experiences an acceleration of $u^2/R$ inwards (i.e. in this case downward towards the aerofoil); the calculation is the same as that of the centripetal force acting on a particle pulled by a string into a circular motion, given in elementary physics texts. Thus the aerofoil creates a force pulling downward on the fluid above it. By the momentum theorem, the fluid correspondingly exerts an upward force on the wing. Since the curvature is less on the lower side, the corresponding forces (now upward on the fluid, downward on the aerofoil) are smaller than those on the upper side. The result is a net upward force, lift, on the aerofoil. This force is manifested by the fluid having a higher pressure below the aerofoil than above it – a result that can be derived from Bernoulli's equation (2.3).

The same principles apply, yet the details are more complicated, when an airplane flies upside down, when a vertical-axis wind turbine blade passes across the wind every 180° of rotation (Figure 9.4b) and when air motion reverses across a Wells turbine in an oscillating-column wave power machine (Figure 12.14).

Stall for an airplane occurs when the lift force is less than the gravitational force. This can occur because the relative speed of the air and the wings has lessened, or because the orientation of all or parts of the wing has changed. Stall is a dangerous condition for an airplane, but can be helpful in a wind turbine to prevent overspeed.

## Problems

2.1 Figure 2.8(a) shows an ideal *Venturi meter* for measuring flow in a pipe.

a Use the equations expressing conservation of mass and conservation of energy to show that the volume of fluid flowing past cross-section 1 of the pipe per unit time is

$$Q = u_1 A_1 = A_1\{(A_1/A_2)^2 - 1\}^{-1/2}\left\{2g\left[\frac{p_1 - p_2}{\rho g} + (z_1 - z_2)\right]\right\}^{\frac{1}{2}}$$

b What is the volume per second flowing past cross-section 2 in Figure 2.8(a)?

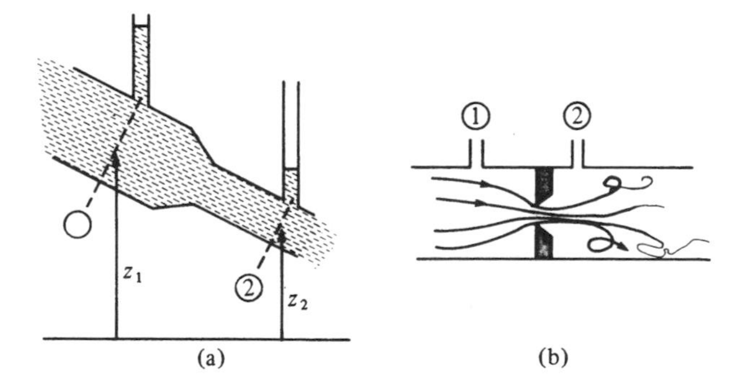


*Figure 2.8* For Problem 2.1. Venturi meters for measuring flow in a pipe. (a) Smooth contraction in diameter along the pipe. (b) Sharp-edged orifice.

c The pressures $p_1$ and $p_2$ are measured by the heights of columns of fluid rising up the side tappings, as shown in Figure 2.8(a). Indicate on the diagram a distance corresponding to the factor in square brackets in the expression in Problem 2.1(a).

d Figure 2.8(b) shows a similar system for measuring flow in a pipe of nominally uniform cross-section. A plate with a sharp-edged hole in it is inserted as shown, and the pressure is measured on either side. Will the flow rate calculated from the equation of Problem 2.1(a) be correct, too high or too low? Why?

*Hint*: Consider the full energy equation (2.1).

2.2 A two-dimensional nozzle discharges water, as a plane horizontal jet of thickness $t$ and width $b$. The jet then strikes a large inclined flat surface as shown in Figure 2.9.

a Neglecting the effects of gravity and viscosity, apply Bernoulli's equation to show that $u_1 = u_2 = u_j$.

*Hint*: The pressure change across a thin layer is negligible.

b What is the component parallel to the plane of the force acting on the fluid? By considering the change in momentum of the fluid in a suitable control volume, and applying conservation of matter, show that the flow rates up and down are, respectively,

$$Q_1 = \frac{1}{2}(1 + \cos\alpha)Q_j$$

$$Q_2 = \frac{1}{2}(1 - \cos\alpha)Q_j$$

c Find also an expression for the force on the plane, and evaluate it for the case $b = 10\,\text{cm}$, $t = 1\,\text{cm}$, $Q_j = 10\,\text{L.s}^{-1}$, $\alpha = 60°$.

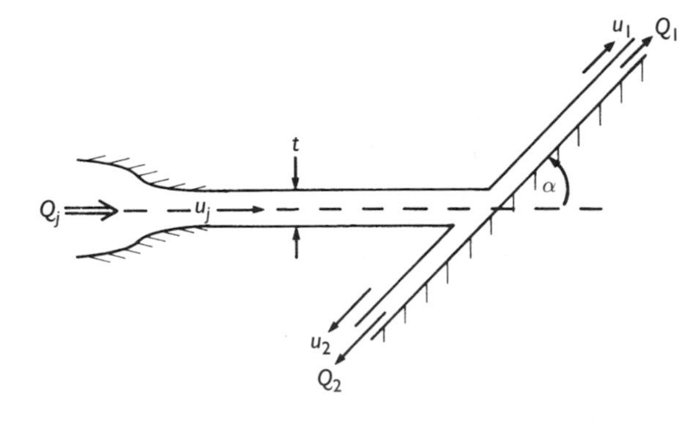


*Figure 2.9* For Problem 2.2. A plane jet strikes an inclined wall.

2.3 Consider a steady *laminar flow between two fixed parallel plates* at $y = 0$ and $y = D$ (cf. Figure 2.4). The fluid is pushed to the right ($x$ increasing) by a constant pressure gradient $\partial p/\partial x (< 0)$.

a What are the forces acting on an element of fluid of length $\Delta x$ and width $\Delta z$, lying between $y$ and $y + \Delta y$? Show that the balance of forces on this fluid element is given by

$$\frac{\partial}{\partial y}\left(\mu \frac{\partial u}{\partial y}\right) = \frac{\partial p}{\partial x}$$

b By integrating the last expression, show that the velocity at a distance $y$ above the lower plate is

$$u(y) = \frac{y}{2\mu}\left(\frac{\partial p}{\partial x}\right)(D - y)$$

c The plates have width $W \gg D$, so that edge effects are negligible. Show that the volumetric flow rate between the plates is

$$Q = \frac{1}{12} WD^3 \left|\frac{\partial p}{\partial x}\right|$$

2.4 *Laminar flow in a pipe*
A constant pressure gradient $(\partial p/\partial x)$ between the ends pushes fluid straight down a circular pipe of diameter $D = 2R$. Let $x$ be the distance along the pipe and $r$ the distance from the axis.

a As in Problem 2.3, show that the balance of forces on an annulus of length $\Delta x$ lying between $r$ and $r + \Delta r$ leads to the equation

$$\frac{\partial}{\partial r}\left(r\mu \frac{\partial u}{\partial r}\right) = -r\frac{\partial p}{\partial x}$$

*Hint*: The shear force at $r$ is $(2\pi r x)\tau(r)$, and $r$ varies across the pipe.

b By integration, show that the velocity at a distance $r$ from the axis is

$$u(r) = \frac{1}{4\mu}\left(\frac{\partial p}{\partial x}\right)(R^2 - r^2)$$

c Hence show that the volume of fluid flowing out of the pipe per unit time is

$$Q = \frac{\pi a^4}{8\mu}\left(\frac{\partial p}{\partial x}\right)$$

d The mean velocity in the pipe is $\overline{u} = Q/A$. By putting the results so far into (2.12) show that $\overline{u}f = 16v/D$, thus verifying (2.15).

2.5 Having been informed of the results of Example 2.1, the accountant of the engineering firm concerned suggests that it would be cheaper to use narrow PVC pipe to carry the flow.

a Disillusion the accountant by calculating the theoretical friction head incurred in passing $0.1\,m^3\,s^{-1}$ of water through 200 m of PVC pipe of diameter 5 cm, and show that it exceeds the available head.
b If gravity is the only force available to move the water, calculate an upper limit to the flow which could in fact be pushed through this pipe.

*Hint*: Take $H_f = L$ (vertical pipe), estimate $f$ and calculate the corresponding $u$.

2.6 A steel pipe of diameter $D$ and length $L$ is to carry a flow $Q$. Assuming that the pipe friction coefficient $f$ varies only slowly with Reynolds number, show that the head loss due to friction is proportional to $D^{-5}$ (for fixed $L$ and $Q$).

## Bibliography

The following selection from the many books on fluid mechanics may prove useful. There are many other good books besides those listed. For work on turbomachinery, books written for engineers are usually more useful than those written for mathematicians, who too often ignore friction and forces. Since the basics of fluid dynamics have not changed, old textbooks can still be useful, especially if they use SI units (which many older books do not).

Batchelor, G.K. (1967) *An Introduction to Fluid Dynamics*, Cambridge University Press. {Classic text, reissued unchanged in 2000. A most precise statement of the foundations, see especially Chapter 3, with many examples. Repays careful reading, but perhaps unsuitable for beginners.}

Francis, J.R. (1974, 4th edn) *A Textbook of Fluid Mechanics*, Edward Arnold, London. {Clear writing makes easy reading for beginners. More engineering detail than Kay and Nedderman.}

Hughes, W.F., Brighton, J.A. and Winowich, N. (1999, 3rd edn) *Theory and Problems of Fluid Dynamics*, Schaum's Outline Series. {Multitude of worked examples.}

Kay, J.M. and Nedderman, R.M. (1985) *Fluid Mechanics and Transfer Processes*, Cambridge University Press. {A concise and wide-ranging introduction.}

Mott, R.L. (2000, 5th edn) *Applied Fluid Mechanics*, Prentice Hall, USA. {Widely used student text for beginners, with exceptionally clear explanations.}

Tritton, D.J. (1988) *Physical Fluid Dynamics*, Oxford Science Publications, Oxford U.P. {Careful mathematical formulation, related carefully to physical reality.}

Webber, N. (1971) *Fluid Mechanics for Civil Engineers*, Chapman and Hall, London. {Delightfully simple but useful introduction for students with little knowledge of physics or engineering.}

Chapter 3

# Heat transfer

## 3.1 Introduction

With direct solar, geothermal and biomass sources, most energy transfer is by heat rather than by mechanical or electrical processes. Heat transfer is a well-established, yet complex, subject. However, we do not need sophisticated detail, which is rarely required to understand and plan renewable energy thermal applications. For instance, as compared with fossil and nuclear fuel engineering plant, temperature differences are often smaller, geometric configurations less complicated and 'most importantly' energy flux densities much lower. Of course, the sophisticated detail is needed for specialised renewables design, as for instance with advanced engines powered by biofuels.

This book uses a unified approach to heat transfer processes, by which several interrelated processes are analysed as one 'heat circuit'. For instance, the solar water heater of Figure 3.1 receives heat by solar radiation at about $1.0\,kW\,m^{-2}$ maximum intensity, producing surfaces about 50 °C higher than the environment. Heat is lost from these surfaces by long wavelength radiation, by conduction and by convection. The useful heat is removed by mass transport. Our recommended method of analysis is to set up a heat transfer circuit of the interconnected processes, such as the one illustrated in Figure 3.2(c), and calculate each transfer process to an accuracy of about 50%.

At this stage, insignificant processes can be neglected and important transfers can be analysed to greater accuracy. Even so, it is unlikely that final accuracies will be better than ±10% of actual performance.

This chapter supports subsequent chapters on individual renewable energy systems. The main formulas needed for practical calculations are summarised in Appendix C. You may find some of this chapter particularly complicated and boring, especially convection; however, be assured, when you use the methods with real hardware, they come alive.

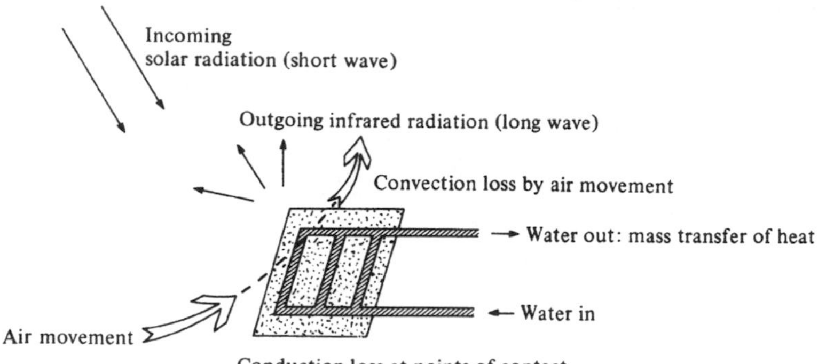


*Figure 3.1* Solar heater (absorbing plate fitted with water pipes): demonstrating types of heat transfer.

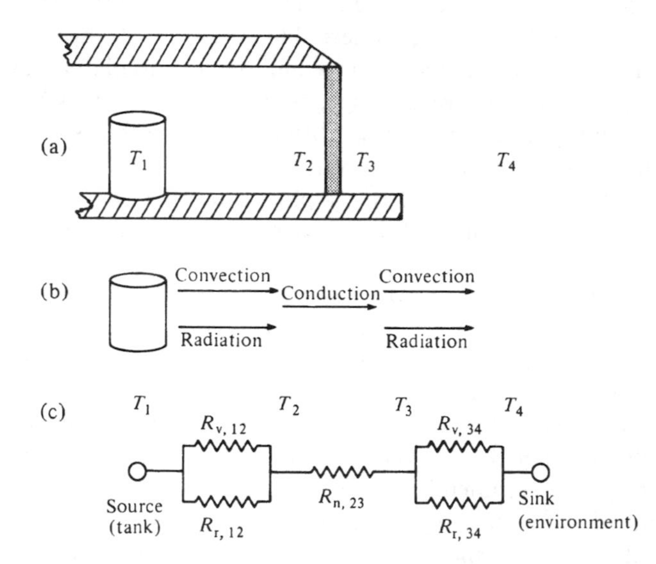


*Figure 3.2* Example of a heat circuit: (a) Physical situation. A hot tank is in a cool room with cold air outside. The roof and the floor are well insulated. $T_1$, $T_2$, $T_3$ and $T_4$ are the temperatures of the tank surface, the wall surfaces and the outside environment respectively. (b) Energy flow mechanisms. (c) Analogue circuit.

## 3.2 Heat circuit analysis and terminology

We introduce our method of heat transfer and circuit analysis with a simple example (which in reality is more complicated). At night, a large tank of hot water stands in a cool enclosed room, with a colder environment outside. The floor and ceiling are so well insulated that heat passes predominantly through the walls. Therefore net heat transfer is down a temperature gradient from the hot tank to the cold outside environment, Figure 3.2(a). Heat is transferred from the tank by radiation and convection to the room walls, by conduction through the walls, and then by radiation and convection to the environment, Figure 3.2(b). This complex transfer of parallel and series connections is described in the heat circuit of Figure 3.2(c).

Each process can be described by an equation of the form

$$P_{ij} = \frac{T_i - T_j}{R_{ij}} \tag{3.1}$$

where the power $P_{ij}$ is the *heat flow* between surfaces at temperatures $T_i$ (hotter) and $T_j$ (colder), and $R_{ij}$ is called the *thermal resistance*. See Appendix A for units. In general $R_{ij}$ is *not* constant with respect to the temperatures, and may well change rapidly and nonlinearly with change in surface temperature and temperature difference. However, for our purposes the heat flows $P_{ij}$ will depend most strongly on the term $(T_i - T_j)$, and the variation of $R_{ij}$ with temperature is always weaker. Thus the concept of thermal resistance is useful. If the direction of heat flow is obvious, then (3.1) may be written as

$$P_{ij} = \frac{\Delta T}{R_{ij}} \tag{3.2}$$

$R$ is called a 'resistance', by analogy with Ohm's law in electricity, because the forcing function is the temperature difference $\Delta T$ (equivalent to electrical potential difference $\Delta V$), and the flow is the heat transfer rate $P$ (equivalent to electrical current $I$)
i.e.

electricity $\quad \Delta V = I \times \text{electrical resistance}$

heat $\quad \Delta T = P \times \text{thermal resistance}$

The thermal resistance method allows each step of a complex of heat transfers to be added together as a set of series and parallel connections, as in electrical circuits. In our example of Figure 3.2

$$P_{14} = \frac{T_1 - T_4}{R_{14}} \tag{3.3}$$

where

$$R_{14} = R_{12} + R_{23} + R_{34}$$

and

$$\frac{1}{R_{12}} = \frac{1}{R_{12}(\text{convection})} + \frac{1}{R_{12}(\text{radiation})}$$

$$R_{23} = R_{23}(\text{conduction})$$

$$\frac{1}{R_{34}} = \frac{1}{R_{34}(\text{convection})} + \frac{1}{R_{34}(\text{radiation})}$$

Even if we use only approximate values of the temperatures, we shall find that the individual resistances can be calculated to obtain the overall resistance $R_{14}$. The heat loss from the tank $P_{14}$ is now calculated in terms of $T_1$ and $T_4$ only. This simplification, together with the diagrammatic quantification of the heat flows, makes thermal resistance a powerful concept.

For general considerations, it is often useful to consider heat flow $q$ across unit area of surface. Then across a surface of area $A$,

$$q = \Delta T/r \tag{3.4}$$

$$P = qA = \Delta T/(r/A) \tag{3.5}$$

so

$$R = r/A \text{ (unit of } R \text{ is K W}^{-1}) \tag{3.6}$$

and

$$r = RA \text{ (unit of } r \text{ is m}^2\text{ K W}^{-1}) \tag{3.7}$$

Here $r$ is the *thermal resistivity of* unit area.

(Note that this is *not* the same as resistance *per* unit area. Since $R$ decreases as the area increases, we *multiply* $R$ by the area to calculate $r$, as with the electrical resistivity of an extended material.)

A common expression for the heat flow per unit area is

$$q = h\Delta T \tag{3.8}$$

where $h$ is the *heat transfer coefficient* (W m$^{-2}$ K$^{-1}$).
Note that

$$h = 1/r \tag{3.9}$$

The heat transfer coefficient $h$ is sometimes called the *thermal conductance* and denoted by symbol $U$. So

$$U = 1/r \tag{3.10}$$

and

$$UA = 1/R \tag{3.11}$$

In this text we use the following subscripts on $R$, $r$ and $h$ to distinguish the various heat transfer mechanisms: $R_n$ for conduction, $R_v$ for convection, $R_r$ for radiation and $R_m$ for mass transfer.

## 3.3 Conduction

Thermal conduction is the transfer of heat by the vibrations of atoms, molecules and electrons without bulk movement. It is the only mechanism of heat transfer in opaque solids, but transparent media also pass heat energy by radiation. Conduction also occurs in liquids and gases, where, however, heat transfer is usually dominated by convection, as heat is carried by the fluid circulating or moving in bulk. Consider the heat flow $P$ by conduction through a slab of material, area $A$, thickness $\Delta x$, surface temperature difference $\Delta T$:

$$P = -kA\frac{\Delta T}{\Delta x} \tag{3.12}$$

$k$ is the *thermal conductivity* (unit $\mathrm{W\,m^{-1}\,K^{-1}}$), and the negative sign indicates that heat flows in the direction of decreasing temperature. By comparison of (3.12) and (3.2), the thermal resistance of conduction is

$$R_n = \frac{\Delta x}{kA} \tag{3.13}$$

and the corresponding thermal resistivity of unit area is

$$r_n = R_n A = \frac{\Delta x}{k} \tag{3.14}$$

The thermal conductivity $k$ of a dry solid is effectively constant over a wide range of temperature, and so the thermal resistance $R_n$ of dry opaque solids is usually considered constant. This is in marked contrast with liquids, gases and vapours, whose thermal resistance varies distinctly with temperature due to convection. Values of thermal conductivity are tabulated in Appendix B.

The inverse of thermal resistivity is thermal conductance (of unit area); in some countries this is called the 'U value' of a building material or component, unit $\mathrm{W\,m^{-2}\,K^{-1}}$. In other countries the inverse of $U$ (i.e. the thermal resistivity of unit area $r$) is called the '*R value*', unit $\mathrm{m^2\,K\,W^{-1}}$. By these definitions, the thermal resistance of a whole extended surface,

$R$, equals the '$R$ value' divided by the area, i.e. the larger the area, the smaller the thermal resistance. If you find this all very muddling, you are not alone. Sadly convention does not allow symmetry or logic in what parameter names end in '-ivity' and what end in '-ance'.

*Example 3.1*
Some values of conductive thermal resistance (using data from Table B.3 of Appendix B)

1 $1.0\,\mathrm{m^2}$ of window glass:

$$R = \frac{5.0\,\mathrm{mm}}{(1.05\,\mathrm{W\,m^{-1}\,K^{-1}})(1.0\,\mathrm{m^2})} = 0.005\,\mathrm{K\,W^{-1}};$$

$$U = 200\,\mathrm{W\,m^{-2}\,K^{-1}};$$

$$r = 0.0050\,\mathrm{m^2\,K\,W^{-1}}$$

2 $5.0\,\mathrm{m^2}$ of the same glass:

$$R_n = 0.0010\,\mathrm{K\,W^{-1}}; \quad U = 200\,\mathrm{W\,m^{-2}\,K^{-1}}; \quad r = 0.0050\,\mathrm{m^2\,K\,W^{-1}}$$

3 $1.0\,\mathrm{m^2}$ of continuous brick wall 220 mm thick:

$$R_n = \frac{220\,\mathrm{mm}}{(0.6\,\mathrm{W\,m^{-1}\,K^{-1}})(1.0\,\mathrm{m^2})} = 0.37\,\mathrm{K\,W^{-1}};$$

$$U = 2.7\,\mathrm{W\,m^{-2}\,K^{-1}};$$

$$r = 0.37\,\mathrm{m^2\,K\,W^{-1}}$$

4 $1.0\,\mathrm{m^2}$ of loosely packed glass fibres ('mineral wool') 80 mm deep as used for ceiling insulation:

$$R = \frac{80\,\mathrm{mm}}{(0.035\,\mathrm{W\,m^{-1}\,K^{-1}})(1.0\,\mathrm{m^2})} = 2.3\,\mathrm{K\,W^{-1}};$$

$$U = 0.4\,\mathrm{W\,m^{-2}\,K^{-1}};$$

$$r = 2.3\,\mathrm{m^2\,K\,W^{-1}}$$

In example 3.1, note the following:

1 The intrinsic conductive resistance of a glass sheet is much less than its effective resistance when in a window. The effective resistance includes

the additional resistances in series due to convection in the air against the glass, which significantly increases the intrinsic resistance, (see Problem 3.7).

2 Since the thermal conductivity of metals is large ($k \sim 100\,\mathrm{W\,m^{-1}\,K^{-1}}$), the conductive thermal resistance of metal components *in series* with other components is often negligible. In parallel with other components, continuous metal provides a thermal short circuit; so avoid metal-framed windows, unless they have a thermal break within the frame.

3 Loosely packed glass fibres have a much higher thermal resistance than pure glass sheet, because the packed fibres incorporate many small pockets of still air. *Still air* is one of the best insulators available ($k \sim 0.03\,\mathrm{W\,m^{-1}\,K^{-1}}$), and all natural and commercial insulating materials rely on it. The thermal resistance of such materials decreases drastically if the material becomes wet, if the material is compressed or if the air pockets are too big (in which case the air carries heat by convection).

4 There are two mechanisms that decrease the conductive thermal resistance of wet or damp materials:

   a Liquid water provides a thermal short circuit within the structure of the material. This is particularly so for biological material, e.g. wood.

   b Evaporation may occur so that vapour diffuses within the structure to regions of condensation (see Section 3.7). This is a dominant aspect of very conductive heat pipes (see Figure 3.18).

Another property closely related to the conductivity is the *thermal diffusivity* $\kappa$, which indicates how quickly changes in temperature diffuse through a material:

$$\kappa = k/(\rho c) \tag{3.15}$$

where $\rho$ is the density and $c$ is the specific heat capacity at constant pressure. $\kappa$ has the unit of $\mathrm{m^2\,s^{-1}}$ as with kinematic viscosity $\nu$ (see (2.10)). The temperature will change quickly only if heat can move easily through the material (large $k$ in the numerator of (3.15)) *and* if a small amount of heat produces a large temperature rise per unit volume (small values $\rho c$ in the denominator).

It takes a time $\sim y^2/\kappa$ for a temperature increase to diffuse along a distance $y$ into a cold mass.

## 3.4 Convection

### 3.4.1 Free and forced convection

Convection is the transfer of heat to or from a moving fluid, which may be liquid or gas. Since the movement continually brings unheated fluid to the source, or sink, of heat, convection produces more rapid heat transfer than conduction through the stationary fluid.

*Figure 3.3* Fluid movement by free convection, away from the hotter surface ($T_2 > T_1$).

In *free convection* (sometimes called *natural convection*) the movement is caused by the heat flow itself. Consider the fluid in contact with the hot surfaces of Figure 3.3; for example, water against the inside surfaces of a boiler or a solar collector. Initially the fluid absorbs energy by conduction from the hot surface, and so the fluid density decreases by volume expansion. The heated portion then rises through the unheated fluid, thereby transporting heat physically upwards, but down the temperature gradient.

In *forced convection* the fluid is moved across a surface by an external agency such as a pump (e.g. water in a solar collector to a storage tank below) or wind (e.g. heat loss from the outside surfaces of a solar collector). The movement occurs independently of the heat transfer (i.e. is not a function of the local temperature gradients). Obviously convection is usually partly forced and partly free, yet usually one process dominates.

### 3.4.2 Nusselt number $\mathcal{N}$

The analysis of convection proceeds from a gross simplification of the processes. We imagine the fluid near the surface to be stationary. We then consider the heat flowing across an idealised boundary layer of stationary fluid of thickness $\delta$ and cross-sectional area $A$ (Figure 3.4). The temperatures

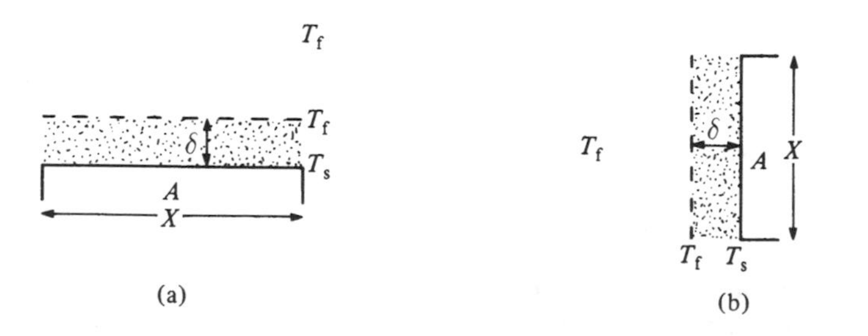


*Figure 3.4* Idealised thermal boundary layer in free convection. (a) Hot surface horizontal. (b) Hot surface vertical.

across the fictitious boundary are $T_f$, the fluid temperature away from the surface, and $T_s$, the surface temperature. This being so, the heat transfer by conduction across unit area of the stationary fluid is

$$q = \frac{P}{A} = \frac{k(T_s - T_f)}{\delta} \tag{3.16}$$

where $k$ is the thermal conductivity of the fluid.

As described here, $\delta$ is fictitious and cannot be measured. We can, however, measure $X$, a 'characteristic dimension' specified rather arbitrarily for each particular surface (see Figure 3.4 and Appendix C).
From (3.13),

$$q = \frac{P}{A} = \frac{k(T_s - T_f)}{\delta} = \frac{X}{\delta}\frac{k(T_s - T_f)}{X} = \mathcal{N}\frac{k(T_s - T_f)}{X} \tag{3.17}$$

$\mathcal{N}$ is the *Nusselt number* for the particular circumstance. It is a dimensionless scaling factor, useful for all bodies of the same shape in equivalent conditions of fluid flow. Tables of values of $\mathcal{N}$ are available for specified conditions, with the appropriate characteristic dimension identified (Appendix C).
From Section 3.2 it follows that:

thermal resistance of convection $$R_v = \frac{X}{\mathcal{N}kA} \tag{3.18}$$

convective thermal resistivity of unit area $$r_v = R_v A = \frac{X}{\mathcal{N}k} \tag{3.19}$$

convective heat transfer coefficient $$h_v = \frac{1}{r_v} = \frac{\mathcal{N}k}{X} \tag{3.20}$$

The amount of heat transferred by convection will depend on three factors:

1 The properties of the fluid
2 The speed of the fluid flow and its characteristics, i.e. laminar or turbulent
3 The shape and size of the surface.

The Nusselt number $\mathcal{N}$ is a dimensionless measure of the heat transfer. Therefore it can depend only on dimensionless measures of the three factors listed. In choosing these measures, it is convenient to separate the cases of forced and free convection.

### 3.4.3 Forced convection

For a given shape of surface, a non-dimensional measure of the speed of the flow is the *Reynolds number*:

$$\mathcal{R} = \frac{uX}{v} \tag{3.21}$$

We saw in Section 2.5 that $\mathcal{R}$ determines the pattern of the flow, and in particular whether it is laminar or turbulent. In flow over a flat plate (Figure 3.5), turbulence occurs when $\mathcal{R} \geq 3 \times 10^5$, with subsequent increase in the heat transfer because of the perpendicular motions involved.

The flow of heat into or from a fluid depends on (a) the thermal diffusivity $\kappa$ of the fluid, and (b) the kinematic viscosity $v$, (2.10), which may be considered 'the diffusivity of momentum', since it affects the Reynolds number and thus the character of the flow. These are the only two properties of the fluid that influence the Nusselt number in forced convection, since the separate effects of $k$, $p$, and $c$ are combined in $\kappa$ (see (3.15)).

A non-dimensional measure of the properties of the fluid is the *Prandtl number*:

$$\mathcal{P} = v/\kappa \tag{3.22}$$

If $\mathcal{P}$ is large, changes in momentum diffuse more quickly through the fluid than do changes in temperature. Many common fluids have $\mathcal{P} \sim 1$, e.g. 7.0 for water at 20 °C. For air at environmental temperatures $\mathcal{P} = 0.7$ (see Appendix B).

Thus, for each shape of surface, the heat transfer by forced convection can be expressed in the form

$$\mathcal{N} = \mathcal{N}(\mathcal{R}, \mathcal{P}) \tag{3.23}$$

That is, for each shape, the Nusselt number $\mathcal{N}$ is a function only of the Reynolds number and the Prandtl number. These relationships may be

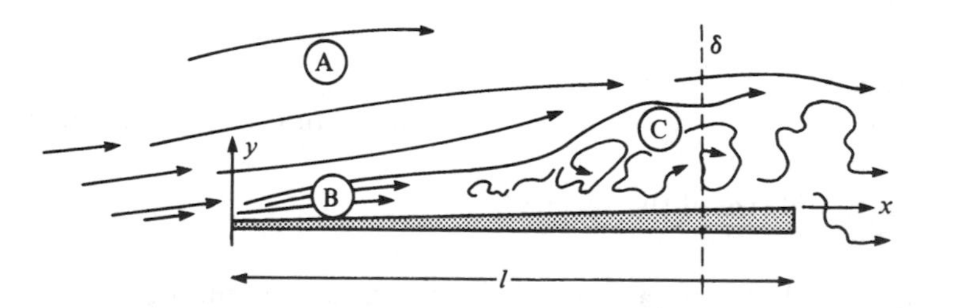


*Figure 3.5* Fluid flow over a hot plate. General view of pathlines, showing regions: (A) well away from the surface; (B) laminar flow near the leading edge; (C) turbulent flow in the downstream region.

expressed with other closely related dimensionless parameters, e.g. the Stanton number $\mathcal{N}/(\mathcal{R}\mathcal{P})$ and the Péclet number $\mathcal{R}/\mathcal{P}$, but neither are used in this book.

Numerical values of $\mathcal{N}$ are determined from experiment, with their method of use explained in Appendix C. An example is given in Section 3.4.5. It is important to realise that these formulas are mostly only accurate to ±10%, partly because they are approximations to the experimental conditions, and partly because the experimental data themselves usually contain both random and systematic errors.

### 3.4.4 Free convection

In free convection, often called 'natural convection', the fluid speed depends on the heat transfer (whereas in forced convection, heat transfer depends on the fluid speed). Analysis still depends on determining the Nusselt number, but as a function of other dimensionless numbers. We replace (3.23) by

$$\mathcal{N} = \mathcal{N}(\mathcal{A}, \mathcal{P}) \tag{3.24}$$

where the *Rayleigh number*

$$\mathcal{A} = \frac{g\beta X^3 \Delta T}{\kappa v} \tag{3.25}$$

These formulae are sometimes expressed in terms of the *Grashof number*

$$\mathcal{G} = \frac{\mathcal{A}}{\mathcal{P}} = g\beta X^3 \frac{\Delta T}{v^2} \tag{3.26}$$

Both the Grashof and Rayleigh numbers are dimensionless measures of the driving temperature difference $\Delta T$; $g$ is the acceleration due to gravity; $\beta$ is the coefficient of thermal expansion and the other symbols are as before.

In this book we prefer to use the Rayleigh number $\mathcal{A}$ because it more directly relates to the physical processes indicated in Figure 3.6. Heated fluid is forced upwards by a buoyancy force proportional to $g\beta\Delta T$, and retarded by a viscous force proportional to $v$. However, the excess temperature (and therefore the buoyancy) is lost at a rate proportional to thermal diffusivity $\kappa$. Therefore the vigour of convection increases with $g\beta\Delta T/(\kappa v)$, i.e. with $\mathcal{A}$. The factor $X^3$ is inserted to make this ratio dimensionless. It is found experimentally that free convection is non-existent if Rayleigh number $\mathcal{A} <\sim 10^3$ and is turbulent if $\mathcal{A} \gtrsim 10^5$.

This argument shows that the Nusselt number in free convection depends mainly on the Rayleigh number. The dimensional argument used in Section 3.4.3 suggests that $\mathcal{N}$ may also depend on the Prandtl number, as indicated in (3.24). Formulas to calculate these Nusselt numbers are given

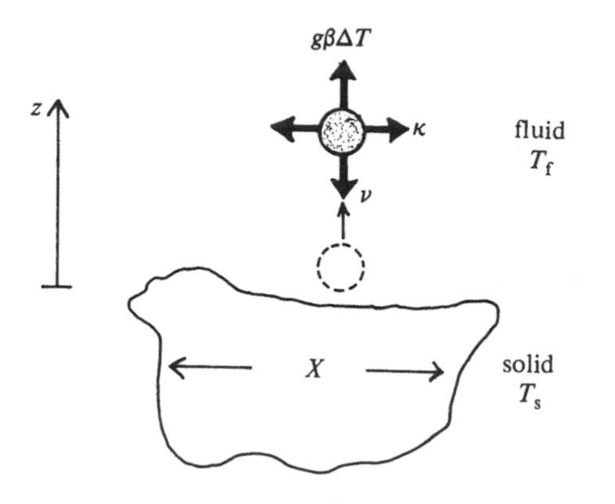


*Figure 3.6* Schematic diagram of a blob of fluid moving upward in free convection. It is subject to an upward buoyancy force, a retarding viscous force, and a sideways temperature loss.

in Appendix C for various geometries. These cannot be expected to give better than 10% accuracy. Note that the Nusselt number (and thus the thermal resistance) in free convection depends on $\Delta T$, through the dependence on $\mathscr{A}$. This is because a larger temperature difference drives a stronger flow, which transfers heat more efficiently. By contrast, in forced convection, the Nusselt number and thermal resistance are virtually independent of $\Delta T$.

### 3.4.5 *Calculation of convective heat transfer*

Because of the complexity of fluid flow, there is no fundamental theory for calculating convective heat transfer. Instead we use experiments on geometrically similar objects. By expressing the results in non-dimensional form, they can be applied to different sizes of the objects and for different fluids. For this, working formulas are given in the tables of Appendix C for shapes common in renewable energy applications; more extensive collections are given in textbooks of heat transfer.

All this seems very confusing. However, when used in earnest for calculating convection, confusion lessens by using the following systematic procedure:

1 Open the tables of heat transfer processes and equations (e.g. Appendix C)
2 Draw a diagram of the heated object.
3 Section the diagram into standard geometries (i.e. parts corresponding to the illustrations in the tables)

4 For each such section:
   a Identify the characteristic dimension ($X$)
   b As required in the tables, calculate $\mathcal{R}$ and/or $\mathcal{A}$ for each section of the object.
   c Choose the formula for $\mathcal{N}$ from tables appropriate to that range of $\mathcal{R}$ or $\mathcal{A}$. (The different formulas usually correspond to laminar or turbulent flow.)
   d Calculate the Nusselt number $\mathcal{N}$ and hence the heat flow across the section $P = qA$.

5 Add the heat flows from each section to obtain the total heat flow from the object.

*Example 3.2 Free convection between parallel plates*
Two flat plates each 1.0 m × 1.0 m are separated by 3.0 cm of air. The lower is at 70 °C and the upper at 45 °C. The edges are sealed together by thermal insulating material acting also as walls to prevent air movement beyond the plates. Calculate the convective thermal resistivity of unit area, $r$, and the heat flux, $P$, between the top and the bottom plate.

*Solution*
Figure 3.7 corresponds to the standard geometry given for (C.7) in Appendix C. Since the edges are sealed, no outside air can enter between the plates and only free convection occurs. Using (3.25) and Table B.1 in Appendix B, for mean temperature 57 °C(= 330 K)

$$\mathcal{A} = \frac{g\beta X^3 \Delta T}{\kappa v} = \left(\frac{g\beta}{\kappa v}\right)(X^3 \Delta T)$$

$$= \frac{(9.8\,\mathrm{m\,s^{-1}})(1/330\,\mathrm{K})}{(2.6 \times 10^{-5}\,\mathrm{m^2\,s^{-1}})(1.8 \times 10^{-5}\,\mathrm{m^2\,s^{-1}})}(0.03\,\mathrm{m})^3(25\,\mathrm{K})$$

$$= 4.1 \times 10^4$$

Using (C.7) a reasonable value for $\mathcal{N}$ can be obtained, although $\mathcal{A}$ is slightly less than $10^5$:

$$\mathcal{N} = 0.062\mathcal{A}^{0.33} = 2.06$$

(From (3.17), this implies the boundary layer is about half way across the gap.)
From (3.19)

$$r_v = \frac{X}{\mathcal{N}k} = \frac{0.03\,\mathrm{m}}{(2.06)(0.028\,\mathrm{W\,m^{-1}\,K^{-1}})} = 0.52\,\mathrm{K\,W^{-1}\,m^2}$$

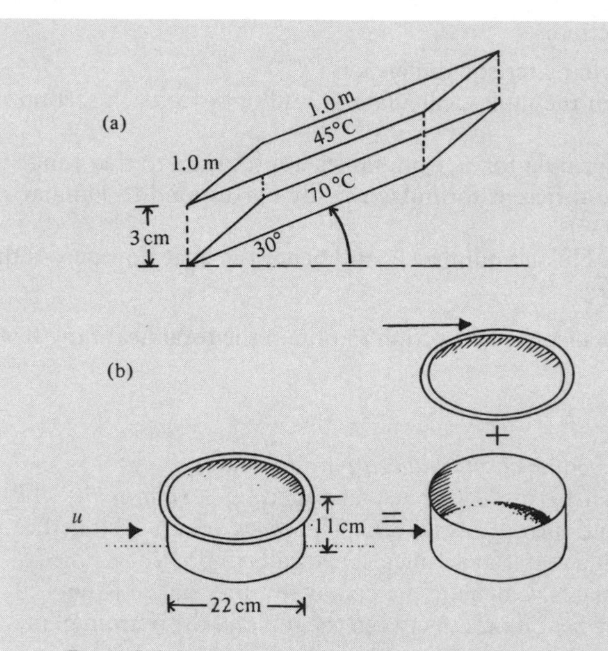


*Figure 3.7* Diagrams for worked examples on convection. (a) Parallel plates, as in Example 3.2 (b) Cooking pot with lid, as in Example 3.3.

From (3.17),

$$P = \frac{A\Delta T}{r} = \frac{(1\,\mathrm{m}^2)(25\,\mathrm{K})}{0.52\,\mathrm{K\,W^{-1}\,m^2}} = 48\,\mathrm{W}$$

Note the following:

1 The factor $(g\beta/\kappa v) = (\mathscr{A}/X^3\,\Delta T)$ is tabulated in Appendix B for air and water.
2 The fluid properties are evaluated at the mean temperature (57 °C in this case).
3 It is *essential* to use consistent units (e.g. SI) in evaluating dimensionless parameters like $\mathscr{A}$.

*Example 3.3 Convective cooling of a cooking pot*
A metal cooking pot with a shiny outside surface, of the dimensions shown in Figure 3.7(b), is filled with food and water and placed on a cooking stove. What is the minimum energy required to maintain it at

boiling temperature for one hour, (1) if it is sheltered from the wind (2) if it is exposed to a breeze of $3.0\,\mathrm{m\,s^{-1}}$?

*Solution*
We assume that the lid is tight, so that there is no heat loss by evaporation. We also neglect heat loss by radiation, as justified in Problem 3.4. Since the conductive resistance of the pot wall is negligible, the problem is then reduced to calculating the convective heat loss from the top and sides of a cylinder with a surface temperature of 100 °C. We shall consider the ambient (air and surrounding walls) temperature to be 20 °C. Therefore heat transfer properties of the air are evaluated at the mean temperature $\bar{T} = 60\,°\mathrm{C}$.

1 *Free convection alone* – For the top (using Table B.1 and (3.25)),

$$\mathcal{A} = (5.8 \times 10^7\,\mathrm{m^{-3}\,K^{-1}})(0.22\,\mathrm{m})^3(80\,\mathrm{K}) = 4.9 \times 10^7$$

and (from (C.2))

$$\mathcal{N} = 0.14\mathcal{A}^{0.33} = 48.4$$

and

$$P_{top} = Ak\mathcal{N}\Delta T/X$$

$$= (\pi/4)(0.22\,\mathrm{m})^2(0.027\,\mathrm{W\,m^{-1}\,K^{-1}})(48.4)\frac{(80\,\mathrm{K})}{0.22\,\mathrm{m}} = 18\,\mathrm{W}$$

For the sides, $X = 0.11\,\mathrm{m}$:

$$\mathcal{A}_{side} = \mathcal{A}_{top}(0.11\,\mathrm{m}/0.22\,\mathrm{m})^3 = 6.1 \times 10^6$$

and (from (C.5))

$$\mathcal{N} = 56\mathcal{A}^{0.25} = 27.8$$

so

$$P_{side} = \pi(0.22\,\mathrm{m})(0.11\,\mathrm{m})(0.027\,\mathrm{W\,m^{-1}\,K^{-1}})(27.8)\frac{(80\,\mathrm{K})}{0.11\,\mathrm{m}} = 41\,\mathrm{W}$$

Hence

$$P_{free} = P_{top} + P_{side} = 59\,\mathrm{W} = (0.059\,\mathrm{kW})(3600\,\mathrm{s\,h^{-1}}) \approx 0.2\,\mathrm{MJ\,h^{-1}}.$$

2 *Forced plus free convection* – Here we calculate the forced convective power losses separately, and add them to those already calculated for free convection, in order to obtain an estimate of the total convective heat loss $P_{total}$.
For the top,

$$\mathcal{R} = \frac{(3\,\mathrm{m\,s^{-1}})(0.22\,\mathrm{m})}{1.9 \times 10^{-5}\,\mathrm{m^2\,s^{-1}}} = 3.5 \times 10^4$$

which suggests the use of (C.8):

$$\mathcal{N} = 0.664\ \mathcal{R}^{0.5}\ \mathcal{P}^{0.33} = 110$$

So

$$P_{top} = Ak\mathcal{N}(\frac{\Delta T}{X}) = 42\,\mathrm{W}$$

For the sides, as for the top,

$$\mathcal{R} = 3.5 \times 10^4$$

which suggests the use of (C.11):

$$\mathcal{N} = 0.26\ \mathcal{R}^{0.6}\ \mathcal{P}^{0.3} = 124$$

and

$$P_{side} = \pi(0.22\,\mathrm{m})(0.11\,\mathrm{m})(0.027\,\mathrm{W\,m^{-1}\,K^{-1}})(124)\frac{(80\,\mathrm{K})}{0.22\,\mathrm{m}}$$
$$= 93\,\mathrm{W}$$

Hence

$$P_{forced} = 93 + 42 = 135\,\mathrm{W}$$

The total estimate is

$$P_{total} = P_{forced} + P_{free} = 194\,\mathrm{W} = (194\,\mathrm{W})(3600\,\mathrm{s\,h^{-1}}) \approx 0.7\,\mathrm{MJ\,h^{-1}}$$

i.e. about 3 times the energy per unit time of the sheltered cooking.

The overall accuracy of calculations like those in Example 3.3 may be no better than ±50%, although the individual formulas are better than this. This is because forced and free convection may both be significant, but their separate contributions do not simply add because the flow induced by free convection may oppose or reinforce the pre-existing flow. Similarly the flows around the 'separate' sections of the object interact with each other.

In Example 3.3 there is an additional confusion about whether the flow is laminar or turbulent. For example, on the top of the pot $\mathcal{A} > 10^5$, suggesting turbulence, but using the external flow speed to calculate $\mathcal{R}$ (as earlier) gives $\mathcal{R} < 10^5$, suggesting laminar flow. In practice such a flow across the top would be turbulent, since it is difficult to smooth out streamlines which have become tangled by turbulence. The only safe way to accurately evaluate a convective heat transfer, allowing for all these interactions, is by experiment! Some formulas, such as (C.15), based on such specialised experiments are available, but have a correspondingly narrow range of applicability. Nevertheless, calculation of convection is essential to give order-of-magnitude understanding of the processes involved.

## 3.5 Radiative heat transfer

### 3.5.1 *Introduction*

Surfaces emit energy by electromagnetic radiation according to fundamental laws of physics. Absorption of radiation is a closely related process. Sadly, the literature and terminology concerning radiative heat transfer are confusing; symbols and names for the same quantities vary, and the same symbol and name may be given for totally different quantities. Here, we have tried to follow the recommendations of the International Solar Energy Society (ISES), whilst maintaining unique symbols throughout the whole book, as on the opening list. In this chapter we consider radiative heat transfer in general. In Chapter 4 we shall consider solar radiation in particular, and in Chapters 5 and 6 heating devices using solar energy.

### 3.5.2 *Radiant flux density (RFD)*

Radiation is energy transported by electromagnetic propagation through space or transparent media. Its properties relate to its wavelength $\lambda$. The named regions of the spectrum are shown in Figure 3.8. The flux of energy per unit area is the *radiant flux density* (abbreviation RFD, unit $\mathrm{W\,m^{-2}}$, symbol $\phi$). The variation of RFD with wavelength is described by the *spectral RFD* (symbol $\phi_\lambda$, unit $(\mathrm{W\,m^{-2}})\,\mathrm{m^{-1}}$ or more usually $\mathrm{W\,m^{-2}\,\mu m^{-1}}$), which is simply the derivative $d\phi/d\lambda$. Thus $\phi_\lambda \Delta\lambda$ gives the power per unit area in a (narrow) wavelength range $\Delta\lambda$, and integration of $\phi_\lambda$ with respect

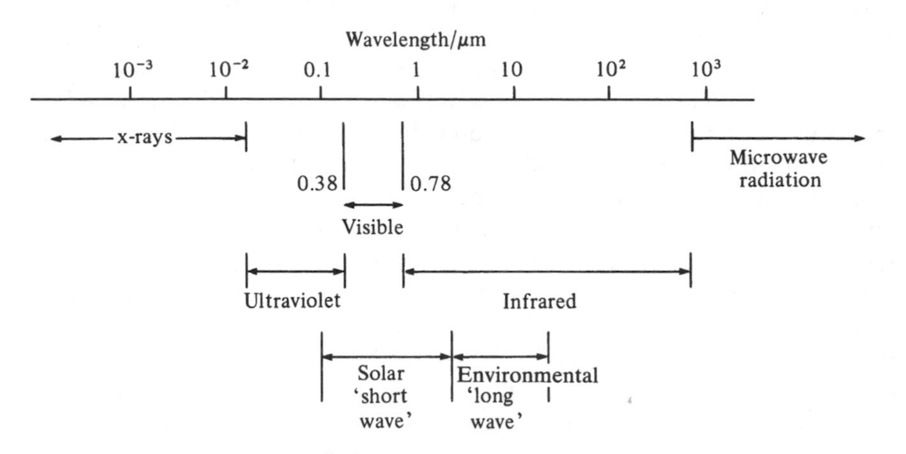


*Figure 3.8* Some of the named portions of the electromagnetic spectrum. (The spectrum extends both 'longwards' and 'shortwards' from that shown.)

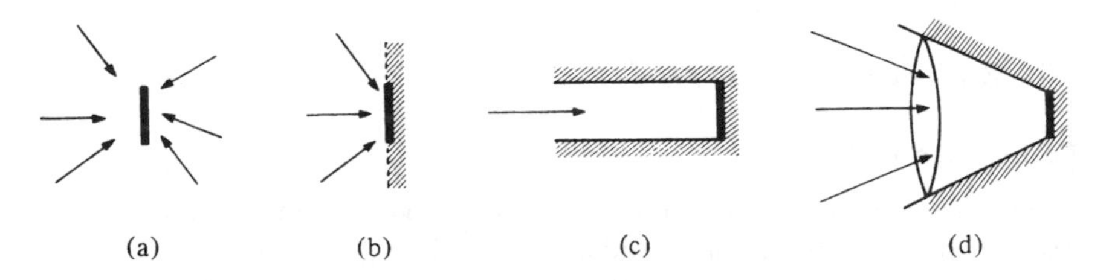


*Figure 3.9* Measurements of various radiation parameters using a small totally absorbing plane. (a) Absorbs all directions. (b) Absorbs from hemisphere above one side only. (c) Absorbs from one direction only. (d) Absorbs from one solid angle only.

to wavelength gives the total RFD, i.e. $\phi = \int \phi_\lambda d\lambda$. Radiation coming *onto* a surface is usually called *irradiance*.

It is obvious that radiation has directional properties, and that these need to be specified. Understanding of this is always helped by:

1 Drawing pictures of the radiant fluxes and the methods of measurement
2 Clarifying the units of the parameters.

Consider a small test instrument for measuring radiation parameters in an ideal manner. This could consist of a small, totally absorbing, black plane (Figure 3.9) that can be adapted to (a) absorb on both sides, (b) absorb on one side only, (c) absorb from one direction only and (d) absorb from one three-dimensional solid angle only.

The energy $\Delta E$ absorbed in time $\Delta t$ could be measured from the temperature rise of the plane of area of one side, $\Delta A$, knowing its thermal capacity. From Figure 3.9(a) the radiant flux density from *all* directions would be $\phi = (\Delta E \Delta t)/(2\Delta A)$. In Figure 3.9(b) the radiation is incident from the hemisphere above one side of the test plane (which may be labelled + or −), so

$$\phi = \frac{\Delta E}{(\Delta A \Delta t)} \tag{3.27}$$

In Figure 3.9(c) a vector quantity is now measured, with the direction of the radiation flux perpendicular to the receiving plane. In Figure 3.9(d) the radiation flux is measured within a solid angle $\Delta\omega$, *centred* perpendicular to the plane of measurement and with the unit of $\mathrm{W\,(m^2\,sr)^{-1}}$.

The wavelength(s) of the received radiation need not be specified, since the absorbing surface is assumed to be totally black. However, if a dispersing device is placed in front of the instrument which passes only a small range of wavelength from $\lambda - (\Delta\lambda/2)$ to $\lambda + (\Delta\lambda/2)$, then the spectral radiant flux density may be measured as

$$\phi_\lambda = \frac{\Delta E}{\Delta A \Delta t \Delta\lambda} \ [\mathrm{unit: W\,m^{-2}\,m^{-1}}] \tag{3.28}$$

This quantity can also be given directional properties per steradian (sr) as with $\phi$. Difficulty may sometimes arise, especially regarding certain measuring instruments. There are two systems of units relating to the measurement of radiation quantities – photometric and radiometric units (see Kaye and Laby 1995). Photometric units have been established to quantify responses as recorded by the human eye, and relate to the SI unit of the candela. Radiometric units quantify total energy effects irrespective of visual response, and relate to the basic energy units of the joule and watt. For our purposes, only radiometric units need be used. As an aside, we note that a similar pair of units exists for noise.

### 3.5.3 Absorption, reflection and transmission of radiation

Radiation incident on matter may be reflected, absorbed or transmitted (Figure 3.10). These interactions will depend on the type of material, the surface properties, the wavelength of the radiation and the angle of incidence $\theta$. Normal incidence ($\theta = 0$) may be inferred if not otherwise mentioned, but at grazing incidence ($90° > \theta \geq 70°$) there are significant changes in the properties.

At wavelength $\lambda$, within wavelength interval $\Delta\lambda$, the *monochromatic absorptance* $\alpha_\lambda$ is the fraction absorbed of the incident flux density $\phi_\lambda \Delta\lambda$. Note that $\alpha_\lambda$ is a property of the surface alone, depending, for example, on the energy levels of the atoms in the surface . It specifies what proportion of

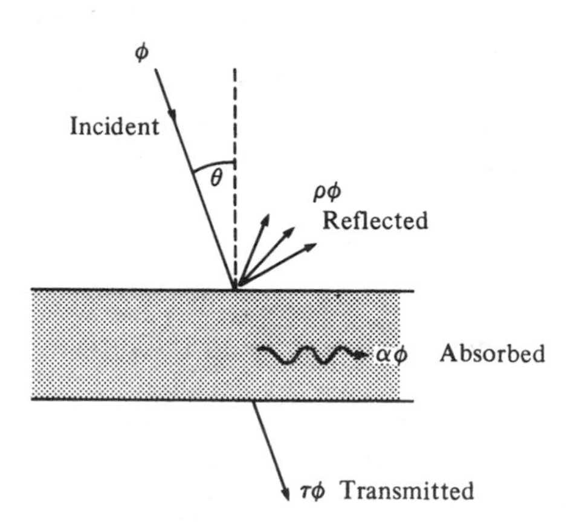


*Figure 3.10* Reflection, absorption and transmission of radiation ($\phi$ is the incident radiation flux density).

radiation at a particular wavelength would be absorbed if that wavelength was present in the incident radiation. The subscript on $\alpha_\lambda$, unlike that on $\phi_\lambda$, does *not* indicate differentiation.

Similarly, we define the *monochromatic reflectance* $\rho_\lambda$ and the *monochromatic transmittance* $\tau_\lambda$.

Conservation of energy implies that

$$\alpha_\lambda + \rho_\lambda + \tau_\lambda = 1 \tag{3.29}$$

and that $0 \le \alpha_\lambda, \rho_\lambda, \tau_\lambda \le 1$. All of these properties are almost independent of the angle of incidence $\theta$, unless $\theta$ is near grazing incidence. In practice, the radiation incident on a surface contains a wide spectrum of wavelengths, and not just one small interval. We define the *absorptance* $\alpha$ to be the absorbed proportion of the total incident radiant flux density:

$$\alpha = \phi_{\text{abs}}/\phi_{\text{in}} \tag{3.30}$$

It follows that

$$\alpha = \frac{\int_{\lambda=0}^{\infty} \alpha_\lambda \phi_{\lambda,\text{in}}\,\mathrm{d}\lambda}{\int_{\lambda=0}^{\infty} \phi_{\lambda,\text{in}}\,\mathrm{d}\lambda} \tag{3.31}$$

Equation (3.31) describes how the total absorptance $\alpha$, unlike $\alpha_\lambda$, does depend on the spectral distribution of the incident radiation. For example, a surface appearing blue in white daylight is black in orange sodium-light. This is because the surface absorbs the photons of orange colour of both lights, and so the remaining reflected daylight appears blue.

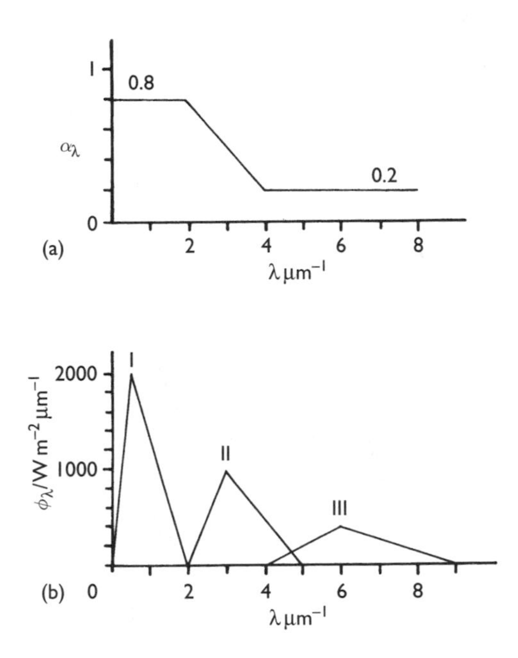


*Figure 3.11* Data for Example 3.4. The maxima of curves I, II, III in (b) are at (0.5, 2000), (3.0, 1000) and (6.0, 400) respectively.

The total reflectance $\rho = \phi_{\text{refl}}/\phi_{\text{in}}$ and the total transmittance $\tau = \tau_{\text{trans}}/\tau_{\text{in}}$ are similarly defined, and again

$$\alpha + \rho + \tau = 1 \tag{3.32}$$

*Example 3.4 Calculation of absorbed radiation*

A certain surface has $\alpha_\lambda$ varying with wavelength as shown in Figure 3.11(a) (this is a typical variation for a 'selective surface', as used on solar collectors, Section 5.6). Calculate the power absorbed by 1.0 m$^2$ of this surface from each of the following incident spectral distributions of RFD:

1 $\phi_\lambda$ given by curve I of Figure 3.11(b) (this approximates a source at 6000 K).
2 $\phi_\lambda$ given by curve II (approximating a source at 1000 K).
3 $\phi_\lambda$ given by curve III (approximating a source at 500 K).

*Solution*

1 Over the entire range of $\lambda$, $\alpha_\lambda = 0.8$. Therefore from (3.31) $\alpha = 0.8$ also, and the absorbed power is

$$P = \alpha(1\,\mathrm{m}^2)\int \phi_{\lambda,\mathrm{in}}\mathrm{d}\lambda$$

$$= (0.8)(1\,\mathrm{m}^2)\left[\left(\frac{1}{2}\right)(2000\,\mathrm{W\,m^{-2}\,\mu m^{-1}})\,(2\,\mu\mathrm{m})\right] = 1600\,\mathrm{W}$$

(The integral is the area under curve I.)

2 Here we have to explicitly calculate the integral $\int \alpha_\lambda \phi_{\lambda,\mathrm{in}}\,\mathrm{d}\lambda$ of (3.31). Tabulate as follows (the interval of $\lambda$ is chosen to match the accuracy of the data; here the spectra are obviously linearised, so an interval $\Delta\lambda \sim 1\,\mu\mathrm{m}$ is adequate):

| $\lambda(\mu\mathrm{m})$ | $\Delta\lambda(\mu\mathrm{m})$ | $\alpha_\lambda$ | $\phi_\lambda(\mathrm{W\,m^{-2}\,\mu m^{-1}})$ | $\alpha_\lambda\phi_\lambda\Delta\lambda(\mathrm{W\,m^{-2}})$ |
|---|---|---|---|---|
| 2.5 | 1 | 0.62 | 500 | 310 |
| 3.5 | 1 | 0.33 | 750 | 250 |
| 4.5 | 1 | 0.2 | 200 | 40 |
| | | | | Total 600 |

Therefore the power absorbed is approximately 600 W.

3 In a manner similar to part 1 of this solution, $\alpha_\lambda = 0.2$ over the relevant wavelength interval. Thus the power absorbed is

$$P = (0.2)(1\,\mathrm{m}^2)\left[\left(\frac{1}{2}\right)(400\,\mathrm{W\,m^{-2}\,\mu m^{-1}})\,(5\,\mu\mathrm{m})\right] = 200\,\mathrm{W}.$$

*Note*: The accuracy of calculations of radiative energy transfer is generally better than for convection. This is because the theory of the physical processes is exactly understood.

### 3.5.4 *Black bodies, emittance and Kirchhoff's laws*

An idealised surface absorbing all incident irradiation, visible and invisible, is named a *black body*. The name is because surfaces having the colour 'black' absorb all visible radiation; note, however, that *black bodies* absorb at all wavelengths, i.e. both visible and invisible radiation. Therefore, a black body has $\alpha_\lambda = 1$ for all $\lambda$, and therefore also has total absorptance $\alpha = 1$. Nothing can absorb more radiation than a similarly dimensioned black body placed in the same incident irradiation.

Kirchhoff proved also that no body can *emit* more radiation than a similarly dimensioned black body at the same temperature.

The *emittance* $\varepsilon$ of a surface is the ratio of the RFD emitted by the surface to the RFD emitted by a black body at the same temperature:

$$\varepsilon = \frac{\phi_{\text{from surface}}(T)}{\phi_{\text{from blackbody}}(T)} \tag{3.33}$$

The *monochromatic emittance*, $\varepsilon_\lambda$, of any real surface is similarly defined by comparison with the ideal black body, as the corresponding ratio of RFD in the wavelength range $\Delta\lambda$ (from $\lambda - (\Delta\lambda/2)$ to $(\lambda + \Delta\lambda/2)$. It follows that

$$0 \leq \varepsilon, \varepsilon_\lambda \leq 1 \tag{3.34}$$

Note that the emittance $\varepsilon$ of a real surface may vary with temperature. Kirchhoff extended his theoretical argument to prove *Kirchhoff's law*: 'for any surface at a specified temperature, and *for the same wavelength*, the monochromatic emittance and monochromatic absorptance are identical,'

$$\alpha_\lambda = \varepsilon_\lambda \tag{3.35}$$

Note that both $\alpha_\lambda$ and $\varepsilon_\lambda$ are characteristics of the surface itself, and not of the surroundings.

For solar energy devices, the incoming radiation is expected from the Sun's surface at a temperature of 5800 K, emitting with peak intensity at $\lambda \sim 0.5\,\mu\text{m}$. However, the receiving surface may be at about 350 K, emitting with peak intensity at about $\lambda \sim 10\,\mu\text{m}$. The dominant monochromatic absorptance is therefore $\alpha_{\lambda=0.5\,\mu\text{m}}$ and the dominant monochromatic emittance is $\varepsilon_\lambda = 10\,\mu\text{m}$. These two coefficients need not be equal, see Section 5.6. Nevertheless, Kirchhoff's Law is important for the determination of such parameters, e.g. at the same wavelength of $10\,\mu\text{m}$, $\varepsilon_{\lambda=10\,\mu\text{m}} = \alpha_{\lambda=10\,\mu\text{m}}$

### 3.5.5 *Radiation emitted by a body*

The monochromatic RFD emitted by a black body of *absolute temperature* $T$, $\phi_{B\lambda}$, is derived from quantum mechanics as *Planck's radiation law*:

$$\phi_{B\lambda} = \frac{C_1}{\lambda^5[\exp(C_2/\lambda T) - 1]} \tag{3.36}$$

where $C_1 = hc^2$, and $C_2 = hc/k$ ($c$, the speed of light in vacuum; $h$, Planck constant; and $k$, Boltzmann constant). Hence $C_1 = 3.74 \times 10^{-16}\,\text{W}\,\text{m}^2$ and $C_2 = 0.0144\,\text{m}\,\text{K}$ are also fundamental constants. Figure 3.12 shows how this spectral distribution $\phi_{B\lambda}$ varies with wavelength $\lambda$ and temperature $T$.

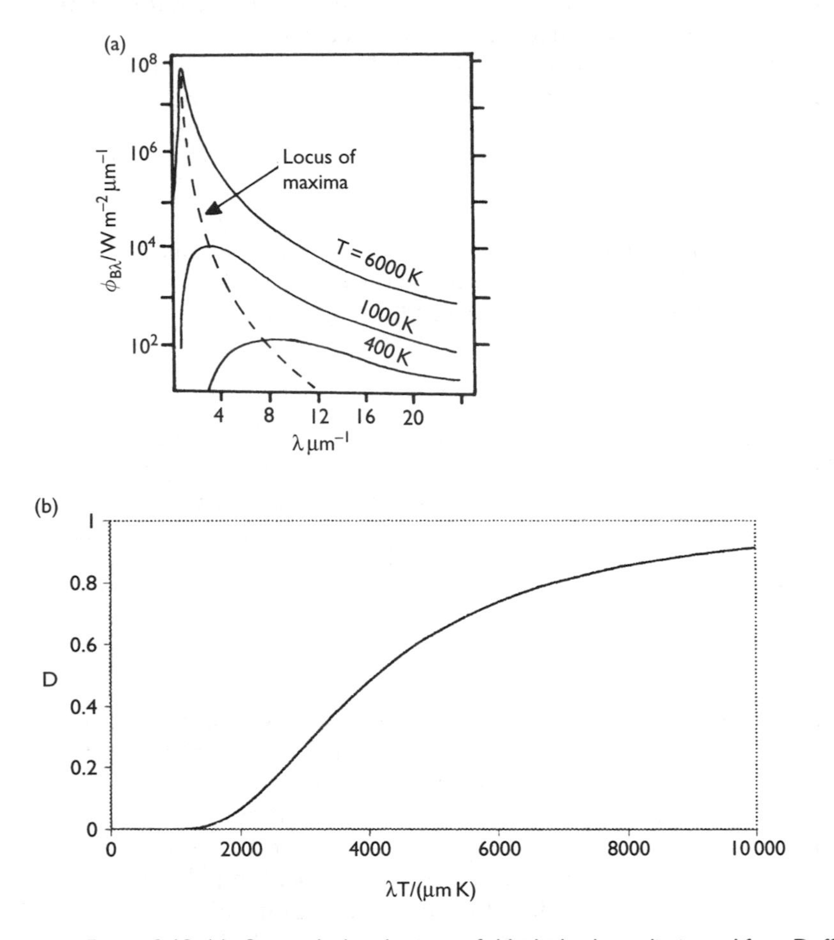


*Figure 3.12* (a) Spectral distribution of black body radiation. After Duffie and Beckman (1991). (b) Cumulative version of (a), in dimensionless form, as in eqn (3.40). Note that $D \rightarrow 1$ as $\lambda T \rightarrow \infty$.

Note that the wavelength $\lambda_m$, at which $\phi_{B\lambda}$ is most intense, increases as $T$ decreases. Indeed, as we know also from experience, when any surface temperature increases above $T \approx 700\,\mathrm{K}(\geq 430\,°\mathrm{C})$ significant radiation is emitted in the visible region and the surface does not appear black, but progresses from red heat to white heat.

By differentiating (3.36) and setting $\mathrm{d}(\phi_{B\lambda})/\mathrm{d}\lambda = 0$, we find that

$$\lambda_m T = 2898\,\mu\mathrm{m\,K} \tag{3.37}$$

This is *Wien's displacement law*. Knowing $T$, it is extremely easy to determine $\lambda_m$, and thence to sketch the form of the spectral distribution $\phi_{B\lambda}$.

From (3.36) the total RFD emitted by a black body is

$$\phi_{\mathrm{B}} = \int_0^\infty \phi_{\mathrm{B}\lambda}\,\mathrm{d}\lambda$$

Standard methods (e.g. see Joos and Freeman, *Theoretical Physics*, p. 616) give the result for this integration as

$$\phi_{\mathrm{B}} = \int_0^\infty \phi_{\mathrm{B}\lambda}\,\mathrm{d}\lambda = \sigma T^4 \tag{3.38}$$

where $\sigma = 5.67 \times 10^{-8}\,\mathrm{W\,m^{-2}\,K^{-4}}$ is the *Stefan–Boltzmann constant*, another fundamental constant.

It follows from (3.33) that the heat flow from a *real* body of emittance $\varepsilon(\varepsilon < 1)$, area $A$ and *absolute* (surface) temperature $T$ is

$$P_{\mathrm{r}} = \varepsilon\sigma A T^4 \tag{3.39}$$

*Note*:

a in using radiation formulae, it is essential to convert surface temperatures in, say, degrees Celsius to absolute temperature, Kelvin; i.e. $x\,°\mathrm{C} = (x + 273)\,\mathrm{K}$,
b the radiant flux dependence on the 4th power of absolute temperature is highly non-linear and causes radiant heat loss to become a dominant heat transfer mode as surface temperatures increase more than $\sim$100 °C.

The Stefan–Boltzmann equation (3.39) gives the radiation emitted *by* the body. The *net* radiative flux away from the body may be much less (e.g. (3.44)). More convenient for calculation than (3.36) is the dimensionless function $D$, where

$$D = \int_0^\lambda \frac{\phi_{\mathrm{B}\lambda}\,\mathrm{d}\lambda}{\sigma T^4} \tag{3.40}$$

which turns out to be a function of the single variable $\lambda T$. This function is graphed in Figure 3.12(b).

### 3.5.6 *Radiative exchange between black surfaces*

All material bodies, including the sky, emit radiation. However, we do not need to calculate how much radiation each body emits individually, but rather what is the *net* gain (or loss) of radiant energy by each body.

Figure 3.13 shows two surfaces 1 and 2, each exchanging radiation. The net rate of exchange depends on the surface properties and on the geometry. In particular we must know the proportion of the radiation emitted by 1 actually reaching 2, and vice versa.

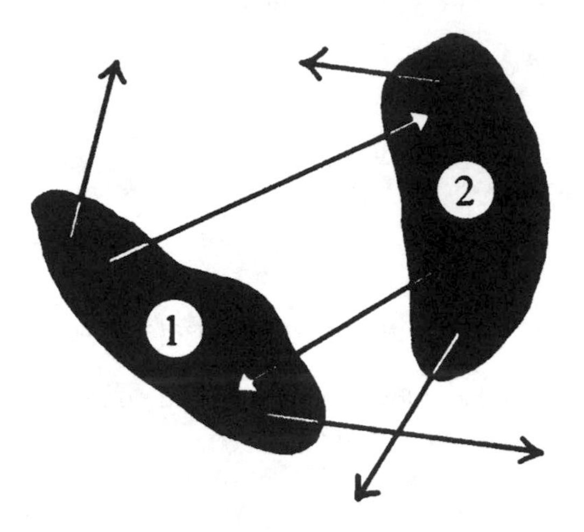


*Figure 3.13* Exchange of radiation between two (black) surfaces.

Consider the simplest case with both surfaces diffuse and black, and with no absorbing medium between them. (A *diffuse* surface is one which emits equally in all directions; its radiation is not concentrated into a beam. Most opaque surfaces, other than mirrors, are diffuse.) The shape factor $F_{ij}$ is the proportion of radiation emitted by surface $i$ reaching surface $j$. It depends only on the geometry and not on the properties of the surfaces. Let $\phi_B$ be the RFD emitted by a black body surface into the hemisphere above it. The radiant power reaching 2 from 1 is

$$P'_{12} = A_1 \phi_{B1} F_{12} \tag{3.41}$$

Similarly the radiant power reaching 1 from 2 is

$$P'_{21} = A_2 \phi_{B2} F_{21} \tag{3.42}$$

If the two surfaces are in thermal equilibrium, $P'_{12} = P'_{21}$ and $T_1 = T_2$: so by (3.38)

$$\phi_{B1} = \sigma T_1^4 = \sigma T_2^4 = \phi_{B2}$$

Therefore

$$A_1 F_{12} = A_2 F_{21} \tag{3.43}$$

This is a geometrical relationship independent of the surface properties and temperature.

If the surfaces are *not* at the same temperature, then the *net* radiative heat flow from 1 to 2, using (3.43), is

$$
\begin{aligned}
P_{12} &= P'_{12} - P'_{21} \\
&= \phi_{B1} A_1 F_{12} - \phi_{B2} A_2 F_{21} \\
&= \sigma T_1^4 A_1 F_{12} - \sigma T_2^4 A_2 F_{21} \\
&= \sigma \left(T_1^4 - T_2^4\right) A_1 F_{12}
\end{aligned} \tag{3.44}
$$

Or, if it is easier to calculate $F_{21}$,

$$P_{12} = \sigma \left(T_1^4 - T_2^4\right) A_2 F_{21} \tag{3.45}$$

In general, the calculation of $F_{ij}$ requires a complicated integration, and results are tabulated in handbooks (e.g. see Wong 1977). Solar collector configurations frequently approximate to Figure 3.14, where the shape factor becomes unity.

### 3.5.7 Radiative exchange between grey surfaces

A *grey body* has a diffuse opaque surface with $\varepsilon = \alpha = 1 - \rho = \text{constant}$, independent of surface temperature, wavelength and angle of incidence.

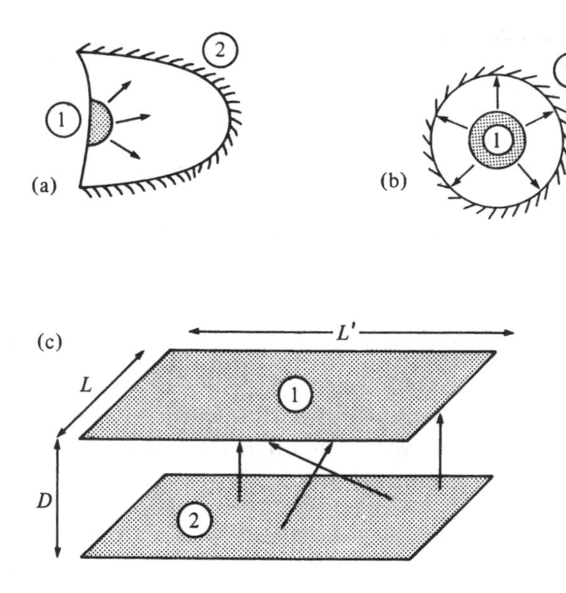


*Figure 3.14* Geometries with shape factor $F_{12} = 1$. (a) convex or flat surface (1) completely surrounded by surface (2). (b) One long cylinder (1) inside another (2). (c) Closely spaced large parallel plates ($L/D, L'/D \gg 1$).

This is a reasonable approximation for most opaque surfaces in common solar energy applications where maximum temperatures are ~200 °C and wavelengths are between 0.3 and 15μm.

The radiation exchange between any number of grey bodies may be analysed allowing for absorption, re-emission and reflection. The resulting system of equations can be solved to yield the heat flow from each body if the temperatures are known, or vice versa. If there are only two bodies, the heat flow from body 1 to body 2 can be expressed in the form

$$P_{12} = \sigma A_1 F'_{12} \left(T_1^4 - T_2^4\right) \tag{3.46}$$

where the *exchange factor* $F'_{12}$ depends on the geometric shape factor $F_{12}$, the area ratio $(A_1/A_2)$ and the surface properties $\varepsilon_1$, $\varepsilon_2$. Comparison with (3.44) shows that for black bodies only, $F'_{12} = F_{12}$.

As in Figure 3.14(c), a common situation is parallel plates with $D \ll L$ and $L'$. In which case $F'_{12} \approx 1/[(1/\varepsilon_1)+(1/\varepsilon_2)-1]$. Such an approximation is acceptable, for instance, in calculating radiative heat exchange in flat plate solar water heaters.

Exchange factors for the most commonly encountered geometries are listed in Appendix C. More exhaustive lists are given in specialised texts (Wong 1977; Rohsenow, Hartnett and Cho 1998).

### 3.5.8 Thermal resistance formulation

Equation (3.46) can be factorised into the form

$$P_{12} = A_1 F'_{12} \sigma \left(T_1^2 + T_2^2\right)(T_1 + T_2)(T_1 - T_2) \tag{3.47}$$

Comparing this with (3.1) we see that the resistance to radiative heat flow from body 1 is

$$R_r = \left[A_1 F'_{12} \sigma \left(T_1^2 + T_2^2\right)(T_1 + T_2)\right]^{-1} \tag{3.48}$$

In general, $R_r$ depends strongly on temperature. However, $T_1$ and $T_2$ in (3.48) are *absolute* temperatures, so that it is often true that $(T_1 - T_2) \ll T_1, T_2$. In this case (3.48) can be simplified to

$$R_r \approx \frac{1}{4\sigma A_1 F'_{12} (\overline{T})^3} \tag{3.49}$$

where $\overline{T} = (T_1 + T_2)/2$ is the mean temperature.

*Example 3.5 Derive typical values of $R_r$, $P_r$*
Two parallel plates of area $1.0\,\mathrm{m}^2$ have emittances of 0.9 and 0.2 respectively. If $T_1 = 350\,\mathrm{K}$ and $T_2 = 300\,\mathrm{K}$ then, using (3.49) and (C.18), Appendix C,

$$R_r = \frac{(1/0.9)+(1/0.2)-1}{4(1\,\mathrm{m}^2)(5.67\times 10^{-8}\,\mathrm{W\,m^{-2}\,K^{-4}})(325\,\mathrm{K})^3} = 0.66\,\mathrm{K\,W^{-1}}$$

This is comparable to the typical convective resistances of Example 3.2. The corresponding heat flow is

$$P_r = \frac{50\,\mathrm{K}}{0.66\,\mathrm{K\,W^{-1}}} = 75\,\mathrm{W}$$

## 3.6 Properties of 'transparent' materials

An ideal transparent material has transmittance $\tau = 1$, reflectance $\rho = 0$ and absorptance $\alpha = 0$. However, in practice 'transparent' materials (such as glass) have $\tau \sim 0.9$ at angles of incidence with the normal of $\leq 70°$, and rapidly reducing $\tau$ and increasing $\rho$ as angles of incidence approach 90°, i.e. the grazing incidence. According to Maxwell's equations of electromagnetism, the reflectance of a material depends on its refractive index and on the angle of incidence with the normal. For most common glasses at angles of incidence less than 40° (the important range in practice) $\rho \approx 0.08$ for visible light. Thus with no absorption, the transmittance would be

$$\tau_r = 1 - \rho \approx 0.92 \tag{3.50}$$

However, some radiation is absorbed as it passes through a partially transparent medium. The proportion reaching a depth $x$ below the surface decreases with $x$ according to the Bouger–Lambert law: the transmitted proportion at $x$ is

$$\tau_{ax} = e^{-Kx} \tag{3.51}$$

where the *extinction coefficient* $K$ varies from about $0.04\,\mathrm{cm}^{-1}$ (for good quality 'water white' glass) to about $0.30\,\mathrm{cm}^{-1}$ (for common window glass with iron impurity, having greenish edges). Iron-free glass has a smaller extinction coefficient than normal window glass, and so is better for solar energy applications. Using the terms $\tau_r$ from (3.50) and $\tau_a$ for $\tau_{a,x}$ when the beam emerges from the material the overall transmittance $\tau$ becomes

$$\tau = \tau_a \tau_r \tag{3.52}$$

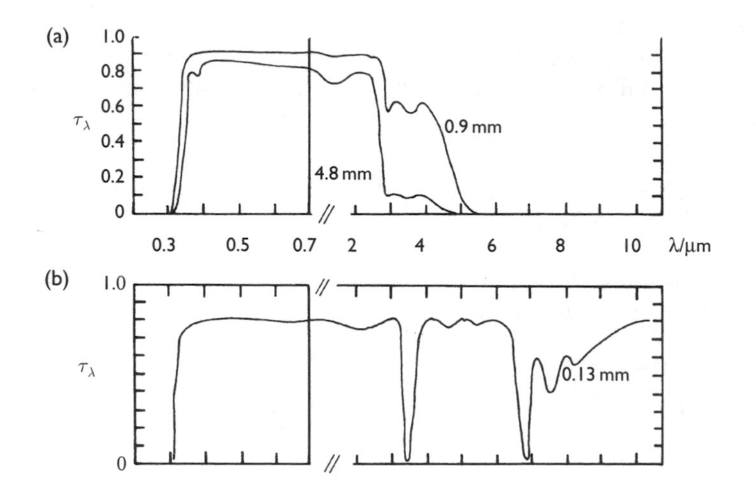


*Figure 3.15* Monochromatic transmittance of: (a) glass (0.15% $Fe_2O_3$) of thickness 4.8 mm and 0.9 mm, (b) polythene thickness 0.13 mm. Note the change of abscissa scale at $\lambda = 0.7\,\mu m$. Data from Dietz (1954) and Meinel and Meinel (1976).

At any particular wavelength $\lambda$, the same reasoning applies to the monochromatic properties, so

$$\tau_\lambda = \tau_{r\lambda}\tau_{a\lambda} \tag{3.53}$$

Figure 3.15(a) shows the variation with wavelength and thickness of the overall monochromatic transmittance, $\tau_\lambda = \tau_{r\lambda}\tau_{a\lambda}$, for a typical glass. Note the very low transmittance in the thermal infrared region ($\lambda > 3\,\mu m$). Glass is a good absorber in this waveband, and hence useful as a greenhouse or solar collector cover to prevent loss of infrared heat. In contrast, Figure 3.15(b) shows that polythene is unusual in being transparent in both the visible and infrared, and hence not a good greenhouse or solar collector cover. Plastics such as Mylar, with greater molecular complication, have transmittance characteristics lying between those of glass and polythene.

## 3.7 Heat transfer by mass transport

Free and forced convection (Section 3.4) is heat transfer by the movement of fluid mass. Analysis proceeds by considering thermal interactions between a (solid) surface and the moving fluid. However, there are frequent practical applications where energy is transported by a moving fluid or solid without considering heat transfer across a surface – for example, when hot water is pumped through a pipe from a solar collector to a storage tank. These systems of heat transfer by mass transport are analysed by considering the fluid alone.

### 3.7.1 *Single phase heat transfer*

Consider the fluid flow through a heated pipe shown in Figure 3.16. According to (2.6), the net heat flow out of the control volume (i.e. out of the pipe) is

$$P_m = \dot{m}c(T_3 - T_1) \tag{3.54}$$

where $\dot{m}$ is the mass flow rate through the pipe (kg/s), $c$ is the specific heat capacity of the fluid ($\text{J kg}^{-1}\,\text{K}^{-1}$) and $T_1$, $T_3$ are the temperatures of the *fluid* on entry and exit respectively. If both $T_1$ and $T_3$ are measured experimentally, $P_m$ may be calculated without knowing the details of the transfer process at the pipe wall. The thermal resistance for this process is defined as

$$R_m = \frac{T_3 - T_1}{P_m} = \frac{1}{\dot{m}c} \tag{3.55}$$

Note here that the heat flow is determined by external factors controlling the rate of mass flow $\dot{m}$, and not by temperature differences. Thus temperature difference is not a driving function here for the mass-flow heat transfer, unlike for conduction, radiation and free convection.

### 3.7.2 *Phase change*

A most effective means of heat transfer is as latent heat of vaporisation/condensation. For example, 2.4 MJ of heat vaporises 1.0 kg of water, which is much greater than the 0.42 MJ to heat 1.0 kg through 100 °C. Heat

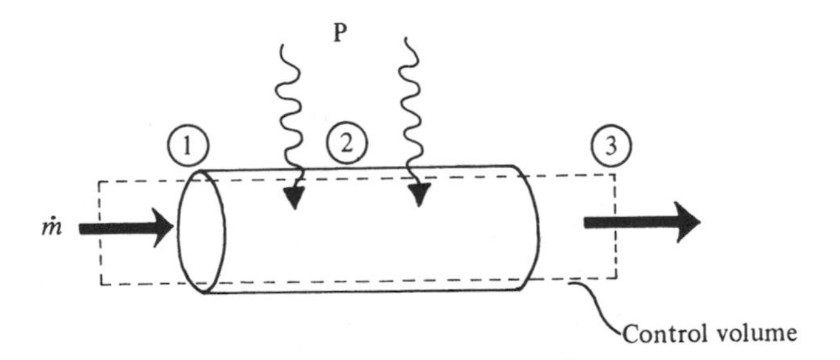


*Figure 3.16* Mass flow through a heated pipe. Heat is taken out by the fluid at a rate $P_m = \dot{m}c(T_3 - T_1)$ regardless of how the heat enters the fluid at (2).

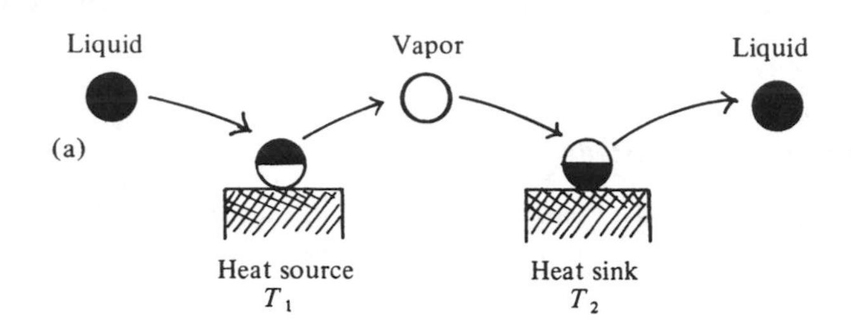


*Figure 3.17* Heat transfer by phase change. Liquid absorbs heat, changes to vapour, then condenses, so releasing heat.

taken from the heat source (as in Figure 3.17) is carried to wherever the vapour condenses (the 'heat sink'). The associated heat flow is

$$P_{\mathrm{m}} = \dot{m}\Lambda \tag{3.56}$$

where $\dot{m}$ is the rate at which fluid is being evaporated (or condensed) and $\Lambda$ is the latent heat of vaporisation. This expression is most useful when $\dot{m}$ is known (e.g. from experiment).

Theoretical prediction of evaporation rates is very difficult, because of the multitude of factors involved, such as (i) the density, viscosity, specific heat and thermal conductivity of both the liquid and the vapour; (ii) the latent heat, the pressure and the temperature difference; and (iii) the size, shape and nucleation properties of the surface. Some guidance and specific empirical formulas are given in the specialised textbooks cited at the end of the chapter.

Since evaporation and condensation are both nearly isothermal processes, the heat flow by this mass transport is not determined directly by the source temperature $T_1$ and the sink temperature $T_2$. The associated thermal resistance can, however, be defined as

$$R_{\mathrm{m}} = \frac{T_1 - T_2}{\dot{m}\Lambda} \tag{3.57}$$

A *heat pipe* is a device for conducting heat efficiently and relatively cheaply for short distances, $<\sim 1\mathrm{m}$, between a separated heat source and heat sink (Figure 3.18). The closed pipe contains a fluid that evaporates in contact with the heat source (at A in the diagram). The vapour rises in the tube (B) and condenses on the upper heat sink (at C). The condensed liquid then diffuses down a cloth wick inside the pipe (at D), to return to the lower end (at E) whence it can continue the cycle. The heat is transferred by mass transfer in the vapour state with very small thermal resistance (high thermal conductance). Many types of evacuated-tube solar water heaters

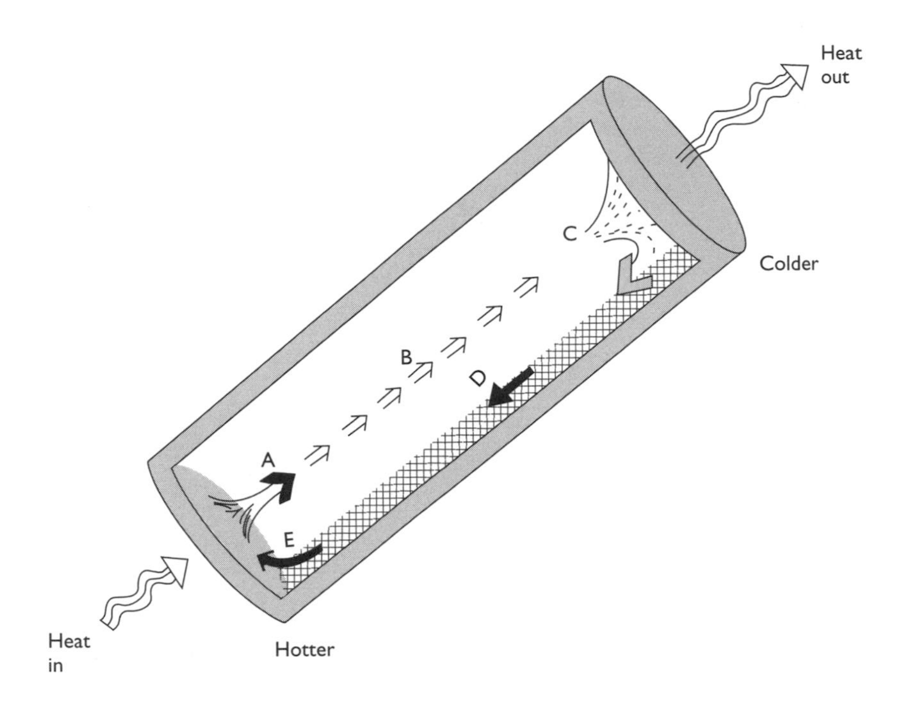


*Figure 3.18* Schematic diagram of a heat pipe (cut-away view). Heat transfer by evaporation and condensation within the closed pipe gives it a very low thermal resistance. See text for further description.

use the heat pipe principle for heat transfer from the collector elements to separately circulating heat transfer fluid.

## 3.8 Multimode transfer and circuit analysis

### *3.8.1 Resistances only*

Section 3.2 showed in general terms how thermal resistances could be combined in series or parallel (or indeed in more complicated networks). The resistances that are combined do not have to refer to the same mode: conduction, convection, radiation and mass transfer can be integrated by this method. Many examples will be found in later chapters, especially Chapter 5.

### *3.8.2 Thermal capacitance*

The circuit analogy can be developed further. Thermal energy can be stored in bulk materials ('bodies') similar to electrical energy stored in capacitors.

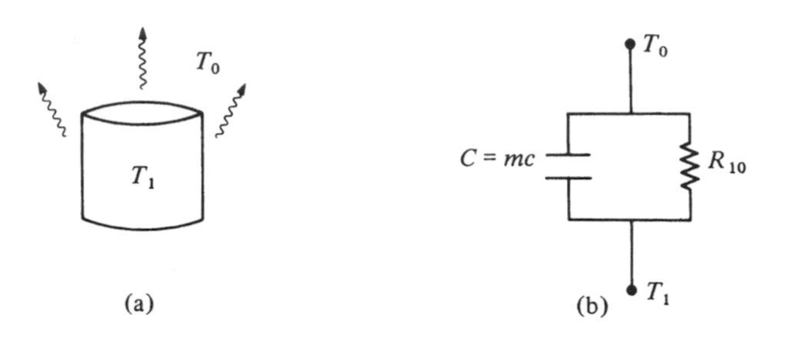




*Figure 3.19* A hot object loses heat to its surroundings. (a) Physical situation. (b) Thermal circuit analogue. (c) Electrical circuit analogue.

For example, consider a tank of hot water standing in a constant temperature environment at $T_0$ (Figure 3.19a). The water (of mass $m$ and specific heat capacity $c$) is at some temperature $T_1$ above the ambient temperature $T_0$. Heat flows from the water to the environment according to the equation

$$-mc\frac{\mathrm{d}}{\mathrm{d}t}(T_1 - T_0) = \frac{T_1 - T_0}{R_{10}} \tag{3.58}$$

where the minus sign indicates that $T_1$ decreases with time when $(T_1 - T_0)$ is positive. $R_{10}$ is the combined thermal resistance of heat loss by convection, radiation and conduction (Figure 3.19(b)). Similarly in the electrical circuit of Figure 3.19(c), electrical current flows from one side of the capacitor (at voltage $V_1$) to the other (at voltage $V_2$) according to the equation

$$\frac{\mathrm{d}q}{\mathrm{d}t} = C_e\frac{\mathrm{d}}{\mathrm{d}t}(V_1 - V_2) = -\frac{V_1 - V_2}{R'_{12}} \tag{3.59}$$

where $q = C_e(V_1 - V_2)$ is the charge held by the capacitor.

Equations (3.58) and (3.59) are exactly analogous; so if the *thermal capacitance* of a body is its heat capacity $C = mc$ (unit $\mathrm{J\,K^{-1}}$), the complete analogy may be listed, as in Table 3.1.

In drawing analogue circuits for thermal systems, care is needed to ensure that the capacitances connect across the correct temperatures (cf. voltages).

*Table 3.1* Comparable electrical and thermal quantities. Note that there is not a 'one to one' correspondence and that much of the terminology is extremely confusing. If in doubt, work out the basic units of the parameter

| *Thermal* | | | *Electrical* | | |
|---|---|---|---|---|---|
| *Quantity* | *Symbol* | *Unit* | *Quantity* | *Symbol* | *Unit* |
| temperature | $T$ | kelvin, K | potential | $V$ | volt, V |
| heat flow | $P$ | watt, W | current | $I$ | ampere, A |
| resistance | $R$ | $\mathrm{K\,W^{-1}}$ | resistance | $R$ | ohm $\Omega = \mathrm{V\,A^{-1}}$ |
| resistivity (of unit area) | $r$ | $\mathrm{m^2 K\,W^{-1}}$ | resistivity | $r$ | $\Omega\,\mathrm{m}$ |
| '$R$ value' $= 1/U = r$ | $r$ | $\mathrm{m^2\,K\,W^{-1}}$. | | | |
| $U$ value (conductance) | $U = 1/r$ | $\mathrm{W\,m^{-2}\,K^{-1}}$ | conductance | $1/R$ | $\Omega^{-1}$ (also *siemens*) |
| conductivity | $k$ | $\mathrm{W\,m^{-1}\,K^{-1}}$ | conductivity | $1/r$ | $\Omega^{-1}\,\mathrm{m^{-1}}$ |
| capacitance | $C$ | $\mathrm{J\,K^{-1}}$ | capacitance | $C$ | farad $\mathrm{F = A\,s\,V^{-1}}$ |

Check that the differential equations, e.g. (3.58), (3.59), correspond exactly with the circuit.

### 3.8.3 Heat exchangers

A heat exchanger transfers heat efficiently from one fluid to another, without allowing them to mix. The so-called 'radiator' in vehicles for extracting heat from the engine cooling-water is probably the most common example. Most solar water heaters have a separate fluid circuit through the collector, with a heat exchanger within the storage tank to transfer the collected heat to the potable water. Figure 3.20 shows the principle of a counter-flow heat exchanger. However, in general there are many different and sophisticated designs, as described in engineering handbooks, e.g. the shell-and-tube design (Figure 14.5).

*Figure 3.20* Sketch of counter-flow heat exchanger principle. Heat is conducted through the wall of the inner tube, thereby cooling the hot inner fluid and heating the cold outer fluid. $T_1 > T_2 > T_4 > T_3$.

In Figure 3.20, consider a fluid, A, losing heat in the inner tube, and fluid, B, gaining heat in the outer tube. Using symbols $\rho$ for density, $c$ for heat capacity and $V$ for rate of volume flow, if these are considered constants with the relatively small changes of temperature:

heat lost by fluid A = heat gained by fluid B + losses

$$\rho_A c_A V_A(T_1 - T_2) = \rho_B c_B V_B(T_4 - T_3) + L \tag{3.60}$$

The efficiency is

$$\eta = \frac{\rho_B c_B V_B(T_4 - T_3)}{\rho_A c_A V_A(T_1 - T_2)} \tag{3.61}$$

The simplest air-to-air heat recovery heat-exchangers operate as ventilation units for rooms in buildings. In which case, usually $V_A = V_B$. With air as the common fluid, and changes in temperature $<50\,°C$, the fluid density and heat capacity are common. So

$$\eta = (T_4 - T_3)/(T_1 - T_2) \tag{3.62}$$

In winter, the incoming fresh air is pre-heated by the outgoing stale air. If the external fresh air is at temperature $T_0$, and the internal stale air at $T_1$, and if, in practice, $T_2 \approx T_3 \approx T_0$, then

$$\eta \approx (T_4 - T_0)/(T_1 - T_0) \tag{3.63}$$

Commonly $T_0 \approx 5\,°C$, $T_1 \approx 22\,°C$ and, in practice, $T_4 \approx 17\,°C$, so

$$\eta \approx (17 - 5)/(22 - 5) \approx 70\% \tag{3.64}$$

In summer, in hot weather, the same flows can pre-cool incoming ventilation air. Such counter-flow heat exchangers are relatively cheap to purchase and to operate as compared with primary energy plant. They provide an excellent example of energy saving and more efficient use of energy.

## Problems

3.1 Show explicitly that for thermal resistances in series, as in Figure 3.2(b), $R_{13} = R_{12} + R_{23}$.
*Hint*: what is the relation between the heat flows in the various resistances?

3.2 Verify from the definitions (3.20) and (3.25) that $\mathcal{N}$ and $\mathcal{A}$ are indeed dimensionless.

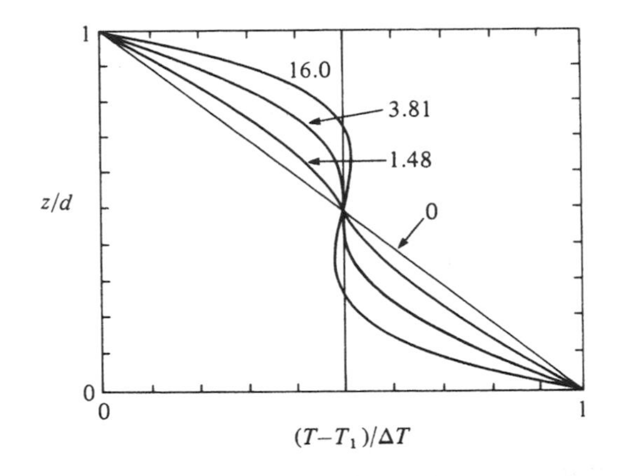


*Figure 3.21* Observed mean temperature profiles in a convecting fluid layer between plates of separation *d*. The curves are labelled by the value of $\mathscr{A}/\mathscr{A}_{crit}$, where $\mathscr{A}_{crit} = 1700$ is the Rayleigh number below which no convection occurs. See (C.7). From Gille (1967).

3.3 A layer of fluid is confined between two horizontal plates, as for (C.7), Appendix C, and Figure 3.21. The hot bottom plate is at height $z = 0$ and temperature $T_1 + \Delta T$, and the cold top plate is at height $z = d$ and temperature $T_1$. Use the measured profiles of Figure 3.21 to measure the thermal thickness $\delta$ at each Rayleigh number shown. Taking $X = d$, find the Nusselt number $\mathscr{N}$ in each case.
*Note*: This is one of the few cases where the thermal thickness $\delta$ of (3.16) can be directly measured. It is the thickness of the layer (near the plate) where the temperature varies almost linearly with $z$. Note that (C.7) does not apply here because $\mathscr{A} \ll 10^5$ and the flow is laminar.

3.4 Calculate the heat lost per hour by radiation from the pot of Example 3.3, and check that it is indeed less than the heat lost by convection.

3.5 Newton's law of cooling (*c*.1666) states that the rate of loss of heat from a body is proportional to the temperature difference between the body and its surroundings. According to this statement, the rate of loss of heat is independent of other variables, such as wind speed.

Calculate the heat flow lost by convection from a flat plate $1\,\text{m}^2$ at $50\,°\text{C}$ to air at $20\,°\text{C}$, at wind speeds of (a) zero (b) $5\,\text{m}\,\text{s}^{-1}$ (c) $10\,\text{m}\,\text{s}^{-1}$. Use (C.15).

What is the percentage error incurred by using a single mean value of heat transfer coefficient $h$, as suggested by Newton's law, to cover the three cases?

3.6 For each of the wind speeds cited in Problem 3.5, calculate the heat flow from the plate of Problem 3.5 by combining separate calculations for

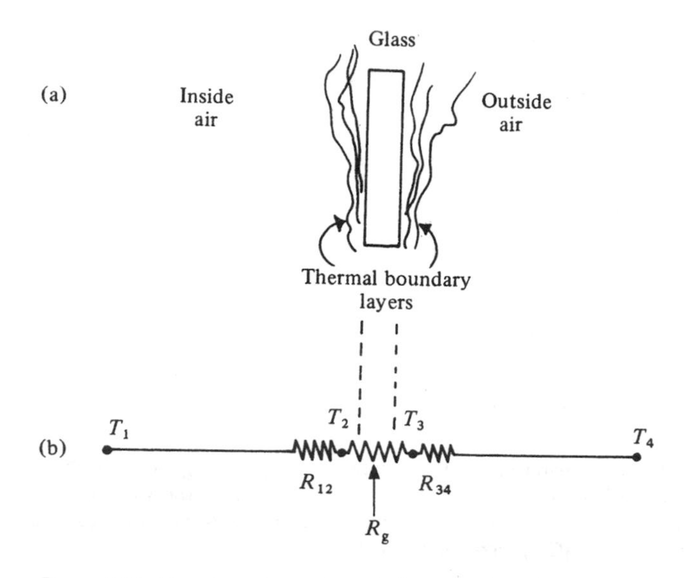


*Figure 3.22* Heat loss through a window; see Problem 3.7.

forced and free convection. Use (C.2) and (C.8). Compare the results and the ease of calculation with those of Problem 3.5.

3.7 *Heat loss through windows*
A room has two glass windows each 1.5 m high, 0.80 m wide and 5.0 mm thick (Figure 3.22). The temperature of the air and wall surface inside the room is 20 °C. The temperature of the outside air is 0 °C. There is no wind. Calculate the heat loss through the glass (a) assuming (falsely) that the only resistance to heat flow is from conduction through the material of the glass, and (b) allowing (correctly) for the thermal resistance of the air boundary layers against the glass, as in Figure 3.22. *Hint*: Assume as a first approximation that $T_2 \approx T_3 \approx (T_1 + T_4)/2$. Justify this assumption afterwards.

## Bibliography

### *General textbooks on heat transfer*

There are many good texts written for engineering students and practising engineers. Without exception these will include heat transfer under large temperature differences using complex situations. Solar energy systems seldom require such complex analysis since temperature differences are relatively small and simple conditions exist. Therefore do not be daunted by the fearsome format of some of these books, of which a small but representative sample follows. The more modern ones have a greater emphasis on computer-aided solutions, but many of the older books are still very useful.

Ede, A.J. (1967) *An Introduction to Heat Transfer Principles and Calculations*, Pergamon Press, Oxford. Clear and simple account for the non-specialist.

Holman, J.P. (1997, 8th edn) *Heat Transfer*, McGraw-Hill, New Jersey. {A widely used text for beginning engineering students, emphasising electrical analogies.}

Incropera, F.P. and Dewitt, D.P. (2002, 5th edn) *Fundamentals of Heat and Mass Transfer*, Wiley. {Includes some solar energy applications as examples.}

Kay, J.M. and Nedderman, R.M. (1985) *Fluid Mechanics and Transfer Processes*, Cambridge University Press. {A terse but clear account for engineering students.}

Kreith, F.R. and Bohn, S. (2000, 6th edn) *Principles of Heat Transfer*, Brooks, New York. {Detailed but clear textbook for would-be specialists.}

Rohsenow, W.R., Hartnett, J.P. and Cho, Y.I. (eds) (1998, 3rd edn) *Handbook of Heat Transfer*, McGraw-Hill, New York. {Comprehensive and detailed handbook for practitioners.}

Wong, H.Y. (1977) *Handbook of Essential Formulae and Data on Heat Transfer for Engineers*, Longmans, London. {A gem of a book, if you can still find it. Easy to use, with comprehensive data expanding Appendices B and C of this book.}

### Heat transfer for solar energy applications

Most books on solar thermal applications include useful chapters of heat transfer formulas. These chapters mostly assume that the basics are known already. For example:

Duffie, J.A. and Beckman, W.A. (1991, 2nd edn) *Solar Engineering of Thermal Processes*, John Wiley and Sons, New York. {Very thorough in a mechanical engineering trasdition and widely used. Uses SI units. Heat transfer is treated from functional relationships, as in many engineering texts, rather than from fundamental physical principles.}

### Specific references in text

De Vries, H.F.W. and Francken, J.C. (1980) 'Simulation of a solar energy system by means of electrical resistance', *Solar Energy*, **25**, 279–341. {Uses an actual physical network.}

Dietz, A.G.H. (1954) 'Diathermanous materials and properties of surfaces', in Hamilton, R.W. (ed.) *Space Heating with Thermal Energy*, MIT Press, Boston.

Gille, J. (1967) 'Interferometric measurement of temperature gradient reversal in a layer of convecting air', *J. Fluid Mech.*, **30**, 371–84.

Joos, G. and Freeman, I.M. (1986, 3rd edn) *Theoretical Physics*, Dover Publications (reprint of 1958 edition).

ISES – International Solar Energy Society (1978) 'Units and symbols in solar energy', *Solar Energy*, **21**, 65–68.

Kaye, H. and Laby, G. (1995) *Tables of Physical and Chemical Constants*, 16th edn, Longmans, London. {Includes an excellent brief account of radiation units.}

Meinel, A.B. and Meinel, M.P. (1976) *Applied Solar Energy*, Addison-Wesley, Massachusetts.

Sargent, S.L. (1972) 'A compact table of black body radiation functions', *Bull. Am. Meteorol. Soc.*, **53**, 360.

Turner, J.S. (1973) *Buoyancy Effects in Fluids*, Cambridge University Press. Especially Chapter 7.

Chapter 4

# Solar radiation

## 4.1 Introduction

Solar radiation reaches the Earth's surface at a maximum flux density of about $1.0\,kW\,m^{-2}$ in a wavelength band between 0.3 and $2.5\,\mu m$. This is called *short wave radiation* and includes the visible spectrum. For inhabited areas, this flux varies from about 3 to $30\,MJ\,m^{-2}\,day^{-1}$, depending on place, time and weather. The spectral distribution is determined by the 6000 K surface temperature of the Sun. This is an energy flux of very high thermodynamic quality, from an accessible source of temperature very much greater than from conventional engineering sources. The flux can be used both thermally (e.g. for heat engines – see Chapters 5 and 6) or, more importantly, for photochemical and photophysical processes (e.g. photovoltaic power and photosynthesis – see Chapters 7 and 10).

The temperatures of the Earth's atmosphere, at about 230 K, and the Earth's surfaces, at about 260–300 K, remain in equilibrium at much less than the 6000 K temperature of the Sun. Therefore the outward radiant energy fluxes emitted by the Earth's atmosphere and surfaces are also of the order of $1\,kW\,m^{-2}$, but occur in an infrared wavelength band between about 5 and $25\,\mu m$, called *long wave radiation*, peaking at about $10\,\mu m$ (see Wien's law, Section 3.5.5). Consequently, the short and long wave radiation regions can be treated as quite distinct from each other, which is a powerful analytical method in environmental science.

The main aim of this chapter is to calculate the solar radiation likely to be available as input to a solar device or crop at a specific location, orientation and time. A secondary aspect is to explain the physical fundamentals associated with the atmospheric greenhouse effect and global climate change; the avoidance of which favours renewable energy. First, we discuss how much radiation is available outside the Earth's atmosphere (Section 4.2). The proportion of this that reaches a device depends on geometric factors, such as latitude (Sections 4.4 and 4.5), and on atmospheric characteristics, such as infrared radiation absorption by water vapour, carbon dioxide and other such molecules (Section 4.6). Two final sections deal briefly with the measurement of solar radiation and with the more difficult problem of how to use other meteorological data to estimate a solar measurement. The most basic information for solar energy devices is contained in Figures 4.7 and 4.15.

## 4.2 Extraterrestrial solar radiation

Nuclear fusion reactions in the active core of the Sun produce inner temperatures of about $10^7$ K and an inner radiation flux of uneven spectral distribution. This internal radiation is absorbed in the outer passive layers which are heated to about 5800 K and so become a source of radiation with a relatively continuous spectral distribution. The radiant flux (W/m$^2$) from the Sun at the Earth's distance varies through the year by ±4% because of the slightly non-circular path of the Earth around the Sun (see (4.25)). The radiance also varies by perhaps ±0.3 per cent per year due to sunspots; over the life of the Earth, there has been probably a natural slow decline of very much less annual significance (see Kyle 1985 or Pap 1997). None of these variations are significant for solar energy applications, for which we consider extra-terrestrial solar irradiance to be constant.

Figure 4.1 shows the spectral distribution of the solar irradiance at the Earth's mean distance, uninfluenced by any atmosphere. Note how similar this distribution is to that from a black body at 5800 K in shape, peak wavelength and total power emitted. (Compare Figure 3.12.) The area beneath this curve is the *solar constant* $G_0^* = 1367\,\mathrm{W\,m^{-2}}$. This is the RFD incident on a plane directly facing the Sun and outside the atmosphere at a distance of $1.496 \times 10^8$ km from the Sun (i.e. at the Earth's mean distance from the Sun).

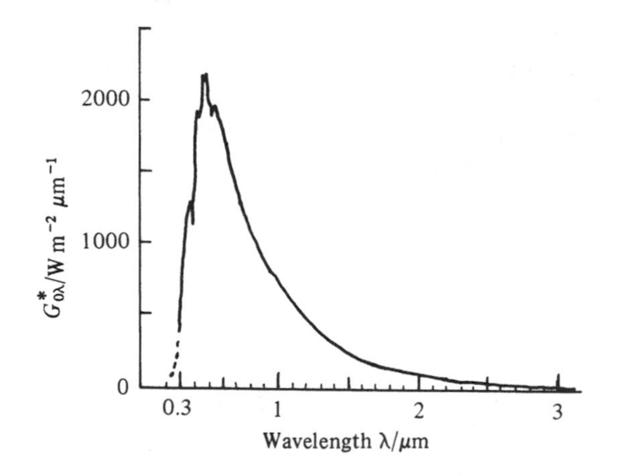


*Figure 4.1* Spectral distribution of extraterrestrial solar irradiance, $G^*_{0\lambda}$. Area under curve equals $1367 \pm 2\,\mathrm{W\,m^{-2}}$ (data source: Gueymard 2004).

The solar spectrum can be divided into three main regions:

1 Ultraviolet region ($\lambda < 0.4\,\mu m$) ~5% of the irradiance
2 Visible region ($0.4\,\mu m < \lambda < 0.7\,\mu m$) ~43% of the irradiance
3 Infrared region ($\lambda > 0.7\,\mu m$) ~52% of the irradiance.

(The proportions given above are as received at the Earth's surface with the Sun incident at about 45°.) The contribution to the solar radiation flux from wavelengths greater than $2.5\,\mu m$ is negligible, and all three regions are classed as solar short wave radiation.

For describing interactions at an atomic level, as in Chapters 7 and 10, it is useful to describe the radiation as individual *photons* of energy $E = hc/\lambda$. Then the range from 0.3 to $2.5\,\mu m$ corresponds to photon energies of 4.1–0.50 eV. (See any textbook on 'modern' physics.)

## 4.3 Components of radiation

Solar radiation incident on the atmosphere from the direction of the Sun is the solar extraterrestrial beam radiation. Beneath the atmosphere, at the Earth's surface, the radiation will be observable from the direction of the Sun's disc in the *direct beam*, and also from other directions as *diffuse radiation*. Figure 4.2 is a sketch of how this happens. Note that even on a cloudless, clear day, there is always at least 10% diffuse irradiance from the molecules in the atmosphere. The practical distinction between the two components is that only the beam radiation can be focused. The ratio between the beam irradiance and the total irradiance thus varies from about 0.9 on a clear day to zero on a completely overcast day.

It is important to identify the various components of solar radiation and to clarify the plane on which the irradiance is being measured. We use subscripts as illustrated in Figure 4.3: $b$ for beam, $d$ for diffuse, $t$ for total, $h$ for the horizontal plane and $c$ for the plane of a collector. The asterisk * denotes the plane perpendicular to the beam. Subscript 0 denotes values outside the atmosphere in space. Subscripts $c$ and $t$ are assumed if no subscripts are given, so that $G(\text{no subscript}) \equiv G_{tc}$.

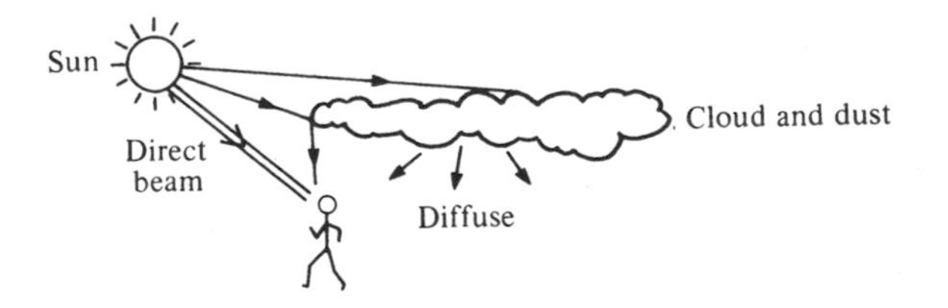


*Figure 4.2* Origin of direct beam and diffuse radiation.

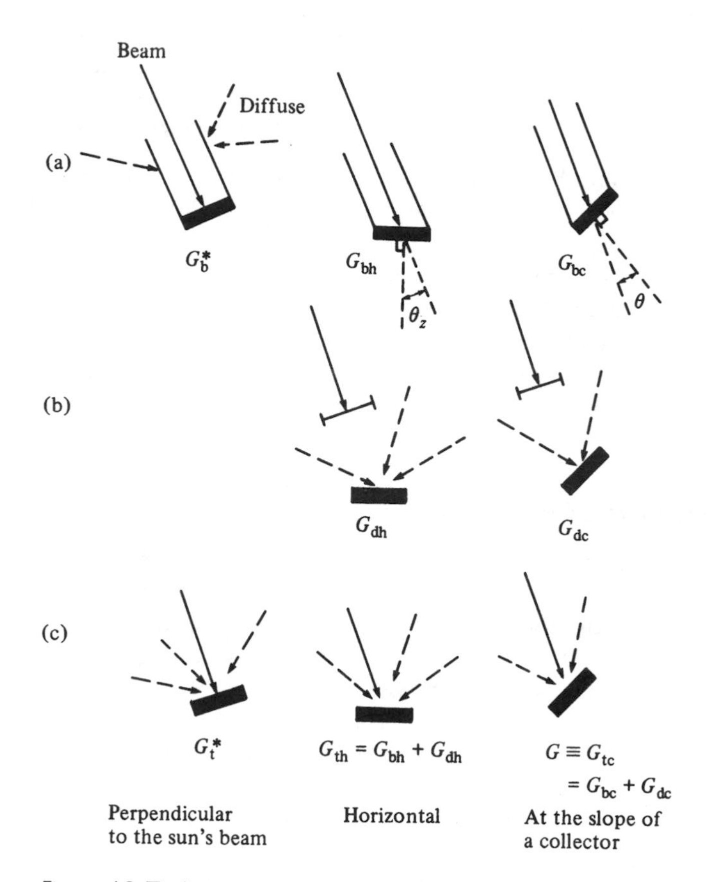


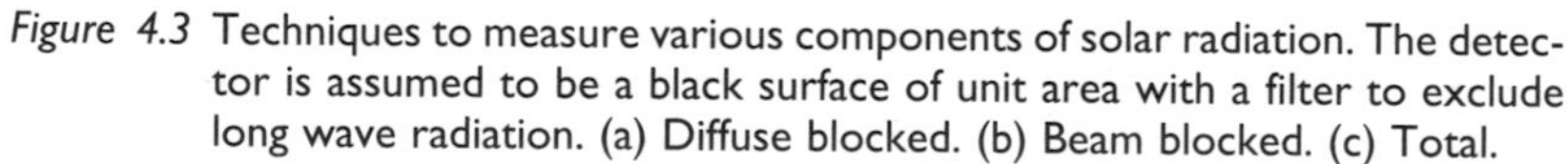
*Figure 4.3* Techniques to measure various components of solar radiation. The detector is assumed to be a black surface of unit area with a filter to exclude long wave radiation. (a) Diffuse blocked. (b) Beam blocked. (c) Total.

Figure 4.3 shows that

$$G_{bc} = G_b^* \cos\theta \tag{4.1}$$

where $\theta$ is the angle between the beam and the normal to the collector surface. In particular,

$$G_{bh} = G_b^* \cos\theta_z \tag{4.2}$$

where $\theta_z$ is the (solar) zenith angle between the beam and the vertical.

The total irradiance on any plane is the sum of the beam and diffuse components, as detailed in Section 4.8.5, so:

$$G_t = G_b + G_d \tag{4.3}$$

See Section 4.8 for more discussion about the ratio of beam and diffuse insolation.

## 4.4 Geometry of the Earth and Sun

### 4.4.1 *Definitions*

You will find it helpful to manipulate a sphere on which you mark the points and planes indicated in Figures 4.4 and 4.5.

Figure 4.4 shows the Earth as it rotates in 24 h about its own axis, which defines the points of the north and south poles N and S. The axis of the poles is normal to the earth's *equatorial plane*. C is the centre of the Earth. The point P on the Earth's surface is determined by its *latitude* $\phi$ and *longitude* $\psi$. Latitude is defined positive for points north of the equator, negative south of the equator. By international agreement longitude $\psi$ is measured positive eastwards from Greenwich, England.[1] The vertical north–south plane through P is the local *meridional plane*. E and G in Figure 4.4 are the points on the equator having the same longitude as P and Greenwich respectively.

*Noon solar time* occurs once every 24 h when the meridional plane CEP includes the Sun, as for all points having that longitude. However, *civil time* is defined so that large parts of a country, covering up to 15° of longitude, share the same official *time zone*. Moreover, resetting clocks for 'summer time' means that solar time and civil time may differ by more than one hour.

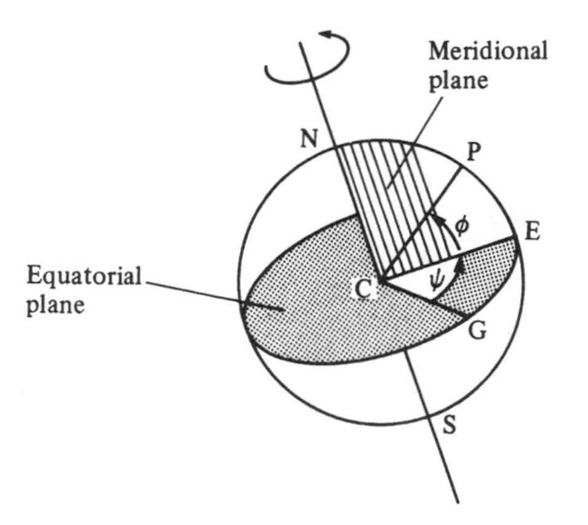


*Figure 4.4* Definition sketch for latitude $\phi$ and longitude $\psi$ (see text for detail).

1 Thereby, the fewest countries are cut by the Date Line at $\phi = 180°$.

The *hour angle* $\omega$ at P is the angle through which the Earth has rotated since solar noon. Since the Earth rotates at $360°/24\,\text{h} = 15°\,\text{h}^{-1}$, the hour angle is given by

$$\begin{aligned}\omega &= (15°\,\text{h}^{-1})(t_{\text{solar}} - 12\,\text{h}) \\ &= (15°\,\text{h}^{-1})(t_{\text{zone}} - 12\,\text{h}) + \omega_{\text{eq}} + (\psi - \psi_{\text{zone}}) \end{aligned} \tag{4.4}$$

where $t_{\text{solar}}$ and $t_{\text{zone}}$ are respectively the local solar and civil times (measured in hours), $\psi_{\text{zone}}$ is the longitude where the Sun is overhead when $t_{\text{zone}}$ is noon (i.e. where solar time and civil time coincide). $\omega$ *is* positive in the evening and negative in the morning. The small correction term $\omega_{\text{eq}}$ is called the equation of time; it never exceeds 15 min and can be neglected for most purposes (see Duffie and Beckman). It occurs because the ellipticity of the Earth's orbit around the Sun means that there are not exactly 24 h between successive solar noons, although the average interval is 24 h.

The Earth orbits the Sun once per year, whilst the direction of its axis remains fixed in space, at an angle $\delta_0 = 23.45°$ away from the normal to the plane of revolution (Figure 4.5). The angle between the Sun's direction and the equatorial plane is called the *declination* $\delta$, relating to seasonal changes. If the line from the centre of the Earth to the Sun cuts the Earth's surface at P in Figure 4.4. then $\delta$ equals $\phi$, i.e. declination is the latitude of the point where the Sun is exactly overhead at solar noon. Therefore in Figure 4.6,

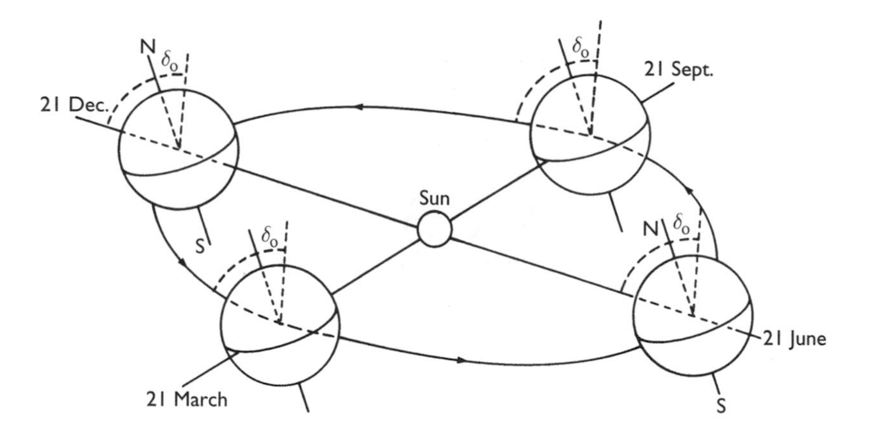


*Figure 4.5* The Earth revolving around the Sun, as viewed from a point obliquely above the orbit (not to scale!). The heavy line on the Earth is the equator. The adjectives 'autumnal, vernal (spring); summer and winter;' may be used to distinguish equinoxes and solstices, as appropriate for the season and hemisphere.

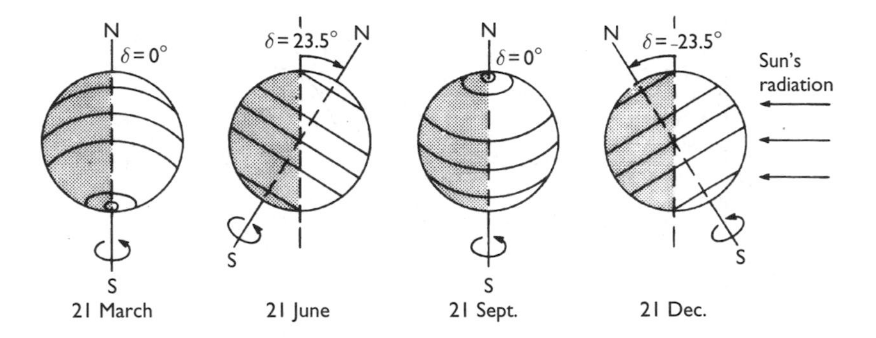


*Figure 4.6* The Earth, as seen from a point further along its orbit. Circles of latitude $0°$, $\pm 23.5°$, $\pm 66.5°$ are shown. Note how the declination $\delta$ varies through the year, equalling extremes at the two solstices and zero when the midday Sun is overhead at the equator for the two equinoxes (equal day and night on the equator).

$\delta$ varies smoothly from $+\delta_0 = +23.45°$ at midsummer in the northern hemisphere, to $-\delta_0 = -23.45°$ at northern midwinter. Analytically,

$$\delta = \delta_0 \sin\left[\frac{360°(284+n)}{365}\right] \tag{4.5}$$

where $n$ is the day in the year ($n = 1$ on 1 January). The error for a leap year is insignificant in practice.

### 4.4.2 *Latitude, season and daily insolation*

The *daily insolation H is* the total energy per unit area received in one day from the sun:

$$H = \int_{t=0\,\mathrm{h}}^{t=24\,\mathrm{h}} G\,\mathrm{d}t \tag{4.6}$$

Figure 4.7 illustrates how the daily insolation varies with latitude and season. The seasonal variation at high latitudes is most significant. The quantity plotted is the clear sky solar radiation on a horizontal plane. Its seasonal variation arises from three main factors:

1 *Variation in the length of the day.* Problem 4.5 shows that the number of hours between sunrise and sunset is

$$N = \frac{2}{15}\cos^{-1}(-\tan\phi\tan\delta) \tag{4.7}$$

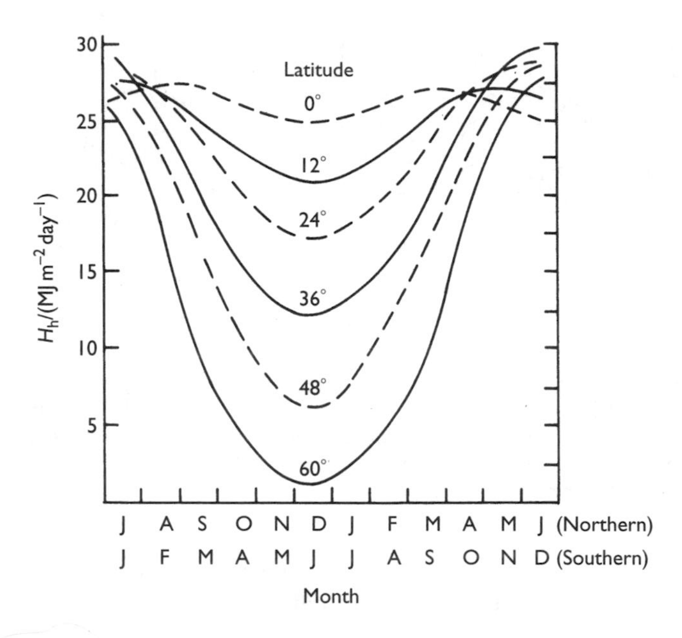


*Figure 4.7* Variation with season and latitude of $H_h$, the daily insolation on a horizontal plane with clear skies. In summer, $H_h$ is about $25\,\text{MJ}\,\text{m}^{-2}\,\text{day}^{-1}$ at all latitudes. In winter, $H_h$ is much less at high latitudes because of shorter day length, more oblique incidence and greater atmospheric attenuation. However, see Figure 4.16 to note how daily insolation varies with the slope of the receiving surface, especially vertical surfaces such as windows.

At latitude $\phi = 48°$, for example, $N$ varies from 16 h in midsummer to 8 h in midwinter. In the polar regions (i.e. where $|\phi| > 66.5°$) $|\tan\varphi \tan\delta|$ may exceed 1. In this case $N = 24$ h (in summer) or $N = 0$ (in winter) (see Figure 4.6).

2 *Orientation of receiving surface.* Figure 4.8 shows that the horizontal plane at a location P is oriented much more towards the solar beam in summer than in winter. Therefore even if $G_b^*$ in (4.2) remains the same, the factor $\cos\theta_z$ reduces $G_{bh}$ in winter, and proportionately reduces $H_h$. Thus the curves of Figure 4.7 are approximately proportional to $\cos\theta_z = \cos(\phi - \delta)$ (Figure 4.8). For the insolation on surfaces of different slopes, see Figure 4.16.
3 *Variation in atmospheric absorption.* The 'clear day' radiation plotted in Figure 4.7 is less than the extraterrestrial radiation because of atmospheric attenuation. This attenuation increases with $\theta_z$, so that $G_b^*$ decreases in winter, thereby the seasonal variation is increased beyond that due to the geometric effects (1) and (2) alone (see Section 4.6).

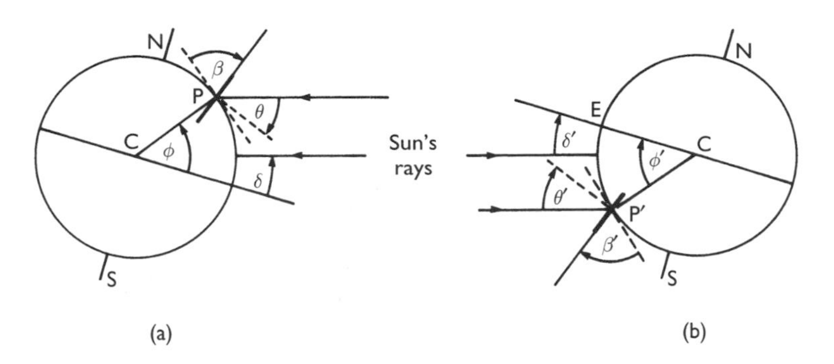


*Figure 4.8* Cross-sections through the earth at solar noon, showing the relation between latitude $\phi$, declination $\delta$, and slope $\beta$ of a collector at P. $\theta$ is the angle of incidence on the north/south-facing collector. (a) Northern hemisphere in summer: $\phi$, $\delta$, $\beta > 0$. (b) 'Symmetrical' example 12 h later in the southern hemisphere ($\phi' = -\phi, \delta' = -\delta, \beta' = \beta, \theta' = \theta$).

In practice 'clear day' radiation is a notional quantity, because actual weather and site conditions vary widely from those assumed in its calculation. Nevertheless, the form of the variations in Figure 4.7 indicates the change in *average* daily insolation *on a horizontal surface* as a function of latitude and season.

Note that for the design of solar buildings, the variation of $H$ on a *vertical surface*, e.g. a window, is significantly different, see Section 4.8.6 and Figure 4.16. Thus, for example, there can be significant solar gain into the windows and conservatories for buildings in regions of middle to high latitude.

## 4.5 Geometry of collector and the solar beam

### 4.5.1 Definitions

For the tilted surface (collector) of Figure 4.9, following Duffie and Beckman, we define:

For the collector surface
*Slope* $\beta$. The angle between the plane surface in question and the horizontal (with $0 < \beta < 90°$ for a surface facing towards the equator; $90° < \beta < 180°$ for a surface facing away from the equator).

*Surface azimuth angle* $\gamma$. Projected on the horizontal plane, $\gamma$ is the angle between the normal to the surface and the local longitude meridian. In either hemisphere, $\gamma$ equals 0° for a surface facing due south, 180° due north, 0° to 180° for a surface facing westwards and, 0° to −180° eastward. For a horizontal surface, $\gamma$ is 0° always.

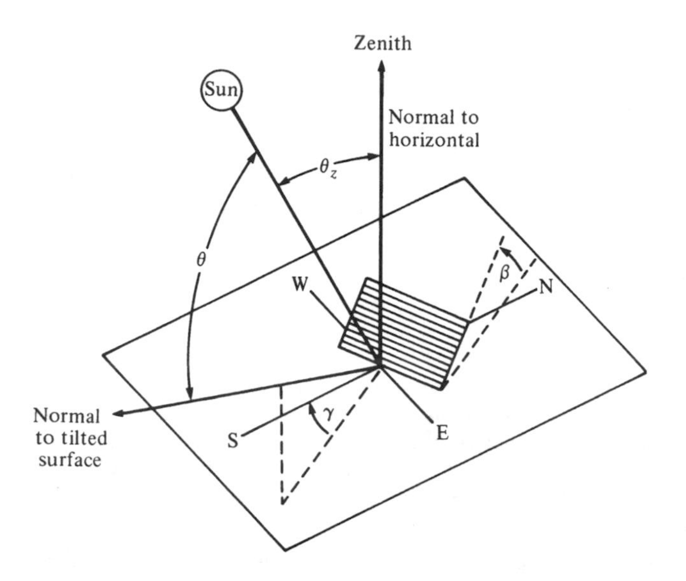


*Figure 4.9* Zenith angle $\theta_z$, angle of incidence $\theta$, slope $\beta$ and azimuth angle $\gamma$ for a tilted surface. (*Note*: for this easterly facing surface $\gamma < 0$.) After Duffie and Beckman.

*Angle of incidence* $\theta$ the angle between solar beam and surface normal.

For the solar beam
*(Solar) zenith angle* $\theta_z$. The angle between the solar beam and the vertical. Note that $\theta_z$ and $\theta$ are not usually in the same plane.

*Solar altitude* $\alpha_s (= 90° - \theta_z)$. The complement to the (solar) zenith angle; angle of solar beam to the horizontal.

*Sun (solar) azimuth angle* $\gamma_s$. Projected on the horizontal plane, the angle between the solar beam and the longitude meridian. Sign convention is as for $\gamma$. Therefore, on the horizontal plane, the angle between the beam and the surface is $(\gamma_s - \gamma)$.

*(Solar) hour angle* $\omega$ (as in (4.4)). The angle Earth has rotated since solar noon (when $\gamma_s = 0$ in the northern hemisphere).

### 4.5.2 Angle between beam and collector

With this sign convention, basic, yet careful, geometry gives equations essential for solar modelling:

$$\cos\theta = (A - B)\sin\delta + [C\sin\omega + (D + E)\cos\omega]\cos\delta \tag{4.8}$$

where

$$A = \sin\phi \cos\beta$$
$$B = \cos\phi \sin\beta \cos\gamma$$
$$C = \sin\beta \sin\gamma$$
$$D = \cos\phi \cos\beta$$
$$E = \sin\phi \sin\beta \cos\gamma$$

and

$$\cos\theta = \cos\theta_z \cos\beta + \sin\theta_z \sin\beta \cos(\gamma_s - \gamma) \tag{4.9}$$

*Example 4.1 Calculation of angle of incidence*
Calculate the angle of incidence of beam radiation on a surface located at Glasgow (56°N, 4°W) at 10 a.m. on 1 February, if the surface is oriented 20° east of south and tilted at 40° to the horizontal.

*Solution*
1 February is day 32 of the year ($n = 32$), so from (4.5)

$$\delta = 23.45^\circ \sin[360^\circ(284+32)/365] = -17.5^\circ$$

Civil time in Glasgow winter is Greenwich Mean Time, which is solar time (±15 min) at longitude $\psi_{zone} = 0$. Hence $t_{solar} \approx 10\,\text{h}$, so (4.4) gives $\omega = -30^\circ$.

We also have $\phi = +56^\circ$, $\gamma = -20^\circ$ and $\beta = +40^\circ$, so that in (4.8)

$$A = \sin 56^\circ \cos 40^\circ = 0.635$$
$$B = \cos 56^\circ \sin 40^\circ \cos(-20^\circ) = 0.338$$
$$C = \sin 40^\circ \sin(-20^\circ) = -0.220$$
$$D = \cos 56^\circ \cos 40^\circ = 0.428$$
$$E = \sin 56^\circ \sin 40^\circ \cos(-20^\circ) = 0.500$$

and so

$$\begin{aligned}\cos\theta &= (0.635 - 0.338)\sin(-17.5^\circ) + [-0.220 \sin(-30^\circ) \\ &\quad + (0.428 + 0.500)\cos(-30^\circ)]\cos(-17.5^\circ) \\ &= 0.783\end{aligned}$$

Thus

$$\theta = 38.5^\circ$$

For several special geometries, the complicated formula (4.8) becomes greatly simplified. For example, Figure 4.8 suggests that a collector oriented towards the equator will directly face the solar beam at noon if its slope $\beta$ is equal to the latitude $\phi$. In this case ($\gamma = 0, \beta = \phi$), (4.8) reduces to

$$\cos \theta = \cos \omega \cos \delta \tag{4.10}$$

For a horizontal plane, $\beta = 0$ and (4.8) reduces to

$$\cos \theta_z = \sin \phi \sin \delta + \cos \phi \cos \omega \cos \delta \tag{4.11}$$

Two cautions should be noted about (4.8), and other formulas similar to it that may be encountered.

1 At higher latitudes in summer, $\theta$ noticeably exceeds 90° in early to mid morning and from mid to late evening, when the sun rises from or falls to the observer's horizon (i.e. $\cos\theta$ negative). When this happens for instance in the northern hemisphere, sunshine is on the north side of buildings and on the rear side of a fixed south-facing collector, not the front.
2 Formulas are normally derived for the case when all angles are positive, and in particular $\phi > 0$. Some northern writers pay insufficient attention to sign, with the result that their formulas do not apply in the southern hemisphere. Southern readers will be wise to check all such formulas, e.g. by constructing complementary diagrams such as Figures 4.8(a,b) in which $\theta' = \theta$ and checking that the formula in question agrees with this.

### 4.5.3 *Optimum orientation of a collector*

A *concentrating collector* (Section 6.8) should always point towards the direction of the solar beam (i.e. $\theta = 0$).

However, the optimum direction of a *fixed flat plate collector* may not be obvious, because the insolation $H_c$ received is the sum of both the beam and the diffuse components:

$$H_c = \int (G_b^* \cos \theta + G_d)\,dt \tag{4.12}$$

A suitable fixed collector orientation for most purposes is facing the equator (e.g. due north in the southern hemisphere) with a slope equal to the latitude, as in (4.10). Other considerations will modify this for particular cases, e.g. the orientation of existing buildings and whether more heat is regularly required (or available) in mornings or afternoons. However, since $\cos\theta \approx 1$ for $\theta < 30°$, variations of $\pm 30°$ in azimuth or slope should have little effect on the total energy collected. Over the course of a year the angle of solar noon varies considerably, however, and it may be sensible to adjust the 'fixed' collector slope month by month.

### 4.5.4 Hourly variation of irradiance

Some examples of the variation of $G_h$ are given in Figure 4.10(a) for clear days and Figure 4.10(b) for a cloudy day. On clear days (following Monteith and Unsworth) the form of Figure 4.10(a) is

$$G_h \approx G_h^{max} \sin\left(\frac{\pi t'}{N}\right) \qquad (4.13)$$

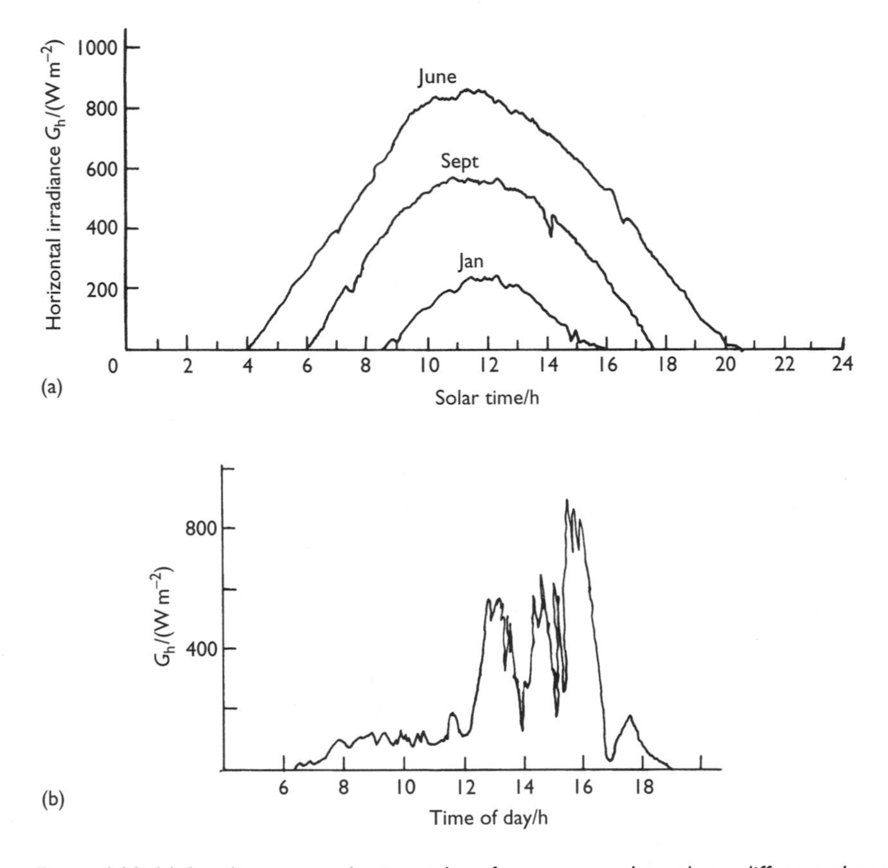


*Figure 4.10* (a) Irradiance on a horizontal surface, measured on three different almost *clear* days at Rothamsted (52°N, 0°W). Note how both the maximum value of $G_h$ and the length of the day are much less in winter than summer. (After Monteith and Unsworth 1990) with permission of Elsevier. (b) Typical variation of irradiance on a horizontal surface for a day of variable cloud. Note the low values during the overcast morning, and the large irregular variations in the afternoon due to scattered cloud.

where $t'$ is the time after sunrise and $N$ is the duration of daylight for the particular clear day (see (4.7) and Figure 4.10(a)). Integrating (4.13) over the daylight period for a clear day,

$$H_{\rm h} \approx (2N/\pi)\, G_{\rm h}^{\max} \tag{4.14}$$

Thus for example at latitude $\pm 50°$ in midsummer, if $G_{\rm h}^{\max} \approx 900\,{\rm W\,m^{-2}}$ and $N \approx 16\,{\rm h}$, then $H_{\rm h} \approx 33\,{\rm MJ\,m^{-2}\,day^{-1}}$. In midwinter at the same latitude, $G_{\rm h}^{\max} \approx 200\,{\rm W\,m^{-2}}$ and $N \approx 8\,{\rm h}$, so $H_{\rm h} \approx 3.7\,{\rm MJ\,m^{-2}\,day^{-1}}$. In the tropics $G_{\rm h}^{\max} \approx 950\,{\rm W\,m^{-2}}$, but the daylight period does not vary greatly from 12 h throughout the year. Thus $H_{\rm h} \approx 26\,{\rm MJ\,m^{-2}\,day^{-1}}$.

These calculations make no allowances for cloud or dust in the atmosphere, and so average measured values of $H_{\rm h}$ are always less than those mentioned. In most regions average values of $H_{\rm h}$ are typically 50–70% of the clear sky value. Only desert areas will have larger averages.

## 4.6 Effects of the Earth's atmosphere

### 4.6.1 *Air-mass-ratio*

The distance travelled through the atmosphere by the direct beam depends on the angle of incidence to the atmosphere (the zenith angle) and the height above sea level of the observer (Figure 4.11). We consider a clear sky, with no cloud, dust or air pollution. Because the top of the atmosphere is not well defined, it is reasonable to consider the mass of atmospheric gases and vapours encountered, rather than the ill-defined distance. For the direct beam at normal incidence passing through the atmosphere at normal pressure, a standard mass of atmosphere will be encountered. If the beam is at zenith angle $\theta_z$, the increased mass encountered compared with the normal path is called the *air-mass-ratio* (or air-mass), with symbol $m$.

The abbreviation AM is also used for air-mass-ratio. AM0 refers to zero atmosphere, i.e. radiation in outer space; AM1 refers to $m = 1$, i.e. sun overhead; AM2 refers to $m = 2$; and so on.

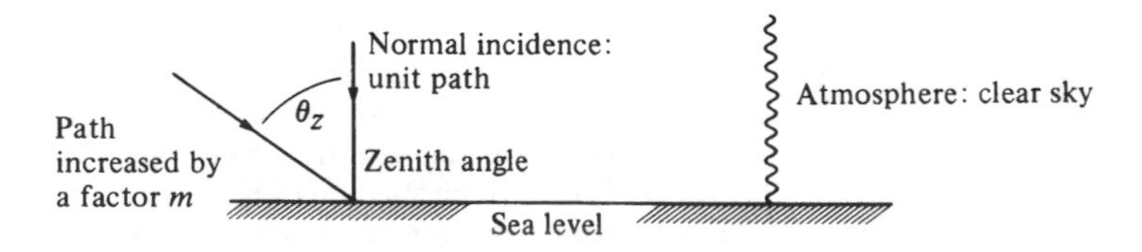


*Figure 4.11* Air-mass-ratio $m = \sec\theta_z$.

From Figure 4.11, since no account is usually taken of the curvature of the earth or of variations with respect to horizontal distance,

$$m = \sec \theta_z \tag{4.15}$$

Changes in air-mass-ratio encountered because of change in atmospheric pressure with time and horizontal distance or with change in height of the observer may be considered separately.

### 4.6.2 Atmospheric absorption and related processes

As the solar short wave radiation passes through the Earth's atmosphere, a complicated set of interactions occurs. The interactions include *absorption*, the conversion of radiant energy to heat and the subsequent re-emission as long wave radiation; *scattering*, the wavelength dependent change in direction, so that usually no extra absorption occurs and the radiation continues at the same frequency; and *reflection*, which is independent of wavelength. These processes are outlined in Figure 4.12.

The effects and interactions that occur may be summarised as follows:

1 *Reflection*. On average, about 30% of the extraterrestrial solar intensity is reflected back into space ($\rho_0 = 0.3$). Most of the reflection occurs from clouds, with a small proportion from the Earth's surface (especially snow and ice). This reflectance is called the *albedo*, and varies with atmospheric conditions and angle of incidence. The continuing short wave solar radiation in clear conditions at midday has flux density $\sim(1-\rho_0)\times 1.3\,\mathrm{kW\,m^{-2}} \approx 1\,\mathrm{kW\,m^{-2}}$.

2 *Greenhouse effect, climate change and long wave radiation*. If the radius of the Earth is $R$, average albedo from space $\rho_0$ and the extraterrestrial solar irradiance (the solar constant) is $G_0$, then the received power is $\pi R^2(1-\rho_0)G_0$. This is equal to the power radiated from the Earth system, of emittance $\varepsilon = 1$ and mean temperature $T_e$, as observed from space. At thermal equilibrium, since geothermal and tidal energy effects are negligible,

$$\pi R^2(1-\rho_0)G_0 = 4\pi R^2\sigma T_e^4 \tag{4.16}$$

and hence, with $\rho_0 = 0.3$,

$$T_e \approx 250\,\mathrm{K}(\text{i.e. } T_e \approx -23\,°\mathrm{C}).$$

Thus, in space, the long wave radiation from the Earth has approximately the spectral distribution of a black body at 250 K. The peak spectral distribution at this temperature occurs at 10 μm, and the distribution does not overlap with the solar distribution (Figure 4.13).

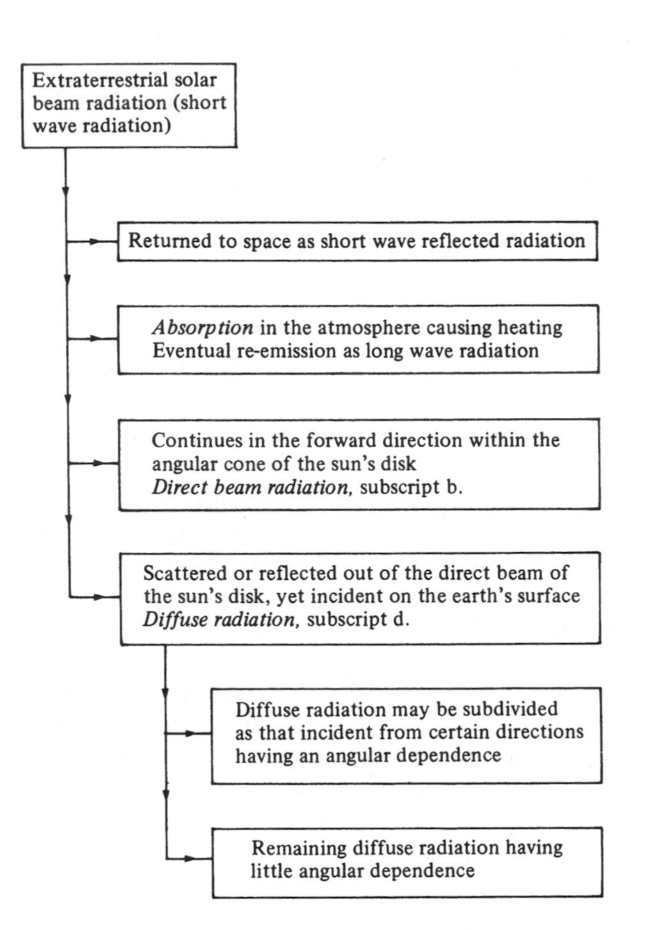


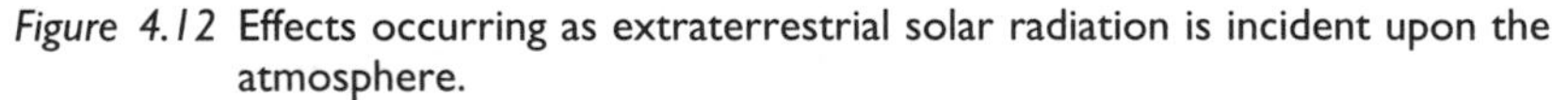

*Figure 4.12* Effects occurring as extraterrestrial solar radiation is incident upon the atmosphere.

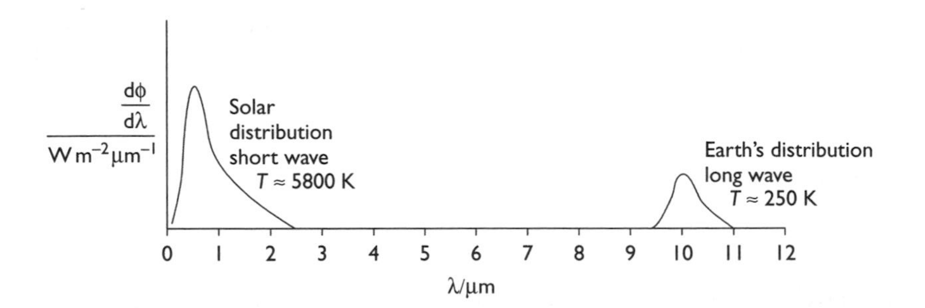


*Figure 4.13* Sketch of the short (including visible) and long wave (far infrared) spectral distributions at the top of the atmosphere. See text and Problem 4.8 for further discussion.

It is obvious from Figure 4.13 that a definite distinction can be made between the spectral distribution (i) of the Sun's radiation (short wave) and (ii) that of the thermal sources from both the Earth's surface and the Earth's atmosphere (long wave). The infrared long wave fluxes at the Earth's surface are themselves complex and large. The atmosphere radiates both down to this surface and up into space. When measuring radiation or when determining the energy balance of an area of ground or a device, it is extremely important to be aware of the invisible infrared fluxes in the environment, which often reach intensities of $\sim 1\,\mathrm{kW\,m^{-2}}$. The black body temperature of the Earth's system in space is effectively that of the outer atmosphere and not of the ground and sea surface. The Earth's average surface temperature, $\sim 14\,°\mathrm{C}$, is about $40\,°\mathrm{C}$ greater than the effective temperature of the outer atmosphere; i.e. about $40\,°\mathrm{C}$ greater than it would be without any atmosphere. In effect, the atmosphere acts as an infrared 'blanket', because some of its gases absorb long wave radiation (see Figure 4.14). This increase in surface temperature (relative to what it would be without the atmosphere) is called the *greenhouse effect*, since the glass of a horticultural glasshouse

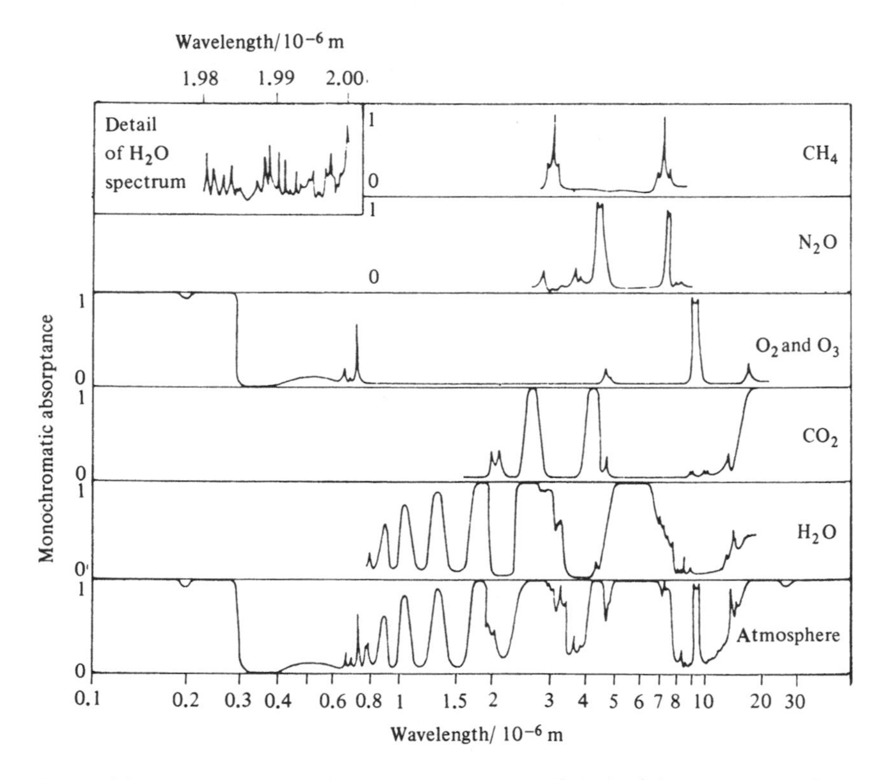


*Figure 4.14* Monochromatic absorptance versus wavelength of the atmosphere. The contributions (not to relative scale) of some main constituents are also shown. From Fleagle and Businger (reprinted by permission of Elsevier).

(a greenhouse) likewise prevents the transmission of infrared radiation from inside to out, but does allow the short wave solar radiation to be transmitted. The gases responsible, notably carbon dioxide ($CO_2$), nitrous oxide ($N_2O$) and methane ($CH_4$), are called *greenhouse gases* (GHG).

Therefore the Earth's atmosphere is not only a source and sink of chemical substances for life; it provides the physical mechanisms for controlling the environmental temperature at which life continues and at which water for life remains liquid.

Measurements of gas trapped in polar ice and the long-term recordings of remote meteorological stations show unequivocally that the concentration of greenhouse gases in the global atmosphere has increased markedly since the industrial revolution of the 18th century. In particular the concentration of $CO_2$ increased from around 280 to 360 ppm by 2000, largely due to the burning of fossil fuels (IPCC 2001). The rate of increase has continued since. The IPCC publications give the theoretical analysis explaining that 'thickening the blanket' in this way increases the average surface temperature of the Earth ('global warming'). The IPCC also give a thorough analysis of the uncertainties involved, since the complexities of atmospheric chemistry, ecology and climate (with its natural variations on timescales of days, seasons, years and centuries) imply that the increase in temperature is unlikely to be directly proportional to the increase in GHG concentration. The authoritative review (IPCC 2001) estimates that collectively the increase in GHG concentrations between the years 1750 and 2000 has had an effect equivalent to an increase of $2.5\,W\,m^{-2}$ in solar irradiance, although some of this effect has been offset by other factors such as an increase in aerosols in the atmosphere, much of which is also due to human activity. The best and easiest to read scientific explanation of this effect and its implications is by Houghton (2004).

Some GHGs contribute more than others to the greenhouse effect. The essential physics is that infrared radiation is absorbed when the electromagnetic radiation resonates with natural mechanical vibrations of the molecules. The more complex the molecules, the more the vibrational modes and the greater the likelihood of absorption at any particular radiation frequency. The impact per unit mass also depends on gaseous density and on secondary reactions and residence time in the atmosphere (Ramaswamy 2001). Thus 1 kg of $CH_4$ (5 atoms per molecule) added to the current atmosphere has as much greenhouse impact over 100 years as 21 kg of $CO_2$ (3 atoms per molecule). This ratio is called the 'global warming potential' (GWP); e.g. the GWP of $CH_4$ is 21. Similarly the GWP of $N_2O$ is 310, while that of most hydrofluorocarbons (used as substitutes for ozone-depleting substances) is over 1000, and that of $CO_2$ is (by definition) 1.000. The measurement of GWP is complex because it depends on the amount of the gases already present and their lifetime in the atmosphere (e.g. methane 'decays' quicker than $CO_2$); the values quoted here are for a 100-year time horizon, and are those used for the purposes of the Kyoto Protocol (see Chapter 17). Allowing for the differing increases in concentrations of the various GHGs, the IPCC find that $CO_2$ is the dominant anthropogenic (human-influenced) greenhouse gas, being responsible for ~60% of the 2.5 $W m^{-2}$ of radiative forcing, with $CH_4$ (at 20%) the next largest contributor.

The IPCC's authoritative review of the relevant scientific literature has concluded that continuing present trends of GHG emissions will lead to an average temperature rise of between 1.5 and 5 °C by 2100, with major consequences for rainfall and sea level. Such man-made *climate change* due to the '*enhanced greenhouse effect*' could have drastic consequences on water supply, the built environment, agriculture, human health and biological ecosystems of all kinds (IPCC 2001). A major motivation for switching from fossil energy sources to renewables is to mitigate these consequences (see Section 1.2).

Since air is nearly transparent, a body on the Earth's surface exchanges radiation not with the air immediately surrounding it, but with the air higher up in the atmosphere, which is cooler. Considering this in terms of Figure 3.14(a), the sky behaves as an enclosure at a temperature $T_s$, the *sky temperature*, which is less than the ambient temperature $T_a$. A common estimate is

$$T_s \approx T_a - 6\,°C \tag{4.17}$$

although in desert regions $(T_a - T_s)$ may be as large as 25 °C.

3 *Absorption in the atmosphere*. Figure 4.14 indicates the relative monochromatic absorption of some main atmospheric components by wavelength. Note the total absorptance (lowest plot) especially. The solar short wave and the atmospheric long wave spectral distributions may be divided into regions to explain the important absorption processes.

a *Short wave ultraviolet region*, $\lambda < 0.3\,\mu m$. Solar radiation is completely removed at sea level by absorption in $O_2$, $O_3$, O and $N_2$ gases and ions.
b *Near ultraviolet region*, $0.3\,\mu m < \lambda < 0.4\,\mu m$. Only a little radiation is transmitted, but enough to cause sunburn.
c *Visible region*, $0.4\,\mu m < \lambda < 0.7\,\mu m$. The pure atmosphere is almost totally transparent to visible radiation, and becomes an open 'window' for solar energy to reach the earth. About half of the solar irradiance is in this spectral region (Figure 4.15). Note, however, that aerosol particulate matter and pollutant gases can cause significant absorption effects.
d *Near infrared (short wave) region*, $0.7\,\mu m < \lambda < 2.5\,\mu m$. Nearly 50% of the extraterrestrial solar radiation is in this region. Up to about 20% of this may be absorbed, mostly by water vapour and also by carbon dioxide in the atmosphere (Figures 4.14 and 4.15). Although the $CO_2$

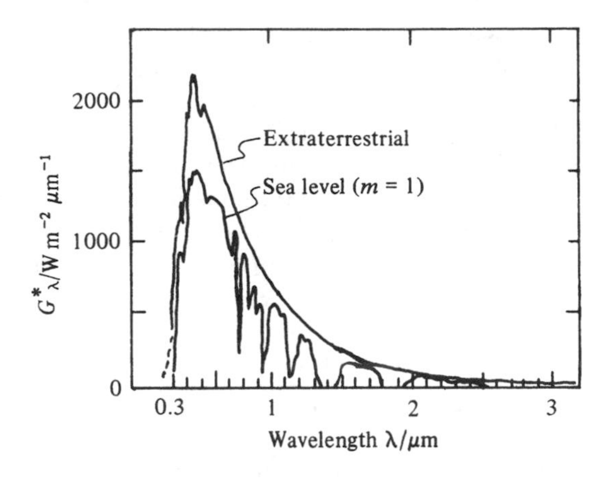


*Figure 4.15* Spectral distributions of solar irradiance received above the atmosphere (upper curve) and at sea level (lower curve). About half the irradiance occurs in the visible region (0.4 – 0.7 μm). There is a gradual decrease of $G_b^*$ as $\lambda$ increases into the infrared, with dips in the sea level spectrum due to absorption by $H_2O$ and $CO_2$. 'Sea level' curve is for air mass $m = 1$.

concentration, now at about 0.04% by volume, is now increasing measurably from year to year, it is relatively constant by month; however, monthly water vapour concentrations may vary significantly to about 4% by volume. Thus fluctuations of absorption by water vapour could be significant in practical applications; however, cloud associated with such increased water vapour is likely to be of far greater significance.

e *Far infrared region*, $\lambda > 12\,\mu m$. The atmosphere is almost completely opaque in this part of the spectrum.

Figure 4.15 shows the cumulative effect on the solar spectrum of these absorptions. The lower curve is the spectrum of the Sun, seen through air-mass-ratio $m = 1$. This represents the radiation received near midday in the tropics (with the Sun vertically above the observer). The spectrum actually received depends on dustiness and humidity, even in the absence of cloud (see Thekaekara 1977 for details).

## 4.7 Measurements of solar radiation

### 4.7.1 Instruments

Table 4.1 lists the commonest instruments used for measuring solar radiation. They are mostly variations on two basic types: a *pyroheliometer*,

*Table 4.1* Classification of instruments for measuring solar radiation

| *Type* | *Measures*[a] | *Stability* | *Absolute output accuracy/%* | *Typical output for 1 kW m$^{-2}$* | *Auxiliary equipment needed* | *Approx. price*[b] | *Notes* |
|---|---|---|---|---|---|---|---|
| | | $\%\,y^{-1}$ | | | | US$ | |
| Active cavity radiometer | Direct irradiance (absolute) | <0–0.1 | <0.3 | – | Varies with use | 15 000 | Used as reference standards and some field observations. Requires tracking mechanism |
| WMO secondary standard pyranometer | Global irradiance | 1 | 3 | 10 mV | Voltage and resistance measurement | 2000 | Thermopile sensor |
| Solar cells | Global irradiance | 2 | 15 | 10 mA | (Milliammeter) current integrator | 200 | Non-uniform spectral response; compact; easily mounted on a collector |
| pyrheliometer (WMO) | Direct irradiance | 1 | 2 | 10 mV | Voltage and resistance | 2000 | Requires tracking mechanism; thermopile |

*Table 4.1* (Continued)

| *Type* | *Measures*[a] | *Stability* %$y^{-1}$ | *Absolute output accuracy/%* | *Typical output for 1 kW m$^{-2}$* | *Auxiliary equipment needed* | *Approx. price*[b] US$ | *Notes* |
|---|---|---|---|---|---|---|---|
| Campbell–Stokes recorder | Sunshine hours | 10 | 20 | Burnt chart | Special cardboard strips | 5000 | Pre-dates pyranometers and satellites |
| Electronic sunshine recorder | Sunshine hours | 2 | 4 | 10 mA | Current integrator | 1000 | Measures difference between shaded and unshaded solar cells |
| Human eye | Cloud fraction | 20 | 20 | Visual scale | Training | – | One site only |
| Satellite estimate[c] | Global insolation | 5 | 10 | Satellite imagery | Radio receiver and special analysis | ?? | Covers whole region – estimates can be derived on a grid spaced across a continent |
| Satellite photo | % cloud | 10 | 20 | Photo | Radio receiver and special analysis | 20 000 | Covers whole region; can be used for forecasting |

Source: Material drawn mainly from WMO, *Guide to meteorological instruments and methods of observation* (1996), esp. Chapter 7 'Measurement of radiation'

Notes:

a Integrating the signal from instruments that measure global irradiance yields insolation.

b Price excluding the ancillary instruments; in most cases, ancillary instruments cost at least as much again.

c In essence, the 'special analysis' compares the radiation coming in from the Sun with that reflected from Earth; the difference is that reaching the Earth's surface. The error in monthly averages (~6%) is about half that for daily averages (~10%). Price to user depends on charging policy of national meteorological service.

which measures the beam irradiance $G_b^*$, and a pyranometer or *solarimeter*, which measures total irradiance $G_{tc}$ (Figure 4.3).

Only the active cavity radiometer (ACR) gives an absolute reading. In this instrument, the solar beam falls on an absorbing surface of area $A$, whose temperature increase is measured and compared with the temperature increase in an identical (shaded) absorber heated electrically. In principle,then,

$$\alpha A G_b^* = P_{elec} \tag{4.18}$$

The geometry of the ACR is designed so that, effectively, $\alpha = 0.999$ (Iqbal 1983).

## 4.8 Estimation of solar radiation

### 4.8.1 Need for estimation

Before installing a solar energy system, it is necessary to predict both the demand and the likely solar energy available, together with their variability. Knowing this and the projected pattern of energy usage from the device, it is possible to calculate the size of collector and storage.

Ideally, the data required to predict the solar input are several years of measurements of irradiance on the proposed collector plane. These are very rarely available, so the required (statistical) measures have to be estimated from meteorological data available either (i) from the site, or (ii) (more likely) from some 'nearby' site having similar irradiance, or (iii) (most likely) from an official solar atlas or database. All such data have systematic error and uncertainty, and natural climatic variability.

### 4.8.2 Statistical variation

In addition to the regular variations depicted in Figures 4.7 and 4.10(a), there are also substantial irregular variations. Of these, perhaps the most significant for engineering purposes are the day-to-day fluctuations (similar to those in Figure 4.10(b)) because they affect the amount of energy storage required within a solar energy system. Thus even a complete record of past irradiance can be used to predict *future* irradiance only in a statistical sense. Therefore design methods usually rely on approximate averages, such as monthly means of daily insolation. To estimate these cruder data from other measurements is easier than to predict a shorter-term pattern of irradiance.

### 4.8.3 Sunshine hours as a measure of insolation

All major meteorological stations measure daily the hours of bright sunshine, $n$. Records of this quantity are available for several decades. It is traditionally measured by a Campbell–Stokes recorder (Table 4.1), which comprises a specially marked card placed behind a magnifying glass. When the sun is 'bright' a hole is burnt in the card. The observer measures $n$ from the total burnt length on each day's card. Sunshine hours are also measured by electronic devices (see Table 4.1), but it is perhaps surprising how often the traditional measurements are continued.

Many attempts have been made to correlate insolation with sunshine hours, usually by an expression of the form

$$H = H_0\,[a + b(n/N)] \tag{4.19}$$

where (for the day in question) $H_0$ is the horizontal irradiance with no atmosphere (i.e. free space equivalent, calculated as in problem 4.6) and $N$ is the length of the day in hours as given by (4.7).

Unfortunately, it has been found that the regression coefficients $a$ and $b$ vary from site to site. Moreover, the correlation coefficient is usually only about 0.7, i.e. the measured data are widely scattered from those predicted from the equation.

Sunshine-hour data give a useful guide to the *variations* in irradiance. For example, it is safe to say that a day with $n < 1$ will contribute no appreciable energy to any solar energy system. The records can also be used to assess whether, for instance, mornings are statistically sunnier than afternoons at the site. The requirement for energy storage can therefore be assessed from the daily data; approximate calculation with some over-design is adequate, but computer modelling gives greater confidence.

Many other climatological correlations with insolation have been proposed, using such variables as latitude, ambient temperature, humidity and cloud cover (see numerous papers in the journal *Solar Energy*). Most such correlations have a limited accuracy and range of applicability.

### 4.8.4 Satellite estimates

Geostationary meteorological satellites operational since about 1990 can produce maps of estimated global insolation across a continent, without using sunshine-hour data as an intermediary. In essence, instruments on the satellite measure separately the radiation coming in from the Sun and that reflected from the Earth; the difference is that reaching the Earth's surface. Many of the data are available from websites, best found with a web search engine.

The main reason for measuring sunshine hours has been to estimate the global insolation (as in Section 4.8.3), but now that regional daily global insolation values are available via satellite-derived analysis, there is little need to monitor ground-based sunshine hours except to calibrate relations like (4.19), which can be used to give a (somewhat inaccurate) record of insolation in the past.

### 4.8.5 *Proportion of beam radiation*

As noted in Section 4.3, the proportion of incoming radiation that is focusable (beam component) depends on the cloudiness and dustiness of the atmosphere. These factors can be measured by the *clearness index* $K_T$, which is the ratio of radiation received on a horizontal surface in a period (usually one day) to the radiation that would have been received on a parallel extraterrestrial surface in the same period:

$$K_T = H_h/H_{oh} \tag{4.20}$$

A clear day may have air-mass-ratio $m = 1$ and therefore $K_T \approx 0.8$. For such days the diffuse fraction is about 0.2; it increases to 1.0 on completely overcast days ($K_T = 0$). On a sunny day with significant aerosol or thin cloud, the diffuse fraction can be as large as 0.5.

The proportion of beam radiation can be found by subtraction:

$$H_{bh}/H_{th} = 1 - (H_{dh}/H_{th}) \tag{4.21}$$

These values of $H_{bh}/H_{th}$ suggest that it is difficult to operate focusing systems successfully in any but the most cloud-free locations. However, notice that such systems track the Sun, and therefore do not collect the horizontal beam component $H_{bh}$, but the larger normal beam component $H_b^*$.

### 4.8.6 *Effect of inclination*

It is straightforward to convert *beam irradiance* measured on one plane (plane 1) to that on another plane (plane 2). This is particularly important for transforming data commonly available for the horizontal plane. Equation (4.8) gives the angle of incidence of the beam to each plane. Then, for the beam component,

$$G_{1b}/\cos\theta_1 = G_{2b}/\cos\theta_2 = G_b^* \tag{4.22}$$

The calculation of the *diffuse irradiance* on another plane, however, cannot be so precise. Consequently the total insolation $H$ on other than the measured surface remains somewhat uncertain.

Duffie and Beckman discuss many refinements for estimating $H$. Although the uncertainty is more than 10%, the results are still instructive. For example, Figure 4.16 shows the variation in estimated daily radiation on various slopes as a function of time of year, at a latitude of 45°N, and with clearness index $K_T = 0.5$. Note that at this latitude, the average insolation on a vertical sun-facing surface varies remarkably little with season, and in winter exceeds $10\,\mathrm{MJ\,m^{-2}\,day^{-1}}$. This is double the insolation on a horizontal surface in winter, and is certainly large enough to provide a

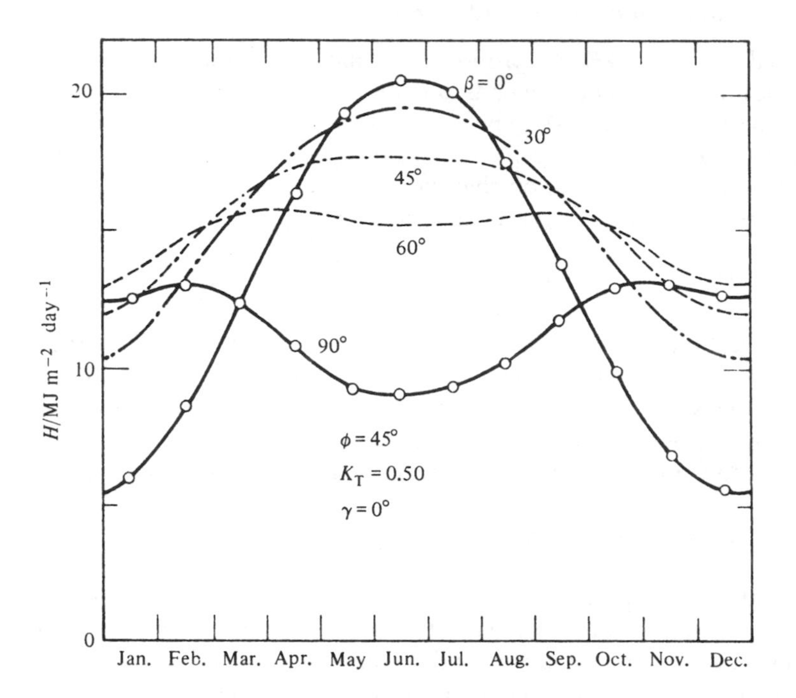


*Figure 4.16* Variation in estimated average daily insolation $H$ on a surface at various slopes, $\beta$, as a function of time of year. For latitude 45°N, with $K_T = 0.50$, $\gamma = 0°$ and ground reflectance 0.20. From Duffie and Beckman (by permission of John Wiley & Sons Inc.).

useful input to passive solar buildings, e.g. through insulating windows, atria and conservatories, and active pre-heating systems (Section 6.3).

## Problems

4.1 a Consider the Sun and Earth to be equivalent to two spheres in space. From the data given below, calculate approximately the solar constant outside the Earth's atmosphere ($W\,m^{-2}$).

b Consider the Earth as apparent from space (i.e. bounded by its atmosphere) to be a black body with surface temperature $T$. Calculate $T$. How does the Earth's surface temperature $T'$ relate to $T$ and what variables control $T'$?

Data:

Diameter of the Sun $2R_S = 1.392 \times 10^9\,m$
Diameter of the Earth $2R_E = 1.278 \times 10^7\,m$
Sun–Earth distance $L = 1.498 \times 10^{11}\,m$
Sun's equivalent black body temperature $= 5780\,K$

4.2 Assume that the sign conventions for $\omega$ (hour angle) in Section 4.4.1 and for $\beta$ (slope) and $\gamma$ (surface azimuth) in Section 4.5.1 are correct for the northern hemisphere. By considering diagrams of appropriate special cases (e.g. Figure 4.8) verify that the conventions are correct also for the southern hemisphere (e.g. a north-facing collector in the southern hemisphere has $\beta > 0$, $\gamma = 180°$).

4.3 At Suva ($\phi = -18°$) at 9 a.m. on 20 May, the irradiance measured on a horizontal plane was $G_h = 1.0\,\mathrm{MJ\,h^{-1}\,m^{-2}}$.

a Determine the angle $\theta_z$ between the beam radiation and the vertical, and hence find the irradiance $G^* = (G_b + G_d)^*$ measured in the beam direction. (Assume that $G_d \ll G_b$, as might be the case on a clear day.)
b Under the same assumptions as in (a), determine the angle $\theta_c$ between the beam and a collector of slope 30° facing due North. Hence find the irradiance $G_c$ on the collector.
c Suppose instead that the diffuse radiation $G_d$ is uniform across the sky, and that $G_{dh} = G_{th}/2$. This is realistic for an overcast day. Recalculate $G^*$ and $G_c$, and comment on the difference between these values and those obtained in (a) and (b).

4.4 Show that the radiative heat loss from a surface at temperature $T$ to the sky (effectively at temperature $T_s$) may be written as

$$P_r = A_1 \varepsilon\sigma \left(T_1^4 - T_s^4\right)$$
$$= A_1 h_r (T_1 - T_a)$$

where

$$h_r = \varepsilon\sigma \left(T_1^2 + T_s^2\right)(T_1 + T_s)\frac{(T_1 - T_s)}{(T_1 - T_a)}$$

4.5 a From (4.11) find the hour angle at sunrise (when the zenith angle $\theta_z = 90°$). Hence show that the number of hours between sunrise and sunset is given by (4.7).
b Calculate the length of the day at midsummer and midwinter at latitudes of (i) 12° and (ii) 60°.

4.6 a If the orbit of the earth were circular, then the irradiance on a horizontal plane outside the atmosphere would be

$$G'_{oh} = G_0^* \cos\theta_z \tag{4.23}$$

where $G_0^*$ is the solar constant.

If $\omega_s$ is the hour angle at sunset (in degrees), show that the integrated daily extraterrestrial radiation on a horizontal surface is

$$H'_{oh} = (G_0^* t_s)[\sin\phi \sin\delta + (180°/\pi\omega_s)\cos\phi \cos\delta \sin\omega_s] \tag{4.24}$$

where $t_s$ is the length of the day.
*Note*: Because of the slight ellipticity of the earth's orbit, the extraterrestrial radiation is not $H'_{oh}$ but

$$H_{oh} = [1 + e\cos(360n/365)]\,H'_{oh} \tag{4.25}$$

where $e = 0.033$ is the eccentricity of the orbit and $n$ is the day number (e.g. $n = 1$ for 1 January).

b Use (4.25) to calculate $H_{oh}$ for $\phi = 48°$ in midsummer and midwinter. Compare your answers with the clear sky radiation given in Figure 4.7.

4.7 Derive (4.10), i.e. $\cos\theta = \cos\omega\cos\delta$, from first principles. (This formula gives the angle $\theta$ between the beam and the normal to a surface having azimuth $\gamma = 0$, slope $\beta = |\text{latitude}|$.
*Hint*: Construct an $(x, y, z)$ co-ordinate system centred on the Earth's centre with the North Pole on $O_z$ and the Sun in the plane $y = 0$, and find the direction cosines of the various directions.
*Note*: The derivation of the full formula (4.8) is similar but complicated. See Coffari (1977) for details.

4.8 In Figure 4.13, should the areas beneath each of the short and long wave distributions be equal? Discuss.

## Bibliography

### General

Duffie, J.A. and Beckman, W.A. (1980, 1st edn; 1991, 2nd edn) *Solar Engineering of Thermal Processes*, Wiley, New York. {Foundation text for serious engineering analysis.}

Iqbal, M. (1983) *An Introduction to Solar Radiation*, Academic Press, New York, reprinted 2004 by Toronto University Press. {Includes a description of the new generation of absolute radiometers and the changes they have made to measurements of the solar constant.}

Monteith, J.L and Unsworth, M. (1990, 2nd edn) *Principles of Environmental Physics*, Edward Arnold, London. {Particularly applied to crop and plant growth, and animal heat balance. Includes a concise description of the radiation environment near the ground.}

Thekaekara, M.P. (1977) 'Solar irradiance, total and spectral' in Sayigh, A.A.M. (ed.) *Solar Energy Engineering*, Academic Press, London. {first hand and first rate account.}

## *Particular*

Coffari, E. (1977) 'The sun and the celestial vault' in Sayigh, A.A.M. (ed.) *Solar Energy Engineering*, Academic Press, London. {Derives the geometric formulae.}

Davies, J.A. and Mackay, D.C. (1989) 'Evaluation of selected models for estimating solar radiation on a horizontal surface', *Solar Energy*, **43**, 153–168.

Dickinson, W.C. and Cheremisinoff, P.N. (eds) (1982) *Solar Energy Technology Handbook*, Butterworths, London. {Clear diagrams of geometry.}

Fleagle, R.C. and Businger, J.A. (1980, 2nd edn) *An Introduction to Atmospheric Physics*, Academic Press, London.

Gueymard, C.A. (2004) 'The sun's total and spectral irradiance for solar energy applications and solar radiation models', *Solar Energy*, **76**, 423–453.

Houghton, John (2004, 3rd edn) *Global Warming – The Complete Briefing*, Cambridge University Press. {An authoritative descriptive text from a Chairman of the Scientific Assessment Working Group of the IPCC – definitely the best explanation of the impact of anthropogenic atmospheric emissions on climate.}

IPCC [Intergovernmental Panel on Climate Change] (2001) *Climate Change 2001: The Scientific Basis*, Cambridge UP. [summary available on the internet at www.ipcc.ch.] {The IPCC is convened by the United Nations to provide an authoritative review on the state of scientific knowledge about climate change. The IPCC produces an updated assessment report about every five years.}

Kyle, H.L. (1985) 'The variation of the Solar constant', *Bul Am Met Soc*, **66**, 1378.

Myers, D.M., Emery, K. and Gueymard, C. (2004) 'Revising and validating spectral irradiance reference standards for photovoltaic performance evaluation,' *J. Solar Engineering* (ASME), **126**, 567–574.

Pap, J.M. (1997) 'Total solar irradiance variability: A review', in *Past and Present Variability of the Solar-terrestrial System: Measurement, Data Analysis and Theoretical Models*, Enrico Fermi international school of physics course CXXXIII, IOS Press, Amsterdam.

Ramaswamy, V. (co-ordinating author) *et al.* (2001) *Radiative Forcing of Climate Change*, Chapter 6 of IPCC (2001). [Understanding how changes in the chemical components of the atmosphere affect the radiation absorption characteristics is an extremely complex subject, fully deserving urgent scientific investigation.]

Renne, D., Perez, R., Zelenka, A., Whitlock, C. and DiPasquale, R. (1999) 'Use of weather and climate research satellites for estimating solar resources', *Advances in Solar Energy*, **13**, 171.

Revfeim, K.J.A. (1981) 'Estimating solar radiation income from "bright" sunshine records', *Q.J. Roy. Met. Soc.*, **107**, 427–435.

World Meteorological Organisation (1996) *Guide to Meteorological Instruments and Methods of Observation*, esp Chapter 7 'Measurement of radiation'.

## Websites

NASA – best updated information from http://solarsystem.nasa.gov/features/planets/sun

www.astm.org – standard reference spectra for solar irradiance at AM0 and AM1.5

Chapter 5

# Solar water heating

## 5.1 Introduction

An obvious use of solar energy is for heating air and water. Dwellings in cold climates need heated air for comfort, and in all countries hot water is used for washing and other domestic purposes. For example, about 30% of the UK's energy consumption is beneficial for heat in buildings and of Australia's energy consumption, about 20% is used for heating fluids to 'low' temperatures (<100 °C). Because of this, the manufacture of solar water heaters has become an established industry in several countries, especially Australia, Greece, Israel, USA, Japan and China. The great majority of solar water heaters are for domestic properties, despite large volumes of hot water being used for process heat in industry.

For solar energy systems, if the insolation is absorbed and utilised without significant mechanical pumping and blowing, the solar system is said to be *passive*. If the solar heat is collected in a fluid, usually water or air, which is then moved by pumps or fans for use, the solar system is said to be *active*. This chapter concentrates on active solar water heaters, since they are common worldwide, they allow practical experiments in teaching and their analysis can provide a step-by-step appreciation of fundamentals for both active and passive applications.

The general principles and analysis that apply to solar water heaters apply also to many other systems which use active and passive mechanisms to absorb the Sun's energy as heat, e.g. air heaters, crop driers, solar 'power towers', solar stills for distilling water, solar buildings. These other applications will be dealt with in Chapter 6. In this chapter we discuss only water heating, starting with essentials and then discussing successively the various refinements depicted in Figure 5.1. These refinements either increase the proportion of radiation absorbed by the heater or decrease the heat lost from the system. Analysis progresses, step by step, to a surprisingly complex heat transfer problem. Table 5.1 shows that although each successive refinement increases efficiency, it also increases the cost. The approximate 'price' in Table 5.1 indicates the cost of manufacture plus some profit. For the institutional reasons discussed in Chapter 17, the monetary cost may

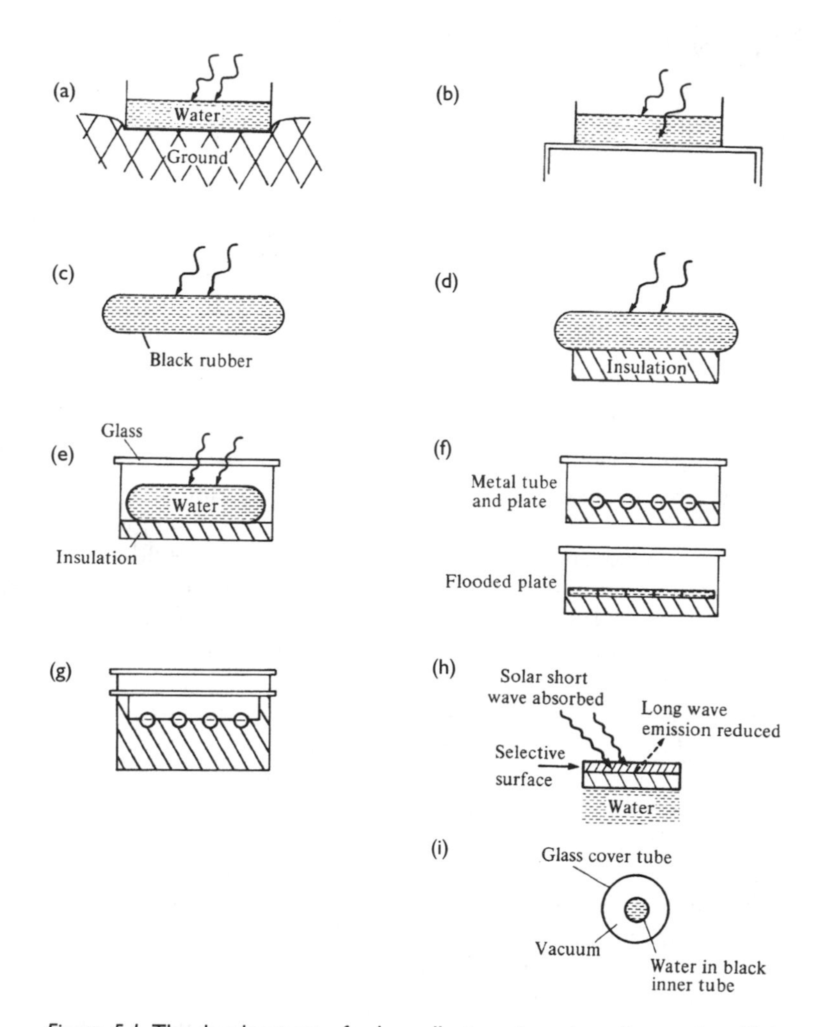


*Figure 5.1* The development of solar collectors, in order of increasing efficiency and cost. See the text for the detailed discussion and analysis of the progression from the most elementary to sophisticated form. (a) Open-container (trough) on ground; heat flows easily to ground. (b) Open trough, off ground. Clear water is not a good absorber; loses heat by evaporation. (c) Black closed container ('tank'); large heat loss, especially to wind; no overnight storage. (d) Black tank, insulated underneath; heat losses confined to top surface, therefore only half those of (c). (e) Sheltered black tank; cheap, but materials degrade. (f) Metal tube and plate collector, and flooded plate. Standard commercial collector; fluid moves through the collector, e.g. to a separate storage tank; flooded plate more efficient than tube and plate. (g) Double glazed flat plate; better insulated version of (f); can operate up to $\sim 100\,°C$; iron-free glass less absorbing than window glass. (h) Selective surface. $\alpha_{short} \gg \varepsilon_{long}$, radiative losses reduced. (i) Evacuated collector. No convection losses to the cover.

*Table 5.1* Summary of the typical performance for different types of collectors

| *Surface* | *Glazing* | *Figure* | $r_{pa}$ ($m^2 KW^{-1}$) | $T_p^{(m)}$ (°C) | *Price* ($\$ m^{-2}$) |
|---|---|---|---|---|---|
| Black | None | 5.1(c) | 0.031 | 40 | 30 |
| Black | Single sheet | 5.1(e), 5.1(f) | 0.13 | 95 | 30–150 |
| Black | Two sheets | 5.1(g) | 0.22 | 140 | 200 |
| Selective | Single sheet | 5.1(f), 5.1(h) | 0.40 | 240 | 200–350 |
| Selective | Two sheets | 5.1(g), 5.1(h) | 0.45 | 270 | 400 |
| Selective | Evacuated tube | 5.1(i) | 0.40 | 300 | 450 |

Notes

1 $r_{pa}$ is the resistance to heat losses through the top of the collector for $T_p = 90\,°C$, $T_a = 20\,°C$, $u = 5\,m\,s^{-1}$
2 $T_p^{(m)}$ is the (stagnation) temperature for which an irradiance of $750\,W\,m^{-2}$ just balances the heat lost through $r_{pa}$. The actual working temperature is substantially less than this (see text).
3 Prices are in US dollars as at 2003, and are very approximate (± factor of 2). They do, however, give some relative indication. Note that 'selective 2-sheet' collectors are no longer commercially produced, and that the price for evacuated collectors may be decreased by mass production.
4 Calculations of $r_{pa}$ and $T_p^{(m)}$ are in Examples 5.1, 5.2 and 5.5 and in Problems 5.3, 5.4, 5.5.

not be the 'true' cost to society or the actual price paid by a consumer in a particular economic framework. Section 5.8 briefly examines such issues, together with social and environmental aspects of the technology.

The main part of a solar heating system is the collector, where solar radiation is absorbed and energy is transferred to the fluid. Collectors considered in this chapter do not concentrate the solar irradiance by mirrors or lenses; they are classed either as *flat plate* or as *evacuated collectors*, in contrast to the focusing collectors discussed in Section 6.8. Non-focusing collectors absorb both beam and diffuse radiation, and therefore still function when beam radiation is cut off by cloud. This advantage, together with their ease of operation and favourable cost (Table 5.1), means that non-focusing collectors are generally preferred for heating fluids to temperatures less than about 80 °C.

We purposely consider the technology step by step for ease of understanding. The simpler collectors (Figure 5.1(a–e)) hold all the water that is to be heated. The more refined collectors, Figure 5.1(f–i), heat only a little water, with the heated water then usually accumulated in a separate storage tank. As discussed in Section 5.5, refinements improve efficiency by reducing the heat losses from the *system* as a whole. Therefore many solar water heaters heat the water indirectly with the collected heat being transferred to potable water in a storage tank through a heat exchanger. A separate fluid in such solar collectors, e.g. an oil or antifreeze solution, is chosen to reduce corrosion, and which does not freeze in winter or boil in normal operation. The analysis of such heaters continues that in Sections 5.4 and 5.5, though with slightly different fluid properties, and is not given separately here.

## 5.2 Calculation of heat balance: general remarks

All solar collectors include an absorbing surface which may be called the *plate*. In Figure 5.2 the radiant flux striking the plate is $\tau_{cov} A_p G$, where $G$ is the irradiance on the collector, $A_p$ is the exposed area of the plate and $T_{cov}$ is the transmittance of any transparent cover that may be used to protect the plate from the wind (e.g. Figure 5.1(e)). The heat transfer terms are all defined in Chapters 3 or 4. Only a fraction $\alpha_p$ of this flux is actually absorbed. Since the plate is hotter than its surroundings, it loses heat at a rate $(T_p - T_a)/R_L$, where $R_L$ is the resistance to heat loss from the plate (temperature $T_p$) to the outside environment (temperature $T_a$). The *net* heat flow into the plate is

$$\begin{aligned} P_{net} &= \tau_{cov}\alpha_p A_p G - [(T_p - T_a)/R_L] \\ &= A_p\left[\tau_{cov}\alpha_p G - U_L(T_p - T_a)\right] \\ &= \eta_{sp} A_p G \end{aligned} \tag{5.1}$$

where $\eta_{sp}$ is the capture efficiency (<1) and $U_L = 1/(R_L A_p)$ is the 'overall heat loss coefficient'. Either of the first two forms of (5.1) is referred to as the Hottel–Whillier–Bliss equation. The parameters of (5.1) for a particular collector are determined experimentally by plotting the collector efficiency as a function of temperature, as in Figure 5.5 (see Section 5.4.3).

It is obvious from the Hottel–Whillier–Bliss equation that the efficiency of solar water heating depends on one set of parameters related to the transmission, reflection and absorption of solar radiation, and another set of parameters related to the retention and movement of heat. In this text we consider each process independently to form a total heat circuit analysis. However, traditional engineering also considers the physical system as a 'black box', to be analysed functionally. For this, practical engineering seeks 'non-dimensional scale-factors' as groups of parameters that, as a group, are independent of particular circumstances; the '*f*-chart' method presented by Duffie and Beckman is a well-used example. Readers are referred to

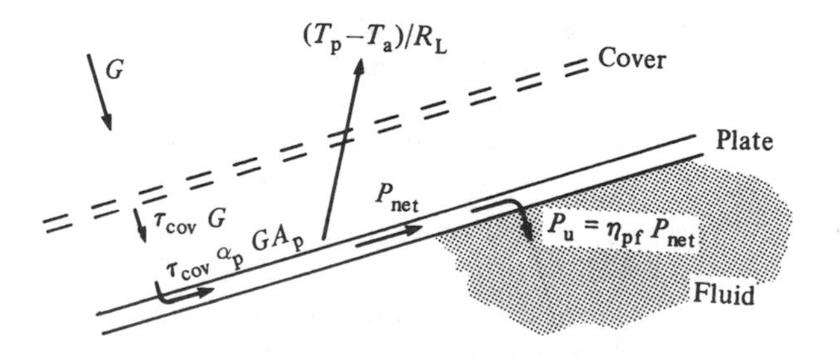


*Figure 5.2* Heat transfer from solar radiation to a fluid in a collector.

Brinkworth (2001) for detailed discussion. However, using such 'lumped parameter' methods may obscure the fundamentals of the heat transfer processes, which are apparent in the 'heat circuit' analysis we use.

In general, only a fraction $\eta_{pf}$ of $P_{net}$ is transferred to the fluid at temperature $T_f$. In a well-designed collector the temperature difference between the plate and the fluid is small, and the *transfer efficiency* $\eta_{pf}$ is only slightly less than 1. Thus the useful output power from the collector is

$$P_u = \eta_{pf} P_{net} \tag{5.2}$$

$$= mc\mathrm{d}T_f/\mathrm{d}t \text{ if a static mass } m \text{ of fluid is being heated} \tag{5.3}$$

$$= \dot{m}c(T_2 - T_1) \text{ if a mass } \dot{m} \text{ flows through the collector in unit time} \tag{5.4}$$

In the third case, (5.4), $T_1$ is the temperature of the fluid as it enters the collector and $T_2$ as it leaves the collector.

These equations are most commonly used to determine the output $P_u$ for a given irradiance $G$. The parameters $A, \tau, \alpha$ of the collector are usually specified, leaving $R_L$ to be calculated using the methods of Chapter 3. Although $T_p$ depends on $P_u$, a reasonable first estimate can be made and then refined later if required. This is illustrated in the following sections.

## 5.3 Uncovered solar water heaters – progressive analysis

### 5.3.1 Uncovered container on the ground

This is the simplest possible water 'heater' (Figure 5.1(a)). An outdoor swimming pool is a common example of a container of water exposed to sunshine, and on, or in, the ground. On a sunny day the water is warmed, but the temperature rise is limited as heat is conducted easily to the ground and also lost by evaporation and convection. Having black surfaces would increase absorption, but obscure cleanliness.

### 5.3.2 Uncovered, open container off the ground

Raising the open container off the ground reduces conductive loss (Figure 5.1(b)), but much of the heat that is retained goes into increased evaporation, thus lessening the temperature increase.

### 5.3.3 Enclosed black container; black tank

Here the water is enclosed in a shallow matt-black tank or bag (Figure 5.1(c)). So no heat is lost by evaporation. The matt-black outer surface absorbs radiation well (typically $\alpha = 0.9$). Some of this absorbed heat is then passed to the water inside by conduction. This type of heater is cheap, easy to make and gives moderately hot water (~20 °C above ambient), but may have a short lifetime. Loss of heat by forced convection from wind severely limits the performance. Despite the simplicity of construction, however, the analysis of the heating is relatively complex, see Example 5.1.

*Example 5.1 The heat balance of an unsheltered black bag*
A rectangular black rubber bag 1 m × 1 m × 0.1 m with walls 5 mm thick is filled with 100 litres of water, supported on a thin, nonconductive, horizontal grid well above the ground, and exposed to a solar irradiance $G = 750\,\mathrm{W\,m^{-2}}$ (Figure 5.3).
The ambient temperature $T_a$ is 20 °C, and the wind speed is $5\,\mathrm{m\,s^{-1}}$. Calculate the resistance to heat losses from the bag. Hence estimate the maximum average temperature of the water, and also the time taken to reach that temperature.

*Solution*
The heat going into or out of the water is conducted through the material skin, which is the 'collector' or 'plate' of this system (see Figure 5.3(a)). Therefore the maximum temperature of the water cannot exceed that of the container. Since the thermal capacity of the thin skin is much less than that of the water and since the conductive resistance of the skin is negligible, we may treat the container and contents as one composite object. This has temperature $T_p = T_f$, absorptance $\alpha = 0.9$ and thermal capacity $C_f = mc$. (In practice, the water at the top of the container will be hotter than $T_f$, and that at the bottom colder; we neglect this for simplicity.) With this approximation, the resistance $R_{pf} = 0$ and $\eta_{pf} = 1$ in (5.2a).

From (5.1) and (5.2b), and with $\tau_{cov} = 1$ since there is no cover,

$$mc\frac{dT_f}{dt} = \alpha AG - (T_f - T_a)/R_L \tag{5.5}$$

In the circuit diagrams of Figure 5.3, $G$ acts as a current source in the analogy with electrical circuits. We shall set up the analysis as in Chapter 3, but here we shall also allow for the environmental temperature to change.

The capacitance $C_f$ *is* shown connected between $T_f$ and a reference $T_{ref}$ in Figure 5.3(b). $T_{ref}$ is an arbitrary but fixed temperature, which is independent of time. This corresponds to the fact that $dT_f/dt$ on the left hand side of (5.5) can be replaced by $d(T_f - T_{ref})/dt$ if $dT_{ref}/dt = 0$. A convenient choice is $T_{ref} = 0\,°C$. Only if the ambient temperature is independent of time

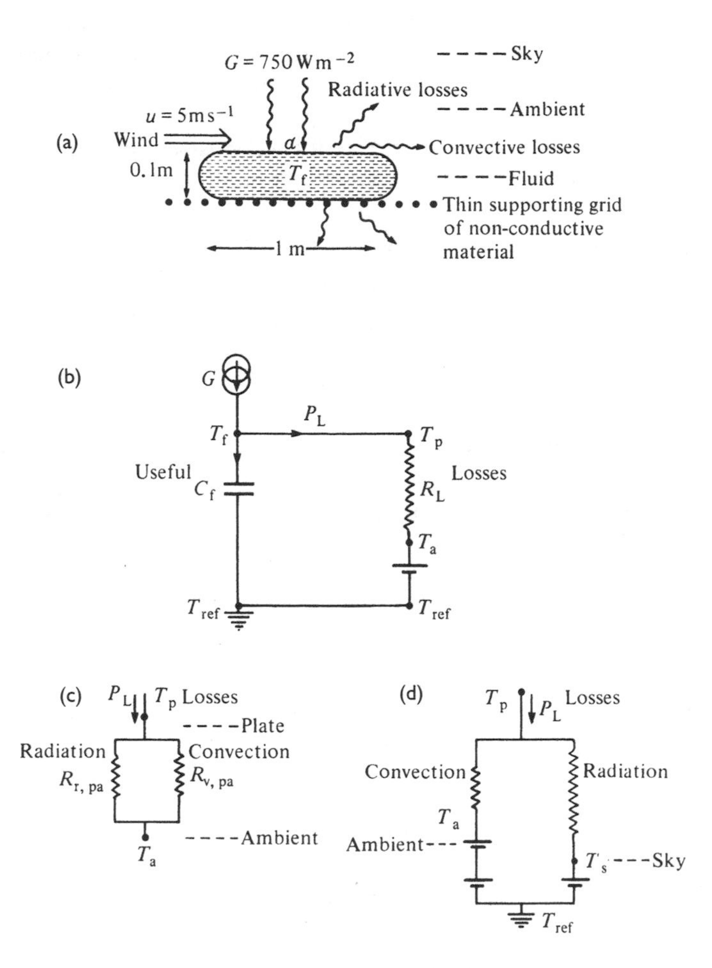


*Figure 5.3* Black bag solar water heater. (a) Physical diagram in section. (b) Simplified circuit analogue. (c) $R_L$ shown as parallel radiation and convection resistances from the plate to the same ambient temperature $T_a$. (d) $R_L$ shown as parallel components losing heat to sinks at different, and possibly changing, temperatures.

can we set $T_{ref} = T_a$ and still preserve the analogy between the circuit and the heat balance equation (5.5). The battery symbol in the right arm of the analogue circuit allows the representation of the ambient temperature $T_a$, as a difference from $T_{ref}$. The resistance $R_L$ between the collector and the environment includes losses from both the top and the bottom. For this system, the top and bottom are similarly exposed to the environment, so that from a total exposed area $A_L = 2\,m^2$ there is a single outward heat flow by convection and radiation in parallel (Figure 5.3(c)).

In many situations, the heat sink temperatures for convection and for radiation are not equal. In general, convective loss is to the ambient air temperature, and radiative loss is to the sky and/or the environment. In Figure 5.3(d) we establish the full circuit diagram for the heat loss component $R_L$. This circuit allows for the different, and possibly changing, heat sink temperatures. In this example, however, $T_{sky}$ and $T_a$ will be treated as constant.

The resistance to convective heat loss is

$$R_{v,pa} = 1/(h_v A_L) \tag{5.6}$$

where $h_v$ is given by (C. 15) of Appendix C as

$$h_v = a + bu = 24.7\,\mathrm{W\,m^{-2}\,K^{-1}} \tag{5.7}$$

for the values given. The radiative heat flow to the sky is given by (C. 17) as

$$P_{r,ps} = \varepsilon_p \sigma A_L (T_p^4 - T_s^4) \tag{5.8}$$

where the effective temperature of the sky $T_s = T_a - 6\,\mathrm{K}$ (see (4.17)).

It is convenient to write the heat flow (5.8) in the form

$$P_{r,ps} = h_{r,pa} A_L (T_p - T_a) \tag{5.9}$$

which will be identically equal to (5.8) if we take

$$\frac{1}{AR_{r,pa}} = h_{r,pa} = \frac{\varepsilon_p \sigma (T_p^2 + T_s^2)(T_p + T_s)(T_p - T_s)}{T_p - T_a} \tag{5.10}$$

We represent the losses as in Figure 5.3(c), where the loss resistance $R_L = (1/R_{v,pa} + 1/R_{r,pa})$ is connected between the plate and ambient, as (5.5) would suggest. It can be verified that $h_{r,pa}$ depends only weakly on $T_p$. Numerically, taking $T_p = 40\,°\mathrm{C}$ as a likely value, we find $h_{r,pa} = 7.2\,\mathrm{W\,m^{-2}\,K^{-1}}$, $r_{pa} = 0.031\,\mathrm{m^2\,K\,W^{-1}}$ and $R_L = 0.015\,\mathrm{K\,W^{-1}}$.
The maximum temperature obtainable occurs when the input balances the losses and (5.5) reduces to (5.1) with $P_{net} = 0$:

$$(T_f - T_a)/R_L = \tau\alpha A_p G$$

Hence $T_f = 31\,°\mathrm{C}$ for this uninsulated bag having $A_p = 1\,\mathrm{m}^2$.

We estimate the time taken to reach this temperature by using (5.5) to find the rate at which $T_f$ is increasing at the halfway temperature $T_f = 25\,°\mathrm{C}$.

Using the value of $R_L$ calculated, $(dT_f/dt)_{25°C} = 8.1 \times 10^{-4}\,K\,s^{-1}$. The time for the temperature to increase by 11 °C is then approximately

$$\Delta t = \Delta T/(dT_f/dt) = 1.3 \times 10^4\,s = 3.7\,h$$

In practice, the irradiance $G$ varies through the day, so the calculations give only very approximate values of $\Delta T$ and $\Delta t$. To obtain a more accurate answer, evaluate (5.5) on an hour-by-hour basis, and also allow for the stratification of water.

### 5.3.4 *Black container with rear insulation*

The heat losses of the system of Figure 5.1(c) can be almost halved simply by insulating the bottom of the container (Figure 5.1(d)). Almost any material that traps air in a matrix of *small* volumes ($\leq$1 mm) is useful as an insulator on this rear side, e.g. fibreglass, expanded polystyrene or wood shavings. The thermal conductivity of all these materials is comparable with that of still air ($k \sim 0.03\,W\,m^{-1}\,K^{-1}$); see Table B.3. The insulating volumes of air must not be too large, since otherwise the air will transfer heat by convection. Also the material must be dry, since water within the matrix is a much better conductor than air (see Appendix B).

Problem 5.2 shows that only a few centimetres of insulation is required to increase the bottom resistance to ten times the resistance of the top. Despite the need for a container to keep the material dry, this is almost always cost-effective for rear insulation.

## 5.4 Improved solar water heaters

### 5.4.1 *Sheltered black container*

The container of Figure 5.1(d) can be sheltered from the wind and so has convective loss reduced by encapsulating it in a covered box with a transparent lid (Figure 5.1(e)). Glass is often the chosen cover material, having small absorptance for the solar short wave irradiation. Clear, i.e. new, polythene sheet also has small short wave absorptance and is cheaper initially, but has to be cleaned and replaced more frequently since it degrades in the open environment. Moreover, glass has a significantly smaller transmittance for infrared radiation than polythene, Figure 3.15, so it absorbs the infrared radiation otherwise lost from the top of the container. This is the 'greenhouse effect' of glass. Polythene is unusual in being transparent to infrared radiation and therefore not good as a cover. However, other types of plastic are available for solar collector covers that have similar properties to glass, but are tougher.

Such a system, with a low total capital cost, may be worthwhile in certain situations where the container is filled by hand.

*Example 5.2 Heat balance of a sheltered collector*

The black container of Example 5.1 is placed inside a box with a glass lid 3.0 cm above it and 10 cm insulation below. For the same external conditions, again calculate the resistance to heat losses from the bag, the theoretical maximum average temperature of the water, and the time taken to attain it.

*Solution*

Figure 5.4(a) shows the physical system and Figure 5.4(b) its circuit analogue. As before, we shall treat the container and contents as a composite system having absorptance $\alpha = 0.90$ and thermal capacity

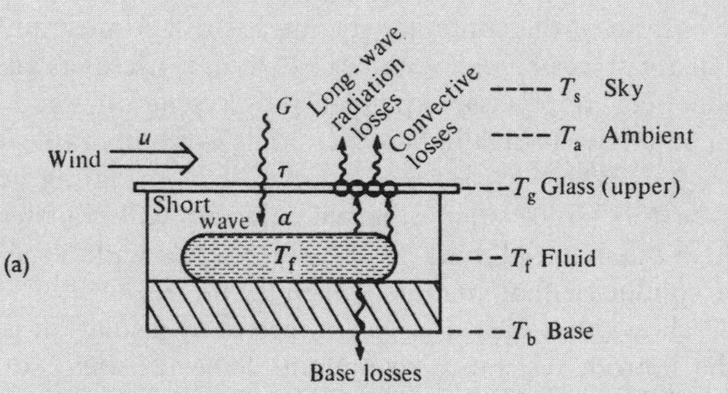


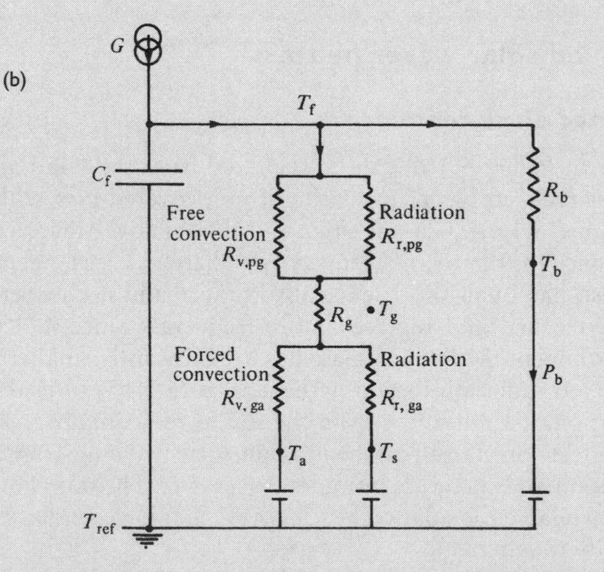


*Figure 5.4* (a) Sheltered black container. (b) Circuit analogue of (a).

$C_f = mc$. The temperature at which heat is lost from the system is the outside temperature of the bag $T_p$. To a first approximation $T_p = T_f$, the mean temperature of the water.

The plate loses heat by conduction through the base, which has an outside temperature $T_b \approx T_a$, so

$$P_b = (T_p - T_b)/R_b \approx (T_p - T_a)/R_b \tag{5.11}$$

Using (3.12) with approximate data,

$$P_b \approx \frac{(T_P - T_a)kA}{x} \approx \frac{(70-20)\mathrm{K}(0.03\,\mathrm{W\,m^{-1}\,K^{-1}})(1\,\mathrm{m^2})}{0.1\,\mathrm{m}} \approx 15\,\mathrm{W} \tag{5.12}$$

which is negligible.

We should incorporate the effective bottom resistance $R_b$ into the loss resistance $R_L$ of (5.1). In practice, as noted earlier, we can usually make $R_b$ great enough so $P_b$ is negligible. Thus the heat balance of the water is given by (5.1) in the form

$$mc\frac{\mathrm{d}T_f}{\mathrm{d}t} = \tau\alpha AG - \frac{T_f - T_a}{R_{pa}} \tag{5.13}$$

The outward heat transfer occurs in the three stages indicated in Figure 5.4(b):

1 Free convection by the air in the gap carries heat to the glass. In parallel with this the plate radiates heat at wavelengths $\sim 10\,\mu\mathrm{m}$. At these wavelengths, glass is not transparent but strongly absorbing (see Figure 3.15). Therefore this radiation is not exchanged directly with the sky but is absorbed by the glass.
2 The heat reaching the glass by these two mechanisms is then conducted to the outer surface of the glass.
3 From here it is transferred to the surroundings by free and/or forced convection, and radiation.

Thus, the overall resistance between the top of the plate and the surroundings is

$$R_{pa} = \left(\frac{1}{R_{v,pg}} + \frac{1}{R_{r,pg}}\right)^{-1} + R_g + \left(\frac{1}{R_{v,ga}} + \frac{1}{R_{r,ga}}\right)^{-1} \tag{5.14}$$

In Figure 5.4(b) the resistance $R_g$ *is* negligible since the glass is thin ($\sim$ 5mm) and a moderately good conductor ($k \approx 1\,\mathrm{W\,m^{-1}\,K^{-1}}$). Therefore the temperature difference across the glass is also negligible. The convective and radiative resistances vary only slowly with the temperatures in the circuit, so the calculation can proceed with initial estimates for these temperatures:

$$T_p = 70\,°\mathrm{C}$$

$$T_g = \frac{1}{2}(T_p + T_a) = 45\,°\mathrm{C} \tag{5.15}$$

For the $1\mathrm{m}^2$ collector, the convective resistance $R_{v,pg}$ follows directly from Example 3.2:

$$R_{v,pg} = 0.52\,\mathrm{K\,W^{-1}}$$

Taking $\varepsilon_p = \varepsilon_g = 0.9$ for long wave radiation, the resistance to radiative heat transfer is, by a calculation similar to that in Example 3.6,

$$R_{r,pg} = 0.16\,\mathrm{K\,W^{-1}}$$

Thus the total plate-to-glass resistance is given by

$$\begin{aligned} R_{pg} &= [(1/R_{v,pg}) + (1/R_{r,pg})]^{-1} \\ &= 0.12\,\mathrm{K\,W^{-1}} \end{aligned} \tag{5.16}$$

The resistance between the outside of the glass and the surroundings is just that already calculated for the unsheltered bag, namely $R_{ga} = 0.031\,\mathrm{K\,W^{-1}}$. Putting these values into (5.14), we obtain $R_{pa} = 0.15\,\mathrm{K\,W^{-1}}$. Then (5.13) with $dT_f/dt = 0, \tau = \alpha = 0.9$ and $G = 750\,\mathrm{W\,m^{-2}}$ implies $T_p^{(m)} = 95\,°\mathrm{C}$. Estimating $(dT/dt)_{60\,°\mathrm{C}}$ as in Example 5.1 gives a theoretical time of 31 hours to reach maximum.

The calculation can be iterated with a better estimate of $T_p$, but the accuracy of the calculation hardly warrants this.

In Example 5.2, the calculated value of the maximum obtainable temperature is over-optimistic because we have neglected the periodicity of the solar radiation, which does not provide sufficient time for the maximum temperature to be reached. Nevertheless we can correctly conclude from Example 5.2 that

1 The presence of a glass cover approximately quadruples the thermal resistance between the hot water and the outside air.
2 A simple sheltered collector can yield water temperatures in excess of 50 °C.

### 5.4.2 Metal plate collectors with moving fluid

We now consider systems of commercial acceptability. In the plate and tube collector (Figure 5.1(f)), water is confined in parallel tubes which are attached to a black metal plate. It is essential to have small thermal resistance between the plate and the tube, and across the plate between the tubes.

Typically the tube diameter is ~2 cm, the tube spacing ~20 cm and the plate thickness ~0.3 cm. The plate and tubes are sheltered from the wind in a framework with a glass top and thick side and rear insulation. This collector has essentially the same circuit analogue as the sheltered black bag (Figure 5.4(b)) and therefore similar resistances to heat loss. Flooded plate collectors are potentially more efficient than tube collectors because of increased thermal contact area. The heated fluid may be used immediately, or it may be stored and/or recirculated, as in Figure 5.6.

### 5.4.3 Efficiency of a flat plate collector

A collector of area $A_p$ exposed to irradiance $G$ (measured in the plane of the collector) gives a useful output

$$P_u = A_P q_u = \eta_c A_p G \tag{5.17}$$

According to (5.1) and (5.3), energy collection is in two sequential stages, so the collector efficiency $\eta_c$ is the product of the capture efficiency $\eta_{sp}$ and the transfer efficiency $\eta_{pf}$:

$$\eta_c = \eta_{sp}\eta_{pf} \tag{5.18}$$

It follows from (5.1) that

$$\eta_{sp} = \tau_{cov}\alpha_p - U_L(T_p - T_a)/G \tag{5.19}$$

which shows that as the plate gets hotter, the losses increase until $\eta_{sp}$ decreases to zero at the 'equilibrium' temperature $T_p^{(m)}$ (also called the stagnation temperature).

As the plate temperature $T_p$ in an operating collector is not usually known, it is more convenient to relate the useful energy gain to the mean fluid temperature $\bar{T}_f$, so that:

$$\eta_c = P_u/(AG) = \eta_{pf}\tau_{cov}\alpha_p - \eta_{pf}U_L(\bar{T}_f - T_a)/G \tag{5.20}$$

In a well-designed collector, the temperature difference between the plate and the fluid is small and the value of $\eta_{pf}$ is nearly one (see Problem 5.8). Typically $\eta_{pf} = 0.85$ and is almost independent of the operating conditions, and, since pipes and storage tanks should be well insulated, $\bar{T}_f \approx T_p$, the collector plate temperature. Hence the $U_L$ in (5.20) is numerically almost the same as that in (5.19). The capture efficiency $\eta_{sp}$ (and therefore also the collector efficiency $\eta_c$) would vary linearly with temperature if $U_L (= 1/R_L)$ were constant in (5.19) and (5.20), but in practice the radiative resistance decreases appreciably as $T_p$ increases. Therefore a plot of $\eta_c$ against operating temperature has a slight curvature, as shown in Figure 5.5.

The performance of a flat plate collector, and in particular its efficiency at high temperatures, can be substantially improved by

1. Reducing the convective transfer between the plate and the outer glass cover by inserting an extra glass cover (see Figure 5.1(g) and Problem 5.3); and/or
2. Reducing the radiative loss from the plate by making its surface not simply black but selective, i.e. strongly absorbing but weakly emitting (see Section 5.6).

The resulting gains in performance are summarised in Table 5.1.

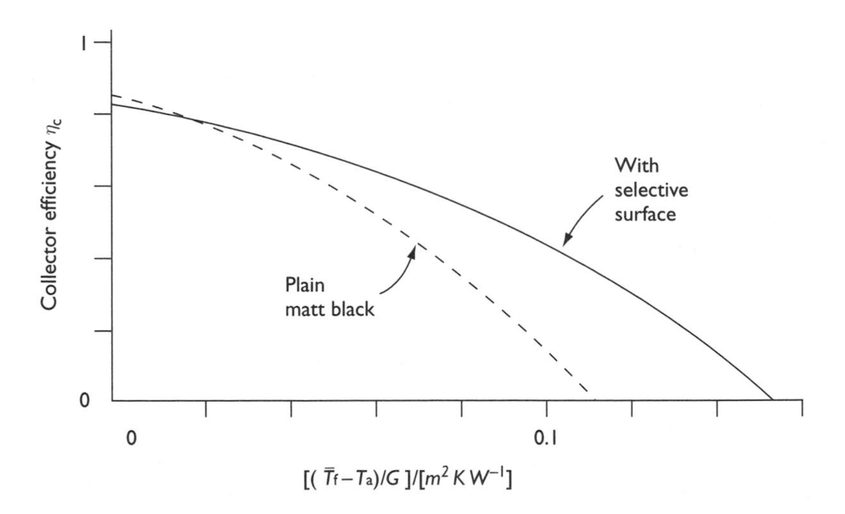


*Figure 5.5* Typical efficiency curves of single-glazed flat plate collectors. $\bar{T}_f$ is the mean temperature of the working fluid and $T_a$ is ambient temperature. After Morrison (2001).

## 5.5 Systems with separate storage

### *5.5.1 Active systems with forced circulation*

The collectors of Figure 5.1(f) can heat only a small volume of water which, therefore, should be passed to an insulated tank for storage (Figure 5.6). For domestic systems, tanks with a volume of about 100–200 litres can store a day's supply of hot water. For forced circulation only a small pump is needed, designed with a pumping rate so the water temperature increases by about 5–10 °C in passing through the collector in sunshine. This incremental temperature increase depends on the solar irradiance $G$ and the inlet temperature $T_1$, so the design temperature rise will be achieved only for one set of conditions if a fixed-speed pump is used. Nevertheless, single-speed pumps are usually used, as they are the cheapest. The pumps are powered either from mains electricity or, in some designs, from a small photovoltaic panel alongside the collector. A simple pump controller switches the pump off if the collector output temperature is less than about 5 °C more than the water in the top of the tank. This prevents needless use of the pump and, in particular, the stupidity of losing heat from the collector in poor sunlight and at night.

*Example 5.3 Temperature rise through a collector*
A flat plate collector measuring $2\,\mathrm{m} \times 0.8\,\mathrm{m}$ has a loss resistance $r_L = 0.13\,\mathrm{m^2\,K\,W^{-1}}$ and a plate transfer efficiency $\eta_{pf} = 0.85$. The glass cover has transmittance $\tau = 0.9$ and the absorptance of the plate is $\alpha = 0.9$. Water enters at a temperature $T_1 = 40\,°\mathrm{C}$. The ambient temperature is $T_a = 20\,°\mathrm{C}$ and the irradiance in the plane of the collector is $G = 750\,\mathrm{W\,m^{-2}}$.

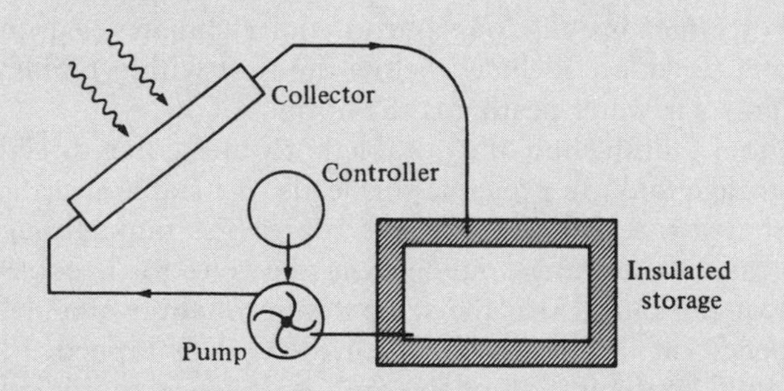


*Figure 5.6* Collector coupled to a separate storage tank by a pump.

a Calculate the flow rate needed to produce a temperature rise of 4 °C.
b Suppose the pump continues to pump at night, when $G = 0$. What will be the temperature *fall* in each passage through the collector? (Assume that $T_1 = 40\,°C$, $T_a = 20\,°C$ still.)

*Solution*
a From (5.1) and (5.2), the useful power per unit area is

$$q_u = (\rho c Q/A)(T_2 - T_1) = \eta_{pf}[\tau\alpha G - (T_p - T_a)/r_L]$$

Assuming $T_p = 42\,°C$ (the mean temperature of the fluid), this yields

$$Q = 3.5 \times 10^{-5}\,m^3\,s^{-1} = 130\,L\,h^{-1}$$

b From (5.15) with $G = 0$, $T_p = 38\,°C$ and with the calculated value of $Q$,

$$T_2 - T_1 = -1.3°C$$

If the collector of Example 5.3 was part of a hot water system with a volume of 130 litres, circulating once per hour, then, if pumping continued without a controller, the water temperature at night would fall by $1.3\,°C\,h^{-1}$ because the collector would lose heat.

An advantage of forced circulation is that an existing water heater system can easily be converted to solar input by adding collectors and a pump. The system is also likely to be more efficient, and the storage tank need not be higher than the collectors. A disadvantage, however, is that the system is dependent on electricity for the pump, which may be expensive or unreliable. For larger installations and in cooler climates, e.g. most of Europe, hot-water tanks are included below the roof within buildings, so forced circulation solar water heating is the norm.

Figures 5.3 and 5.7 both show the potable hot water going directly into the top of the storage tank. In principle this leads to a stable stratification, with the hottest (least dense) water at the top of the tank, though this will not be the case if the water coming from the collector is cooler than that at the top of the tank. Also the temperature of the water delivered to the user depends on the height at which the tank is tapped. In some systems the internal configuration of the tank is designed to minimise the stratification, by promoting mixing of the warmer and cooler water; in this way the water obtained is always 'warm' provided that the extraction rate

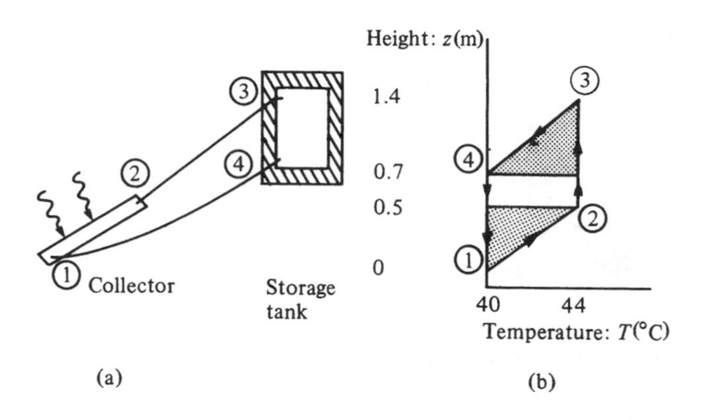


*Figure 5.7* Collector and storage tank with thermosyphon circulation. (a) Physical diagram. (b) Temperature distribution (see Example 5.4).

is not too large. Other systems are designed to promote stratification, so that the hottest water available is drawn off, but this is desirable only if the volume drawn off is significantly less than the total volume of the tank; this may be desirable in colder climates. One ingenious way to achieve this is to have the hot water enter through a vertical pipe with temperature-sensitive valves distributed vertically up it; water then flows into the tank only at the level at which its temperature exceeds that of the water already in the tank.

If the collector circuit gives heat to the tank through an internal heat exchanger, then it need not contain potable water and the fluid can be non-potable and inhibited against freezing. Such an internal heat exchanger, e.g. a coiled pipe, may pass heat to the coldest water at the bottom of the tank, so 'preheating' the hot water supply and reducing the extent of other heating, e.g. thermal boilers and electric 'immerser' heating.

### 5.5.2 *Passive systems with thermosyphon circulation*

Combining the water storage with the collector in one unit at roof height and with no external pump, is common for domestic use in countries with a generally hot climate, e.g. Africa and Australia. The water circulation in such a *thermosyphon* system (Figure 5.7), with the storage tank above the collector as in a roof-top unit, is driven by the density difference between hot and cold water. Consider the simple system shown in Figure 5.8, a closed vertical loop of pipe filled with fluid.

At the section aa′,

$$\int_{\mathrm{a(left)}}^{\mathrm{b}} \rho g \mathrm{d}z - \int_{\mathrm{a(right)}}^{\mathrm{b}} \rho g \mathrm{d}z > 0 \qquad (5.21)$$

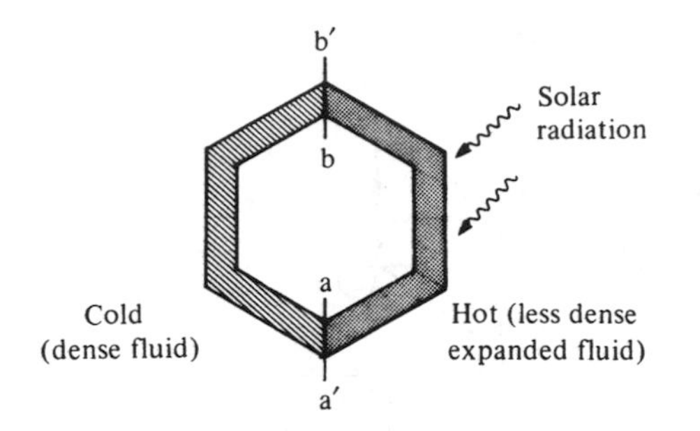


*Figure 5.8* Principle of thermosyphon flow.

The left column of fluid is exerting a greater pressure at aa′ than the right column, thus setting the whole loop of fluid in motion.

The driving pressure, which is precisely the left hand side of (5.21), can be expressed more generally as

$$p_{\mathrm{th}} = \oint \rho g \mathrm{d}z \tag{5.22}$$

where the circle denotes that the integral is taken around a *closed* loop. Note that d$z$ in (5.22) is the vertical increment, and not the increment of length along the pipe. Equation (5.22) can be rewritten as

$$p_{\mathrm{th}} = \rho_0 g H_{\mathrm{th}} \tag{5.23}$$

where the *thermosyphon head*

$$H_{\mathrm{th}} = \oint (\rho/\rho_0 - 1)\mathrm{d}z \tag{5.24}$$

represents the energy gain per unit weight of the fluid and $\rho_0$ is any convenient reference density. This energy gain of the fluid can be lost by other processes and, in particular, by pipe friction represented by the friction head $H_{\mathrm{f}}$ of (2.14).

The expansion coefficient $\beta$ is usually constant,

$$\beta = -(1/\rho)\mathrm{d}\rho/\mathrm{d}T$$

Then (5.24) reduces to

$$H_{\mathrm{th}} = -\beta I_{\mathrm{T}} = -\beta \oint (T - T_0)\mathrm{d}z \tag{5.25}$$

where $T_0$ is a reference temperature. Flow is in the direction for which $I_{\mathrm{T}}$ is positive.

*Example 5.4 Calculation of thermosyphon flow*
In the heating system of Figure 5.7, water enters the collector at temperature $T_1 = 40\,°C$, is heated by $4\,°C$, and goes into the top of the tank without loss of heat at $T_3 = T_2 = 44\,°C$. If the system holds 100 litres of water, calculate the time for all the water to circulate once round the system. Assume the tank has time to achieve stable stratification.

*Solution*
The circulation and insulation insure that the coldest water at the bottom of the tank is at the same temperature as the inlet to the collector (i.e. $T_4 = T_1$). The integral $\oint (T - T_0)dz$ around the contour 1234 is just the area inside the curve (Figure 5.7(b)). This area is the sum of the shaded triangles plus the middle rectangle, i.e.

$$I_T = \frac{1}{2}(0.5\,\text{m})(4\,°\text{C}) + (0.2\,\text{m})(4\,°\text{C}) + \frac{1}{2}(0.7\,\text{m})(4\,°\text{C}) = +3.2\,\text{m K}$$

and is positive since the portion 123 ($z$ increasing) lies to the right of the portion 341 ($z$ decreasing). Therefore flow goes in the direction 1234. Taking a mean value $\beta = 3.5 \times 10^{-4}\,\text{K}^{-1}$ in (5.25) gives $H_{th} = -0.0010\,\text{m}$. This value will be sufficiently accurate for most purposes, but a more accurate value could be derived by plotting a contour of $\rho(z)$, using Table B.2 for $\rho(T)$, and evaluating (5.24) directly.

To calculate the flow speed, we equate the thermosyphon head to the friction head (2.14) opposing it. Most of the friction will be in the thinnest pipes, namely the riser tubes in the collector. Suppose there are four tubes, each of length $L = 2\,\text{m}$ and diameter $D = 12\,\text{mm}$. Then in each tube, using the symbols of Chapter 2,

$$H_{th} = 2fLu^2/Dg$$

where $u$ is the flow speed in the tube and $f = 16v/(uD)$ for laminar flow.

Hence

$$\begin{aligned} u &= \frac{gD^2H_{th}}{32Lv} \\ &= \frac{(1.0 \times 10^{-3}\,\text{m})(12 \times 10^{-3}\,\text{m})^2(9.8\,\text{m}\,\text{s}^{-2})}{(32)(2\,\text{m})(0.7 \times 10^{-6}\,\text{m}^2\,\text{s}^{-1})} \\ &= 0.031\,\text{m}\,\text{s}^{-1} \end{aligned}$$

Checking for consistency, we find the Reynolds number $uD/v = 540$, so that the flow is laminar as assumed.

The volume flow rate through the four tubes is

$$Q = 4(u\pi D^2/4) = 1.4 \times 10^{-5}\,\mathrm{m^3\,s^{-1}}$$

Thus, if the system holds 100 litres of water, the whole volume circulates in a time of

$$(100)(10^{-3}\,\mathrm{m^{-3}})\left(\frac{1}{1.4 \times 10^{-5}\,\mathrm{m^3\,s^{-1}}}\right)\left(\frac{1\,\mathrm{h}}{3.6 \times 10^3\,\mathrm{s}}\right) = 2.0\,\mathrm{h}$$

## 5.6 Selective surfaces

### 5.6.1 *Ideal*

A solar collector absorbs radiation at wavelengths around $0.5\,\mu\mathrm{m}$ (from a source at $6000\,\mathrm{K}$) and emits radiation at wavelengths around $10\,\mu\mathrm{m}$ (from a source at $\sim 350\,\mathrm{K}$). Therefore an ideal surface for a collector would maximise its energy gain and minimise its energy loss, by having a large monochromatic absorptance $\alpha_\lambda$ at $\lambda \sim 0.5\,\mu\mathrm{m}$ and small monochromatic emittance $\varepsilon_\lambda$ at $\lambda \sim 10\,\mu\mathrm{m}$, as indicated schematically in Figure 5.9. Such a surface has $\alpha_{\mathrm{short}} \gg \varepsilon_{\mathrm{long}}$. With a selective surface, $\alpha$ and $\varepsilon$ are weighted means of $\alpha_\lambda$ and $\varepsilon_\lambda$ respectively, over *different* wavelength ranges, cf. (3.31).

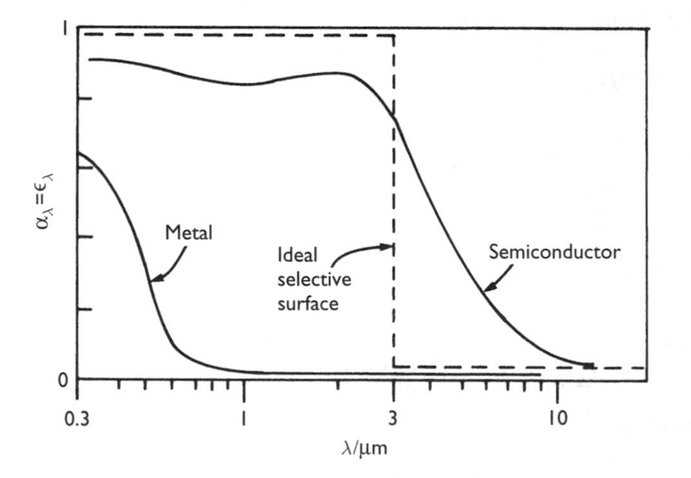


*Figure 5.9* Spectral characteristics of various surfaces. The metal shown is Cu, the semiconductor is $Cu_2O$.

### 5.6.2 *Metal–semiconductor stack*

Some semiconductors have $\alpha_\lambda/\varepsilon_\lambda$ characteristics which resemble those of an ideal selective surface. A semiconductor absorbs only those photons with energies greater than $E_g$, the energy needed to promote an electron from the valence to the conduction band, (see Chapter 7). The critical energy $E_g$ corresponds to a wavelength of 1.1 μm for silicon and 2.0 μm for $Cu_2O$; shorter wavelengths are strongly absorbed, Figure 5.9. However, the poor mechanical strength, small thermal conductivity and relatively large cost of semiconductor surfaces make them unsuitable for the entire collector material.

Metals, on the other hand, are usually mechanically strong, good conductors and relatively cheap. They are also unfortunately good reflectors (i.e. poor absorbers) in the visible and infrared. When light (or other electromagnetic radiation) is incident on a metal, the free electrons near the surface vibrate rapidly in response to the varying electromagnetic field. Consequently, the electrons constitute a varying current, which radiates electromagnetic waves, as in a radio aerial. It appears to an outside observer that the incident radiation has been reflected. The power of the reflected wave is only slightly less than that of the incident wave (Born and Wolf, 1999), so for $\lambda \geq 1\,\mu\text{m}$, $\rho_\lambda \approx 0.97$ (i.e. $\alpha_\lambda = \varepsilon_\lambda \approx 0.03$, see Figure 5.9).

Some metals exhibit an increase in absorptance below a short wavelength $\lambda_p$. For copper $\lambda_p \approx 0.5\,\mu\text{m}$ (see Figure 5.9). Therefore, copper absorbs blue light more than red and appears reddish in colour. The wavelength $\lambda_p$ corresponds to the 'plasma frequency' $f_p = c/\lambda_p$, which is the natural frequency of oscillation of an electron displaced about a positive ion. Net energy has to be fed to the electrons to make them oscillate faster than this frequency, so $\alpha_\lambda$ increases to about 0.5 for frequencies more than $f_p$ (i.e. wavelengths less than $\lambda_p$).

By placing a thin layer of semiconductor over a metal, we can combine the desirable characteristics of both. Figure 5.10 shows how the incoming short wave radiation is absorbed by the semiconductor. The absorbed heat is then passed by conduction to the underlying metal. Since the thermal conductivity of a semiconductor is small, the semiconductor layer should be thin to ensure efficient transfer to the metal. Nevertheless, it should not be too thin: otherwise, some of the radiation would reach the metal and be reflected.

Fortunately the absorption length of a semiconductor at $\lambda = 0.6\,\mu\text{m}$ is typically only $\sim 1\,\mu\text{m}$, i.e. 63% of the incoming radiation is absorbed in the top 1 μm, and 95% in the top 3 μm (see Section 3.6). Therefore, the absorptance for solar radiation is large. The emitted radiation is at wavelengths $\sim 10\,\mu\text{m}$ for which the emittance of both the metal and the semiconductor is small ($\varepsilon \approx 0.1$, as in Figure 5.10).

The result is a composite surface which has much lower radiative loss than a simple black-painted surface (which is black to both visible and infrared

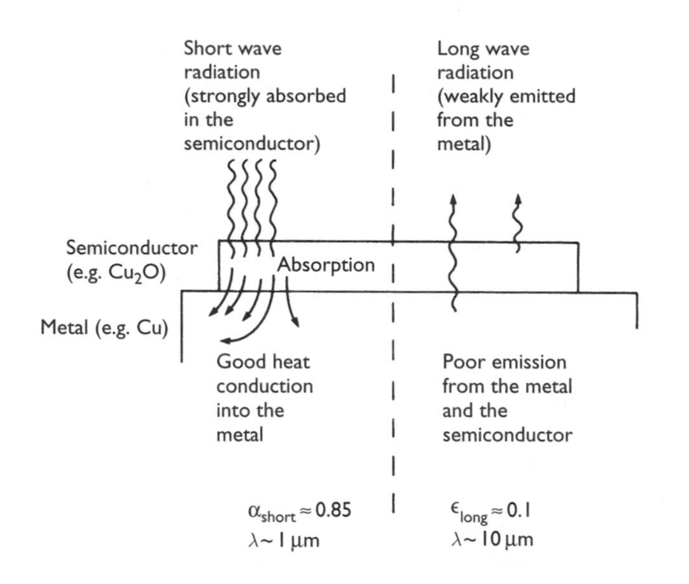


*Figure 5.10* Heat flow in one type of selective surface. Here a semiconductor (which strongly absorbs solar short wave radiation) is deposited on a metal (which is a weak emitter of thermal long wave radiation).

radiation, and therefore has $\alpha = \varepsilon \approx 0.9$). The absorptance is not quite as large as that of a pure black surface, because $\alpha_\lambda$ of the selective surface decreases for $\lambda \geq 1\,\mu m$ (see Figure 5.9), and 30% of the solar radiation is at wavelengths greater than $1\,\mu m$ (see Figure 4.1).

The small emittance of the selective surface becomes more of an advantage as the working temperature increases, since the radiative losses increase as $\varepsilon T^4$. For example, at a plate temperature of 40 °C with $\varepsilon > 0.9$, radiative losses are typically only 20% of the total (e.g. calculate these in Example 5.1); however, at a plate temperature of 400 °C they would be 50% if $\varepsilon = 0.9$ but only 10% if $\varepsilon = 0.1$ (but see caution after (6.17) for $T >$ 1000 °C).

One method for preparing an actual selective surface involves dipping a sheet of copper into an alkaline solution, so that a film of $Cu_2O$ (which is a semiconductor) is formed on it. Many other surface coating types have been successfully developed, including black chrome ($Cr/CrO_x$), metal-pigmented aluminium oxide (e.g. $Ni/Al_2O_3$) and oxidised stainless steel. Most commercial production of selective surfaces is now by sputtering, rather than by electrochemical dipping. Sputtering allows the preparation of water-free composite coatings within which chemical composition, compositional grading, metal-particle size and volume fill-factor can be carefully controlled. Such selective absorbers readily achieve $\alpha > 0.95$ and $\varepsilon < 0.10$.

The absorbing thin-film layer is usually a metal: dielectric composite, often with graded refractive index increasing with depth. A favoured composition is a fine-grained dispersion of submicron-sized conducting particles embedded in an insulating matrix of low dielectric constant which is transparent to infrared radiation. Many physical processes can contribute to the large solar absorptance, e.g. plasma resonance of free electrons (as in Cu), resonant scattering by discrete conducting particles, textural discontinuities and surface roughness, interband transitions (as in semiconductors) and interference effects. Theoretical models of such dispersions using Maxwell's equations go back to 1904, but have recently been refined into 'effective medium theories' which allow computer modelling to be used to evaluate candidate media and optimise designs. (Wackelgard *et al.* 2001, Hutchins 2003).

Advanced uses of solar heat (e.g. the power towers described in Section 6.9.2) work at temperatures of several hundred degrees Celsius, and require selective surfaces capable of withstanding years of fluctuating high temperatures while retaining $\alpha_{short}/\varepsilon_{long}$ as large as possible (e.g. ~30).

## 5.7 Evacuated collectors

Using a selective absorbing surface substantially reduces the radiative losses from a collector. To obtain yet larger temperature differences, (e.g. to *deliver* heat at temperatures around or greater than 100 °C, for which there is substantial industrial demand), it is necessary to reduce the convective losses as well. One way is to use extra layers of glass above a flat plate collector ('double glazing': see Figure 5.1(g) and Problem 5.3). A method that gives better efficiency but is technically more difficult is to evacuate the space between the plate and its glass cover. This requires a very strong structural configuration to prevent the large air pressure forces breaking the glass cover; such a configuration is an outer tube of circular cross-section. Within this evacuated tube is placed the absorbing tube.

One type of evacuated collector uses a double tube, as shown in Figure 5.11(a), with the inner tube containing either the potable water to be heated directly or another heat transfer fluid. The outer tube is made of glass because it is transparent to solar short wave radiation but not to thermal, long wave, radiation, and because glass is relatively strong compared with transparent plastic materials. Both tubes are usually made of glass since glass holds a vacuum better than most other materials. The outgassing rate from baked Pyrex glass is such that the pressure can be held less than $0.1\,N\,m^{-2}$ for 300 years, which is about $10^{12}$ times longer than for a copper tube. The inner tube has a circular cross-section. This helps the weak glass withstand the tension forces produced in it by the pressure difference between the fluid inside and the vacuum outside. Typically the tubes have outer diameter $D = 5\,cm$ and inner diameter $d = 4\,cm$. By suitably connecting an array of these tubes, collectors may receive both direct and diffuse

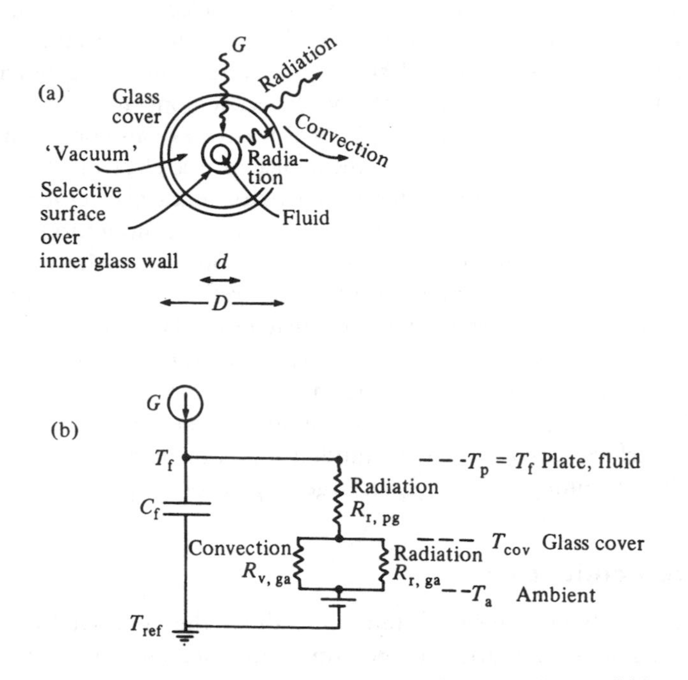


*Figure 5.11* (a) Evacuated collector. (b) Circuit analogue of (a).

solar radiation. Other variations on the basic geometry are also marketed successfully, but that of Figure 5.11(a) is perhaps the simplest to analyse.

*Example 5.5 Heat balance of an evacuated collector*

Calculate the loss resistance of the evacuated collector of Figure 5.11(a) and estimate its stagnation temperature. Take $D = 5.0\,\text{cm}$, $d = 4.0\,\text{cm}$, length of tube 1.0 m; long-wave (infrared) emittances $\varepsilon_p = 0.10$, $\varepsilon_g = 1.0$, $\varepsilon_{air} = 1.0$; short-wave (solar) absorptance of plate $\alpha_p = 0.85$, short-wave transmittance of glass $\tau_g = 0.90$, $G = 750\,\text{W}\,\text{m}^{-2}$, $T_a = 20\,°\text{C}$, $T_{cov} = T_g = 40\,°\text{C}$; $T_p = 100\,°\text{C}$, $u = 5.0\,\text{m}\,\text{s}^{-1}$

*Solution*

The symbols and methods of Chapter 3 are used, together with information in Appendices B (Table B.5) and C. The circuit analogue is shown in Figure 5.11(b), with no convective pathway between the 'plate' (inner tube) and glass (outer tube) because of the vacuum. The only convection is from the outer glass to the environment. Consider a unit length of tube. $T_p = 100\,°\text{C} = 373\,\text{K}$, $T_g = 40\,°\text{C} = 313\,\text{K}$. Treating the two tubes as close parallel surfaces, then by (3.1), (3.6), and (C.18) we obtain by algebraic factorisation

$$1/r_{pg} = \sigma\varepsilon_p\varepsilon_g(T_p^2 + T_g^2)(T_p + T_g) = 0.92\,\mathrm{W\,m^{-2}\,K^{-1}}$$

Taking the characteristic area $A_{pg}$ to be that of a cylinder of length 1 m and *mean* diameter 4.5 cm, we find

$$A_{pg} = \pi(4.5 \times 10^{-2}\mathrm{m})(1.0\,\mathrm{m}) = 0.14\,\mathrm{m}^2$$

hence

$$R_{pg} = r_{pg}/A_{pg} = 7.7\,\mathrm{K\,W^{-1}}$$

For the outside surface of area $A_g = \pi(5.0\,\mathrm{cm})(1.0\,\mathrm{m}) = 0.157\,\mathrm{m}^2$, and the convective loss coefficient is approximately, by (C.15),

$$h_{v,ga} = a + bu = (5.7 + (3.8 \times 5.0)) = 24.7\,\mathrm{W\,m^{-2}\,K^{-1}}$$

By (3.6), (3.9) and (3.49), and, since $\varepsilon_g = \varepsilon_{air} = 1.0$, giving $F'_{12} = 1.0$ in (3.44), the radiative loss coefficient for the outer surface is

$$h_{r,ga} = 4\sigma[(T_g + T_a)/2]^3 = 6.2\,\mathrm{W\,m^{-2}\,K^{-1}}$$

The losses by convection and radiation from the external glass to the environment are in parallel, and since by (3.9) $h = 1/r$, the combined thermal resistance is

$$R_{ga} = 1/[(h_{v,ga} + h_{r,ga})A_g] = 0.21\,\mathrm{K\,W^{-1}}$$

and

$$R_{pa} = R_{ga} + R_{pg} = (0.21 + 7.7)\mathrm{K\,W^{-1}} = 7.9\,\mathrm{K\,W^{-1}}$$

Note how the radiation resistance $R_{pg}$ dominates, since there is no convection to 'short circuit' it. It does not matter that the mixed convection formula (C.15) applies to a flat surface, since it will underestimate the resistance from a curved surface.

Since each 1 m of tube occupies the same collector area as a flat plate of area $0.05\,\mathrm{m}^2$, we could say that the equivalent resistance of unit area of this collector is $r_{pa} = 0.40\,\mathrm{m^2\,K\,W^{-1}}$, although this figure does not have the same significance as for true flat plates.

To calculate the heat balance on a single tube, we note that the heat input is to the projected area of the inner tube, whereas the losses are from the entire outside of the larger outer tube. With no heat removed by a stagnant fluid, input solar energy equals output from losses, so

$$\tau_g \alpha_p G d(1.0\,\mathrm{m}) = (T_\mathrm{p} - T_\mathrm{a})/R_\mathrm{pa}$$

so

$$T_\mathrm{p} - T_\mathrm{a} = 0.90 \times 0.85 \times 750\,\mathrm{W\,m^{-2}} \times 0.04\,\mathrm{m} \times 1.0\,\mathrm{m} \times 7.9\,\mathrm{K\,W^{-1}} = 180\,\mathrm{K}$$

giving, 200 °C for the maximum (stagnation) temperature.

*Note*

This temperature is less than that listed for the double glazed flat plate in Table 5.1. However $T_\mathrm{p}^{(\mathrm{m})}$ and, more importantly, the outlet temperature $T_2$ when there is flow in the tubes, can be increased by increasing the energy input into each tube, e.g. by placing a white surface behind the tubes, which helps by reflecting short wave radiation and reducing the effect of wind.

From the 1990s, evacuated tube collectors have been mass-produced in China (mostly for domestic consumption) and, of a more sophisticated design using a central heat pipe within a central metal strip collector, in the UK (mostly for export). The manufacturing process, especially with automatic equipment, is sophisticated. The tubes should have a long lifetime, but are susceptible to damage from hailstones and vandalism.

## 5.8 Social and environmental aspects

Solar water heating is an extremely benign and acceptable technology. The collectors are not obtrusive, especially when integrated into roof design. There are no harmful emissions in operation and manufacture involves no especially dangerous materials or techniques. Installation requires the operatives to be trained conventionally in plumbing and construction, and to have had a short course in the solar-related principles. The technology is now developed and commercial in most countries, either extensively (e.g. Greece, Cyprus, Israel and Jordan) or without widespread deployment (e.g. USA, France and the UK). It works best everywhere in summer and especially in sunny climates, e.g. the Mediterranean, and where alternatives, such as gas or electricity, are most expensive e.g. northern Australia. Moreover in every climate, solar water heaters have pre-heating value. In the UK for instance, a 4 $m^2$ collector is sufficient for nearly 100% supply to a family of 2–4, with careful use, from mid-April to late-September, and will pre-heat in other months.

In almost all cases, using solar energy for water heating in practice replaces brown (fossil) energy that would otherwise be used for the same purpose. This gives the benefits of improved sustainability and less greenhouse gas emissions, as described in Section 1.2. For this reason, some governments partially subsidise household purchase of solar water heaters, in an attempt to offset the 'external costs' of brown energy (see Chapter 17 for a general discussion of external costs and policy tools). The fossil fuel use might be direct (e.g. gas heating) or indirect (e.g. gas- or coal-fired electricity). Especially in colder countries, the replacement is likely to be seasonal, with the 'solar-deficit' in the cooler months being supplied by electric heating, central-heating boilers or district heating. Depending on the source of these other supplies, which may be from renewable energy but is frequently from fossil fuels, this may reduce the greenhouse gas savings. Installing a solar water heating system can be undertaken by a practical householder, although most people employ a properly trained tradesperson. The collectors (and for some systems the water tank also) are usually fixed on roofs of sufficient strength. In most situations, a 'conventional' water heater is available either as a back-up or as an alternative new installation. Nevertheless, the payback time against the running cost of a conventional system is usually 5–10 years, which is substantially less than the life of the solar system, (see Examples 17.1 and 17.2).

Solar water heaters, even relatively sophisticated ones, can be manufactured almost anywhere on a small or medium scale, thus giving employment. They do not need to be imported and there is a market, especially among the middle class and members of 'green' organisations. The technology is modular and can be scaled up for commercial uses, such as laundries and hotels. Thus by far the largest national production of solar water heaters is in China, where even basic cheap units can provide domestic hot water, even if only for half the year in the winter climate and high latitude of China. Many of these units are single glazed or even unglazed, often with relatively poor thermal connection between the plate and the tubes, but their price/performance ratio is acceptable.

All these features are examples of the benefits of renewable energy systems generally, as set out in Chapter 1.

## Problems

5.1 In a sheltered flat plate collector, the heat transfer between the plate and the outside air above it can be represented by the network of Figure 5.12, where $T_p$, $T_g$ and $T_a$ are the mean temperatures of plate, glass and air respectively.

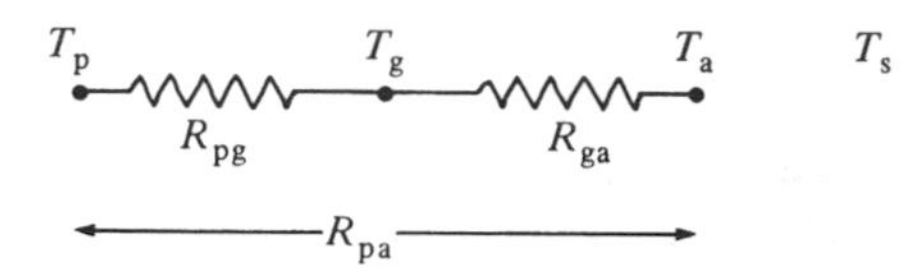


*Figure 5.12* Thermal resistances for Problem 5.1.

a Show that

$$T_g = T_a + (R_{ga}/R_{pa})(T_p - T_a)$$

Verify that, for $T_p = 70\,°C$ and the resistances calculated in Example 5.2, this implies $T_g = 32\,°C$.

b Recalculate the resistances involved, using this second approximation for $T_g$ instead of the first approximation of $\frac{1}{2}(T_p + T_a) = 45\,°C$ used in the example, and verify that the effect on the overall resistance $r_{pa}$ is small.

c Use the resistances calculated in (b) in the formula in (a) to calculate a third approximation for $T_g$. Is a further iteration justified?

5.2 The collector of Example 5.2 had a resistivity to losses from the top of $r_{pa} = 0.13\,m^2\,K\,W^{-1}$. Suppose the *bottom* of the plate is insulated from the ambient (still) air by glass wool insulation with $k = 0.034\,W\,m^{-1}\,K^{-1}$. What thickness of insulation is required to insure that the resistance to heat loss at the bottom is (a) equal to (b) 10 times the resistance of the top?

5.3 A certain flat plate collector has *two* glass covers. Draw a resistance diagram showing how heat is lost from the plate to the surroundings, and calculate the resistance (for unit area) $r_{pa}$ for losses through the covers. (Assume the standard conditions of Example 5.2.) Why will this collector need thicker *rear* insulation than a single glazed collector?

5.4 Calculate the top resistance $r_{pa}$ of a flat plate collector with a single glass cover and a *selective* surface. (Assume the standard conditions of Example 5.2.) See Figure 5.1(h).

5.5 Calculate the top resistance of a flat plate collector with double glazing and a selective surface. (Again assume the standard conditions.) See Figure 5.1(g).

5.6 Bottled beer is pasteurised by passing 50 litres of hot water (at 70 °C) over each bottle for 10 min. The water is recycled, so that its minimum temperature is 40 °C.

a A brewery in Kenya proposes to use solar energy to heat this water. What form of collector would be most suitable for this purpose? Given that the brewery produces 65 000 filled bottles in an 8 h working day, and that the irradiance at the brewery can be assumed to be always at least $20\,\mathrm{MJ\,m^{-2}\,day^{-1}}$ (on a horizontal surface), calculate the minimum collector area required, assuming no heat supply losses.

b Refine your estimate of the required collector area by allowing for the usual losses from a single glazed flat plate collector. (Make suitable estimates for $G$, $T_a$, $u$.)

c For this application, would it be worthwhile using collectors with (i) double glazing (ii) selective surface?

Justify your case as quantitatively as you can.
*Hint:* Use the results summarised in Table 5.1.

5.7 Some of the radiation reaching the plate of a glazed flat plate collector is reflected from the plate to the glass and back to the plate, where a fraction $\alpha$ of that is absorbed, as shown in Figure 5.13.

a Allowing for multiple reflections, show that the product $\tau\alpha$ in (5.1) and (5.13) should be replaced by

$$(\tau\alpha)_{\mathrm{eff}} = \frac{\tau\alpha}{1-(1-\alpha)\rho_{\mathrm{d}}}$$

where $\rho_d$ is the reflectance of the cover system for diffuse light.

b The reflectance of a glass sheet increases noticeably for angles of incidence greater than about 45° (why?). The reflectance $\rho_d$ can be estimated as the value for incidence of 60°; typically $\rho_d \approx 0.7$. For $\tau = \alpha = 0.9$, calculate the ratio $(\tau\alpha)_{\mathrm{eff}}/\tau\alpha$, and comment on its effect on the heat balance of the plate.

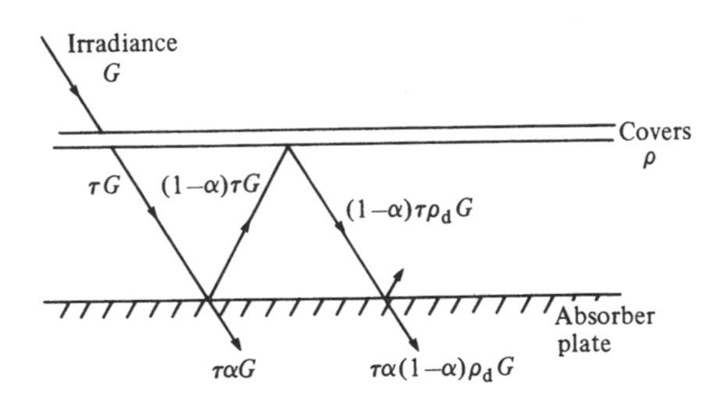


*Figure 5.13* Multiple reflections between collector cover(s) and plate (for Problem 5.7).

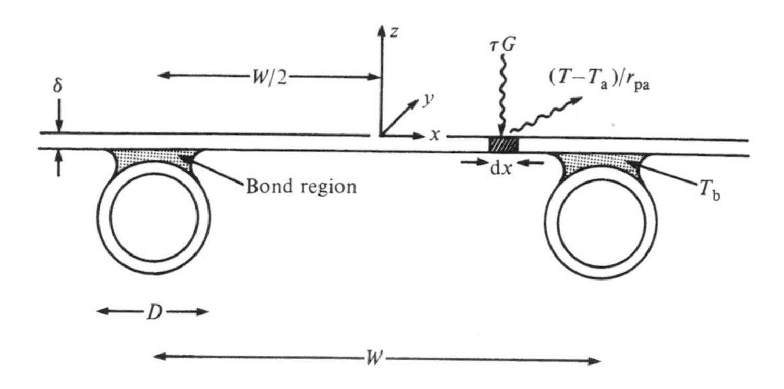


*Figure 5.14* Cross-section of a tube and plate collector (for Problem 5.8).

5.8 *Fin efficiency*

Figure 5.14 shows a tube and plate collector. An element of the plate, area $dxdy$, absorbs some of the heat reaching it from the sun, loses some to the surroundings, and passes the rest by conduction along the plate (in the $x$ direction) to the bond region above the tube. Suppose the plate has conductivity $k$ and thickness $\delta$, and the section of plate above the tube is at constant temperature $T_b$.

a Show that in equilibrium the energy balance on the element of the plate can be written $k\delta \dfrac{d^2T}{dx^2} = (T - T_a - \tau G r_{pa})/r_{pa}$

b Justify the boundary conditions

$$\frac{dT}{dx} = 0 \text{ at } x = 0$$

$$T = T_b \text{ at } x = (W - D)/2$$

c Show that the solution of (a), (b) is

$$\frac{T - T_a - \tau G r_{pa}}{T_b - T_a - \tau G r_{pa}} = \frac{\cosh mx}{\cosh m(W - D)/2}$$

where $m^2 = 1/(k\delta r_{pa})$, and that the *heat* flowing into the bond region from the side is

$$(W - D)F[\tau G - (T_b - T_a)/r_{pa}]$$

where the fin efficiency is given by

$$F = \frac{\tanh m(W - D)/2}{m(W - D)/2}$$

d Evaluate $F$ for $k = 385\,\mathrm{W\,m^{-1}\,K^{-1}}$, $\delta = 1\,\mathrm{mm}$, $W = 100\,\mathrm{mm}$, $D = 10\,\mathrm{mm}$.

5.9 What happens to a thermosyphon system at night? Show that if the tank is wholly above the collector, then the system can stabilise with $H_{th} = 0$, but that a system with the tank lower (in parts) will have a reverse circulation.
*Hint*: Construct temperature–height diagrams like Figure 5.7(b).

## Bibliography

### *General*

Duffie, J.A. and Beckman, W.A. (1991, 2nd edn) *Solar Engineering of Thermal Processes*. John Wiley and Sons, New York. {The standard work on this subject, including not just the collectors but also the systems of which they form part.}

Gordon J. (ed.) (2001) *Solar Energy – The State of the Art*, James & James, London. {10 chapters by solar thermal, photovoltaic and glazing experts; plus single chapters on policy and wind power.} See in particular: Wackelgard, E., Niklasson, G. and Granqvist, C., on 'Selectively solar-absorbing coatings', and Morrison, G.L., on 'Solar collectors' and 'Solar water heating'.

### *Specialised references*

Born, M. and Wolf, W. (1999, 7th edn) *Principles of Optics*, Cambridge UP. {Electromagnetic theory of absorption etc. Heavy going!}

Brinkworth, B.J. (2001) 'Solar DHW system performance correlation revisited', *Solar Energy*, **71** (6), pp. 377–387. {A thorough review of 'black box' comparative analysis and standards for domestic hot water (DHW) systems, including storage; based on the search for comprehensive non-dimension groups of parameters which provide generalised reference methods of performance.}

Close, D.J. (1962) 'The performance of solar water heaters with natural circulation', *Solar Energy*, 5, 33–40. {A seminal paper of theory and experiments on thermosyphon systems. Many ongoing articles in the same journal elaborate on this.}

Hutchins, M.G. (2003) 'Spectrally selective materials for efficient visible, solar and thermal radiation control' in M. Santamouris (ed.), *Solar thermal technologies for buildings*, James & James, London.

Morrison, G.L. (2001) 'Solar collectors' in Gordon (2001), pp. 145–221.

Peuser, F.A., Remmmers, K.-H. and Schnauss, M. (2002) *Solar Thermal Systems*, James & James, London, with Solarpraxis (Berlin). {Predominantly considers large solar water heating plant, this book demonstrates the complex learning-curve of commercial experience in Germany; read this to appreciate the engineering demands of successful large installations.}

Chapter 6

# Buildings and other solar thermal applications

## 6.1 Introduction

Solar heating has many other applications than the heating of water; this chapter reviews some of the most important, using the theory of heat transfer and storage already considered in Chapters 4 and 5. We introduce main concepts only and give guidance to specialist literature for detailed knowledge.

Keeping *buildings* warm in winter, and cool in summer, accounts for up to half of the energy requirements of many countries (see Figure 16.2). Even a partial contribution to this load, by designing or redesigning buildings to make use of solar energy, abates nationally significant amounts of fuel per year. Section 6.3 considers the design and construction of energy-efficient, solar-friendly buildings that has become an important aspect of modern architecture (yet sadly some very energy-inefficient buildings are still constructed!). For best results, the design requires an integrated approach, taking account of not only the solar inputs and their interaction with the building envelope, but also the internal heat transfers in the building, not least those gains arising from the activities, equipment, plant and machines of the occupants.

Solar heat can also be used to heat air for *drying crops* (Section 6.4). Much of the present world grain harvest is lost to fungal attack, which could be prevented by proper drying. Crop drying requires the transfer not only of heat but also of water vapour. This is even more so in the solar *desalination* systems discussed in Section 6.6, including the use of solar heat to distil fresh (potable) water from saline or brackish impure water.

Heat *engines* convert heat into work (which may in turn be converted to electricity), and can be powered by solar radiation. Indeed, since the potential efficiency of heat engines increases with their working temperature, there are theoretical advantages in using solar radiation, which arrives at a thermodynamic temperature of 6000 K, as discussed in Section 6.8. High temperatures are obtained by concentrating clear sky insolation on a surface of area much less than that of the concentrating mirror. Indeed, if the concentrators are large and the area is shielded in a cavity, temperatures approaching but not equalling 6000 K can be obtained. Such devices are treated in Section 6.8, and their application for *thermal electricity generation* and other applications in Section 6.9. Generating electricity from solar heat in this way has been developed to commercial practice. However, *focusing collectors* have numerous other uses, not least in connection with photovoltaics (Chapter 7).

Also discussed briefly in this chapter are two other applications of solar heat: absorption *refrigerators* (Section 6.5) and *solar ponds* (Section 6.7). The chapter concludes with a brief review of some of the *social and environmental aspects* of the technologies discussed.

## 6.2 Air heaters

Hot air is required for two main purposes: warming people (Section 6.3) and drying crops (Section 6.4). Solar air heaters are similar to the solar water heaters of Chapter 5 in that the fluid is warmed by contact with a radiation absorbing surface. In particular, the effects on their performance of orientation and heat loss by wind etc. are very similar for both types.

Two typical designs are shown in Figure 6.1. Note that air heaters are cheap because they do not have to contain a heavy fluid, can be built of light, local materials, and do not require frost protection.

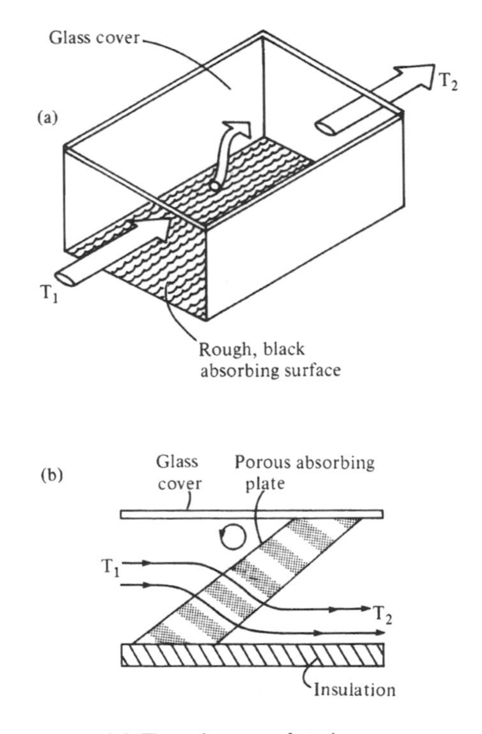


*Figure 6.1* Two designs of air heater.

Equation (2.6) gives the useful heat flow into the air:

$$P_u = \rho c Q (T_2 - T_1) \tag{6.1}$$

The density of air is 1/1000 that of water, and so for the same energy input, air can be given a much greater volumetric flow rate $Q$. However, since the thermal conductivity of air is much less than that of water for similar circumstances, the heat transfer from the plate to the fluid is much reduced. Therefore air heaters of the type shown in Figure 6.1(a) should be built with roughened or grooved plates, to increase the surface area and turbulence available for heat transfer to the air. An alternative strategy is to increase the contact area by using porous or grid collectors (Figure 6.1(b)).

A full analysis of internal heat transfer in an air heater is complicated, because the same molecules carry the useful heat and the convective heat loss, i.e. the flow 'within' the plate and from the plate to the cover are coupled, as indicated in Figure 6.2. The usual first approximation is to ignore this coupling and to use (5.1) as for other solar collector devices (see Sections 5.2 and 5.4.3). If the component of solar irradiance incident perpendicular to the collector is $G_c$ on area A, the collector efficiency is

$$\eta_c = \rho c Q \frac{(T_2 - T_1)}{G_c A} \tag{6.2}$$

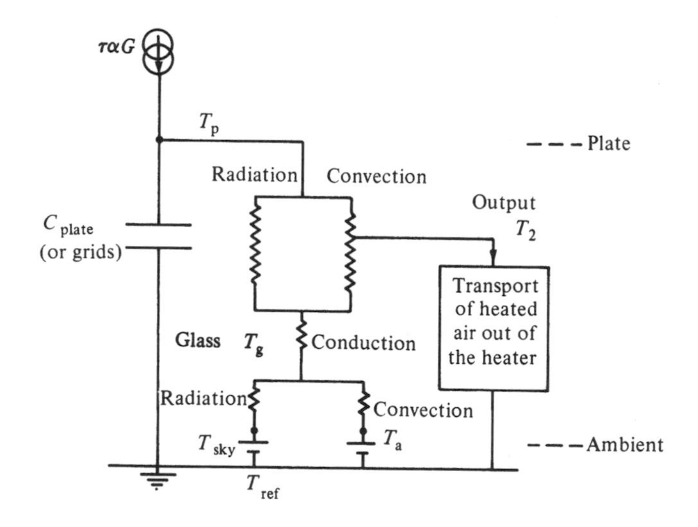


*Figure 6.2* Heat circuit for the air heater of Figure 6.1(a). Note how air circulation within the heater makes the exit temperature $T_2$ less than the plate temperature $T_p$. Symbols are as in Chapter 5.

This efficiency can also be defined using an overall heat loss factor $U_c$. The useful heat is the difference between the absorbed heat and the heat losses. The absorbed heat is a fraction $f$ of the irradiance on the collector:

$$P_u = fAG_c\tau_{cov}\alpha_p - U_cA(T_p - T_a) \tag{6.3}$$

where we assume a single value for the collector temperature (otherwise the collector has to be zoned).

The standard practical evaluation of system characteristics is to measure $\eta$ from (6.2) and then plot $\eta$ against $(T_p - T_a)/G_c$, as in Figure 5.5. If the material properties $\tau_{cov}$ and $\alpha_p$ are known, then the overall loss factor $U_c$ and the collection fraction $f$ are obtained from the slope and ordinate intercept.

## 6.3 Energy-efficient buildings

A major use of energy in colder climates is to heat buildings, especially in winter. What a person considers a comfortable air temperature depends on the humidity, the received radiation flux, the wind speed, clothing and that person's activity, metabolism and life-style. Consequently, inside (room) temperature $T_r$ may be considered comfortable in the range of about 15–22 °C. The internal built environment should be at such a 'comfort temperature', whilst using the minimum artificial heating or cooling ($P_{boost}$), even when the external (ambient) temperature $T_a$ is well outside the comfort range. The heat balance of the inside of a building with solar input is described by equations similar to (5.1). The simplest formulation considers lumped parameters, such that

$$mc\frac{dT_r}{dt} = \tau\alpha GA + P_{boost} - \frac{(T_r - T_a)}{R} \tag{6.4}$$

Detailed mathematical modelling of a building is most complex and is undertaken with specialist software packages. Nevertheless, (6.4) contains the basis of all such modelling, namely energy fluxes and heat capacities.

Note also that the energy required to make the materials and for the construction of the building is also an 'energy expenditure' that should be considered. This is called the 'embodied energy', which is determined from specialist data resources.

### *6.3.1 Passive solar systems*

Passive solar design in all climates consists of arranging the lumped building mass $m$, the sun-facing area $A$ and the loss resistance $R$ to achieve optimum solar benefit, by structural design. The first step is to insulate the building properly (large $R$), including draught prevention and, if necessary, controlled ventilation with heat recovery. The orientation, size and position of windows should allow a sufficient product of $GA$ (perpendicular to the glazing) for significant solar heating in winter, with shading preventing overheating in summer. The windows themselves should have an advanced, multi-surface, construction so their resistance to heat transfer, other than short wave solar radiation, is large.

For passive solar buildings at higher latitudes, solar heat gain in winter is possible because the insolation on vertical sun-facing windows and walls is significantly more than on horizontal surfaces. The sun-facing internal mass surfaces should have a dark colour with $\alpha > 0.8$ (Figure (a)6.3), and the building should be designed to have large mass of interior walls and floors (large $m$) for heat storage within the insulation, thereby limiting the variations in $T_r$. Overheating can of course be prevented by fitting external shades and shutters, which also provide extra thermal insulation at night.

*Example 6.1 Solar heat gain of a house*

The Solar Black House shown in Figure 6.3(a) was designed as a demonstration for Washington DC (latitude 38°N), with a large window on the south side and a massive blackened wall on the north. Assuming that the roof and walls are so well insulated that all heat loss is through the window, calculate the solar irradiance required so that direct solar heating alone maintains room temperature 20 °C above ambient.

*Solution*

If the room temperature is steady, (6.4) reduces to

$$\tau\alpha G = \frac{T_r - T_a}{r}$$

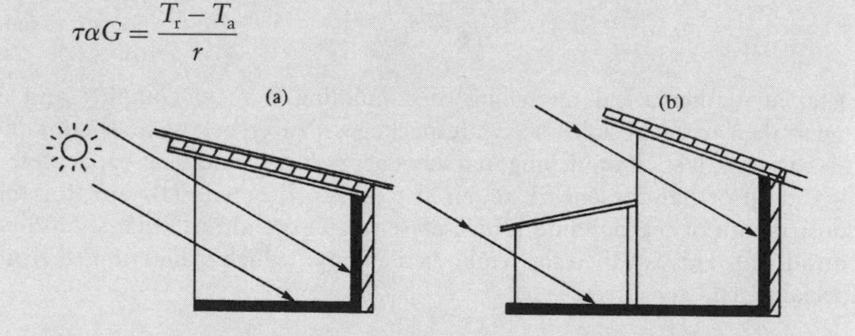


*Figure 6.3* Direct gain passive solar heating. (a) Basic system. (b) Clerestory window (to give direct gain on the back wall of the house). Note the use of massive, dark coloured, rear insulated surfaces to absorb and to store the radiation.

where $r$ is the thermal resistivity from room to outside of a vertical window, single glazed. By the methods of Chapters 3 and 5,

$$r = 0.07\,\text{m}^2\,\text{K}\,\text{W}^{-1}$$

Take glass transmittance $\tau = 0.9$ and wall absorptance $\alpha = 0.8$, then

$$G = \frac{20\,^\circ\text{C}}{(0.07\,\text{m}^2\,\text{K}\,\text{W}^{-1})(0.9)(0.8)} = 400\,\text{W}\,\text{m}^{-2}$$

This irradiance may be expected on a vertical sun-facing window on a clear day in winter.

Example 6.1 correctly suggests that most of the heating load of a well-designed house can be contributed by solar energy, but the design of practical passive solar systems is more difficult than the above example would suggest. For example, the calculation shows only that the Solar Black House will be adequately heated in the middle of the day. But the heat must also be retained at night and there must be an exchange of air for ventilation.

*Example 6.2 Heat loss of a house*

The Solar Black House of the previous example measures 2.0 m high by 5.0 m wide by 4.0 m deep. The interior temperature is 20 °C at 4.00 p.m. Calculate the interior temperature at 8.00 a.m. the next day for the following cases:

a Absorbing wall 10 cm thick, single window as before;
b Absorbing wall 50 cm thick, thick curtain covering the inside of the window.

*Solution*

With $G = P_{boost} = 0$, (6.4) reduces to

$$\frac{dT_r}{dt} = -\frac{(T_r - T_a)}{RC}$$

with $C = mc$.
The solution is

$$T_r - T_a = (T_r - T_a)_{t=0}\exp[-t/(RC)]$$

assuming $T_a$ *is* constant (cf. Section 16.4). As before, assume all heat loss is through the window, of area $10\,m^2$. Assume the absorbing wall is made of concrete.

a

$$R = rA^{-1} = 0.007\,K\,W^{-1}$$

$$C = mc = [(2.4)(10^3\,kg\,m^{-3})(2\,m)(5\,m)(0.1\,m)](0.84 \times 10^3\,kg^{-1}\,K^{-1})$$
$$= 2.0 \times 10^6\,J\,K^{-1}$$

$$RC = 14 \times 10^3\,s = 4.0\,h$$

After 16 hours, the temperature excess above ambient is

$$(20\,°C)\exp(-16/4) = 0.4\,°C.$$

b A curtain is roughly equivalent to double glazing. Therefore take $r \approx 0.2\,m^2\,K\,W^{-1}$ (from Table 5.1). Hence

$$R = 0.02\,K\,W^{-1}$$

$$C = 10 \times 10^6\,J\,K^{-1}$$

$$RC = 2.0 \times 10^5\,s = 55\,h$$

$$T_r - T_a = (20\,°C)\exp(-16/55) = 15\,°C$$

Example 6.2 shows the importance of $m$ and $R$ in the heat balance, and also the importance of having parts of the house adjustable to admit heat by day while shutting it in at night (e.g. curtains, shutters).

One drawback of simple direct gain systems is that the building can be too hot during the day, especially during the summer, although this discomfort can be reduced by suitably large roof overhangs as shades. Improved comfort and better use of the solar heat can be achieved by increasing the heat storage of the building within the insulation by increasing the internal 'thermal-mass' (strictly, the thermal capacitance $mc$) with thick walls and floors of rock or dense concrete. If solar and other heat flows are controlled appropriately, large interior thermal mass is always beneficial for comfort in both cold and hot climates.

### 6.3.2 *Active solar systems*

An alternative space heating method is to use external (separate) collectors, heating either air (Section 6.2) or water (Chapter 5) in an active solar system. Such systems are easier to control than purely passive systems and can be fitted to existing houses. However, the collectors have to be large, and retrofitting is usually far less satisfactory than correct passive design at the initial construction stage. In either case a large storage system is needed (e.g. the building fabric, or a rockbed in the basement, or a large tank of water; see Section 16.4). Water-based systems require heat exchangers, (e.g. 'radiators') to heat the rooms, and air-based systems need substantial ducting. A system of pumps or fans is needed to circulate the working fluid.

Like passive systems, active solar systems will work well only if heat losses have been minimised. In practice so-called passive houses are much improved with electric fans, controlled to pass air between rooms and heat stores. Thus the term 'passive' tends to be used when the Sun's heat is first trapped in rooms or conservatories behind windows, even if controlled ventilation is used in the building. 'Active' tends to be used if the heat is first trapped in a purpose-built exterior collector.

### 6.3.3 *Integrated energy-efficient buildings*

The analysis for real houses is complicated because of the complex absorber geometry, heat transfer through the walls, the presence of people in the house and the considerable 'free gains' from lighting etc. People make independent adjustments, such as opening windows or drawing the curtains that cannot be easily predicted. Also their metabolism contributes appreciably to the heat balance of an 'energy conscious' building with 100–150 W per person in the term $P_{boost}$ of (6.4). A reasonable number of air changes (between one and three per hour) is required for ventilation, and this will usually produce significant heat loss unless heat exchangers are fitted. Computer programs such as ENERGY-10 (USA) and BREEAM (UK) are designed to model the interactions between all the factors affecting the energy performance of a building and are widely used, but it is still essential for analysts to appreciate the importance of the individual effects through simplified, order of magnitude, calculations such as those in Examples 6.1 and 6.2.

Figures 6.4 and 6.5 show some actual buildings, designed and constructed for energy efficiency. The design principles of many such buildings go beyond just energy efficiency and 'solar architecture' to *sustainable design*, which also considers the sustainability of the materials used in construction and the impacts of the building and its use on other environmental flows such as clean water. Such considerations would include the 'embodied energy' of the building and also the extent of use of non-renewable resources, e.g. rain-forest timber. Although experienced architects can produce such buildings with a cost little greater than those of 'conventional' buildings, it is unfortunate that far too many modern buildings fall far short of these standards. Indeed in most countries, other than in northern Europe,

(a)

(b)

*Figure 6.4* Two energy-efficient residential buildings, whose key features are described in the text. (a) A cool-climate house in Canberra, Australia (latitude 35 °S) with good solar input and thermal mass. (b) A house for the humid semi-tropical climate of the Gold Coast, Australia (latitude 28 °S) with good natural ventilation, daylighting and small embodied energy. [Photos by courtesy of Australian Building Energy Council and the Australian Greenhouse Office.]

the building regulations do not require more than minimal energy efficiency in buildings; the shortfall is made up at great cost to the occupants and the environment by overly large energy inputs in the form of heating and air conditioning from non-renewable energy sources.

The Canberra house, Figure 6.4(a), is designed for winters which are cool (minimum temperatures approximately −5 °C) but sunny. This means that orientation is important, as are overhangs to keep out direct sunlight in summer. The house is designed to optimise (passive) solar performance while being of similar construction cost to a conventional house of the same size and similar materials (brick); this was achieved, with energy running costs only about 10% of those prevalent in the area. Glazing is maximised on the northern 'equatorward' side shown in the figure, and minimised on the southern side. The external walls and ceiling are insulated (though not as much as required in Scandinavia), the floor and internal walls are of

(a)

*Figure 6.5* Two energy-efficient institutional buildings, whose key features are described in the text. (a) The Center for Environmental Studies at Oberlin College, Ohio, USA, (latitude 40° N) demonstrates passive solar design for a cold climate, and incorporates many other 'green' features. [Photo by Ron Judkoff, courtesy of (US) National Renewable Energy Laboratory.] (b – *next page*) The Environment and Information Sciences building of Charles Sturt University at Thurgoona, Australia (latitude 37°S), has passive solar features to cope with a hot dry summer, including shading, natural ventilation and thermal mass. [Photo by courtesy of Australian Building Energy Council and the Australian Greenhouse Office.]

(b)

*Figure 6.5* (Continued).

concrete slab, and there is a short (750 mm) extra wall for heat storage set just inside the main windows.

The Oberlin College building, Figure 6.5(a), is designed for a more severe cold climate, and incorporates many features similar to the Canberra house. Appropriately for its function as a teaching centre for environmental studies, the building is designed to optimise passive solar performance and daylighting, and witness the prominent glazed atrium. Thermal mass in the floors and walls retains and radiates heat. Energy-efficient ventilation (e.g. the clerestory windows, which open in summer), insulated roof construction and walls, and building controls for lighting and glazing are used. There are water-source heat pumps for heating, cooling and ventilation. The building also incorporates photovoltaic panels in the roof (not visible in this photo), a wastewater treatment system mimicking a natural wetland, and building materials chosen with regard to sustainability of the building and the sources of materials.

By contrast, the Gold Coast house, Figure 6.4(b), resembles a classic 'Queenslander' design, which are lightweight buildings set high to catch the breeze to ameliorate the hot humid climate. It is a light timber frame dwelling that makes maximum use of passive environmental control features such as beneficial daylight, controlled solar gain, cross ventilation and stack ventilation. Where possible, low embodied energy, low toxicity and recycled materials have been specified. The impact of household operations has also been considered; appliances have been selected to conserve energy, and electricity is supplied by photovoltaic panels. (The photovoltaic panels are not shown in the photo; their use in buildings is discussed in Section 7.10). Rainwater is stored and treated on site for household use, and waste 'greywater' (not sewage) is treated for use in the garden.

The building in Figure 6.5(b), like that in Figure 6.5(a), is intended to teach good environmental practice by example, but in this case in a hot dry climate (inland Australia). It features rammed earth walls, with thermal chimneys, which double as skylights. The building is oriented with a long north–south axis, with openable equatorward windows protected by sunshading. There is natural ventilation, with night purging by automatically opening low and high level louvers. An active system circulates water through floor and ceiling slabs; the roof-mounted solar collectors (visible in the photo) take heat in during winter and out during summer. Self-sufficiency in water (with 'natural wetland' or 'reed bed' cleansing) is a feature of the campus.

## 6.4 Crop driers

Grains and many other agricultural products have to be dried before being stored. Otherwise, insects and fungi, which thrive in moist conditions, render them unusable. Examples include wheat, rice, coffee, copra (coconut flesh), certain fruits and, indeed, timber. We shall consider grain drying, but the other cases are similar. All forms of crop drying involve transfer of water from the crop to the surrounding air, so we must first determine how much water the air can accept as water vapour.

### 6.4.1 *Water vapour and air*

The *absolute humidity* (or 'vapour concentration') $\chi$ is the mass of water vapour present in $1.0\,m^3$ of the air at specified temperature and pressure. This reaches a practical maximum at saturation, so if we try to increase $\chi$ beyond saturation (e.g. with steam), liquid water condenses. The saturation humidity $\chi_s$ depends strongly on temperature (Table B.2(b)). A plot of $\chi$ (or some related measure of humidity) against $T$ is called a *psychometric chart* (Figure 6.6). The ratio $\chi/\chi_s$ is called the *relative humidity*, and ranges from 0% (completely dry air) to 100% (saturated air). Many other measures of humidity are also used (Monteith and Unsworth 1990).

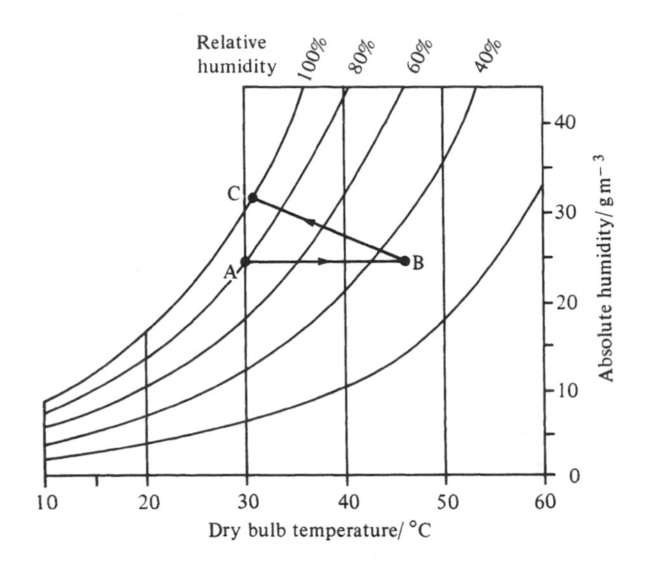


*Figure 6.6* Psychrometric chart (for standard pressure 101.3 kN m$^{-2}$).

Consider air with the composition of point B in Figure 6.6. If it is cooled without change in moisture content, its representative point moves horizontally to A. Alternatively the air can be cooled by evaporating liquid water. If this happens in a closed system with no other heat transfer (i.e. the air and water cool adiabatically), the humidity of the air increases, and its representative point moves diagonally upwards (BC).

### 6.4.2 *Water content of crop*

The percentage *moisture content* (dry basis) $w$ of a sample of grain is defined by

$$w = (m - m_0)/m_0 \tag{6.5}$$

where $m$ *is* the total mass of the sample 'as is' and $m_0$ is the mass of the dry matter in the sample ($m_0$ can be determined for wood by drying the sample in an oven at 105 °C for 24 hours). We shall always use this definition of moisture content ('dry weight' basis), which is standard in forestry. In other areas of agriculture, moisture content on a 'wet weight' basis may be used:

$$\begin{aligned} w' &= (m - m_0)/m \\ &= w/(w+1) \end{aligned} \tag{6.6}$$

The determination of $m_0$ requires care, and should be measured in a laboratory according to the standard procedures for each crop or product. The temperature and time for drying to determine 'oven dry mass' is limited so that other chemical changes do not occur. Some chemically bound water may remain after this process. It is also important to realise that there are limiting temperatures for drying crops for storage, so the product does not crack and allow bacterial attack. Further detail is available in the references listed at the end of this chapter.

If left for long enough, a moist grain will give up water to the surrounding air until the grain reaches its *equilibrium moisture content* $w_e(\%)$. The value of $w_e$ depends on the crop, and especially on the temperature and humidity of the surrounding air. For rice in air at 30 °C and 80% relative humidity (typical of rice growing areas), $w_e \approx 16\%$.

Note that the drying process is not uniform. Much of the moisture present in a crop is 'free water', which is only loosely held in the cell pores, and is therefore quickly lost after harvest. All other parameters remaining constant, the moisture content reduces at a constant rate as this loosely held water is removed. The remaining water (usually 30–40%) is bound to the cell walls by hydrogen bonds, and is therefore harder to remove; this moisture is lost at a decreasing rate. It is important that the grains be dried quickly, i.e. within a few days of harvest, to about 14% moisture content to prevent the growth of fungi that thrive in moist or partly moist grain. Even if the fungi die, the waste chemicals that remain can be poisonous to cattle and humans. Once dried, the grain has to remain dry in ventilated storage.

### 6.4.3 Energy balance and temperature for drying

If unsaturated air is passed over wet material, the air will take up water from the material as described in the previous section. This water has to be evaporated, and the heat to do this comes from the air and the material. The air is thereby cooled. In particular, if a volume $V$ of air is cooled from $T_1$ to $T_2$ in the process of evaporating a mass $m_w$ of water, then

$$m_w \Lambda = \rho c (T_1 - T_2) \tag{6.7}$$

where $\Lambda$ is the latent heat of vaporisation of water and $\rho$ and $c$ are the density and specific heat of the 'air' (i.e. including the water vapour) at constant pressure at the mean temperature, for moderate temperature differences.

The basic problem in designing a crop drier is therefore to determine a suitable $T_1$ and $V$ to remove a specified amount of water $m_w$. The temperature $T_1$ must not be too large, because this would make the grain crack and so allow bacteria and parasites to enter, and high humidity at elevated temperature for extended time will encourage microbial growth.

*Example 6.3*
Rice is harvested at a moisture content $w = 0.28$. Ambient conditions are 30 °C and 80% relative humidity, at which $w_e = 0.16$ for rice. Calculate how much air at 45 °C is required for drying 1000 kg of rice if the conditions are as in Figure 6.6.

*Solution*
From (6.5), $m/m_0 = w + 1 = 1.28$, so the dry mass of rice is $m_0 =$ 780 kg. The mass of water to be evaporated is therefore

$$m_w = (0.28 - 0.16)(780\,\text{kg}) = 94\,\text{kg}$$

i.e. $94/220 = 42\%$ of the total water present

For the moist air leaving the drier, the exit temperature is found from the humidity data (Table B.2(b)) as follows. 1 $m^3$ of air at 30 °C and 80% relative humidity has absolute humidity

$$0.8 \times 30.3\,\text{g}\,\text{m}^{-3} = 24.2\,\text{g}\,\text{m}^{-3}$$

(point A in Figure 6.6).

Note that moist air is less dense than drier air at the same temperature and pressure. (Water molecules have less mass than either oxygen or nitrogen molecules; that is partly why moist air rises to form clouds.) Nevertheless, if the small change in density is neglected, this value will also be the absolute humidity of the same air after heating to 45 °C (point B). (The relative humidity will of course be reduced.) After passing through the rice, the exit air will become more moist. If the conditions are according to Figure 6.6, the exit air is at C, and its temperature will be about 30 °C. Then (6.7) gives

$$V = \frac{(94\,\text{kg})(2.4\,\text{MJ}\,\text{kg}^{-1})}{(1.15\,\text{kg}\,\text{m}^{-3})(1.0\,\text{kJ}\,\text{kg}^{-1}\,\text{K}^{-1})(45 - 30)\,°\text{C}}$$

$$= 13 \times 10^3\,\text{m}^3$$

where the other data come from Appendix B.
More exact calculations would allow for the variations in latent heat, density and humidity of the exit air, but the conclusion would be the same: large scale drying requires the passage of large volumes of warm dry air. Drying with forced convection is an established and complex subject. Drying without forced air flow is even more complex, especially if drying times and temperatures are limited.

## 6.5 Space cooling

Solar heat can be used not only to heat but also to cool. A mechanical device capable of doing this is the *absorption refrigerator* (Figure 6.7). All refrigerators depend on the surroundings giving up heat to evaporate a working fluid. In a conventional electrical (or compression) refrigerator, the working fluid is recondensed by heat exchange at increased pressure applied by a motor. In an absorption refrigerator, the required pressure rise is obtained from the difference in vapour pressure of the refrigerant between (1) a part containing refrigerant vapour above a concentrated solution of refrigerant liquid (the generator), and (2) a part containing refrigerant vapour above a dilute solution (the absorber). Instead of an external input of work, as in a compression cycle, the absorption cycle requires an external input of heat. This heat is applied to the generator in order to maintain its temperature such that the vapour pressure of the fluid equals the saturation pressure in the condenser.

A suitable combination of chemicals is water as refrigerant and lithium bromide as absorbent. The heat can be applied either by a flame, by waste heat or by solar energy. Although systems are commercially available for use with flat plate collectors, operating at ~80 °C, they are handicapped by mechanical complexity and poor coefficient of performance $\eta$, where

$$\eta = \frac{\text{heat removed from cool space}}{\text{heat applied to generator}} \approx 0.7 \qquad (6.8)$$

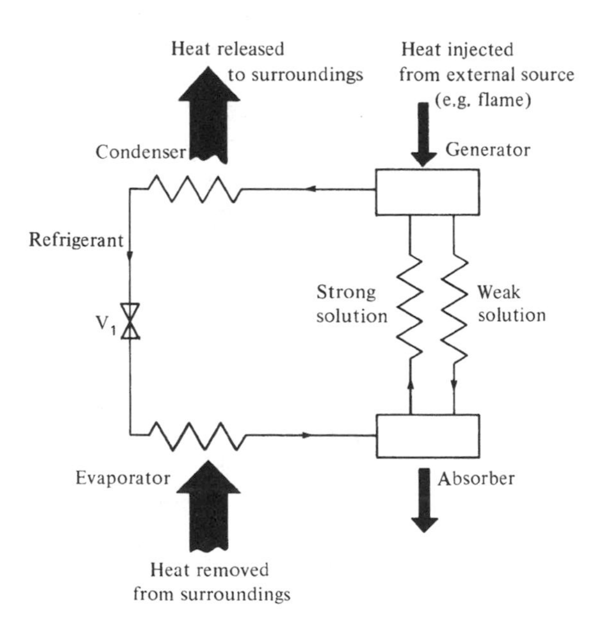


*Figure 6.7* Schematic diagram of an absorption refrigerator. Zigzags represent heat exchangers.

There is a great variety of solar vapour cycle refrigerators, including some straightforward designs working on a 24 h cycle, but their use is not widespread.

A better way to cool *buildings* in hot climates is again to use passive designs (cf. Section 6.3). These either harness the natural flows of cooling air (in humid areas), store coolness from night time or winter (in dry areas) or in some cases automatically generate a cooling flow by convection. A comprehensive account of the relevant design principles, with examples, is given in the *Manual of Tropical Building* (Koenigsberger *et al.* 1974). For cooling foodstuffs etc. offgrid, at least in small quantities, commercial compression refrigerators and freezers are available, powered by solar cells (see Section 7.9). At present these are economically attractive only in areas remote from conventional electricity supplies.

## 6.6 Water desalination

To support a community in arid or desert conditions, potable, i.e. fit to drink, water must be supplied for domestic use, and other water for crops and general purposes. Many desert regions (e.g. central Australia) have regions of salt or brackish water underground, and it is usually much cheaper to purify this water than to transport fresh water from afar. Since deserts usually have large insolation, it is reasonable to use solar energy to perform this purification by distillation.

The most straightforward approach is to use a *basin solar still* (Figure 6.8). This is an internally blackened basin containing a shallow depth of impure water. Over this is a transparent, vapour-tight cover that completely encloses the space above the basin. The cover is sloped towards the collection channel. In operation, solar radiation passes through the cover and warms the water, some of which then evaporates. The water vapour diffuses and moves convectively upwards, where it condenses on the cooler cover. The condensed drops of water then slide down the cover into the catchment trough.

Example 6.4 shows that substantial areas of glass are required to produce enough fresh water for even a small community.

*Example 6.4 Output from an ideal solar still*

The insolation in a dry sunny area is typically $20\,\mathrm{MJ\,m^{-2}\,day^{-1}}$. The latent heat of evaporation of water is $2.4\,\mathrm{MJ\,kg^{-1}}$. Therefore if all the solar heat absorbed by the evaporation, and all the evaporated water, is collected, the output from the still is

$$\frac{20\,\mathrm{MJ\,m^{-2}\,day^{-1}}}{2.4\,\mathrm{MJ\,kg^{-1}}} = 8.3\,\mathrm{kg\,day^{-1}\,m^{-2}} \tag{6.9}$$

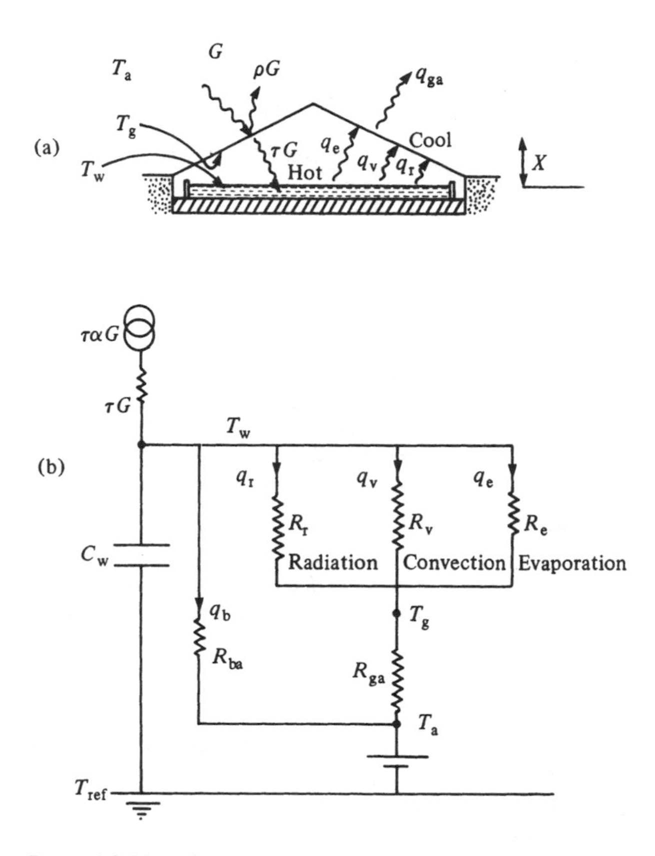


*Figure 6.8* Heat flows in a solar still. (a) Schematic. (b) Heat circuit. Symbols as before, with subscripts: *b* base, e evaporation, *v* convection, *r* radiation, *w* water and *a* ambient.

To calculate the output of a real solar still, we have to determine the proportion of the input solar energy that causes evaporation. Figure 6.9 shows the results of calculations allowing for the transfer of heat and solute (water vapour) by (air) convection inside the still, and for the relatively high reflectance of the glass top (caused by condensed water droplets). We see that the fraction of heat going into evaporation is almost independent of $(T_w - T_g)$ but increases strongly with the water temperature $T_w$. This is to be expected, since the vapour concentration $\chi(T_w)$ increases non-linearly with $T_w$ (see Figure 6.6). The results also show that the maximum production achievable with this type of still with basin water at ~50 °C is 60% of that calculated in Example 6.4, i.e. $5\,\text{kg}\,\text{day}^{-1}\,\text{m}^{-2}(= 5\,\text{litre/day}\,\text{m}^2)$

An alternative approach is the *multiple effect still* in which the heat given off by the condensation of the distilled fresh water is used to evaporate a

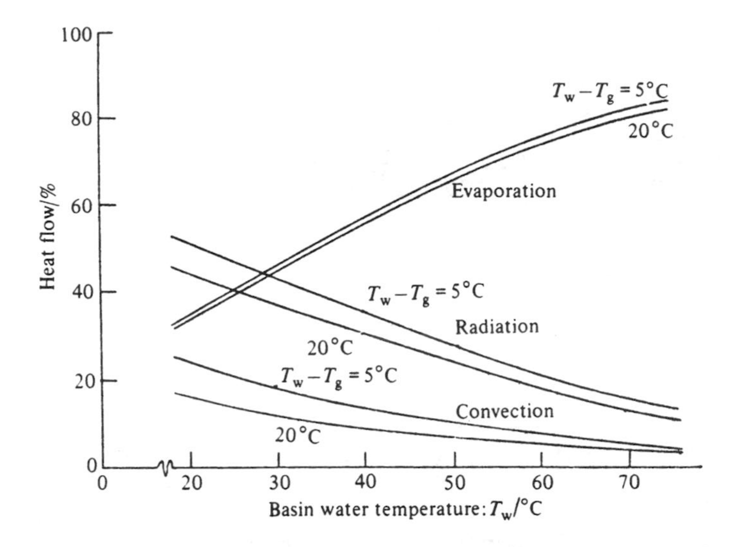


*Figure 6.9* Effect of water temperature on the effectiveness of a basin solar still, as calculated by Cooper and Read (1974). [Copyright 1974, reprinted with permission from Elsevier.]

second mass of saline water. The heat given off by the condensation of the second mass can in turn be used to evaporate a third mass of saline water, and so on. Practical performance is limited by imperfections in heat transfer and by the complexity of the system.

The economics of desalination depends on the price of alternative sources of fresh water. In an area of large or moderate rainfall ($>40\,cm\,y^{-1}$), it is almost certainly cheaper to build a water storage system than any solar device. If remote desalination is necessary (i.e. in very dry areas), then an alternative approach using photovoltaics is now usually cheaper than solar distillation. In this approach, water is purified by reverse osmosis, with the water pumped against the osmotic pressure across special membranes which prevent the flow of dissolved material; solar energy (photovoltaics) can be used to drive the pumps, including any needed to raise water from underground.

## 6.7 Solar ponds

In applications calling for large amounts of low temperature heat ($<100\,°C$), the conventional collectors described in Chapter 5 are often too expensive. A solar pond is an ingenious collector, which uses water as its top cover. Consequently a large 'pond', of surface area perhaps $10^4\,m^2$ and containing

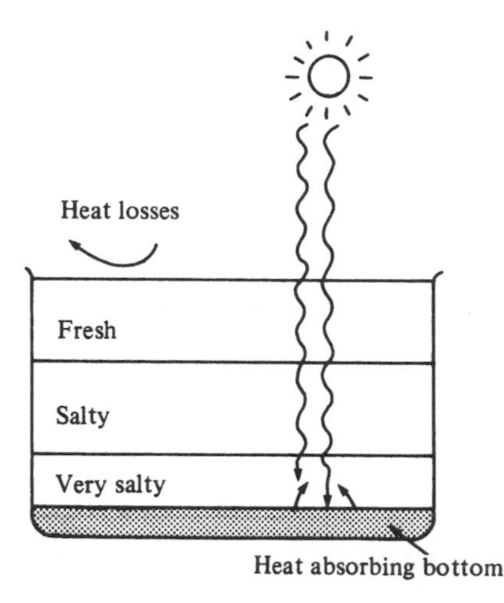


*Figure 6.10* In a solar pond, convection is suppressed and the bottom layer retains the heat from the sun.

$10^4$ m$^3$ of water, can be constructed with simple earthworks at low cost. Moreover, it incorporates its own heat storage, which extends its range of uses.

A solar pond comprises several layers of salty water, with the saltiest layer on the bottom (Figure 6.10), at about 1.5 m deep. Sunshine is absorbed at the bottom of the pond, so the lowest layer of water is heated the most. In an ordinary homogeneous pond, this warm water would then be lighter than its surroundings and would rise, thus carrying its heat to the air above by free convection (cf. Section 3.4). But in the solar pond, the bottom layer was initially made so much saltier than the one above that, even though its density decreases as it warms, it still remains denser than the layer above. Thus convection is suppressed, and the bottom layer remains at the bottom getting hotter and hotter. Indeed there are other liquid solutions that increase density with increase in temperature, so producing very stable solar ponds.

Of course, the bottom layer does not heat up indefinitely but settles to a temperature determined by the heat lost by conduction through the stationary water above. Calculation shows that the resistance to this heat loss is comparable to that in a conventional plate collector (Problem 6.3). Lowest layer equilibrium temperatures of 90 °C or more have been achieved, with boiling being observed in some exceptionally efficient solar ponds. Note that to set up such a solar pond in practice takes up to several months, because if the upper layers are added too quickly, the resulting turbulence stirs up the lower layers and destroys the desired stratification.

In a large solar pond, the thermal capacitance and resistance can be made large enough to retain the heat in the bottom layer from summer to winter (Problem 6.3). The pond can therefore be used for heating buildings in the winter. The pond has also many potential applications in industry, as a steady source of heat at a moderately high temperature. It is also possible to produce electricity from a solar pond by using a special 'low temperature' heat engine coupled to an electric generator. Such systems are conceptually very similar to OTEC systems (Chapter 14). A solar pond at Beit Ha'Harava in Israel produced a steady and reliable 5 MW(e) at a levelised cost of around 30 USc/kWh) (Tabor and Doron, 1990).

## 6.8 Solar concentrators

### 6.8.1 Basics

Many potential applications of solar heat require larger temperatures than those achievable by even the best flat plate collectors. In particular a working fluid at 500 °C can drive a conventional heat engine to produce mechanical work and thence (if required) electricity. Even larger temperatures (~2000 °C) are useful in the production and purification of refractory materials.

A concentrating *collector* comprises a *receiver*, where the radiation is absorbed and converted to some other energy form, and a *concentrator*, which is the optical system that directs beam radiation onto the receiver, (e.g. Figure 6.11). Therefore it is usually necessary to continually orientate the concentrator so that it faces the solar beam. (Section 6.8.4 considers a non-tracking case.)

The aperture of the system $A_a$ is the projected area of the concentrator facing the beam. We define the *concentration ratio* $X$ to be the ratio of the area of aperture to the area of the receiver:

$$X = A_a/A_r \tag{6.10}$$

For an ideal collector, $X$ would be the ratio of the flux density at the receiver to that at the concentrator, but in practice the flux density varies greatly across the receiver. The temperature of the receiver cannot be increased indefinitely by simply increasing $X$, since by Kirchhoff's laws (Section 3.5.4) the receiver temperature $T_r$ cannot exceed the equivalent temperature $T_s$ of the Sun. Moreover the Sun (radius $R_s$, distance $L$) subtends a finite angle at the Earth which limits the achievable concentration ratio to

$$X < (L/R_s)^2 = 45\,000 \tag{6.11}$$

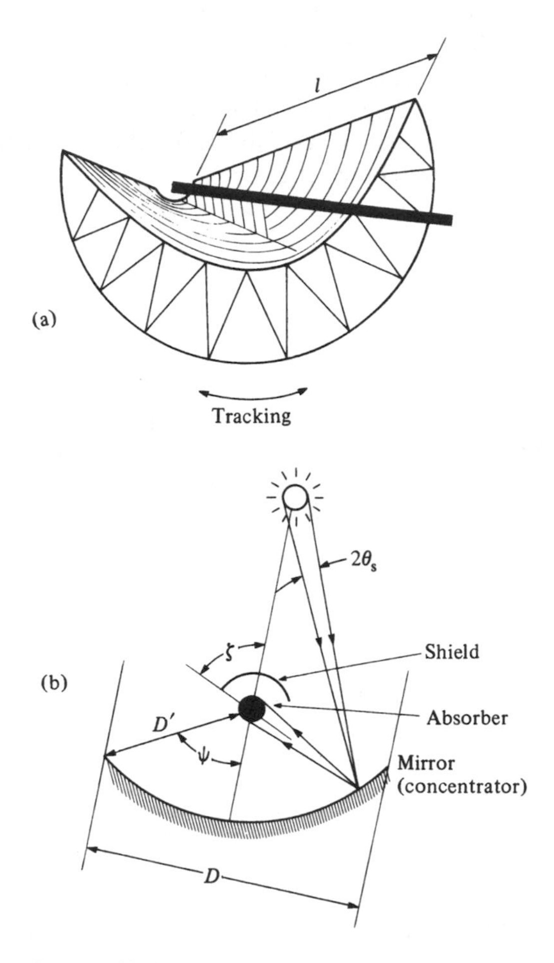


*Figure 6.11* A parabolic trough concentrator. (a) General view, showing the receiver running along the axis. Support struts for the receiver and mirror are also drawn. (b) End view of the design discussed in the text (not to scale).

(See Problem 6.4.) In the next section we shall see how closely these limits on $T_r$ and $X$ can be approached in practice. In Chapter 7 concentrators are discussed for solar cell arrays (see Figure 7.22).

### 6.8.2 Parabolic trough concentrator

Figure 6.11(a) shows a typical trough collector. The concentrator is a parabolic mirror of length $l$ with the receiver running along its axis. This gives concentration only in one dimension, so that the concentration factor is less than for a paraboloid dish. On the other hand, the one-dimensional arrangement is mechanically simpler. Similarly, it is usual to have a trough collector track the Sun only in one dimension. The axis is aligned north–south, and the trough rotated (automatically) about its axis to follow the Sun in tilt only. The power absorbed by the absorbing tube is

$$P_{abs} = \rho_c \alpha l D G_b \tag{6.12}$$

where $\rho_c$ is the reflectance of the concentrator, $\alpha$ is the absorptance of the absorber, $lD$ *is* the area and $G_b$ is the averaged beam irradiance on the trough.

The shield shown in Figure 6.11(b) is intended to cut down heat losses from the absorber. It also cuts out some unconcentrated direct radiation, but this is insignificant compared with the concentrated radiation coming from the other side. The absorber loses radiation only in directions unprotected by the shield. Therefore its radiated power is

$$P_{rad} = \varepsilon \left(\sigma T_r^4\right)(2\pi r l)(1 - \zeta/\pi) \tag{6.13}$$

where $T_r$, $\varepsilon$ and $r$ are respectively the temperature, emittance and radius of the absorber tube. To minimise the losses we want $r$ small, but to gain the full power $P_{abs}$ the tube must be at least as big as the Sun's image. Therefore for large temperature we choose

$$r = D'\theta_s \tag{6.14}$$

in the notation of Figure 6.11(b). In principle, other heat losses can be eliminated, but radiative losses cannot. Therefore by setting $P_{rad} = P_{abs}$ we find the stagnation temperature $T_r$:

$$T_r = \left[\frac{\alpha \rho_c \tau_a G_0 \cos\omega}{\varepsilon\sigma}\right]^{\frac{1}{4}} \left[\frac{D}{2\pi r (1 - \zeta/\pi)}\right]^{\frac{1}{4}} \tag{6.15}$$

This will be a maximum when the shield allows outward radiation only to the mirror, i.e. $\zeta \to \pi - \psi$. By trigonometry the geometric term inside the second bracket can be simplified to $1/\theta_s$, so that the maximum obtainable temperature is

$$T_r^{(max)} = \left[\frac{\alpha \rho_c \tau_a G_0 \cos\omega}{\varepsilon\sigma\theta_s}\right]^{\frac{1}{4}} = 1160\,\mathrm{K} \tag{6.16}$$

for the typical conditions $G_0 = 600\,\mathrm{W\,m^{-2}}$, $\rho_c = 0.8$, $\alpha/\varepsilon = 1$, $\theta_s = R_s/L = 4.6 \times 10^{-3}$ rad and $\sigma = 5.67 \times 10^{-8}\,\mathrm{W\,m^{-2}\,K^{-4}}$, $T_r = 1160\,\mathrm{K}$ is a much larger temperature than that obtainable from flat plate collectors (cf. Table 5.1). Practically obtainable temperatures are less than $T_r^{(\max)}$ for two main reasons:

1 Practical troughs are not perfectly parabolic, so that the solar image subtends angle $\theta_s' > \theta_s = R_s/L$.
2 Useful heat $P_u$ is removed by passing a fluid through the absorber, so

$$T_r^4 \propto P_{rad} = (P_{abs} - P_u) < P_{abs}$$

Nevertheless useful heat can be obtained at ~700 °C under good conditions (see Problem 6.5).

Although (6.15) suggests that $T_r$ could be raised even further by using a selective surface with $\alpha/\varepsilon > 1$, this approach yields only limited returns because the selectivity of the surface depends on the fact that $\alpha$ and $\varepsilon$ are averages over different regions of the spectrum (cf. Section 5.6).

Indeed, according to the definitions (3.31),

$$\alpha = \frac{\int_0^\infty \alpha_\lambda \phi_{\lambda,\mathrm{in}}\,\mathrm{d}\lambda}{\int_0^\infty \phi_{\lambda,\mathrm{in}}\,\mathrm{d}\lambda}; \; \varepsilon = \frac{\int_0^\infty \varepsilon_\lambda \phi_{\lambda,\mathrm{B}}\,\mathrm{d}\lambda}{\int_0^\infty \phi_{\lambda,\mathrm{B}}\,\mathrm{d}\lambda} \tag{6.17}$$

As $T_r$ increases, the corresponding black body spectrum $\phi_{\lambda,\mathrm{B}}(T_r)$ of the emitter becomes more like the equivalent black body spectrum of the sun, $\phi_{\lambda,\mathrm{in}} = \phi_{\lambda,\mathrm{B}}(T_s)$. Since Kirchhoff's law (3.35) states that $\alpha_\lambda = \varepsilon_\lambda$ for each $\lambda$, (6.17) implies that as $T_r \to T_s$, $\alpha/\varepsilon \to 1$.

### 6.8.3 *Parabolic bowl concentrator*

Concentration can be achieved in two dimensions by using a bowl shaped concentrator. This requires a more complicated tracking arrangement than the one-dimensional trough, similar to that required for the 'equatorial mounting' of an astronomical telescope. As before, the best focusing is obtained with a parabolic shape, in this case a paraboloid of revolution. Its performance can be found by repeating the calculations of Section 6.8.2, but this time let Figure 6.11(b) represent a section through the paraboloid. The absorber is assumed to be spherical. The maximum absorber temperature is found in the limit $\zeta \to 0$, $\psi \to \pi/2$ and becomes

$$T_r^{(\max)} = \left[\frac{\alpha \rho_c \tau_a G_0 \sin^2 \psi}{4\varepsilon\sigma\theta_s^2}\right]^{\frac{1}{4}} \tag{6.18}$$

Comparing this with (6.16) we see that the concentrator now fully tracks the Sun, and $\theta_s$ has been replaced by $(2\theta_s/\sin\psi)^2$. Thus $T_r^{(max)}$ increases substantially. Indeed for the ideal case, $\sin\psi = \alpha = \rho_c = \tau_a = \varepsilon = 1$, we recover the limiting temperature $T_r = T_s$. Even allowing for imperfections in the tracking and in the shaping of the mirror, and especially for difficulties in designing the receiver, temperatures of up to 3000 K can be achieved in practice.

### 6.8.4 *Nontracking concentrators*

The previous sections described how large concentration ratios can be achieved with geometric precision and accurate tracking. Nevertheless cheap concentrators of small concentration ratio are useful (e.g. Winston 2000). For example, it may be more cost-effective and equally satisfactory to use a 5 m$^2$ area concentrator of concentration ratio 5 coupled to a 1 m$^2$ solar photovoltaic cell array, than to use 5 m$^2$ of photovoltaic cells with no concentration. Such economy may perhaps be achieved more readily if the concentrator does not track the Sun. However, with photovoltaic cells, care has to be taken to avoid unequal illumination across the array (see Chapter 7).

## 6.9 Solar thermal electric power systems

Collectors with concentrators can achieve temperatures large enough ($\geq$700 °C) to operate a heat engine at reasonable efficiency, which can be used to generate electricity. However, there are considerable engineering difficulties in building a single tracking bowl with a diameter exceeding 30 m. A single bowl of that size could receive at most a peak thermal power of $\pi(15\,\text{m})^2(1\,\text{kW}\,\text{m}^{-2}) = 700\,\text{kW}$, with subsequent electricity generation of perhaps 200 kW. This would be useful for a small local electricity network, but not for established utility networks.

How then can a solar power station be built large enough to make an appreciable contribution to a local grid, say 10 MW (electric)? Two possible approaches are illustrated in Figures 6.12 and 6.14, namely distributed collectors and a central 'power tower'.

### 6.9.1 *Distributed collectors*

In Figure 6.12 many (small) concentrating collectors, each individually track the Sun. The collectors can be two-dimentional parabolic troughs as shown here or three-dimentional parabolic bowls, which provide larger temperatures but at the cost of greater engineering complexity. Each collector transfers solar heat to a heat-transfer fluid, and this hot fluid is then

*Figure 6.12* Electricity generation from distributed parabolic collectors. The photo shows part of the facility at Kramer Junction, California, which has a total capacity of 150 MW (elec) and covers more than 400 ha (1000 acres). Working fluid is heated in the pipe at the focus of each parabolic trough. [Photo by Martin Bond, courtesy of Kramer Junction Company.]

gathered from all the collectors at a central 'power station'. The transfer fluid could be steam, to be used directly in a steam turbine, or it could be a special mineral oil, which heats the steam indirectly, as in the Kramer Junction system of Figure 6.12, which also maintains generation at night and

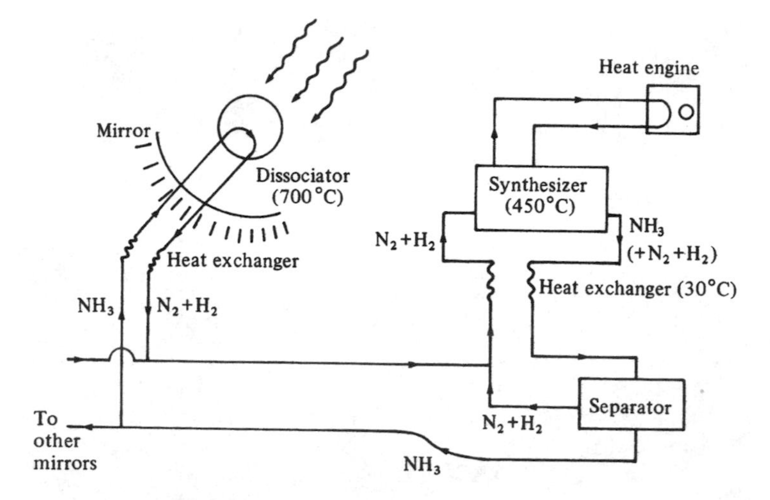


*Figure 6.13* Dissociation and synthesis of ammonia as a storage medium for solar energy. After Carden (1977).

in cloud for contractual reasons by combusting natural gas. Alternatively the transfer fluid could be some *thermochemical storage medium*, such as dissociated ammonia, as illustrated in Figure 6.13.

The advantage of the latter system, initially proposed by Carden (1977) and with later development, is that no energy stored in the chemical is lost between the collectors and the heat engine, so that transmission can be over a long distance or a long time (e.g. overnight) thus allowing continuous power generation. In the original system, the Sun's rays are focused on to a receiver in which ammonia gas (at high pressure, ~300 atmospheres) is dissociated into hydrogen and nitrogen. (This reaction is endothermic with $\Delta H = -46\,\text{kJ}(\text{mol NH}_3)^{-1}$; the heat of reaction is provided by the solar energy.) Within the central plant, the $N_2$ and $H_2$ are partially recombined in the synthesis chamber, using a catalyst. The heat from this reaction can be used to drive an external heat engine or other device. The outflow from the synthesis chamber is separated by cooling it, so that the ammonia liquefies. Numerical details are worked out in Problems 6.6, 6.7.

### 6.9.2 Power tower

An alternative approach is to use a large field of sun-tracking plane mirrors, which focus on to a large central receiver. In Figure 6.14 the mirrors focus beam radiation on to the receiver on top of the tower (hence the name 'power tower').

*Figure 6.14* Solar 'power tower' for generating steam for electricity production. The photograph shows the 7 MW (thermal) CESA-1 facility at Almeria, Spain. The facility has 330 mirrors, each of 39.6 $m^2$. At a typical irradiance of 950 $MW/m^2$ an intensity of 3.3 $MW/m^2$ is obtained on a 4 m-diameter circle. [Photo by courtesy of Plataforma Solar de Almeria.]

## 6.10 Social and environmental aspects

Keeping buildings warm in winter, and cool in summer, accounts for up to half of the energy requirements of many countries (see Figure 16.2). Even a partial contribution to this load, by designing or redesigning buildings to make use of solar energy, abates nationally significant amounts of fuel per year, thus also abating the related greenhouse gas emissions.

The best results are achieved by allowing for energy considerations at the design and construction stage–not least by suitable orientation of the building (facing equatorwards to catch the sunshine in winter but with shades to mitigate the more vertical solar input in summer). Incorporation of site-specific features in this way also makes the buildings architecturally interesting (see for example Figures 6.4 and 6.5). The marginal cost of such solar features is minimal at construction of new buildings. Nevertheless, very significant energy gains can be made by retrofitting insulation, shading, curtains, skylights, etc. to existing buildings, with the savings in fuel costs accruing to the householder. This is nearly always a paying proposition, with payback time often only a few months, and is therefore one way in which individual householders can both benefit themselves and contribute to national greenhouse gas strategies and to sustainable development through the abatement of non-renewable energy use.

The paybacks for commercial buildings are often equally short, but the 'landlord–tenant problem' intrudes: the landlord pays the (extra) cost of insulation or control equipment for heating and air-conditioning so that operating the building is more energy efficient (e.g. heating or cooling specific spaces and not treating the whole building as one unit), but the benefits accrue to the tenant. This is a case where intervention by government regulation is warranted, so mandating appropriate minimum standards for energy performance, as happens in a few countries. (See Chapter 17 for further discussion of such institutional factors.)

Sustainable development aspects of buildings go well beyond energy use within the building envelope, for example to the embodied energy used in the construction process (including the energy and other resources embodied in the building materials). (See discussions of sustainable development in Chapter 1 and Section 17.3.2). One can even incorporate electricity production into the fabric of the building (e.g. by photovoltaic arrays on the walls or roof, e.g. Figure 7.29) as a step towards making the building self sufficient (i.e. not depleting outside resources).

The other technologies examined in this chapter (solar crop driers, solar distillation, absorption refrigerators, solar ponds and solar thermal power systems), although not nearly so widely applicable as energy-efficient buildings, can all make a positive social and environmental contribution locally. Although solar ponds collect large quantities of salt, they are only likely to be used in areas where salt (or salty water) is already abundant, and are therefore unlikely to contribute appreciably to worsening salination of agricultural land. Solar thermal electric power systems necessarily involve strong beam radiation, which can be a hazard for the eyes of people and birds, but this is easily accommodated within normal safety standards.

Mention should also be made here of two age-old solar thermal technologies which both alleviate the need for considerable fossil fuel use: clothes drying and salt production (by evaporation of salt water in large salt pans). It is particularly distressing that in many affluent households and organisations, washed clothes are always dried by heat from electricity in clothes driers, rather than using sunshine whenever possible. Such practices are examples of non-sustainable development.

## Problems

6.1 *Theory of the chimney*

A vertical chimney of height $h$ takes away hot air at temperature $T_h$ from a heat source. By evaluating the integral (5.22) inside and outside the chimney, calculate the thermosyphon pressure $p_{th}$ for the following conditions:

a $T_a = 30\,°C$, $T_h = 45\,°C$, $h = 4\,m$ (corresponding to a solar crop drier)

b $T_a = 5\,°C$, $T_h = 300\,°C$, $h = 100\,m$ (corresponding to an industrial chimney).

6.2 *Flow through a bed of grain*

Flow of air through a bed of grain is analogous to fluid flow through a network of pipes.

a Figure 6.15(a) shows a cross-section of a solid block pierced by $n$ parallel tubes each of radius $a$. According to Poiseuille's law, the volume of fluid flowing through *each* tube is

$$Q_1 = \frac{\pi a^4}{8\mu}\left(\frac{dp}{dx}\right)$$

where $\mu$ is the dynamic viscosity (see Chapter 2) and $dp/dx$ is the pressure gradient driving the flow. Show that the bulk fluid flow speed through the solid block of cross-section $A_0$ is

$$\overline{v} = \frac{Q_{total}}{A_0} = \frac{\varepsilon a^2}{8\mu}\frac{dp}{dx}$$

where the porosity $\varepsilon$ is the fraction of the volume of the block which is occupied by fluid, and $Q_{total}$ is the total volume flow through the block.

b The bed of grain in a solar drier has a total volume $V_{bed} = A_0 \Delta x$ (Figure 6.15(c)). The drier is to be designed to hold 1000 kg of grain of bulk volume $V_{bed} = 1.3\,m^3$. The grain is to be dried in four days (= 30 hours of operation). Show that this requires an air flow of at least $Q = 0.12\,m^2\,s^{-1}$. (Refer to Example 6.1.)

c Figure 6.15(b) shows how a bed of grain can likewise be regarded as a block of area $A_0$ pierced by tubes whose diameter is comparable to (or smaller than) the radius of the grains. The bulk flow velocity is reduced by a factor $k(<1)$ from that predicted by (a), because of the irregular and tortuous tubes. If the driving pressure is $\Delta p$,

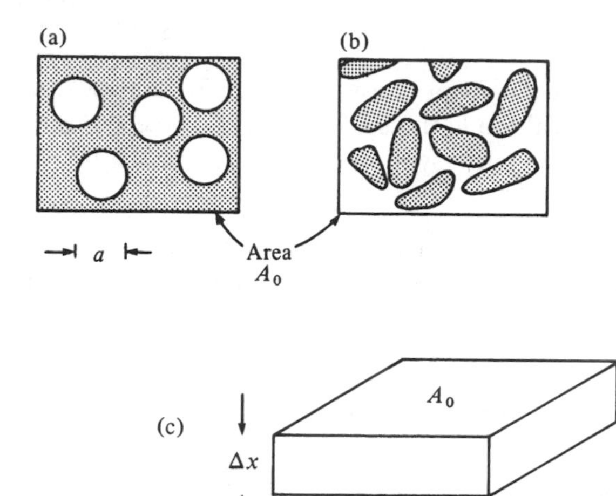


*Figure 6.15* For Problem 6.2. (a) Block pierced by parallel tubes. (b) Pores in a bed of grain. (c) Volume of grain bed.

show that the thickness $\Delta x$ through which the flow $Q$ can be maintained is

$$\Delta x = \left( \frac{k\varepsilon a^2}{8\mu} \frac{V\Delta p}{Q} \right)^{\frac{1}{2}}$$

For a bed of rice, $\varepsilon = 0.2$, $a = 1$ mm, $k = 0.5$ approximately. Taking $Q$ from (b), and $\Delta p$ from Problem 6.1(a), calculate $\Delta x$ and $A_0$.

6.3 *The solar pond*

An idealised solar pond measures 100 m × 100 m × 1.2 m. The bottom 20 cm (the storage layer) has an effective absorptance $\alpha = 0.7$. The 1.0 m of water above (the insulating layer) has a transmittance $\tau = 0.7$, and its density increases downwards so that convection does not occur. The designer hopes to maintain the storage layer at 80 °C. The temperature at the surface of the pond is 27 °C (day and night).

a Calculate the thermal resistance of the insulating layer, and compare it with the top resistance of a typical flat plate collector.

b Calculate the thermal resistance of a similar layer of fresh water, subject to free convection. Compare this value with that in (a), and comment on any improvement.

c The density of NaCl solution increases by 0.75 g per litre for every 1.0 g of NaCl added to 1.0 kg $H_2O$. A saturated solution of NaCl contains about 370 g of NaCl per kg $H_2O$. The volumetric coefficient of thermal expansion of NaCl solution is about $4 \times 10^{-4}\,K^{-1}$.

Calculate the minimum concentration $C_{min}$ of NaCl required in the storage layer to suppress convection, assuming the water layer at the top of the pond contains no salt. How easy is it to achieve this concentration in practice?

d Calculate the characteristic time scale for heat loss from the storage layer, through the resistance of the insulating layer. If the temperature of the storage layer is 80 °C at sunset (6 p.m.) what is its temperature at sunrise (6 a.m.)?

e The molecular diffusivity of NaCl in water is $1.5 \times 10^{-5}\,\mathrm{cm^2\,s^{-1}}$. The pond is set up with the storage layer having twice the critical concentration of NaCl, i.e. double the value $C_{min}$ calculated in (c). Estimate the time for molecular diffusion to lower this concentration to $C_{min}$.

f According to your answers to (c)–(e), discuss the practicability of building such a pond, and the possible uses to which it could be put.

6.4 The limiting concentrating system of Section 6.8.3 has as receiver a black body of area $A_r$, which is in radiative equilibrium at temperature $T_r = T_s$, the equivalent temperature of the Sun. By considering the energy balance of the receiver, show that these conditions correspond to a limiting concentration ratio

$$X^{(m)} = (L/R_s)^2$$

where the Sun has radius $R_s$ and distance $L$ from the Earth.

6.5 Figure 6.16 shows the key feature of a system for the large scale use of solar energy similar to one implemented in California in the 1980s. Sunlight is concentrated on a pipe perpendicular to the plane of the diagram and is absorbed by the selective surface on the outside of the pipe. The fluid within the pipe is thereby heated to a temperature $T_f$ of about 500 °C. The fluid then passes through a heat exchanger where it produces steam to drive a conventional steam turbine, which in turn drives an electrical generator.

a Why is it desirable to make $T_f$ as large as possible?

b Suppose the inner pipe is 10 m long and 2.0 mm thick and has a diameter of 50 mm, and that the fluid is required to supply 12 kW of thermal power to the heat exchanger. If the pipe is made of copper, show that the temperature difference across the pipe is less than 0.1 °C. (Assume that the temperature of the fluid is uniform.)

c Suppose the selective surface has $\alpha/\varepsilon = 10$ at the operating temperature of 500 °C. What is the concentration factor required of the lens (or mirror) to achieve this temperature using the evacuated collector shown? Is this technically feasible from a two-dimensional system?

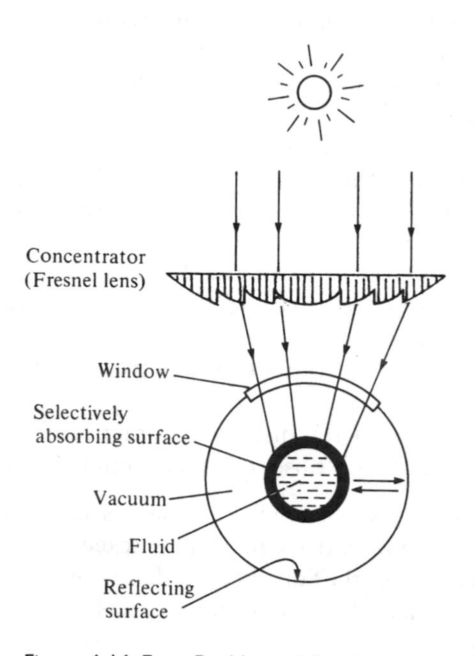


*Figure 6.16* For Problem 6.5. A proposed concentrator system for power generation.

d Suppose the copper pipe was not shielded by the vacuum system but was exposed directly to the air. Assuming zero wind speed, calculate the convective heat loss per second from the pipe.

e Suppose the whole system is to generate 50 MW of *electrical* power. Estimate the collector area required.

f Briefly discuss the advantages and disadvantages of such a scheme, compared with (i) an oil-fired power station of similar capacity, and (ii) small scale uses of solar energy, such as domestic water heaters.

6.6 Suppose the system of Figure 6.13 is to be used to supply an average of 10 MW of electricity.

a Estimate the total collector area this will require. Compare this with a system using photovoltaic cells.

b Briefly explain why a chemical (or other) energy store is required, and why the mirrors have to be pointed at the Sun. How might this be arranged?

c To insure a suitably high rate of dissociation, the dissociator is to be maintained at 700 °C. Plumbing considerations (Problem 6.7) require that the dissociator has a diameter of about 15 cm. Assuming (for simplicity) that it is spherical in shape, calculate the power lost from each dissociator by radiation.

d Each mirror has an aperture of $10\,m^2$. In a solar irradiance of $1\,kW\,m^{-2}$, what is the irradiance at the receiver? Show that about $2.5\,g\,s^{-1}$ of $NH_3$ can be dissociated under these conditions.

6.7 The system of Figure 6.13 requires $2.5\,g\,s^{-1}$ of $NH_3$ to pass to each concentrator (see Problem 6.6). Suppose the ammonia is at a pressure of 300 atmospheres, where it has density $\rho = 600\,kg\,m^{-3}$ and viscosity $\mu = 1.5 \times 10^{-4}\,kg\,m^{-1}\,s^{-1}$. The ammonia passes through a pipe of length $L$ and diameter $d$. To keep friction to an acceptable low level, it is required to keep the Reynolds number $\mathcal{R} < 6000$.

a Calculate (i) the diameter $d$ (ii) the energy lost to friction in pumping 2.5 g of ammonia over a distance $L = 50\,m$.

b Compare this energy loss with the energy carried. Why is the ammonia kept at a pressure of ~300 atmospheres rather than ~1 atmosphere? (*Hint*: Estimate the dimensions of a system working at ~1 atmosphere.)

## Bibliography

### General

Duffie, J.A. and Beckman, W.A. (1991, 2nd edn) *Solar Engineering of Thermal Processes*, John Wiley and Sons. {The classic text, especially for solar thermal theory and application. Covers most of the topics of this chapter by empirical engineering analysis.}

Gordon, J. (ed.) (2001) *Solar Energy – The State of the Art*, James & James, London. {10 chapters by experts in solar thermal, photovoltaic and glazing.}

Yogi Goswami, D., Frank Kreith, Jan F. Kreider (2000, 2nd edn) *Principles of Solar Engineering*, Taylor and Francis, Philadephia. {Another standard textbook on solar thermal systems.}

### Journals

The most established journal, covering all aspects of solar (sunshine) energy is *Solar Energy*, published by Elsevier in co-operation with the International Solar Energy Society, ISES.

### Air heaters and crop-drying

Brenndorfer, B., Kennedy, L., Bateman, C.O., Mrena, G.C. and Wereko-Brobby C. (1985) *Solar Dryers: Their Role in Post-harvest Processing*, Commonwealth Secretariat, London.

Monteith, J. and Unsworth, K. (1990, 2nd edn) *Principles of Environmental Physics*, Edward Arnold. {Includes a full discussion of humidity.}

### Buildings

Balcomb, J.D. (ed.) (1991) *Passive Solar Buildings*, MIT Press. {One of a series on 'solar heat technologies' summarising US research in the 1970s and 1980s.}

Eicker, U. (2003) *Solar Technologies for Buildings*, Wiley. {Translated from a German original of 2001. Includes chapters on solar heating and cooling, and on absorption cooling.}

Koenigsberger, O.H., Ingersol, T.G., Mayhew, A. and Szokolay, S.V. (1974) *Manual of Tropical Housing and Building. Part I: Climatic Design*, Longmans, London. {A guide for architects, containing much relevant physics and data. Stresses passive design.}

Mobbs, M. (1998) *Sustainable House*, University of Otago Press. {Describes the author's autonomous house in Sydney, Australia.}

Santamouris, M. (ed.) (2003) *Solar Thermal Technologies for Buildings: The State of the Art*, James & James, London. {Part of a series on buildings, energy and solar technology.}

Vale, Brenda and Vale, Robert (2000) *The New Autonomous House*, Thames & Hudson, ISBN 0-500-34176-1. {Design and construction of low energy, solar conscious and sustainable-materials housing, with specific UK example. A serious study of a common subject.}

Weiss, W. (ed.) (2003) *Solar Heating Systems for Houses*, James and James, London. {One of a series of publications emerging from the Solar Heating and Cooling Program of the International Energy Agency. This book focuses on combisystems (i.e. the use of solar water heaters integrated with other heating for buildings).}

### Water desalination

Cooper, P.I. and Read, W. (1974) Design philosophy and operating experience for Australian solar stills, *Solar Energy*, **16**, 1–8. {Summarises much earlier work.}

Delyannis, E. and Belessiotis, V. (2001) Solar energy and desalination, *Advances in Solar Energy*, **14**, 287. {Useful review with basic physics displayed; notes that 'almost all large state-of-the-art stills have been dismantled'.}

Howe, E.W. and Tleimat, B. (1977) 'Fundamentals of water desalination', in A. Sayigh, (ed.) *Solar Energy Engineering*, Academic Press, London. {Useful review with basic physics displayed.}

### Solar absorption cooling

Wang, R.Z. (2003) Solar refrigeration and air conditioning research in China, *Advances in Solar Energy*, **15**, 261. {Clear explanation of principles; notes that there have been few commercial applications as yet.}

### Solar ponds

Tabor, H. (1981) 'Solar ponds', *Solar Energy*, **27**, 181–194. {Reviews practical details and costs.}

Tabor, H. and Doron, B. (1990) 'The Beit Ha'Harava 5 MW(e) solar pond', *Solar Energy*, **45**, 247–253. {Describes the largest working solar pond yet built.}

### *Concentrators*

Welford, W. and Winston, R. (1989) *High Collection Non-imaging Optics*, Academic Press, New York.
Winston, R. (1974) 'Solar concentrators of novel design', *Solar Energy*, **16**, 89.
Winston, R. (2000) *Solar Concentrators*, in Gordon (2001), pp. 357–436.

### *Solar thermal electricity generation*

Carden, P.O. (1977) Energy co-radiation using the reversible ammonia reaction, *Solar Energy*, **19**, 365–378. {First of a long series of articles}; see also Luzzi, A. and Lovegrove, K. (1997) A solar thermochemical power plant using ammonia as an attractive option for greenhouse gas abatement, *Energy*, **22**, 317–325.
Mills, D.R. (2001) *Solar Thermal Electricity*, in Gordon (2001), pp. 577–651.
Winter, C.-J., Sizmann, R.L. and Vant-Hull, L.L. (eds) (1991) *Solar Power Plants: Fundamentals, Technology, Systems, Economics*, Springer Verlag, Berlin. {Detailed engineering review.}

### *Web sites*

[There are countless web sites dealing with applications in solar energy, some excellent and many of dubious academic value. Use a search engine to locate these and give most credence to the sites of official organisations, as with the examples cited below.]
International Solar Energy Society, ISES. www.ises.org. The largest, oldest and most authoritative professional organisation dealing with the technology and implementation of solar energy.
Association of Environment Conscious Buildings (UK). www.aecb.net Literature & practical information.
IDEA (Interactive Database for Energy-efficient Architecture). http://nesa1.uni-siegen.de/projekte/idea/idea_1_e.htm. Numerous case studies from Europe, both residential and commercial buildings.
Australian Green Development Forum. http://www.agdf.com.au/ showcase.asp (formerly Australian Building Energy Council. http://www.netspeed.com.au/abeccs/). Case studies from Australia.

Chapter 7

# Photovoltaic generation

## 7.1 Introduction

Only two methods are used to generate significant electric power. The first, discovered by Michael Faraday in 1821 and in commercial production by 1885, requires the relative movement of a magnetic field and a conductor and hence an external engine or turbine. The second is photovoltaic generation using *solar cells* (more technically called *photovoltaic cells*). These devices produce electricity directly from electromagnetic radiation, especially light, without any moving parts. The photovoltaic effect was discovered by Becquerel in 1839 but not developed as a power source until 1954 by Chapin, Fuller and Pearson using doped semiconductor silicon. Photovoltaic power has been one of the fastest growing renewable energy technologies: annual production of cells grew tenfold from about 50 MW in 1990 to more than 500 MW by 2003 (see Figure 7.31), with this growth continuing since. Demand has been driven by the modular character, standalone and grid-linked opportunities, reliability, ease of use, lack of noise and emissions, and reducing cost per unit energy produced.

Photovoltaic generation of power is caused by electromagnetic radiation separating positive and negative charge carriers in absorbing material. If an electric field is present, these charges can produce a current for use in an external circuit. Such fields exist permanently at junctions or inhomogeneities in photovoltaic (PV) cells as 'built-in' electrostatic fields and provide the electromotive force (EMF) for useful power production. Power generation is obtained from cells matched to radiation with wavelengths from the infrared ($\sim 10\,\mu m$) to the ultraviolet ($\sim 0.3\,\mu m$), however, unless otherwise stated, we consider cells matched to solar short wave radiation ($\sim 0.5\,\mu m$). The built-in fields of most semiconductor/semiconductor and metal/semiconductor cells produce potential differences of about 0.5 V and current densities of about $400\,A\,m^{-2}$ in clear sky solar radiation of $1.0\,kW\,m^{-2}$. Commercial photovoltaic cells, dependent on price, have efficiencies of 10–22% in ordinary sunshine, although laboratory specimens and arrangements reach greater efficiency. The cells are usually linked in series and fixed within weather-proof modules, with most modules producing about 15 V. The current from the cell or module is inherently direct current, DC. For a given module in an optimum fixed position, daily output depends on the climate, but can be expected to be about 0.5–1.0 kWh /($m^2$ day$^{-1}$). Output can be increased using tracking devices and solar concentrators.

Junction devices are usually named 'photovoltaic cells', although the solar radiation photons produce the current, since the voltage is already present across the junction. The cell itself provides the source of EMF. It is important to appreciate that photovoltaic devices are *electrical current sources* driven by a flux of radiation. Efficient power utilization depends not only on efficient generation in the cell, but also on dynamic load matching in the external circuit. In this respect, photovoltaic devices are similar to other renewable sources of power, although the precise methods may vary (e.g. by using DC-to-DC converters as 'maximum power tracking' interfaces, Section 7.9.2).

On a reasonably sunny site of insolation 20 MJ/ ($m^2$ day)$^{-1}$, power can be produced that is significantly cheaper over extended use than that from diesel generators, especially in remote areas where fuel supply and maintenance costs may be large. The eventual aim is to be competitive with some daytime peak grid-electricity prices, which is most likely if the polluting forms of generation are charged for external costs.

Sections 7.2–7.5 outline the basic science and technology of photovoltaic cells. Preliminary analysis will always refer to the silicon p–n junction single crystal solar cell (Figure 7.1) since this is the most common and well-established type. Section 7.6 considers how cells are constructed, and how this has been improved over the years to overcome many of the limitations of such cells. Variations, including the development of cells of materials other than Si, are discussed in Section 7.7. Sections 7.8 and 7.9 examine and illustrate the circuits and systems in which photovoltaics are actually used. Readers whose main interest is in applications may wish to read these sections first, along with Section 7.10, which examines economic, social

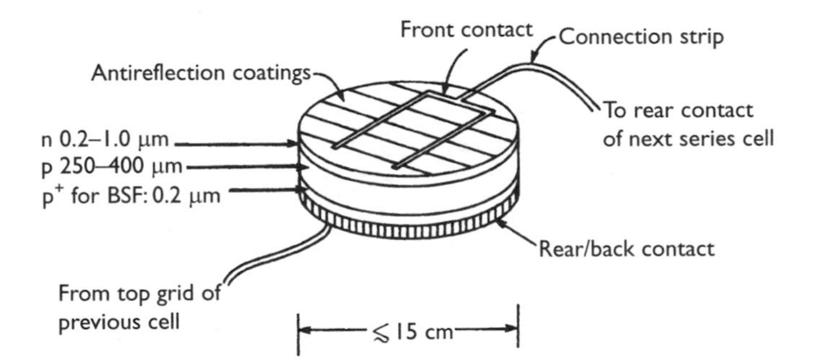


*Figure 7.1* Basic structure of *p–n* junction solar cell. The cover (glass or plastic) above the cell and the filler between the cover and the cell are not shown. BSF: back surface field.

and environmental aspects of the use and production of photovoltaics. Before 2000, most photovoltaics were in stand-alone systems, progressing from space satellites to lighting, water pumping, refrigeration, telecommunications, solar homes, proprietary goods and mobile or remotely isolated equipment (e.g. small boats, warning lights, parking meters). Grid connected PV power, e.g. incorporated with buildings, has become a major activity for the 21st century.

## 7.2 The silicon p–n junction

The properties of semiconductor materials are described in an ample range of solid state physics and electronics texts, almost all of which include the properties of the p–n junction without illumination, because of its centrality to microelectronics and the vast industry that springs from that. This theory is summarised in this section, and extended to the illuminated junction for solar applications in Sections 7.3 and 7.4.

### 7.2.1 *Silicon*

Commercially pure (intrinsic) silicon, Si, has concentrations of impurity atoms of $<10^{18}\,\mathrm{m}^{-3}$ (by atom, <1 in $10^9$) and electrical resistivity $\rho_e \approx 2500\,\Omega\,\mathrm{m}$. As the basis of the microelectronics industry it is of great commercial importance, and its properties and ways to manipulate them have been extensively studied.

The electrical properties of Si are described by the theory of the band gap between conduction and valence bands (Figure 7.2). The density of charge carrier electrons in the conduction band and holes in the valence band of pure intrinsic material is proportional to $\exp(-E_g/(2\,kT))$ if impurity atoms have no effect. This is equivalent to there being no electron or hole charge carriers with an energy state within the forbidden band gap. Table 7.1 gives basic data of silicon.

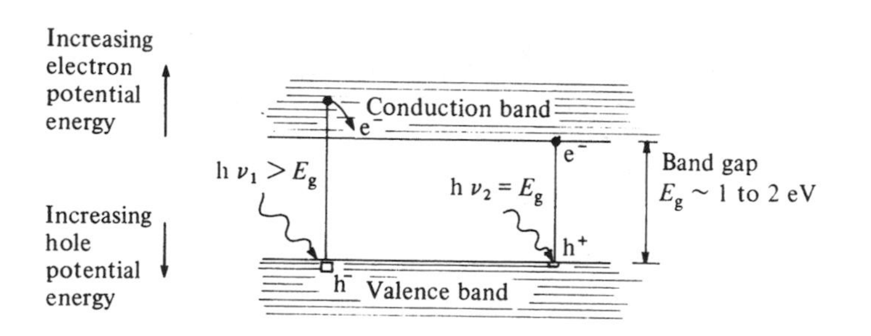


*Figure 7.2* Semiconductor band structure, intrinsic pure material. Photon absorption; $h\nu_2 = E_g$, photon energy equals band gap; $h\nu < E_g$, no photovoltaic absorption; $h\nu_1 > E_g$, excess energy dissipated as heat.

*Table 7.1* Solar cell–related properties of silicon

*Intrinsic, pure material*
Band gap $E_g$ (27 °C): 1.14 eV(corresponding $\lambda = 1.09\,\mu m$)
(~200 °C): 1.11 eV(corresponding $\lambda = 1.12\,\mu m$)
Carrier mobility $\mu$, electron $0.14\,m^2\,V^{-1}\,s^{-1}$, hole $0.048\,m^2\,V^{-1}$
Carrier diffusion constant $D_e = 35 \times 10^{-4}\,m^2\,s^{-1}$, $D_h = 12 \times 10^{-4}\,m^2\,s^{-1}$
Refractive index at $\lambda = 6\,\mu m$ $n = 3.42$
Extinction coefficient $K$ at $\lambda = 1\,\mu m$ $K = 10^4\,m^{-1}$
at $\lambda = 0.4\,\mu m$ $K = 10^5\,m^{-1}$
Thermal conductivity $157\,W\,m^{-1}\,K^{-1}$
Specific heat capacity $694\,J\,kg^{-1}\,K^{-1}$
Density; atoms $5.0 \times 10^{28}\,m^{-3}$; $2329\,kg\,m^{-3}$

*Typical Si homojunction n–p/p$^+$ solar cell*
n layer, thickness 0.25–0.5 μm; dopant conc. $\lesssim 10^{26}\,m^{-3}$
p layer, thickness 250–350 μm; dopant conc. $\lesssim 10^{24}\,m^{-3}$
p$^+$ layer, thickness 0.5 μm; dopant conc. $\lesssim 10^{24}\,m^{-3}$
Surface recombination velocity $10\,m\,s^{-1}$
Minority carrier
diffusion constant $D \sim 10^{-3}\,m^2\,s^{-1}$
path length $L \sim 100\,\mu m$
lifetime $\tau \sim 10\,\mu s$
Dopant concentration about 1 in 10 000 atoms

*Guide to silicon crystal solar cell efficiencies*

| | |
|---|---|
| laboratory, single crystal | ~25% |
| laboratory, polycrystalline | ~20% |
| best commercial single crystal (e.g. for space) | ~19% |
| general commercial, single crystal | ~15% |
| concentrated insolation on general commercial | ~20% |

### 7.2.2 *Doping*

Controlled quantities of specific impurity ions are added to the very pure (intrinsic) material to produce doped (extrinsic) semiconductors. Si is tetravalent, in group IV of the periodic table. Impurity dopant ions of less valency (e.g. boron, group III) enter the solid Si lattice and become electron *acceptor* sites which trap free electrons. These traps have an energy level within the band gap, but near to the valence band. The absence of the free electrons produces positively charged states called *holes* that move through the material as free carriers. With such electron acceptor impurity ions, the semiconductor is called *p (positive) type material*, having holes as *majority carriers*.

Conversely, atoms of greater valency (e.g. phosphorus, group V) are electron *donors*, producing *n (negative) type material* with an excess of conduction electrons as the majority carriers. A useful mnemonic is 'acceptor–*p*-type', 'donor–*n*-type'.

In each case, however, charge carriers of the complementary polarity also exist in much smaller numbers and are called *minority carriers* (electrons in p-type, holes in n-type). Holes and electrons may recombine when they meet freely in the lattice or at a defect site. Both p- and n-type extrinsic material have higher electrical conductivity than the intrinsic basic material. Indeed the resistivity $\rho_e$ is used to define the material. Common values for silicon photovoltaics range between $\rho_e \approx 0.010\,\Omega\,\mathrm{m} = 1.0\,\Omega\,\mathrm{cm}(N_d \approx 10^{22}\,\mathrm{m}^{-3})$ and $\rho_e \sim 0.10\,\Omega\,\mathrm{m} = 10\,\Omega\,\mathrm{cm}(N_d \approx 10^{21}\,\mathrm{m}^{-3})$, where we use the symbol $N_d$ for dopantion concentration.

### 7.2.3 Fermi level

The n-type material has greater conductivity than intrinsic material because electrons easily enter the conduction band by thermal excitation. Likewise p-type has holes that easily enter the valence band. The Fermi level is a descriptive and analytical method of explaining this (Figure 7.3). It is the apparent energy level within the forbidden band gap from which majority carriers (electrons in n-type and holes in p-type) are excited to become charge carriers. The probability for this varies as $\exp[-e\phi/(kT)]$, where e is the charge of the electron and hole, $e = 1.6 \times 10^{-19}$ C, and $\phi$ is the electric potential difference between the Fermi level and the valence or conduction bands as appropriate; $\phi \ll E_g$. Note that electrons are excited 'up' into the conduction band, and holes are excited 'down' into the valence band. Potential energy increases upwards for electrons and downwards for holes on the conventional diagram.

### 7.2.4 Junctions

The p-type material can have excess donor impurities added to specified regions so that these become n-type in the otherwise continuous material, and vice versa. The region of such a dopant change is a *junction* (which is

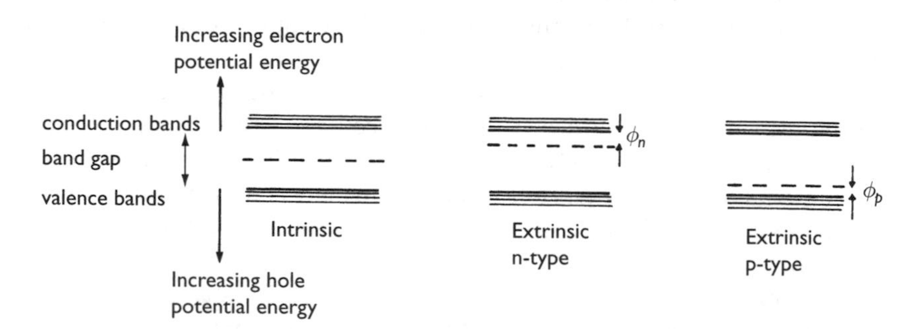


*Figure 7.3* Fermi level in semiconductors (shown by broken line) – this describes the potential energy level for calculations of electron and hole excitation.

not formed by physically pushing two separate pieces of material together!). Imagine, however, that the junction has been formed instantaneously in the otherwise isolated material (Figure 7.4(a)). Excess donor electrons from the n-type material cross to the acceptor p-type material, and vice versa for holes. A steady state is eventually reached. The electric field caused by the accumulation of charges of opposite sign on each side of the junction balances the diffusive forces arising from the different concentrations of free electrons and holes. As a result the Fermi level is at constant potential throughout the whole material. However, a net movement of charge has occurred at the junction, with excess negative charge now on the p side and positive on the n side.

The band gap $E_g$ still exists throughout the material, and so the conduction and valence bands have a step at the junction as drawn in Figure 7.4(b). The depth of the step is $eV_B$ in energy and $V_B$ in electric potential difference (voltage). $V_{B(I=O)}$ is the band step potential at zero current through the material and is the built-in field potential of the isolated junction. Note that $V_B < E_g/e$ because

$$V_{B(I=0)} = E_g/e - (\phi_n + \phi_p) \tag{7.1}$$

$(\phi_n + \phi_p)$ decreases with increase in dopant concentration. For a heavily doped Si p–n junction (dopant ions $\sim 10^{22}\,\mathrm{m}^{-3}$), $E_g = 1.11\,\mathrm{eV}$, and $(\phi_n + \phi_p) \approx 0.3\,\mathrm{V}$. So in the dark, with no current flowing,

$$V_{B(I=0)} \approx 0.8\,\mathrm{V} \tag{7.2}$$

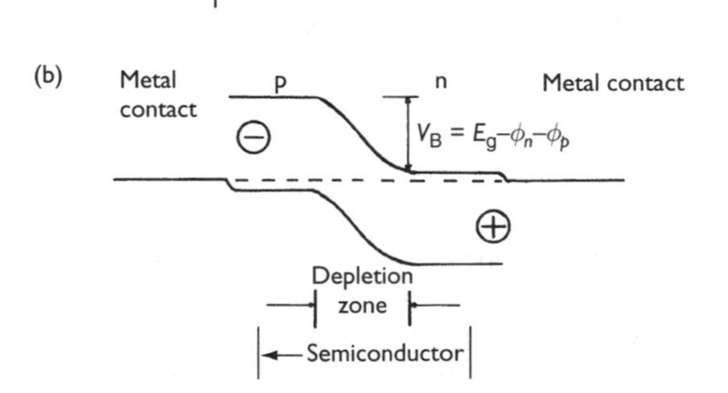


*Figure 7.4* (a) Diagrammatic 'formation' of a p–n homojunction cell with metal connectors. Fermi levels of isolated components shown by broken line. (b) Energy level diagram of a p–n homojunction with metal non-rectifying (ohmic) contacts. Electrons and holes have diffused to reach an equilibrium.

### 7.2.5 Depletion zone

The potential energy balance of carriers from each side of the junction (represented by the constancy of the Fermi level across the junction) results in the p-type region having a net negative charge ('up' on the energy diagram) and vice versa for the donor region. The net effect is to draw electron and hole carriers out of the junction, leaving it greatly depleted in total carrier density. Let $n$ and $p$ be the electron and hole carrier densities. Then the product $np = C$, a constant, throughout the material. For example,

1 p region:

$$np = C = (10^{10}\,\mathrm{m}^{-3})(10^{22}\,\mathrm{m}^{-3}) = 10^{32}\,\mathrm{m}^{-6} \tag{7.3}$$

2 n region:

$$np = C = (10^{22}\,\mathrm{m}^{-3})(10^{10}\,\mathrm{m}^{-3}) = 10^{32}\,\mathrm{m}^{-6} \tag{7.4}$$

$$n + p = 10^{22}\,\mathrm{m}^{-3}$$

3 Depletion zone: $n = p$ by definition. So

$$n^2 = p^2 = C = 10^{32}\,\mathrm{m}^{-6}$$

$$n = p = 10^{16}\,\mathrm{m}^{-3}$$

$$n + p = 2 \times 10^{16}\,\mathrm{m}^{-3} \tag{7.5}$$

The typical data of this example show that the total charge carrier density at the depletion zone is reduced by at least $\sim 10^5$ as compared with the n and p regions each side.

The width $w$ of the junction can be approximated to

$$w \approx \left[\left(\frac{2\varepsilon_0 \varepsilon_r V_B}{e\sqrt{(np)}}\right)\right]^{\frac{1}{2}} \tag{7.6}$$

where $\varepsilon_0$ is the permittivity of free space, $\varepsilon_r$ is the relative permittivity of the material, and the other terms have been defined previously.

For Si at $10^{22}\,\mathrm{m}^{-3}$ doping concentration and $w \approx 0.5\,\mu\mathrm{m}$, the electric field intensity $V_B/w$ is $\sim 2 \times 10^6\,\mathrm{V\,m}^{-1}$. The current carrying properties of the junction depend on minority carriers being able to diffuse to the depletion zone and then be pulled across in the large electric field. This demands that $w < L$, where $L$ is the diffusion length for minority carriers, and this is a criterion easily met in solar cell p–n junctions, see (7.11).

### 7.2.6 *Biasing*

The p–n junction may be fitted with metal contacts connected to a battery (Figure 7.5). The contacts are called 'ohmic' contacts – non-rectifying junctions of low resistance compared with the bulk material. In 'forward bias' the positive conventional circuit current passes from the p to the n material across a reduced band potential difference $V_B$. In 'reverse bias', the external battery opposes the internal potential difference $V_B$ and so the current is reduced. Thus the junction acts as a rectifying diode with an $I$–$V$ characteristic that will be described later (Figure 7.8).

### 7.2.7 *Carrier generation*

At an atomic scale, matter is in a continuous state of motion. The atoms in a solid oscillate in vibrational modes with quantised energy (phonons). In semiconductor material, electrons and holes are spontaneously generated from bound states for release into the conduction and valence bands as charge carriers. This is a thermal excitation process with the dominant temperature variation given by the Boltzmann probability factor $\exp[-E/(kT)]$, where $E$ is the energy needed to separate the electrons and holes from their particular bound states, k is the Boltzmann constant and $T$ is the absolute temperature. For pure intrinsic material $2E = E_g$, the band gap. For doped extrinsic material $|E| = |e\phi|$, where $\phi$ is the potential difference needed to excite electrons in n-type material into the conduction band, or holes in $p$-type material into the valence band, Figure 7.4(a). Note that $\phi$ is determined locally at the dopant site and $|e\phi| < E_g$. In general, $\phi$ decreases with increase in dopant concentration. For heavily doped Si ($\rho_e \approx 0.01\,\Omega\,\text{m}$, $N_d \approx 10^{22}\,\text{m}^{-3}$), $|e\phi| \approx 0.2\,\text{eV}$.

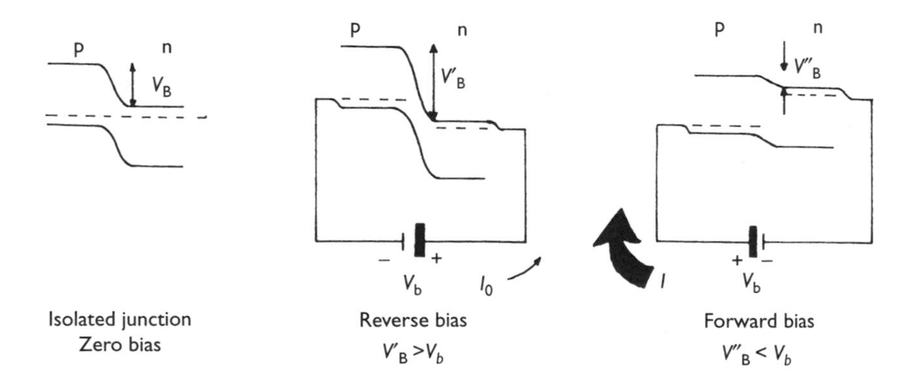


*Figure 7.5* Reverse and forward biasing of a p–n junction. $I_0 \ll I$, conventional current. *Note*: Conventional current direction is opposite to electron current direction.

### 7.2.8 Relaxation (recombination) time and diffusion length

Thermally or otherwise generated electron and hole carriers recombine after a typical relaxation time $\tau$, having moved a typical diffusion length $L$ through the lattice. In very pure intrinsic material, relaxation times can be long ($\tau \sim 1$ s), but for commercial doped material, relaxation times are much shorter ($\tau \sim 10^{-2}$ to $10^{-8}$ s). Lifetime is limited by recombination at sites of impurities, crystal imperfections, surface irregularities and other defects. Thus highly doped material tends to have short relaxation times. Surface recombination is troublesome in solar cells because of the large area and constructional techniques. It is characterised by the surface recombination velocity $S_v$, typically $\sim 10\,\mathrm{m\,s^{-1}}$ for Si, as defined by

$$J = S_v N \tag{7.7}$$

where $J$ is the recombination current number density perpendicular to the surface ($\mathrm{m^{-2}\,s^{-1}}$) and $N$ is the carrier concentration in the material ($\mathrm{m^{-3}}$).

The probability per unit time of a carrier recombining is $1/\tau$. For $n$ electrons the number of recombinations per unit time is $n/\tau_n$, and for $p$ holes it is $p/\tau_p$. In the same material at equilibrium these must be equal, so

$$\frac{n}{\tau_n} = \frac{p}{\tau_p}, \quad \tau_n = \frac{n}{p}\tau_p, \quad \tau_p = \frac{p}{n}\tau_n \tag{7.8}$$

In $p$ material, if $p \sim 10^{22}\,\mathrm{m^{-3}}$ and $n \sim 10^{11}\,\mathrm{m^{-3}}$, then $\tau_n \ll \tau_p$ and vice versa. Therefore in solar cell materials, minority carrier lifetimes are many orders of magnitude shorter than majority carrier lifetimes (i.e. minority carriers have many majority carriers to recombine with).

Thermally generated carriers diffuse through the lattice down a concentration gradient $\mathrm{d}N/\mathrm{d}x$ to produce a number current density (in the $x$ direction) of

$$J_x = -D\left(\frac{\mathrm{d}N}{\mathrm{d}x}\right) \tag{7.9}$$

where $D$ *is* the diffusion constant. A typical value for Si is $35 \times 10^{-4}\,\mathrm{m^2\,s^{-1}}$ for electrons, $12 \times 10^{-4}\,\mathrm{m^2\,s^{-1}}$ for holes.

Within the relaxation time $\tau$, the diffusion distance $L$ is given by Einstein's relationship

$$L = (D\tau)^{1/2} \tag{7.10}$$

Therefore a typical diffusion length for minority carriers in $p$-type Si ($D \sim 10^{-3}\,\mathrm{m^2\,s^{-1}}$, $\tau \sim 10^{-5}$ s) is

$$L \approx (10^{-3} 10^{-5})^{1/2}\mathrm{m} \approx 100\,\mu\mathrm{m} \tag{7.11}$$

Note that $L \gg w$, the junction width of a typical p–n junction (7.6).

### 7.2.9 *Junction currents*

Electrons and holes may be generated thermally or by light, and so become carriers in the material. Minority carriers, once in the built-in field of the depletion zone, are pulled across electrostatically down their respective potential gradients. Thus minority carriers that cross the zone become majority carriers in the adjacent layer (consider Figures 7.5 and 7.6). The passage of these carriers becomes the *generation current* $I_g$, which is predominantly controlled by temperature in a given junction without illumination. In an isolated junction there can be no overall imbalance of current across the depletion zone. A reverse *recombination current* $I_r$ of equal magnitude occurs from the bulk material. This restores the normal internal electric field. Also the band potential $V_B$ is slightly reduced by $I_r$. Increase in temperature gives increased $I_g$ and so decreased $V_B$ (leading to reduced photovoltaic open circuit voltage $V_{oc}$ with increase in temperature, see later). For a given material, the generation current $I_g$ is controlled by the temperature. However, the recombination current $I_r$ can be varied by external bias as explained in Section 7.2.6 and in Figures 7.5 and 7.7. Without illumination, $I_g$ is given by

$$I_g = eN_i^2\left(\frac{1}{p}\frac{L_p}{\tau_p} + \frac{1}{n}\frac{L_n}{\tau_n}\right) \tag{7.12}$$

where $N_i$ is the intrinsic carrier concentration and the other quantities have been defined before. In practice the control of material growth and dopant

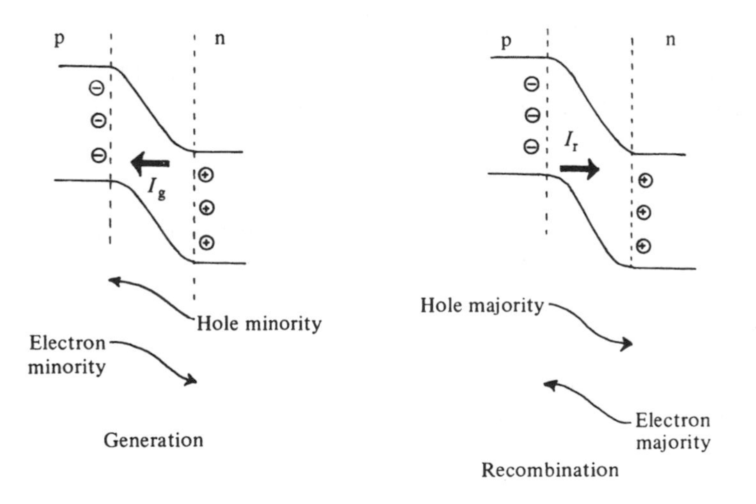


*Figure 7.6* Generation and recombination currents at a p–n junction.

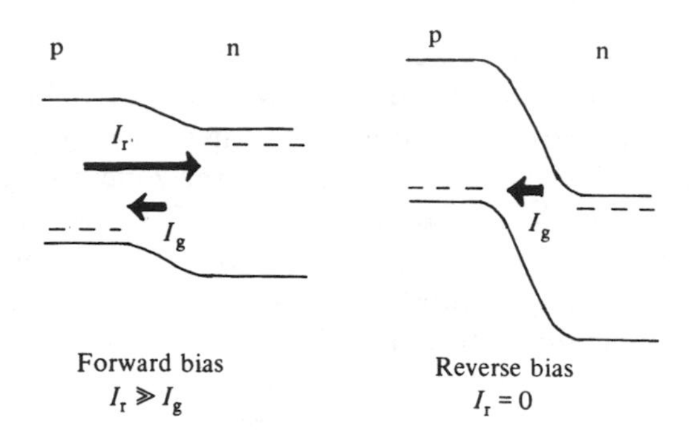


*Figure 7.7* Recombination and generation junction currents with externally applied bias.

concentration is not exact enough to predict how $L$ and $\tau$ will vary with material properties and so $I_g$ is not controlled.

Note that recombination is unlikely to occur in the depletion zone, since the transit time across the zone is

$$t \approx \frac{w}{u} = \frac{w}{\mu(V_B/w)} = \frac{w^2}{\mu V_B} \sim 10^{-12}\,\text{s} \tag{7.13}$$

where $u$ is the carrier drift velocity and $\mu$ is the mobility ($\sim 0.1\,\text{m}^2\,\text{V}^{-1}\,\text{s}^{-1}$) in the electric field $V_B/w(V_B \sim 0.6\,\text{V}, w \sim 0.5\,\mu\text{m})$.
Thus $t \ll \tau_r$, (where $\tau_r$ is the recombination time that varies from $\sim 10^{-2}$ to $10^{-8}$ s).

### *7.2.10 Circuit characteristics*

The p–n junction characteristic (no illumination) is explained by the previous discussion and shown in Figure 7.7. With no external bias ($V_b = 0$),

$$I_r = I_g \tag{7.14}$$

With a positive, forward, external bias across the junction of $V_b$, the recombination current becomes an increased forward current:

$$I_r = I_g \exp[eV_b/(kT)] \tag{7.15}$$

as explained in basic solid state physics texts.
The net current (in the dark, no illumination) is

$$\begin{aligned} I_D &= I_r - I_g \\ &= I_g \{\exp[eV_b/(kT)] - 1\} \end{aligned} \tag{7.16}$$

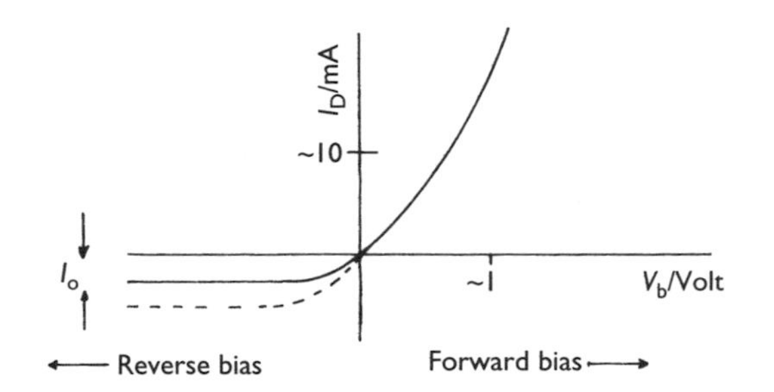


*Figure 7.8* p–n junction dark characteristic. Plot of diode junction current $I_D$ versus external voltage bias $V_b$ (see equation 7.17). Note how the magnitude of the saturation current $I_0$ increases with temperature (- - -).

This is the Shockley equation for the junction diode, usually written

$$I_D = I_0 \{\exp[eV_b/(kT)] - 1\} \tag{7.17}$$

where $I_0(= I_g)$ is the *saturation current* under full reverse bias before avalanche breakdown. It is also called the leakage or diffusion current. For good solar cells $I_0 \sim 10^{-8}\,\text{A}\,\text{m}^{-2}$.

## 7.3 Photon absorption at the junction

So far, we have considered the junction 'in the dark'; now let light appear. The dominant process causing the absorption of electromagnetic radiation in semiconductors is the generation of electron–hole pairs. This occurs in direct transitions of electrons across the band gap $E_g$ when

$$h\nu \geq E_g \tag{7.18}$$

where $h$ is the Planck constant ($6.63 \times 10^{-34}\,\text{J s}$) and $\nu$ is the radiation frequency. The semiconductor material of solar cells has $E_g \approx 1\,\text{eV}$.

Absorption of photons near this condition occurs in *indirect band gap transitions* (e.g. in silicon) because of interaction within the crystal lattice with a lattice vibration *phonon* of energy $h\Omega \sim 0.02\,\text{eV}$, where $\Omega$ is the phonon frequency. In this case the radiation absorption is not 'sharp' (see also Section 7.7.2(6)) because the condition for photon absorption is

$$h\nu \pm h\Omega \geq E_g \tag{7.19}$$

*Direct band gap semiconductors* (e.g. GaAs) absorb photons without lattice phonon interaction. They therefore have sharp absorption band transitions

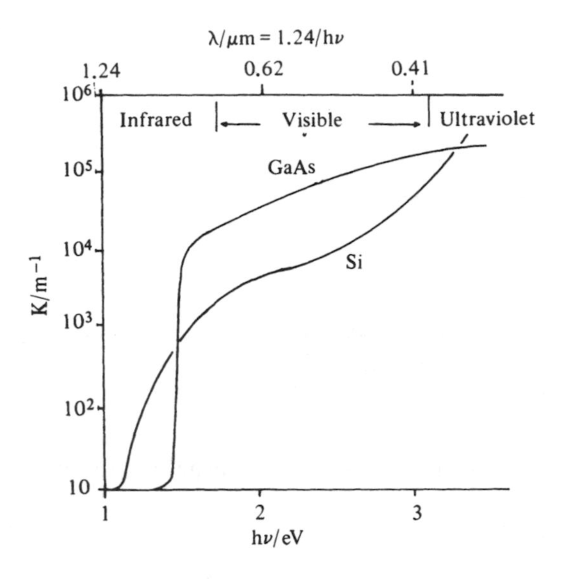


*Figure 7.9* Extinction coefficient $K$ of materials with a direct (GaAs) and indirect (Si) band gap. Radiant flux density varies as $G(x) = G_0 \exp(-Kx)$ where $x$ is the depth into the surface. Note the logarithmic scale, which masks the sharpness of the band gap absorption. After Wilson (1979).

with relatively large values of extinction coefficient ($v > E_s/h$). This contrasts with the indirect band gap semiconductors (e.g. Si) that have less sharp absorption bands and smaller extinction coefficients $K$ (Figure 7.9). Band gap absorption for semiconductors occurs at frequencies within the solar spectrum; for Si this occurs when

$$v > E_g/h \approx \frac{(1.1\,\mathrm{eV})(1.6 \times 10^{-19}\,\mathrm{J\,eV^{-1}})}{6.63 \times 10^{-34}\,\mathrm{J\,s}} = 0.27 \times 10^{15}\,\mathrm{Hz} \qquad (7.20)$$

and

$$\lambda \approx \frac{3.0 \times 10^{8}\,\mathrm{m\,s^{-1}}}{0.27 \times 10^{15}\,\mathrm{s^{-1}}} = 1.1\,\mu\mathrm{m} \qquad (7.21)$$

The number flux of photons in the solar spectrum is large ($\sim 1\,\mathrm{kW\,m^{-2}}/[(2\,\mathrm{eV})(1.6 \times 10^{-19}\,\mathrm{J\,eV^{-1}})] \approx 3 \times 10^{21}$ photon $\mathrm{m^{-2}\,s^{-1}}$). So the absorption of solar radiation in semiconductors can greatly increase electron–hole generation apart from thermal generation. If this charge carrier creation occurs near a p–n junction, the built-in field across the depletion zone can be the EMF to maintain charge separation and produce currents in an externally connected circuit (Figure 7.10). Thus the photon

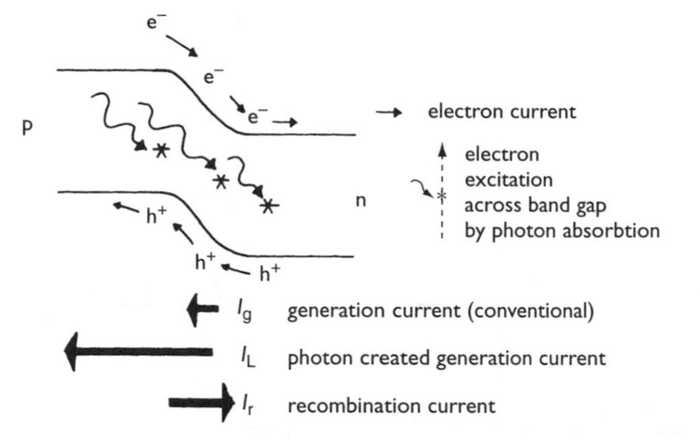


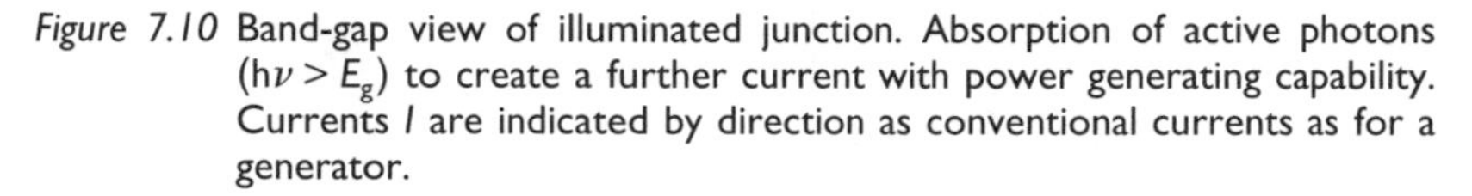


*Figure 7.10* Band-gap view of illuminated junction. Absorption of active photons ($h\nu > E_g$) to create a further current with power generating capability. Currents $I$ are indicated by direction as conventional currents as for a generator.

generation of carriers in sunlight adds to, and dominates, the thermal generation already present. In dark conditions, of course, only the totally negligible thermal generation occurs.

The p–n junction with photon absorption is therefore a DC source of current and power, with positive polarity at the p-type material. Power generation from a solar cell corresponds to conditions of diode forward bias, as illustrated in Figure 7.11.

The solar cell current $I$ is determined by subtracting the photon-generated current $I_L$ from the diode dark current $I_D$ (Figure 7.11).

$$I = I_D - I_L \tag{7.22}$$

So from (7.17),

$$I = I_0[\exp(eV_b/kT) - 1] - I_L \tag{7.23}$$

In the sign convention used so far for rectifying diodes, $I_D$ is positive, so $I$ is negative in the power production quadrant; therefore under illumination the current flows into an externally connected battery to charge it, as in Figure 7.11(b). Section 7.8 gives more detail on these characteristics and their implications for practical photovoltaic power systems. (*Caution*: In discussing photovoltaic power systems (Section 7.8), the solar cells are considered as generators of positive current, so the somewhat contorted diode definitions are not used. Further confusion may occur, since arrays of cells may be connected to conventional rectifying diodes to prevent the solar

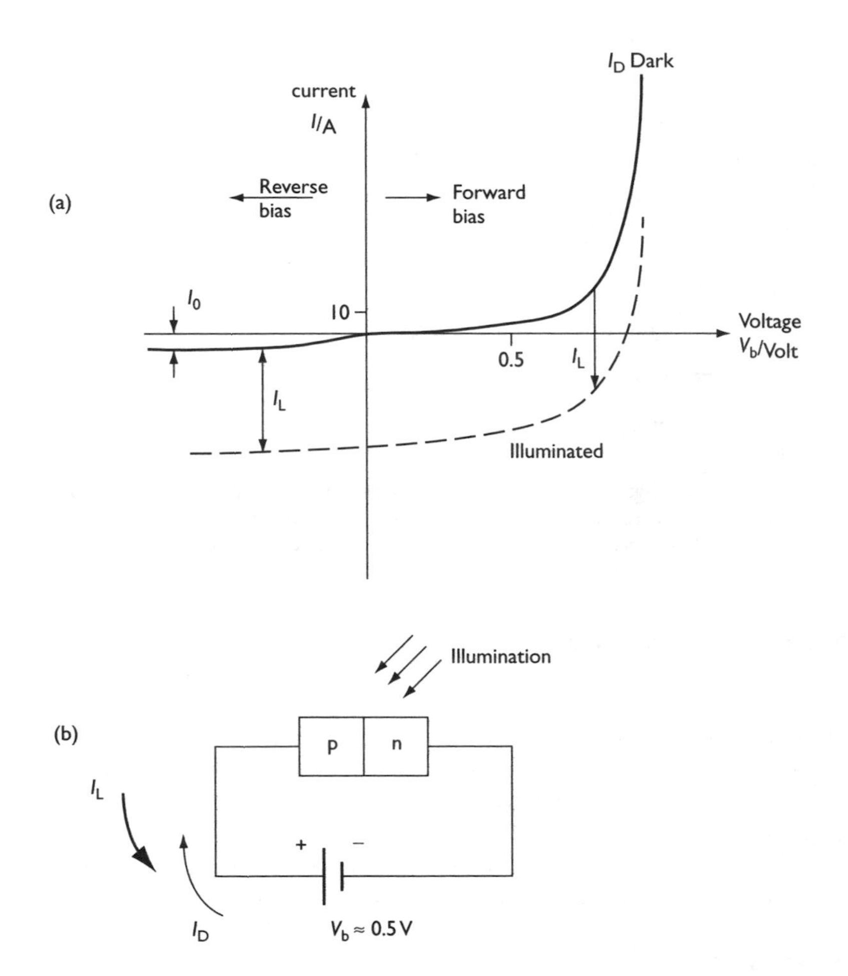


*Figure 7.11* Sketch diagrams of the p-n diode operating as a solar cell. (a) *I–V* characteristic of the p–n junction solar cell without illumination ( ____ ), as in Figure 7.8, and with illumination (- - -). Without illumination, $I = I_D$. However, with illumination, the light-generated current $I_L$ is superimposed on the dark current $I_D$ of Figure 7.8 to give a net current $I = I_D - I_L$. This results in a region in the lower right quadrant where power can be generated with the p–n junction as a solar cell and forced into a battery or grid line. This figure and Figure 7.8 are both drawn in the manner of rectifying diodes. (b) Corresponding physical set-up of the device as a solar cell, connected so the battery is charged by the light-generated current $I_L$. Such a connection would be the 'forward biased' configuration for a solar cell as a diode, c.f. Figure 7.5 for a diode in the dark.

cells passing a reversed current in the dark. Anyway, it is only the physicists who become bothered by such conventions; electrical power engineers just make the systems work!).

Photocurrent generation depends on photon absorption near the junction region. If the incident solar radiant flux density is $G_0$, then at depth $x$, the absorbed power per unit area, $G_{abc}$ is

$$G = G_0 - G_x = G_0\{1 - \exp[G_{abc} - K(v)x]\} \tag{7.24}$$

where

$$\frac{dG}{dx} = K(v)G_x \tag{7.25}$$

$K(v)$ is the extinction coefficient of Figure 7.9, and is critically dependent on frequency. Photons of energy less than the band gap are transmitted with zero or very little absorption. At depths of $1/K$, absorption is 63%, at $2/K$ it is 86% and at $3/K$ it is 95%. For Si at frequencies greater than the band gap, $2/K$ equals $\sim 400\,\mu m$, which gives approximately the minimum thickness for solar cell material, unless thin wall reflection (light trapping) techniques are used (Section 7.5.5).

## 7.4 Solar radiation absorption

Detailed properties of solar radiation were considered fully in Chapter 4. Figure 7.12 shows the spectral distribution of irradiance (Figure 4.15) replotted in terms of photon energy $h\nu$ rather than wavelength $\lambda$. This mathematical transformation shifts the peak of the curve, though not of course the area under it (which is the total irradiance). Figure 7.12(b) also smooths out the prominent dips (due to atmospheric absorption) in the lower curve of Figure 4.15.

For photovoltaic power generation in a typical solar cell, e.g. Si material, the essential factors indicated in Figure 7.12 are:

1 The solar spectrum includes frequencies too small for photovoltaic generation ($hv < E_g$) (region A). Absorption of these low frequency (long wavelength) photons produces heat, but no electricity.
2 At frequencies of band gap absorption ($hv > E_g$), the excess photon energy ($hv - E_g$) is wasted as heat (region C).
3 Therefore there is an optimum band gap absorption to fit a solar spectrum for maximum electricity production (Figure 7.13). Note that the spectral distribution of the received solar radiation varies with penetration through the atmosphere so the curves of Figure 7.12(a) and (b) peak at different energies. The spectral distribution (and total irradiance) also varies of course with cloudiness, humidity, pollution, etc.

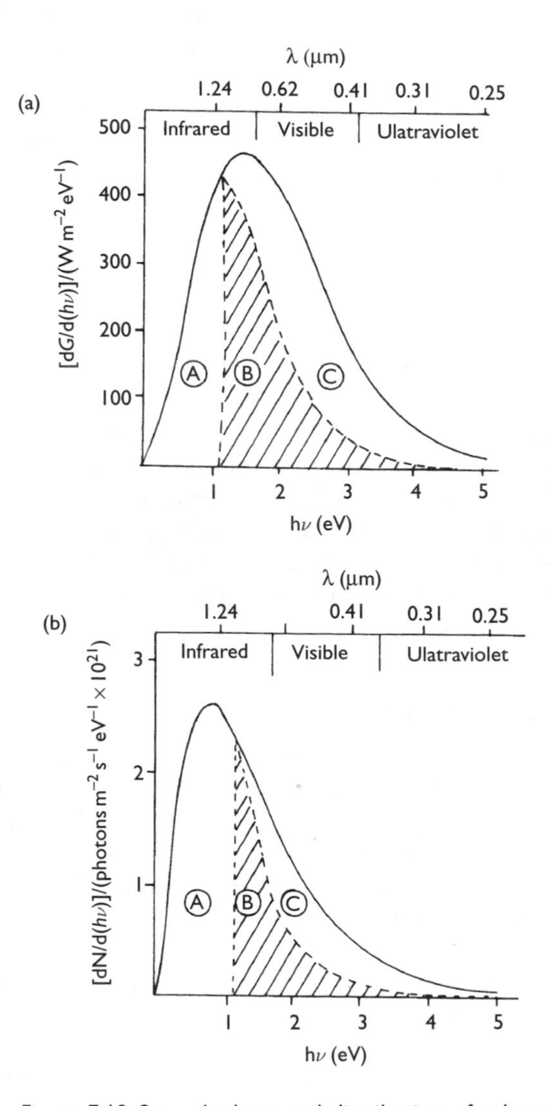


*Figure 7.12* Smoothed spectral distribution of solar radiation. (a) Irradiation above the Earth's atmosphere (AM0). Abscissa (lower scale): photon energy in eV; ordinate: irradiance per photon energy. Also shown (on upper scale) are corresponding wavelengths of radiation; if $h\nu$ has the unit of electron volt (eV), then $\lambda = (1.24\,\mu\text{m})/(h\nu/\text{eV})$. (b) Irradiation onto the Earth's surface in clear sky conditions (AM1). Abscissa as for (a); ordinate is the distribution of irradiance but expressed as *number* of photons. In each chart, region (B) represents energy actually available for power production by a Si solar cell – see text for further detail.

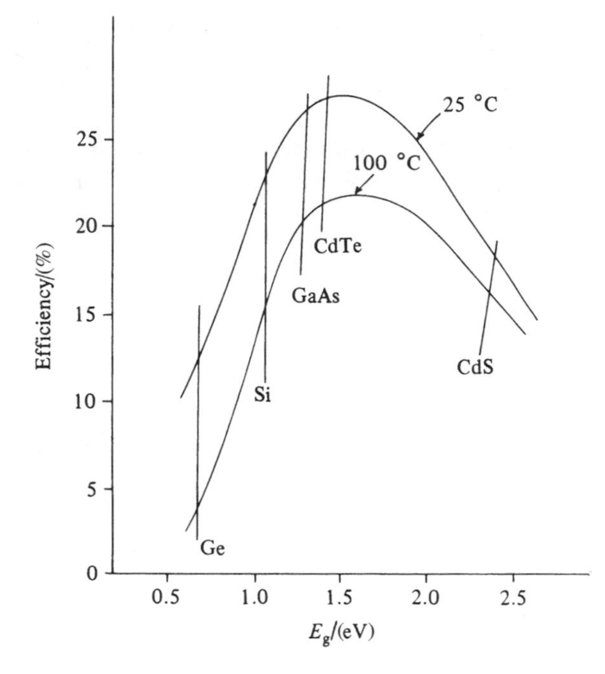


*Figure 7.13* Indicative theoretical efficiency of homojunction solar cells as a function of band gap. Note the decrease in performance with increase in temperature. Band gaps of some semiconductor materials are indicated.

(See Section 4.6.1 concerning air-mass-ratio, i.e. AM0 in space, AM1 at zenith, AM2 at zenith angle 60°; AM1.5 conditions are usually considered as standard for solar cell design.)

4 Only the energy in region B of Figure 7.12 is potentially available for photovoltaic power in a solar cell with a single band gap. The maximum proportion of total energy $[B/(A+B+C)]$, where $A$, $B$, $C$ are the areas of regions A, B, C, is about 47%, but the exact amount varies slightly with spectral distribution. Note that not all this energy can be generated as *useful* power, due to the cell voltage $V_B$ being less than the band gap $E_g$ (see Figure 7.4 and Section 7.5.7). The useful power, at current $I$, is $V_B I$, not $E_g I$. Therefore, in practice, with $V_B/E_g \approx 0.75$, *only a maximum of about 35% (=75% of 47%) of the solar irradiance is potentially available for conversion to electrical power.*

Points (1)–(3) above explain the peak in Figure 7.13 in terms of the incoming photons. Alternatively one can consider the output of a solar cell. If the material has a large band gap, the output with the same irradiance will have larger voltage but smaller current, because there are fewer photons available with the requisite energy, and thus there will be less power. Conversely, if the material has a small band gap, the output will have larger current (many photons qualify) but at smaller voltage. Somewhere in between, the power output will be a maximum. For the solar spectrum at AM1, this peak is at a band gap of about 1.6 eV.

For photons with energy greater than the band gap, if the distribution of photon number ($N$) with photon energy ($E = h\nu$) is $\mathrm{d}N/\mathrm{d}E$, then the maximum theoretical power produced is

$$P = \int_{Eg}^{\infty} \left(\frac{\mathrm{d}N}{\mathrm{d}E}\right) E_g \mathrm{d}E \tag{7.26}$$

but

$$\mathrm{d}P = \mathrm{h}\nu \mathrm{d}N = E\mathrm{d}N \tag{7.27}$$

so

$$P = \int_{Eg}^{\infty} \left(\frac{\mathrm{d}P}{\mathrm{d}E}\right) \frac{E_g}{E} \mathrm{d}E \tag{7.28}$$

## 7.5 Maximising cell efficiency

Photovoltaic cells are limited in efficiency by many losses; some of these are avoidable but others are intrinsic to the system. Some limits are obvious and may be controlled independently, but others are complex and cannot be controlled without producing interrelated effects. For instance, increasing dopant concentration can have both advantageous and harmful effects. Table 7.2 portrays typical losses for commercial Si p–n junction single-crystal solar cells in AM1 irradiance. Unfortunately there is no standard convention for the names of the loss factors. Also shown are corresponding figures for the PERL cell (Figure 7.21), which is one of the most efficient cells yet made from Si. Such increased efficiency comes at a cost of complexity and consequent cost of manufacturing. Nevertheless laboratory-scale 'champion' cells (which also often lack durability) show the potential for improvement in a variety of solar cell technologies (see also Table 7.3). The balance between cost, complexity and efficiency is a delicate commercial judgement for both manufacturers and users of solar cells. In general, there is a bias to greater efficiency, since installations of a given total power will have smaller area and less transportation and placement cost. There are a few applications, such as solar car racing or in space, where users seek the largest efficiency with reasonable durability, almost regardless of cost. But for most terrestrial applications with ample area, cost per kWh generated is a key criterion.

*Table 7.2* Limits to efficiency in Si solar cells. Refer to Section 7.5 for explanation of each process

| *Text section for process* | *Data for common commercial Si cells (c.2003)* | — | — | | *Data for champion cells (c.2003)* | — | — |
|---|---|---|---|---|---|---|---|
| | *Energy remaining after process loss/ %* | *Power loss/ %* | *Efficiency factor* | Notes | *Efficiency factor* | *Power loss/ %* | *Energy remaining after process loss/ %* |
| 7.5.3 | 80 | 20 | 0.8 | No photovoltaic absorption: $h\nu < E_g$ | 0.8 | 20 | 80 |
| 7.5.4 | 48 | 32 | 0.6 | Excess photon energy lost as heat: $h\nu - E_g$ | 0.6 | 32 | 48 |
| 7.5.2 | 46 | 2 | 0.95 | Surface reflection | 0.99 | 0.5 | 48 |
| 7.5.5 | 45 | 0.5 | 0.99 | Quantum efficiency | 0.99 | 0.5 | 47 |
| 7.5.1 | 42 | 4 | 0.92 | Top surface contact grid obstruction | 0.98 | 1 | 46 |
| 7.5.7 | 25 | 17 | 0.6 | Voltage factor $eV_B < E_g$ | 0.65 | 16 | 30 |
| 7.5.8 | 20 | 5 | 0.81 | Curve factor = (max. power)$/I_{sc}V_{oc}$ | 0.84 | 5 | 25 |
| 7.5.9 | 16 | 4 | 0.8 | Additional curve factor *A*, recombination collection losses | 0.99 | 0.3 | 25 |
| 7.5.10 | 16 | 0.5 | 0.97 | Series resistance | 0.99 | 0.2 | 25 |
| 7.5.11 | 16 | 0.2 | 0.99 | Shunt resistance | 0.99 | 0.2 | 24 |
| 7.5.12 | 16 | | | Delivered power | | | 24 |

*Table 7.3 Solar cell base material parameters, AM1 conditions**. The 'champion cell' is the PERL cell (Figure 7.21)

| *Material base* | *Band gap* $E_g^{\dagger}$ eV | *Direct D or Indirect (I)* | *Example of cell* | *Efficiency of commercial cells (c.2003)/ %* | *Efficiency of 'champion' (laboratory) cells (c.2003)/ %* |
|---|---|---|---|---|---|
| Ge | 0.6 | I | Not used | N/a | |
| Si (single crystal) | 1.1 | I | p/n | 15 | 25 |
| Si (multicrystal) | | | Commonest commercial cell | 13 | 20 |
| Si (amorphous) | | | Thin film or ribbon | ~8 | 12 |
| GaAs | 1.4 | D | p/n | N/a | 25 |
| CdTe | 1.4 | D | Thin film (heterojunction with CdS) | ~8 | 16 |
| $Ga_{1-x}Al_xA_s(0 < x < 0.34)$ | 1.9–2.2 | I | Heterojunction with GaAs base | N/a | ~18 |
| $Ga_{1-x}Al_xA_s(0.34 < x < 1)$ | 1.9–2.2 | I | Heterojunction with GaAs base | N/a | ~18 |
| CdS | 2.4 | D | Only in heterojunctions | N/a | |
| $Cu(InGa)Se_2$ | ~1.2# | D | Thin film | ~10 | 19 |

* The optimum band gap in AM1 radiation is between 1.4 and 1.5 eV (Figure 7.13)

† Data here for ambient temperature (~25 °C). Band gap decreases with temperature increase [e.g. Si 1.14 eV (30 °C), 1.09 eV (130 °C)]

# Composition is actually $Cu(In_{1-x}Ga_x)Se_2$, with band gap (from 1.1 to 1.6 V) and efficiency depending on $x$. Most commercial CIGS cells in 2003 have $x \approx 0.3$, $Eg \approx 1.2$ V.

In the following sections, the losses are given as a percentage of total incident irradiance, AM1 = 100%, and are listed from the top to the base of the cell. The efficiency factors in Table 7.2 refer to the proportion of the remaining irradiance that is usefully absorbed at that stage in the photovoltaic generation of electricity.

### 7.5.1 Top surface contact obstruction (loss ~3%)

The electric current leaves the top surface by a web of metal contacts arranged to reduce series resistance losses in the surface (see Section 7.5.10). These contacts have a finite top surface area and so they cover part of the otherwise active surface; this loss of area is not always accounted for in efficiency calculations.

### 7.5.2 Reflection at top surface (loss ~1%)

Without special precautions, the reflectance from semiconductors is large, ~40% of the incident solar radiation. Fortunately this may be dramatically reduced to 3% or less by thin film surface or other treatment.

Consider three materials (air, cover, semiconductor) of refractive index of $n_0$, $n_1$ and $n_2$. For dielectric, electrically insulating materials, the reflectance between two media is

$$\rho_{\text{ref}} = \frac{(n_0 - n_1)^2}{(n_0 + n_1)^2} \tag{7.29}$$

For the reflectance between air ($n_0 = 1$) and plastic (say $n_1 = 1.6$), $\rho_{\text{ref}} = 5.3\%$.

Semiconductors have a refractive index represented by a complex number (since they are partly conducting) which is frequency dependent and averages about 3.5 in magnitude over the active spectrum. Considering the radiation frequency in electron volts, the reflectance in air varies from $\rho_{\text{ref}}$ (1.1 eV) = 34% to $\rho_{\text{ref}}(5\,\text{eV}) = 54\%$.

A thin film (thickness $t$) of appropriate material placed between air and a semiconductor can largely prevent reflection (Figure 7.14) if, for normal incidence, the main reflected components $a$ and $b$ are of equal intensity and differ in phase by $\pi$ radians ($\lambda/2$ path difference). For this to occur the reflectance at each surface is equal, so $n_1 = \surd(n_0 n_2)$ and also $t = \lambda/(4n_1)$. There is only one wavelength for which this condition is met exactly; however, broad band reflectance is considerably reduced. For Si (if $n_1 = 1.9$, thickness $t = 0.08\,\mu\text{m}$) the broad band reflectance is reduced to ~6%. Multiple thin layers can reduce broad band reflectance to <3%.

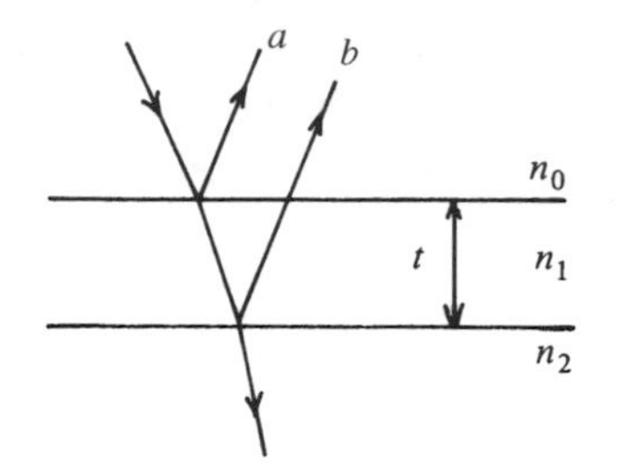


*Figure 7.14* Antireflection thin film.

*Figure 7.15* Top surfaces for increased absorption; scale of 10–100 μm. (a) Idealised textured shape, e.g. by chemical etching. (b) Structured shape, e.g. by laser machining.

Reflection losses can also be reduced by top surface configurations that reflect the beam for a second opportunity of absorption (Figure 7.15). The *textured surface* of Figure 7.15(a) can be produced by chemical etching on Si.

### *7.5.3 Photon energy less than band gap (loss ~23%)*

Photons of quantum energy $h\nu < E_g$ cannot contribute to photovoltaic current generation. For Si ($E_g \approx 1.1\,\text{eV}$) the inactive wavelengths have $\lambda > 1.1\,\mu\text{m}$ and include 23% of AM1 irradiance. If this energy is absorbed it causes heating with a temperature rise that lowers power production still further. These photons can theoretically be removed by filters. However, a more 'energy efficient' strategy is to make use of the heat in a combined heat and power system, such as cooling the pv module as part of a solar water heater.

### *7.5.4 Excess photon energy (loss ~33%)*

The excess energy of active photons ($h\nu - E_g$) also appears as heat.

### 7.5.5 *Quantum efficiency (loss ~0.4%)*

Quantum efficiency is the fraction of incident absorbed active photons producing electron–hole pairs, which usually approaches 100%. Thus the design of the cell has to ensure that the material is thick enough for at least 95% absorption, as explained in Section 7.3. Reflecting layers at the rear of the cell can return transmitted radiation for a second, and many more, passes. This is the principle of light trapping, as in thin-walled silicon cells deposited on a supporting glass substrate.

### 7.5.6 *Collection efficiency*

Collection efficiency is a vague term used differently by different authors. It may be applied to include the losses described in Sections 7.5.3 and 7.5.4 or usually, as here, to electrical collection of charges after carrier generation. Collection efficiency is therefore defined as the proportion of radiation generated electron–hole pairs that produce current in the external circuit. For 10% overall efficiency cells, the collection efficiency factor is usually about 0.7. Increasing this to about 0.9 would produce >20% overall efficiency cells, and so collection efficiency improvement is a major design target.

There are many factors affecting collection efficiency, as outlined in the following. One improvement not otherwise mentioned is back surface field (BSF). A layer of increased dopant concentration is formed as a further layer beyond the p–n junction, e.g. $1\,\mu m$ of $p^+$ on p to produce a further junction of $\sim 200\,kV\,m^{-1}$ (Figure 7.16). Electron minority carriers formed in the p layer near this $p^+$ region are 'reflected' down a potential gradient back towards the main p–n junction rather than up the gradient to the rear metal contact. Electron–hole recombination at the rear contact is therefore reduced. Similar diode-like layers, shown here as an n on p cell, can be added to the front surface (e.g. $n^+$ on n) to produce

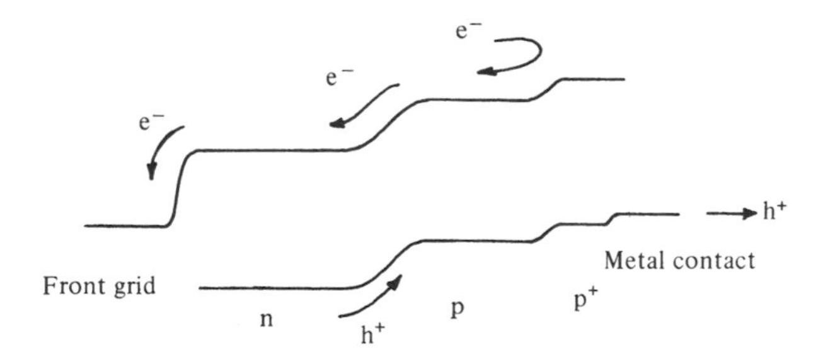


*Figure 7.16* Back surface field (BSF) to lessen diffusion leakage of electron current carriers at the rear of cells, shown here as an n on p cell.

the same effect for holes, providing optical absorption is not significantly increased.

### 7.5.7 *Voltage factor $F_v$ (loss ~20%)*

Each absorbed photon produces electron–hole pairs with an electric potential difference of $E_g/e$ (1.1 V in Si). However, only part ($V_B$) of this potential is available for the EMF of an external circuit. This is made clear in Figure 7.4, where the displacement of the bands across the junction in open circuit produces the band potential $V_B$. The voltage factor is $F_v = eV_B/E_g$. In Si, $F_v$ ranges from ~0.6 (for 0.01 Ωm material) to ~0.5 (for 0.1 Ωm material), so $V_B \approx 0.66–0.55$ V.

The 'missing' EMF ($\phi_n + \phi_1$) is because in open circuit the Fermi level across the junction equates at the dopant n and p levels, and not at the displaced conduction to valence band levels. Increased dopant concentration increases $F_v$(0.01 Ω m Si has greater $V_B$ and $V_{oc}$ than 0.1 Ωm Si), but other effects limit the maximum dopant concentrations in Si to $\sim 10^{22}\,m^{-3}$ of 0.01 Ωm materials. In GaAs, $F_v$ is ~0.8.

When producing current on load, the movement of carriers under forward bias produces heat as resistive internal impedance heating. This may be included as voltage factor loss, as $A$ factor loss (Section 7.5.9) or, as here, by series resistance heating (Section 7.5.10).

### 7.5.8 *Curve factor $F_c$ (loss ~4%)*

The solar cell $I$–$V$ characteristic is strongly influenced by the p–n diode characteristic (Figure 7.7). Thus as the solar cell output voltage is raised towards $V_{oc}$ the diode becomes increasingly forward biased, so increasing the internal recombination current $I_r$ across the junction. This necessary behaviour is treated as a fundamental loss in the system. Peak power $P_{max}$ is less than the product $I_{sc}V_{oc}$ owing to the exponential form of the characteristic (7.23). The curve factor (also called the fill factor) is $F_c = P_{max}/(I_{sc}V_{oc})$. The maximum value in Si is 0.88.

### 7.5.9 *Additional curve factor A (loss ~5%)*

In practice the cell characteristic does not follow (7.23) and is better represented by

$$I = I_0[\exp(eV/AkT) - 1] - I_L \qquad (7.30)$$

The factor $A$ (>2 for many commercial cells) results from increased recombination in the junction. This effect also tends to change $V_{oc}$ and $I_0$, so in general optimum output would occur if $A = 1$.

Recombination has already been mentioned for back surface field (Section 7.5.6). Within the cell, recombination is lessened if:

1 Diffusion paths are long (in Si > 50 to 100 μm). This requires long minority carrier lifetimes (in Si up to 100 μs).
2 The junction is near the top surface (within 0.15 μm, as in 'violet' cells, rather than 0.35 μm as in normal Si cells).
3 The material has few defects other than the dopant.

Surface recombination effects are influential owing to defects and imperfections introduced at crystal slicing or at material deposition.

### 7.5.10 Series resistance (loss ~0.3%)

The solar cell current has to pass through the bulk material to the contact leads. At the rear the contact can cover the whole cell and the contribution to series, electrical 'ohmic', resistance is very small. The top surface, however, must be exposed to the maximum amount of light. It should therefore be covered by the minimum area of contacts. This, however, produces relatively long current path lengths with noticeable series resistance. Significant improvements have now been made in forming these contacts, for instance forming laser-cut channels into which thin contacts can be formed, and arranging the surface layout to minimise the top surface series resistance to ~0.1 Ω in a cell resistance of ~20 Ω at peak power.

### 7.5.11 Shunt resistance (negligible, ~0.1%)

Shunt resistance is caused by structural defects across and at the edge of the cell. Present technology has reduced these to a negligible effect, so shunt resistance may be considered infinite in single crystal Si cells. This may not be so in polycrystalline cells, however.

### 7.5.12 Delivered power (Si cell 10 to 14%)

Table 7.2 shows the remaining power, after the losses, as the delivered power. This assumes optimum load matching at full insolation, without overheating, to produce peak power on the $I$–$V$ characteristic. Table 7.2 (right hand side) also shows corresponding data of one of the highest-efficiency Si single crystal cells to date, with efficiency ~25%. Note that the spectrum of incident solar radiation sets an absolute limit for the efficiency (Sections 7.4, 7.5.3, 7.5.4). For Si cells this limit is ~47%.

## 7.6 Solar cell construction

We shall describe briefly the construction of a standard single-crystal Si photovoltaic cell. There are many variations, and commercial competition produces continued revision of cell type and fabrication method. A general design is shown in Figure 7.1.

### 7.6.1 *General design criteria*

1. Initial materials have to be of excellent chemical purity with consistent properties.
2. The cells must be mass-produced with the minimum cost, but total control of the processes and high levels of precision must be maintained.
3. The final products are usually guaranteed for at least 20 years, despite exposure in hostile environments and significant changes of temperature. For even without concentration of the insolation, the cell temperature may range between −30 and +100 °C. Electrical contacts must be maintained and all forms of corrosion avoided. In particular water must not be able to enter the module.
4. The design must allow for some faults to occur without failure of the complete system. Thus redundant electrical contacts are useful and modules are connected in parallel strings. The parallel and series connections between the cells must allow for some cells to become faulty without causing an avalanche of further faults.
5. The completed modules have to be safely transported, often to inaccessible and remote areas.

### 7.6.2 *Crystal growth*

Very pure, electronic grade, base material is obtained in polycrystalline ingots. Impurities should be less than 1 atom in $10^9$, i.e. less than $10^{18}$, atoms per $m^3$. This starter material has to be made into large single crystals.

1. *Czochralski technique*. This well-established crystal growing technique consists of dipping a small seed crystal into molten material, Figure 7.17(a). Dopant (e.g. boron acceptors for p-type) is added to the melt. Slowly the crystal is mechanically pulled upwards with a large (~15 cm diameter) crystal growing from the seed. As with other single crystal growth techniques, the crystal is then sliced ~300 μm thick with highly accurate diamond saws. About 40–50% of crystalline material may be lost during this process, which represents a most serious loss.
2. *Zone refining*. Polycrystalline material is formed as a rod. A molten zone is passed along the rod by heating with a radio frequency coil or

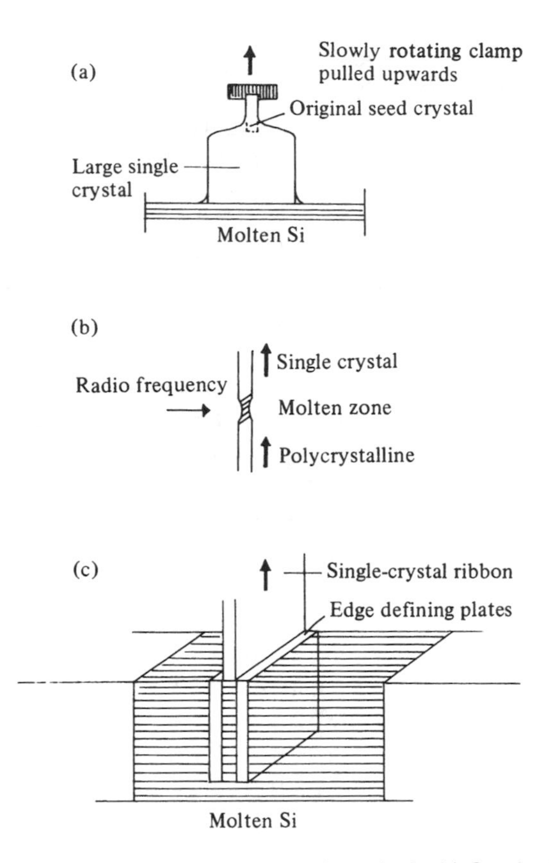


*Figure 7.17* Some crystal growth methods. (a) Czochralski. (b) Zone recrystallization or laser heating. (c) Ribbon.

with lasers (Figure 7.17(b)). This process both purifies the material and forms a single crystal. The single crystal has to be sliced and treated as for other techniques.

3 *Ribbon growth*. This method avoids slicing and the consequent wastes by growing a continuous thin strip of single crystal up to 10 cm wide and 300 μm thick, as shown in Figure 7.17(c). The ribbon may be stored on large diameter rolls, and then cut for surface treatment to form cells.
4 *Vacuum deposition*. This technique may be used at different stages of construction, e.g. for the top metal layer of a Schottky diode. Vacuum deposition of Si is difficult and not usually successful.
5 *Casting*. Polycrystalline material is formed. The cheapness of the process may compensate for the lower efficiency cells produced.

### 7.6.3 *Slice treatment*

The 200–400 μm thick slices are then chemically etched. A very thin layer of n-type material is formed by diffusion of donors (e.g. phosphorus) for the top surface. One method is to heat the slices to 1000 °C in a vacuum chamber, into which is passed $P_2O_5$, but more often the slices are heated in nitrogen with the addition of $POCl_3$.

Photolithographic methods may be used to form the grid of electrical contacts. First, Ti may be deposited to form a low resistance contact with the Si; then a very thin Pd layer to prevent chemical reaction of Ti with Ag; and then the final Ag deposit for the current carrying grid. Other methods depend on screen printing and electroplating.

The important antireflection layers are afterwards carefully deposited by vacuum techniques; however, the similar properties of textured surfaces are produced merely by chemical etching. The rear surface may be diffused with aluminium to make a BSF of $p^+$ on p (see Section 7.5.6). Onto this is laid the rear electrical metal contact as a relatively thick overall layer.

### 7.6.4 *Modules and arrays*

The individual cells, of size ~ (10 cm × 10 cm), are then connected and fitted into modules. Traditionally, most modules had about 36 cells in series to provide an overvoltage to charge nominally 12 V batteries. But many later types of module have greater numbers of cells in series for larger voltages more compatible with efficient inverters for AC grid connected systems. The cells are sandwiched in an inert filler between a clear front cover, usually ultraviolet resistant plastic, and a backing plate. The cover sealing must be watertight under all conditions including thermal stress. The rear plate must be strong and yet present a small thermal resistance for cooling.

## 7.7 Types and adaptations of photovoltaics

Although the flat plate Si solar cell has been the dominant commercial product, there is a great variety of alternative types and constructions. These seek to improve efficiency and/or to decrease the cost of the power produced by reducing capital cost. This section is a brief summary of a complex and continually changing scene (Table 7.3).

### 7.7.1 *Variations in Si material*

1 *Single crystal.* The cells described so far assume single-crystal base material. This can be produced by most of the methods of Section 7.6.2. In particular, offcuts of best-grade Si material (usually Czochralski)

have been available relatively cheaply from the microelectronics industry, which has encouraged their use in solar cells. However, the ever increasing growth of pv manufacture means that this resource is not in itself sufficient.

2 *Polycrystalline.* Considerable savings in production costs can be avoided by using polycrystalline material (also called multicrystalline material). This is not necessarily structurally weak, but the presence of boundaries between the crystal grains increases recombination of electron–hole pairs. Consequently polycrystalline solar cells have smaller efficiencies than single crystal material. However, R&D diminished the gap in efficiency so half the total commercial production of solar cells were polycrystalline Si by 2004. Polycrystalline cells may be made by less costly thin-film techniques, described later in Section 7.7.5(1), which may, however, alter the solid state properties significantly.

3 *Amorphous.* Amorphous materials are solids that have only a short range of order (glass has an amorphous structure). Materials that are considered semiconductors (e.g. Si) may under particular conditions remain with semiconductor properties in the amorphous state. Electrical resistivity may be similar, and in particular n- and p-type dopants may be added to produce similar effects to that in crystalline material. Normally, however, the amorphous structure produces a very large proportion of unsatisfied 'dangling' chemical bonds that readily act as electron or hole traps in an uncontrollable manner. However, if a high proportion of hydrogen is present as the material is formed, the number of such bonds is dramatically reduced. Amorphous Si (a-Si) can be used in thin-film solar cells (see below), in which the total thickness of semiconductor is only about $1\,\mu m$ (i.e. $\sim 1/100$ of the thickness of a single-crystal cell). Development of amorphous cells, with multiple junctions, within that $1\,\mu m$ has increased efficiency to about 10% and allows reduced cost installations. A practical difficulty is reduced efficiency with age, especially in the first few years of operation. An advantage is that the output of a-Si cells does not diminish as temperature increases.

### 7.7.2 *Variations in junction type*

4 *Homojunctions.* If the base semiconductor material remains the same across the p–n junction, and only changes in type or concentration of dopant, it is a homojunction. The Si cells so far discussed are homojunctions. The band gap is constant across the junction, Figure 7.18(a).

5 *Heterojunctions.* The band gap changes across the junction because of a significant change in base material, Figure 7.18(b). The advantage is

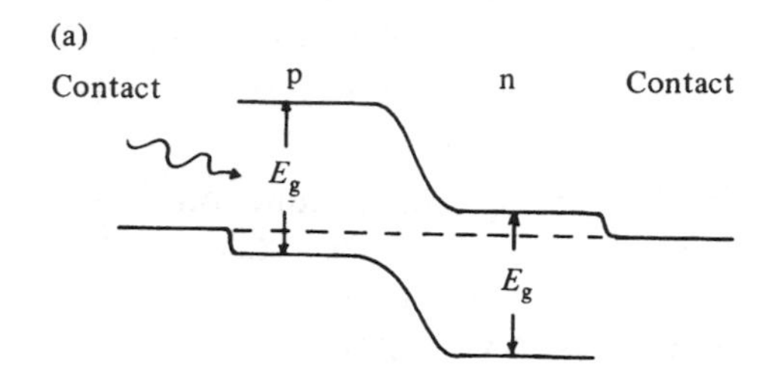


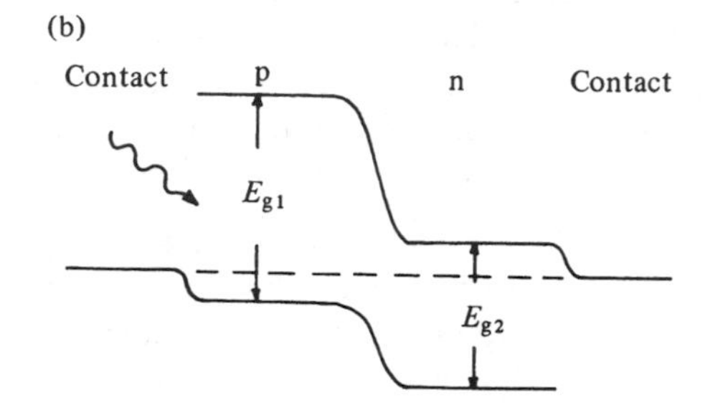


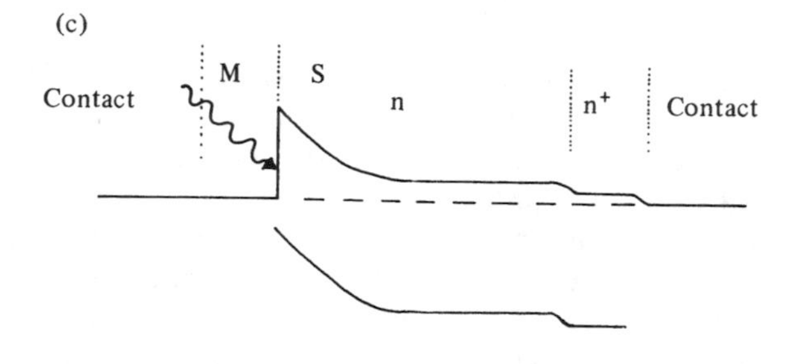


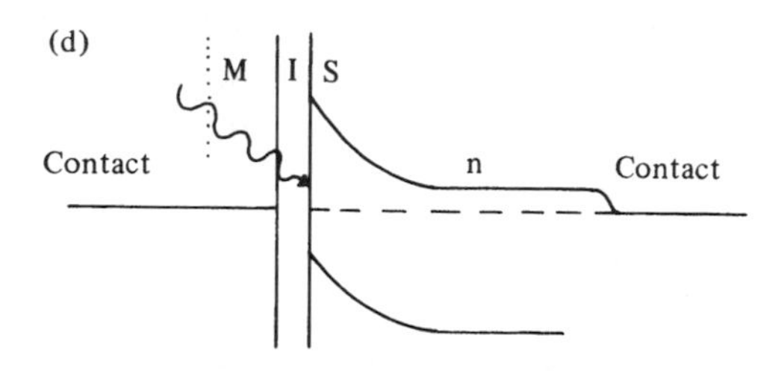


*Figure 7.18* Solar cell junction types. (a) Homojunction: base material and band gap constant across junction. (b) Heterojunction: base material and band gap change across junction. (c) Schottky metal semiconductor (MS) cell, e.g. Au/Si, shown here with $n/n^+$ back surface field. The important antireflection layer is not shown. (d) Schottky metal insulator semiconductor (MIS) cell.

that band gap photon absorption is now possible at two frequencies. This increases the total proportion of photons that may be absorbed, and so decreases the excess photon energy loss ($h\nu - E_g$). Normally the wider band gap material is on the top surface, so the less energetic photons cross to the narrower band gap material. The lattice structure on each side of the heterojunction has to be compatible for continuous growth of the complete cell. Examples are $Ga_{1-x}Al_xAs$ on GaAs (operational efficiency 12%), and $SnO_2$ on n-type Si (operational efficiency 10%).

A continuously decreasing band gap (the graded band gap cell) has been proposed. Manufacture is difficult yet possible (e.g. with $Ga_{1-x}Al_x$ material), but $V_B$ would be small.

6 *Direct and indirect band gap.* Consider the Brillouin energy ($E$), momentum ($k$) diagrams of Figure 7.19. We wish photons to be able to cause electron–hole transitions across the minimum band gap. With direct band gap materials, the transition can be caused by photons alone, with $h\nu = E_g$. However, with indirect band gap materials, the minimum gap can only be crossed with low energy photons if the absorption coincides with phonon absorption. Thus if the phonon energy is $h\Omega$, the condition for absorption at the minimum between the bands is

$$h\nu \pm h\Omega = E_g \tag{7.31}$$

As explained in section 7.3, indirect band gap material (e.g. Si) has a smaller extinction coefficient than direct band gap material (e.g. GaAs), so requiring thicker cells.

7 *Schottky barrier, MS and MIS.* A p–n junction may be formed at a metal–semiconductor (MS) interface as shown in Figure 7.18(c). The

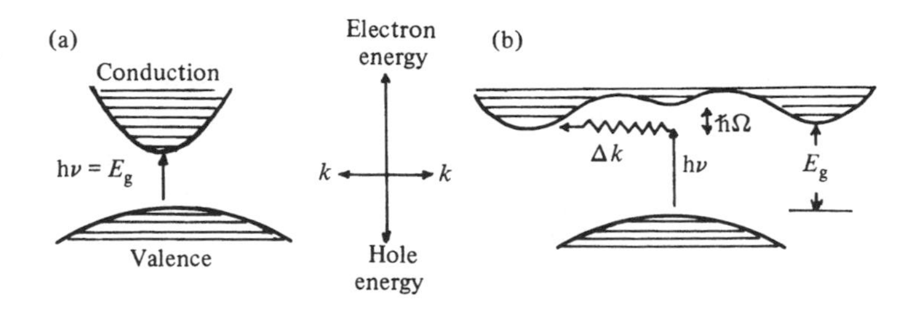


*Figure 7.19* Brillouin zone, energy band/lattice vibration momentum diagrams. Absorption near the band gap requires the creation of a lattice phonon (energy $\pm h\Omega$) with indirect band gap semiconductors. (a) Direct band gap e.g. GaAs. (b) Indirect band gap e.g. Si. See device texts for further explanation.

advantage is simplicity of construction, since the metal can be deposited as a thin film on the one base material. The disadvantages are the increased reflectance of the metal (and hence loss of input) and the increased recombination losses at the junctions. In fabrication it is difficult to avoid a thin insulating layer of oxide forming between the metal and the semiconductor (MIS cell). Control of this insulating layer can, however, lead to improved cells, Figure 7.18(d), through suppression of surface recombination.

### 7.7.3 Variations in semiconductor mechanisms

The following have been the subject of laboratory experiments or theoretical investigation, but by 2004 had not yet progressed significantly.

8 *Liquid interface.* The top surface of a solar cell may be a liquid electrolyte. The advantages are potentially good electrical contact, and the possibility of chemical change in the liquid for energy storage. The disadvantages are the operational difficulties, generally low efficiency and easy contamination.

9 *Organic material.* Carbon-based organic materials may be semiconductors (carbon is in the same periodic group as silicon but not itself a semiconductor). The extensive knowledge of organic chemistry and the possible cheapness of organic materials make developments in this area of great interest. Efficiencies of 1% have been achieved, which should be improved to about 10% with present research.

10 *Intermediate transitions (phosphors).* In principle, the front surface of a photovoltaic cell could be coated with a fluorescent or phosphorescent layer to absorb photons of energy significantly greater than the band gap ($h\nu_1 \gg E_g$). The emitted photons would have to be actively absorbed ($h\nu_2 \geqslant E_g$). Thus the excess energy of the original photons ($h\nu_1 - h\nu_2$) would be dissipated in the surface, hopefully with less temperature increase of the cell. Other similar ideas have been put forward for either releasing two active photons from each original photon or absorbing two inactive photons ($h\nu < E_g$) to produce one active photon in a manner reminiscent of photosynthesis. Commercial exploitation of these ideas has not been tried.

### 7.7.4 Other substrate materials

There are a great number of multi-element materials having semiconductor junction properties for use in solar cells. Specialist books, journals and websites should be used to search for these. Table 7.3 lists important parameters for a small range of solar cell materials, and gives some typical properties of p–n junction cells made with the material as bottom layer. Si is commercially and historically the most important. However, as an indirect-gap semiconductor, crystalline Si has a relatively small extinction coefficient (Table 7.1), so that at least ten times as much crystalline Si is required to absorb a given fraction of sunlight compared to other semiconductors such as GaAs or even amorphous Si. Therefore it needs only a much thinner layer of these other semiconductor materials to make a solar cell, with potentially much smaller cost of manufacture. The materials that have been most intensively developed for use in *thin-film solar* cells include amorphous Si (Section 7.7.1), CdTe and Cu(InGa)$Se_2$; many more variations are under investigation, particularly based on mixed elements of groups III and V of the periodic table like GaAs. For none of them are the solid-state properties as well understood as for Si.

1 *Copper indium gallium selenide, Cu(InGa)*$Se_2$ *(CIGS)*. Thin-film cells based on this alloy have achieved close to 20% efficiency in the laboratory. Earlier development was based on CIS (no gallium), but its performance was limited by its low band gap of 1.0 eV. One way to form such a polycrystalline thin cell is by simultaneous evaporation of Cu, Ga, In and Se onto a neutral substrate such as Mo-coated glass. The alloy film is p-type, with the p–n junction formed by depositing an n-type layer of CdS , ZnO or other suitable and stable material.
2 *Cadmium telluride (CdTe)*. CdTe is a direct band gap semiconductor with $E_g = 1.5$ eV, which is near the optimum band gap for a solar cell in AM1 insolation (see Figure 7.13). It can be deposited in thin polycrystalline films by electrodeposition or other means, and a heterojunction formed with CdS. Efficiencies of 16% have been reported, but performance of CdTe cells is sensitive in ways not yet fully understood to the precise conditions of manufacture, with some cells degrading badly over time, though others do not.
3 *Gallium arsenide (GaAs)*. Heterojunctions with $Ga_{1-x}Al_xAs$ can be made commercially. Theoretical target efficiencies for cells are high at about 25%, and GaAs devices have reached practical efficiencies of 16%. The high extinction coefficient necessitates accurate control of layer depths, and surface recombination can be high.

### 7.7.5 Variations in cell construction

1 *Thin-film cells*. The advantages of thin-film cells, and the most promising materials for them, have been explained above. In principle many types of cells could be manufactured by thin-film techniques, but the

difficulties of forming efficient cells seem very great. However, technical improvements should occur in the near future to make such cells commercially viable.

2 *Stacked cells*. These are a series of physically separated p–n junctions of decreasing band gap, with light incident on the largest band gap material. Photons of energy less than the first gap ($h\nu < E_{g1}$) are transmitted to the next junction, and so on. Efficiencies ~35% have been obtained in the laboratory from triple-junction cells, especially under concentrated radiation – i.e. considerably higher than any single-junction cell can obtain. Further improvement is anticipated.

3 *Vertical multijunction cells (VMJ)*

   a *Series linked* A series of perhaps hundred similar p–n junctions are made in a pile (Figure 7.20(a)). Light is incident through the edges of the junctions, so the output potential (~50 V) is the series sum of many junctions. The current is related only to the absorbed radiation flux per junction edge area, and so is not large.

   b Parallel linked arrangements are also possible. This is a form of *grating cell*, usually made with the aim of absorbing photons more efficiently in the region of the junction (Figure 7.20(b)).

4 *Reflecting or textured surfaces*. The top surface of the solar cells can be designed to pass reflected radiation back into the surface (see Figure 7.15). Some systems have to be made mechanically; others may be 'textured' by etching with chemicals that give pitted surface structure according to crystal axis orientation.

5 *PERL cell (Passivated Emitter, Rear Locally diffused)* (Figure 7.21). This type of cell, developed at the University of New South Wales, is one of the most efficient using crystalline Si, with an efficiency of 24% or more. Cells of this and similar structure have been made in semi-commercial quantities for specialised applications. Its intricate

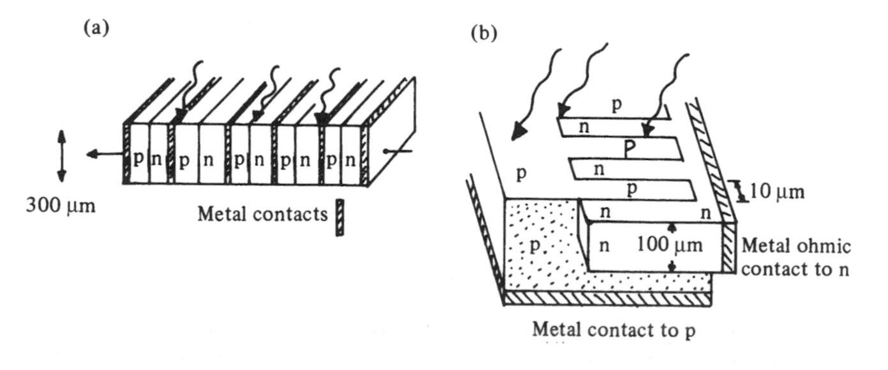


*Figure 7.20* Vertical multijunction cells (VMJ). (a) Series linked. (b) Parallel linked.

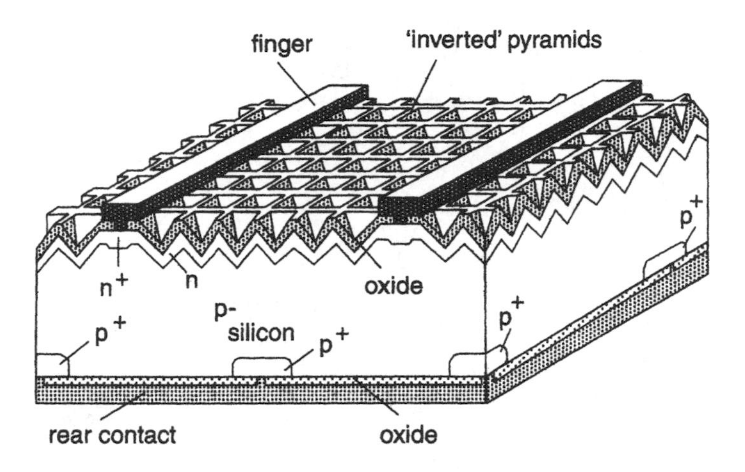


*Figure 7.21* An ultra-high efficiency crystalline Si cell (the PERL cell of UNSW). Reproduced with permission from Green (2001).

structure illustrates the complexity (and thus the cost) of achieving high efficiency. It features detailed attention to maximising the absorption of light by careful manufacture of a textured top surface in the form of inverted pyramids with width ~10 μm. The oxide layer at the rear reflects most of the weakly absorbed infrared light back into the cell, thus further increasing the absorption, as does an antireflection coating (see Section 7.5.2). In addition, the oxide layers at top and bottom 'passivate' the carriers, giving small recombination rates at the surfaces with minimal doping. Electrical contacts use the laser-grove technique.

6 *Sliver cells.* This novel manufacturing approach allows ~30 times more active area than usual from a Si wafer by micromachining the wafer (of thickness ~1 mm) perpendicular to its surface, thereby producing hundreds of strips ('slivers') of Si, each about 50 mm long, 1 mm wide and 60 μm thick. These slivers are treated (doped, etc.) while still held together by the rest of the wafer, so that each sliver becomes a cell at an oblique orientation with the active face being 50 mm × 1 mm (Blakers *et al.*, 2003).

### *7.7.6 Variation in system arrangement*

1 *Concentrators* (see Section 6.8). Since the active solar cell material is usually the most expensive component of an array, it is sensible strategy to concentrate the insolation onto the cell (Figure 7.22). Performance is not impaired in the increased radiation flux if the cell

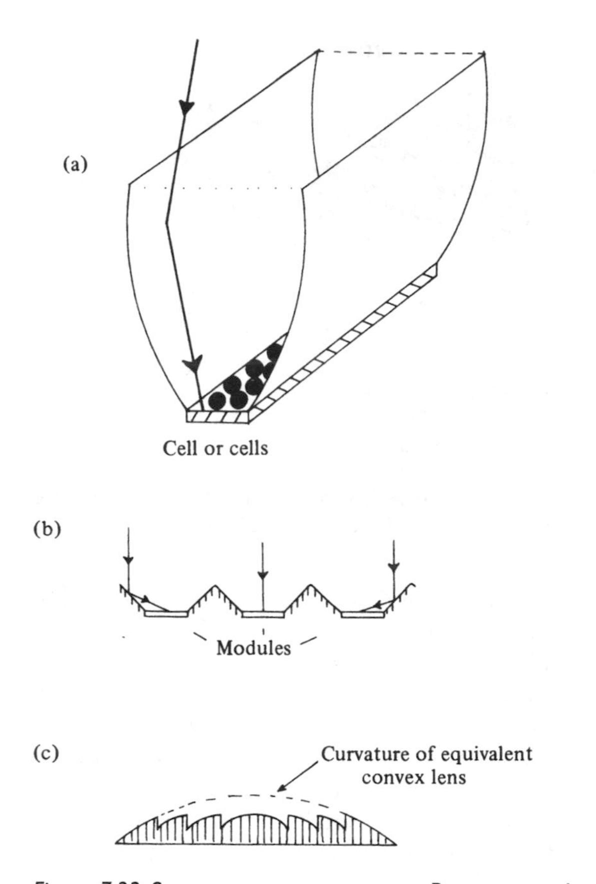


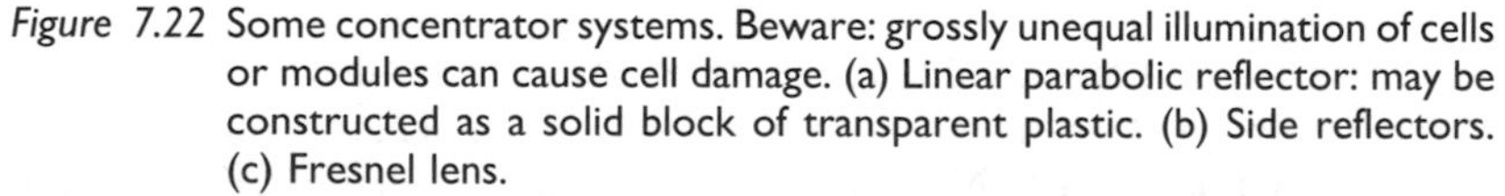


*Figure 7.22* Some concentrator systems. Beware: grossly unequal illumination of cells or modules can cause cell damage. (a) Linear parabolic reflector: may be constructed as a solid block of transparent plastic. (b) Side reflectors. (c) Fresnel lens.

temperature remains near ambient, by active or passive cooling. Indeed the cell efficiency increases with concentration. The heat removed in active cooling may be used to increase the total energy efficiency of the system.

The *concentration ratio* $X$ of a perfectly focused system is the ratio of the concentrator input aperture to the surface area of the cell. In practice, energy concentration reaches 80–90% of this geometric factor. Systems with small values of $X (X \leq 5)$ do not need to be oriented through the day to follow the Sun and so make use of some diffuse as well as direct radiation. However, systems with larger values of $X$ should track the Sun, and are only sensible in places with a large proportion (>70%) of direct radiation.

There is a wide variety of concentrator systems based on lenses (usually Fresnel flat plane lenses), mirrors and various novelties such as prism internal reflection (Figure 7.22). (see also Section 6.8.) Multi-junction cells under concentration have achieved the highest efficiencies of any solar cells: up to 35% in laboratory conditions.

2 *Spectral splitting*. Separate solar cells with increasing band gap could be laid along a solar spectrum (say from a prism, and ranging from infrared to ultraviolet) to obtain excellent frequency matching. The dominant losses (totalling ~50%) from the mismatch of photon energy and band gap could therefore be greatly decreased. Final efficiencies of about 60% might be possible. Economically this would only be possible if combined with a high concentration ratio before dispersion. Systems of this principle using three dichroic mirrors have been made.
3 *Thermophotovoltaic*. Solar radiation may be highly concentrated on an absorbing surface that then re-radiates according to its temperature. Effectively the peak frequency of the radiation is shifted into the infrared to obtain a better match with a low band gap photovoltaic cell. Efficiencies of 40% have been claimed in laboratory systems. The same cells are able to generate electricity from the otherwise wasted infrared heat radiation from surfaces in industrial processes, so enabling a form of energy recovery.
4 *Quantum-dot devices*. Luminescence occurs when high energy photons are absorbed, leading to the emission of another photon with less energy at longer wavelength. By containing the luminescent material in a thin glass 'tank', most of the emitted photons may be internally reflected onto an end wall covered by a photovoltaic cell. The system therefore becomes a photovoltaic concentrator, with the possibility of increased electrical output per unit area of collector and of cheaper cost per unit of electrical energy produced.
5 *Dye sensitive cells*. A new form of solar cell under active development resembles photosynthesis in its operation. Rather than the sunlight being absorbed in a semiconductor, the cell absorbs light in dye molecules containing ruthenium ions. These dye molecules are coated onto the whole outside surfaces of nanocrystals of the wide-band gap semiconductor $TiO_2$, as in Figure 7.23. The mechanism of photon absorption and subsequent electron 'exciton' transfer to a 'processing centre' resembles the photosynthetic process, see Chapter 10 (e.g. see Figure 10.6). Light photon absorption through the sun-facing surface of transparent conductive oxide (TCO) excites electrons in the dye to an energy where they are injected into the conduction band of the adjacent n-type $TiO_2$ and thence to the front surface and the external circuit. The electron current passes through the external load to the back electrode, where it reduces tri-iodide to iodide, which then diffuses through

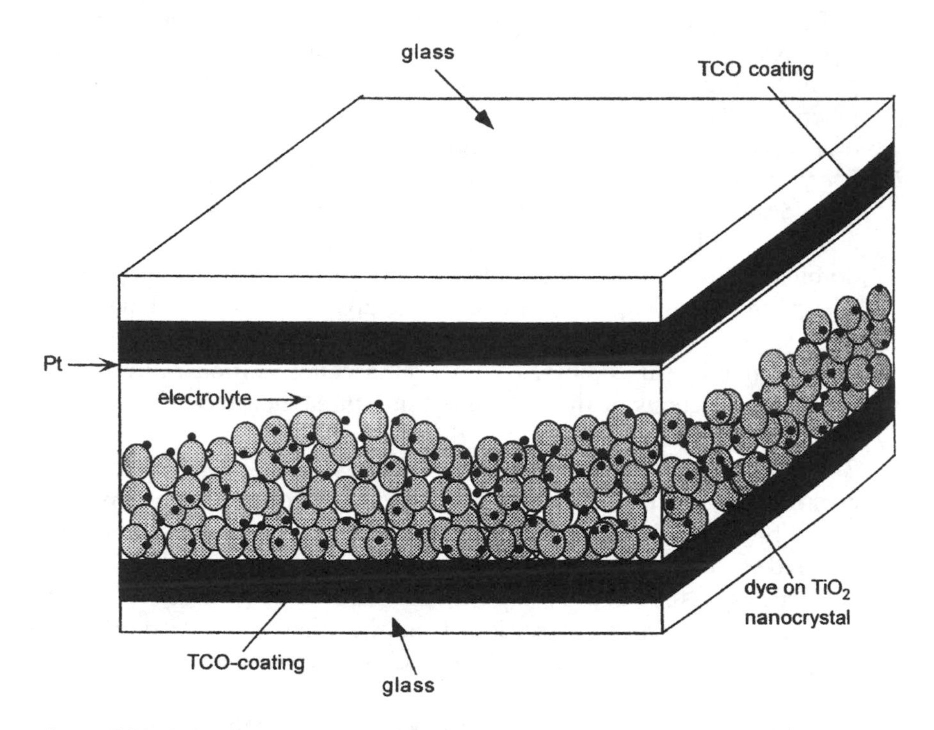


*Figure 7.23* A dye-sensitive solar cell. The dye covers the surfaces of the $TiO_2$ nanocrystals. TCO – transparent conductive oxide. Reproduced with permission from Green (2001).

the electrolyte to reduce the photo-oxidised dye molecules back to their original state. Efficiencies over 10% have been achieved in the laboratory. Technologies of this type, but using infrared absorbing dyes, have the potential to produce 'visually transparent' modules which would be of great commercial interest as electricity-generating windows in buildings. Processes like this based on liquids have the prospect of large-scale and relatively cheap mass production.

## 7.8 Photovoltaic circuit properties

With photovoltaic cells, as with all renewable energy devices, the environmental conditions provide a *current source* of energy. The equivalent circuit portrays the essential macroscopic characteristics for power generation (Figure 7.24).

From the equivalent circuit (Figure 7.24),

$$I = I_{\mathrm{L}} - I_{\mathrm{D}} - \frac{(V + IR_{\mathrm{s}})}{R_{\mathrm{sh}}} \tag{7.32}$$

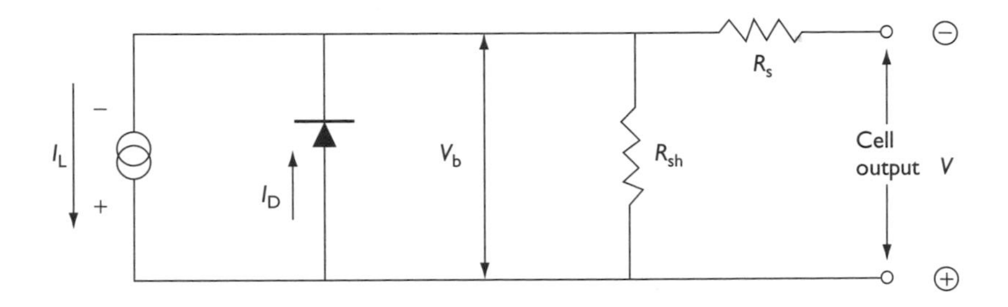


*Figure 7.24* Equivalent circuit of a solar cell.

In discussing photovoltaic cells as power generators in operation (i.e. when illuminated), it is usual to take the device current $I$ as positive when flowing from the positive terminal of the cell (i.e. the generator) through the external load. This is the convention with all DC generators, but is opposite to the convention used in Sections 7.2–7.4 where the physics was derived from a simple diode. Thus (7.32) has the reverse form of (7.22). Equation (7.32) also allows for the internal resistances $R_s$ and $R_{sh}$ shown in Figure 7.24.

Therefore the I–V characteristics of Figure 7.25 have $I$ as positive; this characteristic is the reflection about the voltage axis, of the lower right (power production) quadrant of Figure 7.11. (The characteristics of a single solar cell have the same shape but the voltages and currents are obviously smaller.) From 0 V (short circuit conditions) to about 10 V, the $I$–$V$ characteristics are dominated by the illumination current $I_L$, which increases in proportion to the irradiance. The sharp curvature at $V \sim 12\,V$ results from the curvature of the diode characteristics; see Figures 7.8 and 7.11. Following (7.30), but now including the voltage drop across the series resistance $R_s$:

$$I_D = I_0\{\exp[e(V + IR_s)/(AkT)] - 1\} \tag{7.33}$$

where $A$ is the *ideality factor*, introduced so the model fits the empirical characteristic. For all cells $A \geq 1$. Larger values of $A$ relate to $I$–$V$ characteristics that are more curved, so reducing maximum power; this effect is called the *additional curvature factor*. It results from increased electron–hole recombination at defects in the junction. Section 7.5 discusses these aspects more fully. For Si material, $I_0 \sim 10^{-7}\,A\,m^{-2}$.

In Figure 7.25, for a given illumination, the curve goes from $V = 0$ (short circuit, with current $I_{sc}$) to $V = V_{oc}$ (open circuit voltage, with $I = 0$). The open circuit voltage $V_{oc}$ increases only slightly with irradiance, unlike the short circuit current $I_{sc}$. Maximum power is transmitted to an external load $R_L$ when $R_L$ equals the internal resistance of the source $R_{int}$. However, $R_{int}$ is regulated by the absorbed photon flux, so good power

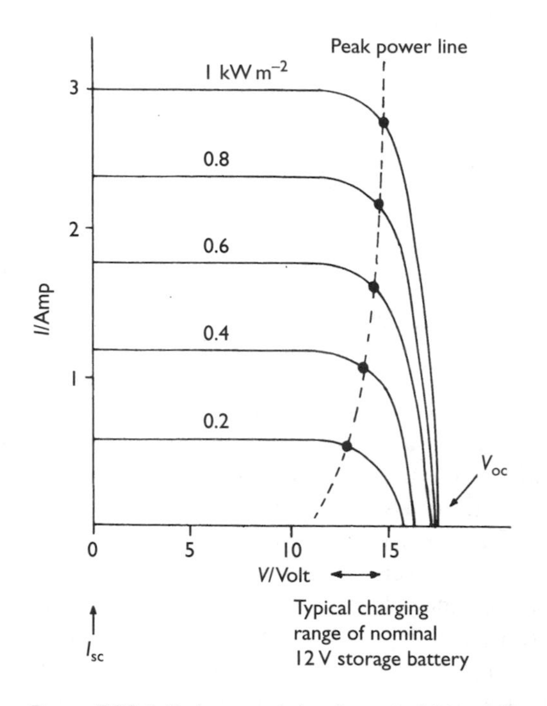


*Figure 7.25* *I–V* characteristic of a typical 36-cell Si module used for battery charging or other loads for various irradiances. Note that even without maximum power load control, the peak power line, of maximum *IV* product, is a good match with the charging voltage range of the nominally 12 V battery.

matching in a solar cell requires $R_L$ to change in relationship to the solar irradiance.

An increase in cell material temperature $\theta$ affects performance by decreasing $V_{oc}$ and increasing $I_{sc}$, with the characteristic changing accordingly. Effectively $R_{sh}$ (often taken as infinitely large) and $R_s$ (made as small as possible) decrease with increase in temperature. Empirical relationships for these effects at $1\,\mathrm{kW\,m^{-2}}$ insolation on Si material are

$$V_{oc}(\theta) = V_{oc}(\theta_1)[1 - a(\theta - \theta_1)] \tag{7.34}$$

$$I_{sc}(\theta) = I_{sc}(\theta_1)[1 + b(\theta - \theta_1)] \tag{7.35}$$

where $\theta_1 = 25\,°\mathrm{C}$ is a convenient reference temperature, $\theta$ is the material temperature, and the temperature coefficients are

$$a = 3.7 \times 10^{-3}(°\mathrm{C})^{-1}, \quad b = 6.4 \times 10^{-4}(°\mathrm{C})^{-1}$$

The net effect of an increase in temperature is to reduce the power $P$ available. An empirical relationship for crystalline Si material is

$$P(\theta) = P(\theta_1)[1 - c(\theta - \theta_1)] \tag{7.36}$$

where $c = 4 \times 10^{-3} (^\circ\text{C})^{-1}$. Thus a module operating at 65 °C (quite possible in a sunny desert environment) loses about 16% of its nominal power. (Amorphous Si is one of the few solar cell materials not to show such a decrease.)

The remaining requirements for good power production are obvious from the equivalent circuit, namely:

1. $I_L$ should be a maximum: for example through minimum photon losses, absorption near the depletion layer, a low surface reflectance, a small top surface electrical contact area, a high dopant concentration and few recombination centres.
2. $I_D$ should be a minimum: for example through high dopant concentration.
3. $R_{sh}$ should be large, such as by careful edge construction.
4. $R_s$ should be small, for example by ensuring a short path for surface currents to electrical contacts, and by using low resistance contacts and leads.
5. $R_{load} = R_{internal} = V/I$ for optimum power matching.

*Solar cell arrays* are often assembled from a combination of individual modules usually connected in series and parallel. Each module is itself a combination of cells in series. Each cell is a set of surface elements connected in parallel (Figure 7.26). Maximum open-circuit potential of common modules is ~22 V, and maximum short-circuit current at the module terminals is ~5.5 A in standard conditions. Such modules were developed originally for charging '12 V' batteries. Larger modules may be more cost-effective, e.g. 72 cell modules for 100–160 W at about 32 V in full sunshine, open circuit.

Since the cells are in series, difficulties will arise if one cell or element of a cell becomes faulty, or if the array is unequally illuminated by shading or by unequal concentration of light. Remember, a cell that is not illuminated properly behaves as a rectifying diode (see Figures 7.8 and 7.11). Therefore, current generated in a properly illuminated PV cell tries to pass to the next shaded cell in the direction that is now blocked, since the shaded cell behaves as a diode 'in the dark'. Thus when cells are connected in series and if only one is shaded, little current passes. (The analogy is stepping on a water hose pipe.) Consequently, shadows should never be allowed to fall on pv modules. If shading is unavoidable then the connected strings of modules should be so arranged that, if possible, each string either remains wholly

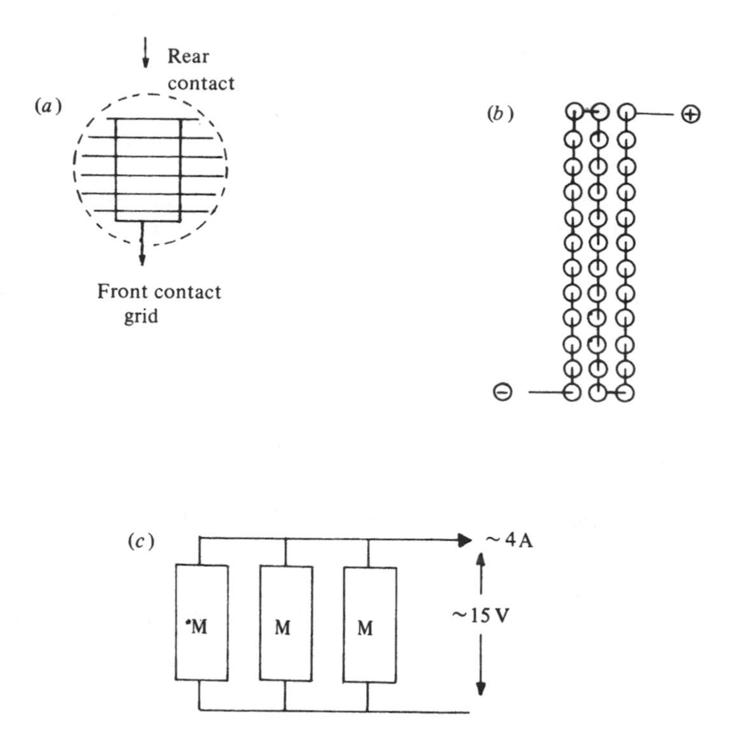


*Figure 7.26* Typical arrangements of commercial Si solar cells. (a) Cell. (b) Module of 36 cells. (c) Array.

in sunshine or is shaded. Moreover, it is possible that a shaded or faulty cell becomes overheated. Such faults may, in principle, avalanche unless protective bypass diodes are set in parallel with each series-linked cell. In practice, such protection is not installed for each cell within a module, but whole modules or lines of modules will be so protected.

Commercial solar modules with proven encapsulation give trouble-free service so long as elementary abuse is avoided. Lifetimes of at least twenty years are guaranteed, with expectation of even longer successful operation.

## 7.9 Applications and systems

### 7.9.1 Stand-alone applications

Photovoltaic modules are very reliable, have no moving parts and require no maintenance or external inputs such as fuel but only a flux of solar energy. Their first significant uses were therefore in applications where a small quantity of electric power was essential but where it was difficult or expensive to bring in fuel for conventional generators. Such applications include space satellites and autonomous, stand-alone systems in remote areas such as meteorological measurement, marine warning lights and telecommunication repeater stations; see Figure 7.27 (the operating cost of a boat to bring top-up fuel to a remote lighthouse every few months was far more than the capital cost of the photovoltaic system, even in the 1980s!). As in Figure 7.28, most stand-alone PV systems use a battery to store and help regulate the power. Storage is needed, since the solar input does not coincide with use, e.g. for lighting at night or for peak power when signals are transmitted. Regulation is needed, usually with the addition of electronic controllers, since otherwise there would be no voltage reference. Stand-alone applications are commonly automatic, being periodically maintained by trained staff and not requiring unskilled people.

Significant application from the 1970s onwards has been by the affluent for lighting and communications in holiday chalets, 'mobile homes' (or 'recreational vehicles'), yachts, etc. Also significant has been PV power for pocket calculators and 'gadgets' of many kinds. Such 'markets for the wealthy' have been alongside remarkable growth in PV stand-alone power for homes ('solar homes'), villages and medical and educational facilities in developing countries, especially for lighting, water pumping, radio and

*Figure 7.27* A typical stand-alone application of photovoltaics: powering a railway signal box in a remote area of Australia. [Photo by courtesy of BP Solar.]

television, where success has depended as much on social and institutional factors (Section 7.10) as on the technology.

As the cost of PV systems has fallen, so has the distance from the electric grid at which the installations are cost competitive. For example, it is now often cheaper, and always safer, to install car-parking meters or lighting for footpaths as stand-alone, solar-powered systems than install a connection and metering from the grid for the small amounts of power required. Moreover, the latest electronic devices, including lighting, always tend to use less power than their predecessors, so PV power is even more likely to be used.

### 7.9.2 *Balance of system components*

Figure 7.28(a) shows schematically how the array of Figure 7.26 can be connected to a DC load in a stand-alone system. For simplicity, the system is shown as operating at a nominal 12 V, but operation of 24 V DC appliances is readily achieved by reconfiguring the modules. The array has an equivalent circuit, as Figure 7.24, and I–V characteristics as Figure 7.25, but with numerical values appropriately scaled up. Examples of a typical appliance load corresponding to the applications described above are fluorescent light(s), a radio transceiver, a charger for portable (NiCd) rechargeable batteries, a small refrigerator or a water pump. All are available commercially in 12 and 24 V DC versions; many are mass-produced for the 'recreational vehicle' market.

Maximum power is obtained by controlling $V$ and $I$ to lie on the *maximum power line*, as the received insolation and load resistance vary (see Figure 7.25). This is nearly at constant voltage, within 25% of $V_{oc}$, even with the cells at constant temperature (the condition for which IV curves

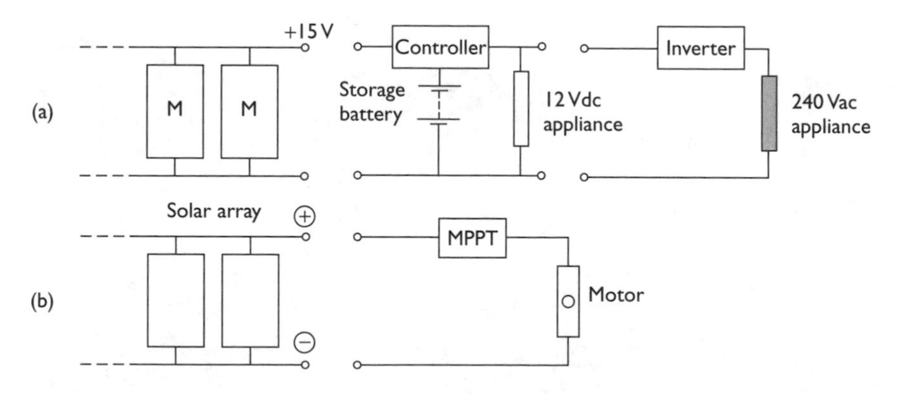


*Figure 7.28* Schematic diagram of a stand-alone photovoltaic system. (a) 12 V DC system with battery charge controller, with possible 240 V (or 110 V) AC appliances. (b) System with maximum power tracker (MPPT).

are usually drawn by manufacturers). In practice, the operating temperature usually rises with irradiance, which reduces the output voltage according to (7.34), so that the peak power line is more nearly vertical than indicated in Figure 7.25. The terminal voltage of an electrical storage battery remains nearly constant whatever the charging current, but increases with increase in state of charge. Therefore the $V/I$ load line for normal battery charging can be matched near to the maximum power line.

This is in marked contrast to a purely resistive load or a DC motor where the terminal voltage varies directly as the current. If such a load is used without a battery (as it may be for a water pump), performance is much enhanced by a maximum power point tracker (MPPT) as illustrated in Figure 7.28(b). This is essentially a DC-to-DC converter which keeps the input voltage close to the maximum power point as the irradiance varies. Such load management control devices are often built into solar pumping systems with names like 'maximiser' or 'linear current booster', and can enable 95% of the maximum output to be used gainfully for a load under varying solar conditions.

Virtually all applications of 12 V PV systems require battery storage of electricity, most obviously for lighting at night but also to cope with load surges such as radio transmission. Batteries are discussed in Section 16.5. The performance and reliability of a PV system is much better using a battery designed for that purpose than using a vehicle battery, whose characteristics are markedly different. A *charge controller* is essential to keep the battery within the limits of charge rate and depth of discharge suggested by the manufacturer. The controller is a relatively simple electronic circuit which switches either the battery or the load in and out of the circuit according to the voltage and current at the battery. Without this, the batteries can fail within days, especially if non-technical users are involved. Even with a controller, the battery lifetime is commonly 3–6 years, which is much shorter than the life of the modules and often shorter than the system designers imply. The charge controller is often incorporated with an MPPT in a single unit.

To operate an AC appliance (240 V/50 Hz or 110 V/60 Hz) in a PV system requires an *inverter*, as shown on the right of Figure 7.28(a). (This should be done only for appropriate appliances; it is not sensible to power an electric heater from a PV system!) A stand-alone inverter uses an internal frequency generator and switching circuitry to transform the low voltage DC power to higher voltage AC power. The shape of the AC waveform may be a square wave (cheap inverter) or an almost pure sine wave (sophisticated solid state electronic inverter). The inverter should be sized for the particular installation so that it can cope with the surge currents associated with motor-starting (if applicable), but not so large that it normally operates at a small fraction of its rated power (say $<15\%$) and therefore at poor efficiency ($<85\%$). Solid state electronic inverters are commercially

*Figure 7.29* Examples of grid-connected photovoltaic installations. (a) An office building at the University of Melbourne, Australia, with an envelope of PVs on its north-facing wall. This building also includes many other features of 'sustainable design' – see Section 6.3. [Photo courtesy of University of Melbourne.] (b) A service station in Australia with PVs on its roof. Note the juxtaposition of renewable and non-renewable energy sources. [Photo by courtesy of BP Solar.]

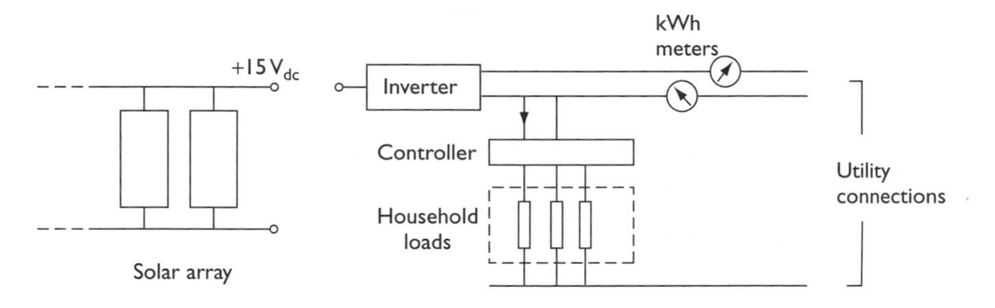


*Figure 7.30* Schematic of a grid-connected photovoltaic system (this is *not* a wiring diagram).

available with excellent reliability and efficiency (~95%), and at reasonable cost.

### 7.9.3 *Grid-connected systems*

Since 2000, distributed grid-connected systems such as those in Figure 7.29 have been the largest, and fastest-growing, use of photovoltaics. All such systems use an inverter to transform the DC electricity from the PV array into AC as used by the grid (see Section 16.9). Grid-connected inverters are different from stand-alone inverters; they use the prevailing line voltage frequency on the utility line as a control parameter to ensure that the PV system output is fully synchronised with the grid. Associated with this property, the PV systems have the necessary ability to disconnect when there are utility power failures, so no unexpected and potentially dangerous 'rogue' voltage appears on the grid line. An MPPT is usually incorporated with the inverter. Normally electricity can flow either out of the building from excess on-site PV power generation, or into the building with inadequate PV generation (Figure 7.30); both the exported and the imported energy can be metered, as indeed can the net energy exchange. The financial arrangements for imported and exported power and for abated-carbon credit vary widely by country and utility.

## 7.10 Social and environmental aspects

Photovoltaic power is a renewable energy technology that has really 'taken off' since the 1980s. Both the reduction in costs and the growth in use have been dramatic (see Figure 7.31). These two effects are closely linked, and are examples of 'learning curves' (compare Figure 17.1). At the high initial price only a few units can be sold, and producers have to recoup their set-up and R&D costs from them. They can then reduce the price, which stimulates new

(a)

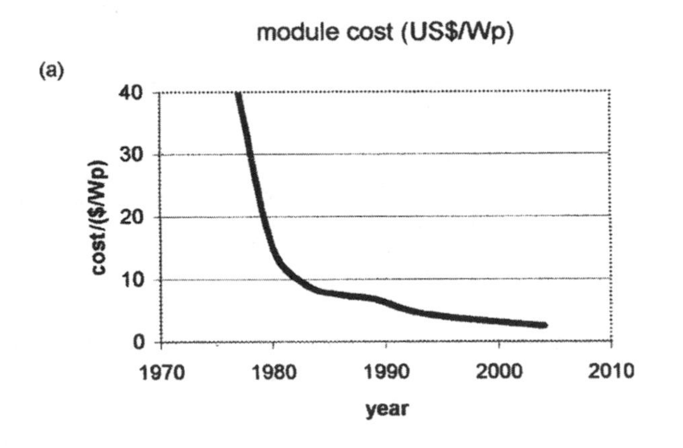


(b)

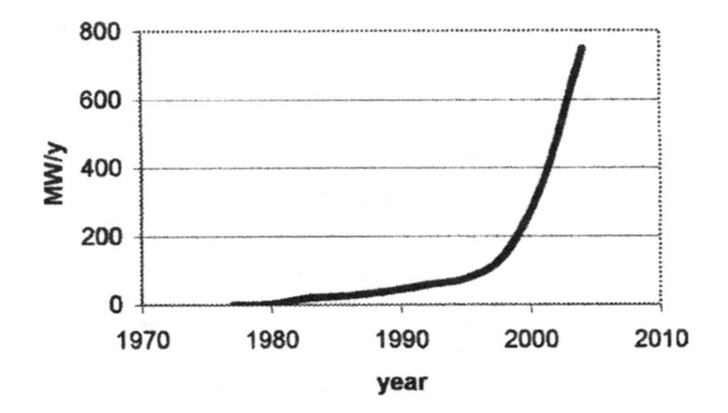


*Figure 7.31* Historical trends of (a) cost and (b) volume of production of solar cells. *Caution*: These costs are for PV modules, not for complete systems, which cost two to four times as much. [Data from Luque & Hegedus (eds) *Handbook of photovoltaic science and engineering* (Wiley) and Solarbuzz.com].

demand, particularly as the price falls below that of particular 'competing' technologies. Of particular importance has been the strong demand for PV module manufacture as supported by the policies of many countries, notably Japan and Germany. Nevertheless, even at these large rates of growth (averaging 30% p.a. over a decade), electricity production from photovoltaics in 2010 may be less than 1% of the then increased world demand.

The costs in Figure 7.31 are expressed as dollars per peak Watt ($/Wp). This is a standard measure of capital cost; it relates to the output of the panel under light of radiant flux density 1000 $W/m^2$ with a standard spectral distribution (corresponding to the Sun at 48 degrees from vertical, i.e. AM1.5) and with the panel temperature fixed at 25 °C. However, a fully illuminated panel rated at (say) 80 Wp may produce less than 80 W because either the irradiance is less than 1 kW $m^{-2}$ or the operating temperature is more than 25 °C (both of which are most likely). Moreover the capital cost per peak Watt of *output to the grid* from a grid-connected system is significantly more, perhaps double per peak Watt from the panel, because of the capital cost of the other components of the system (Figure 7.30) and the cost of installation. These 'other costs' are the 'balance of system costs'. If the system uses storage batteries, the cost per peak Watt available for use is at least four times that per peak Watt from the panel. Usually as important in practice as capital cost per peak Watt of a new system is the cost per kWh of electricity produced. At an unshaded fixed location in California, an efficient array rated at 1 kWp may produce 1800 kWh $y^{-1}$, yet in parts of the UK this output may require 3 kWp. For stand-alone applications, the most important measure is cost of *service delivered at a particular site* (e.g. comparing a PV-powered light of a certain light intensity with a kerosene-fuelled light of similar intensity). Thus there is a trade-off between system components, e.g. better energy-efficient appliances require fewer panels and fewer batteries, so usually the savings are greater than the extra cost of improved efficiency appliances.

The major component of the growth in demand for PV has been for grid-connected systems. This has been accelerated by government programmes, such as the solar rooftops programmes in Japan and Germany, which have induced a large market, especially for local manufacturers of PV systems. The sun-facing roof area of a majority of suburban houses in Europe, if covered in photovoltaics, can be expected to generate an amount of electricity in the year equal to 50–100% of the annual electricity demand of the household. Householders with PV use their PV-generated electricity in the daytime (selling any excess to the local utility) and buy electricity from the utility at night. The grid thus acts as their 'storage'. The government programmes referred to have made this a reasonable financial proposition for the householder by one or more of: (a) mandating that the utility must buy such electricity at a preferential price (feed laws), (b) subsidising the initial capital cost of the solar array and (c) establishing payments for carbon-abatement 'credits' obtained in proportion to the renewable energy generated. The modular nature of PV generation and the light weight of the modules make such distributed (embedded) generation relatively easy technologically. The economics and ease of construction are improved by the development of 'structural' PV panels that can be incorporated on building facades and roofs, in place of conventional wall and roof materials.

Substantial use and demand has continued in developing countries for programmes of rural electrification – a very important part of social and economic development. Most by number of the PV installations in developing countries have been very small in capacity (~40 W per 'solar home'). Initially, many such programmes failed, though rarely because of technical failure of the photovoltaic modules. Successful projects have been distinguished by their attention to institutional and social factors, such as provision of technical infrastructure, training, cultural understanding and design of the payment and institutional structure (see Section 17.2.2). An application of particular social importance has been in powering small refrigerators in rural health clinics, as part of the 'cold chain' for vaccines. Because it is linked to an existing (or emerging) technical infrastructure, this has usually been successful.

In operation, photovoltaics are environmentally benign, with no emissions and no noise. However, their manufacture involves some noxious chemicals and uses energy. The time for a given PV module to generate electricity equal in energy to that used in its manufacture depends on the site insolation and the method of manufacture. For a typical temperate climate, the 'simple' *energy payback* time for single-crystal silicon encapsulated modules is about three to four years (see Problem 7.5); for thin-film technologies and for sunnier locations, it is less. The guaranteed life of a module is typically ~20 years, but most modules will last much longer in practice, so that PV easily pays for itself in energy terms.

*Figure 7.32* Training local people to install and maintain photovoltaic systems is critical to their success. Photo shows proud householders with a small house-lighting system in Bhutan. [Photo by courtesy of BP Solar.]

## Problems

7.1 The band gap of GaAs is 1.4 eV. Calculate the optimum wavelength of light for photovoltaic generation in a GaAs solar cell.

7.2 a Give the equation for the I–V characteristic of a p–n junction diode in the dark.

b If the saturation current is $10^{-8}\,A\,m^{-2}$, calculate and draw the I–V characteristic as a graph to 0.2 V.

7.3 a What is the *approximate* photon flux density (photon $s^{-1}\,m^{-2}$) for AM1 solar radiation at $0.8\,kW\,m^{-2}$?

b AM1 insolation of $0.8\,kW\,m^{-2}$ is incident on a single Si solar cell of area $100\,cm^2$. Assume 10% of photons cause electron–hole separation across the junction leading to an external current. What is the short circuit current $I_{sc}$ of the cell? Sketch the I–V characteristic for the cell.

7.4 A small household-lighting system is powered from a nominally 8 V (i.e. four cells at 2 V) storage battery having a 30 Ah supply when charged. The lighting is used for 4.0 h each night at 3.0 A.

Design a suitable photovoltaic power system that will charge the battery from an arrangement of Si solar cells.

a How will you arrange the cells?

b How will the circuit be connected?

c How will you test the circuit and performance?

7.5 a Calculate the approximate time for a single-crystalline PV module to generate electricity equal in energy terms the primary energy used in its manufacture. Consider a typical Japanese climate with $1450\,kWh\,(m^2y)^{-1}$ of insolation, modules of 15% efficiency and 1.0 kWp peak power of modules. The total direct manufacturing and processing energy (includes refining, etc.) is $1350\,kWh/kW_p$, plus a further $1350\,kWh/kW_p$ of factory 'overhead energy', with 90% of both from electricity (generated thermally at 55% efficiency from fuels in combined cycle generation).

b Straightforward energy-payback of (a) assumes that electricity has the same 'worth' or 'value' as heat. Is this correct? What result would be obtained if PV cells and modules were manufactured entirely from, say, hydroelectricity? How can manufacture be more energy-efficient?

7.6 What is the best fixed orientation for power production from a photovoltaic module located at the South Pole?

7.7 a The band gap of intrinsic Si at 29 °C is 1.14 eV. Calculate the probability function $\exp(-E_g/(2kT))$ for electrons to cross the full band gap by thermal excitation.

c If the Fermi level in n-type Si is about 0.1 eV below the conduction band, calculate the probability function for electrons to be thermally excited into the conduction band.

Compare your answers for (a) and (b).

7.8 Einstein won the Nobel Physics Prize in 1905 for explaining the *photoelectric effect*, in which light incident on a surface can lead to the emission of an electron from that surface with energy

$$E = h\nu - \Phi$$

where $h\nu$ is the energy of a photon of light and $\Phi$ is a property of the surface.

a What are the main differences and similarities between the photoelectric effect and the photovoltaic effect?

b Discuss how, if at all, the photoelectric effect could be used to yield useful energy.

7.9 A Si photovoltaic module is rated at 50 W with insolation 1000 W m$^{-2}$, as for peak insolation on Earth. What would be its peak output on Mars? (*Data*: Mean distance of the Sun from Earth is $1.50 \times 10^{11}$ m; from Mars $2.28 \times 10^{11}$ m. You may also like to consider the implications of there being no significant atmosphere on Mars.)

## Bibliography

### *Comprehensive books*

Luque, A. and Hegedus, S. (eds) (2003) *Handbook of photovoltaic science and engineering*, Wiley, 1138 pp. {Comprehensive reference covering physics, construction, testing, systems, applications, economics and implications for rural development. Includes an excellent and readable overview by the editors of the state of the art in all the above, and in-depth chapters on each of the major device technologies.}

Lorenzo, E. (1994) *Solar Electricity – engineering of photovoltaic systems*, PROGENSA, Seila, Spain. In English, *Electicidad Solar* in Spanish. {Thorough text, with electrical characteristics and circuitry emphasised. Nevertheless covers solid state properties. Underpins practical implementation with necessary theory.}

Markvart, T. (ed.) (1994) *Solar Electricity*, UNESCO series, John Wiley, UK. {Chapters drafted by different experts; with end-of-chapter set problems. Includes scientific, commercial and institutional factors.}

Messenger, R. and Ventre, J. (1999) *Photovoltaic Systems Engineering*, CRC Press, Boca Raton, Florida. {Wide ranging and concisely written text, with basics and fundamentals, but also covers practical engineering aspects in a professional manner.}

Partan, L.D. (ed.) (1995) *Solar Cells and their Applications*, John Wiley. {An edited review, at advanced level, by experts in the science, manufacture, commerce, policy and futures of PV.}

### Principally device physics

Green, M.A. (1998) *Solar Cells: Operating Principles, Technology and System Application*, Prentice-Hall. Reprinted by the University of New South Wales, Australia. {A basic text from nearly first principles. Excellent text, with later revisions, by an outstanding researcher. See Wenham *et al.* for a companion applied-text.}

Green, M.A. (2001) Photovoltaic physics and devices, in Gordon, J.E. (ed.) *Solar Energy: The State of the Art*, James & James, London, pp. 291–355. {Concise and comprehensive review in an excellent general solar text.}

### Principally applications

Hankins, M. (1995) *Solar Electric Systems for Africa*, Commonwealth Science Council, London. {Straightforward, practical guide for technicians and users.}

Loos, G. and van Hemert, B. (eds) (1999) *Stand-alone Photovoltaic Applications: Lessons Learned*, International Energy Agency, Paris, and James & James, (London).

Liebenthal, A., Mathur, S. and Wade, H. (1994) *Solar Energy: Lessons from the Pacific Island Experience*, World Bank, Washington, DC. {Practical experiences of success and failure, especially with small household systems, covering both technical and institutional factors.}

Roberts, S. (1991) *Solar Electricity – A Practical Guide to Designing and Installing Small Photovoltaic Systems*, Prentice Hall, UK. {A thorough and general 'manual' for PV applications, especially stand-alone systems; e.g. includes battery, inverter and water pumping detail.}

Sick, F. and Erge, T. (eds) (1996) *Photovoltaics in Buildings: A Design Handbook for Architects and Engineers*, International Energy Agency, Paris, and James & James, (London). {Good on ancillary systems and case studies.}

Strong, S. (1987) *The Solar Electric House*. Rodale Press, Emmaus. {Practical application.}

Wenham, S.R., Green, M.A. and Watt, M.E. (~1998) *Applied Photovoltaics*, Centre for Photovoltaic, Devices and Systems, University of New South Wales, Australia. {Written for university and college students, but with basic theory.}

### Specific references

Blakers, A., Weber, K., Stocks, M., Babeaei, J., Everett, V., Kerr, M. and Verlinden, P. (2003) Sliver solar cells, in *Destination Renewables*, Australia and New Zealand Solar Energy Society, Melbourne.

Knapp, K. and Jester, T. (2001) Empirical investigation of the energy payback time for photovoltaic modules, *Solar Energy*, **71**, pp. 165–172.
Wilson, J.I.B. (1979) *Solar Energy*, Wykeham, London. {Introductory college level text, somewhat dated but still useful on fundamentals.}

### Journals and websites

*Progress in Photovoltaics*, bimonthly by Wiley, Chichester. {An important journal with world leaders in photovoltaics on the Board of Editors.}
www.solarbuzz.com Includes industry statistics and news, outlines of technologies and their status, etc.

Chapter 8

# Hydro-power

## 8.1 Introduction

The term hydro-power is usually restricted to the generation of shaft power from falling water. The power is then used for direct mechanical purposes or, more frequently, for generating electricity. Other sources of water power are waves and tides (Chapters 12 and 13).

Hydro-power is by far the most established and widely used renewable resource for electricity generation and commercial investment. The early generation of electricity from about 1880 often derived from hydro-turbines, and the capacity of total worldwide installations has grown at about 5% per year since. Hydro-power now accounts for about 20% of world's electric generation. Output depends on rainfall and the terrain. Table 8.1 reviews the importance of hydroelectric generation for various countries and regions, while Figure 8.1 indicates the global increase. In about one-third of the world countries, hydro-power produces more than half the total electricity. In general the best sites are developed first on a national scale, so the rate of exploitation of total generating capacity tends to diminish with time. By the 1940s, most of the best sites in industrialised countries had already been exploited – hence the large 'fraction harnessed' percentages shown in Table 8.1. Almost all the increase in Figure 8.1 is in developing countries, notably India, China and Brazil, as reflected in the 'under construction' column in Table 8.1.

However, global estimates can be misleading for local hydro-power planning, since small-scale (1 MW to ~10 kW) applications are often neglected, despite the sites for such installations being the most numerous. This may be because the large surveys have not recognised the benefits perceived by the site owners, such as self-sufficiency or long-term capital assets. Thus the potential for hydro generation from run-of-river schemes (i.e. with only very small dams) is often underestimated. Social and environmental factors are also important, and these too cannot be judged by global surveys but only by evaluating local conditions. Coupled with the direct construction costs, these factors account for the 'economic potential' for the global study

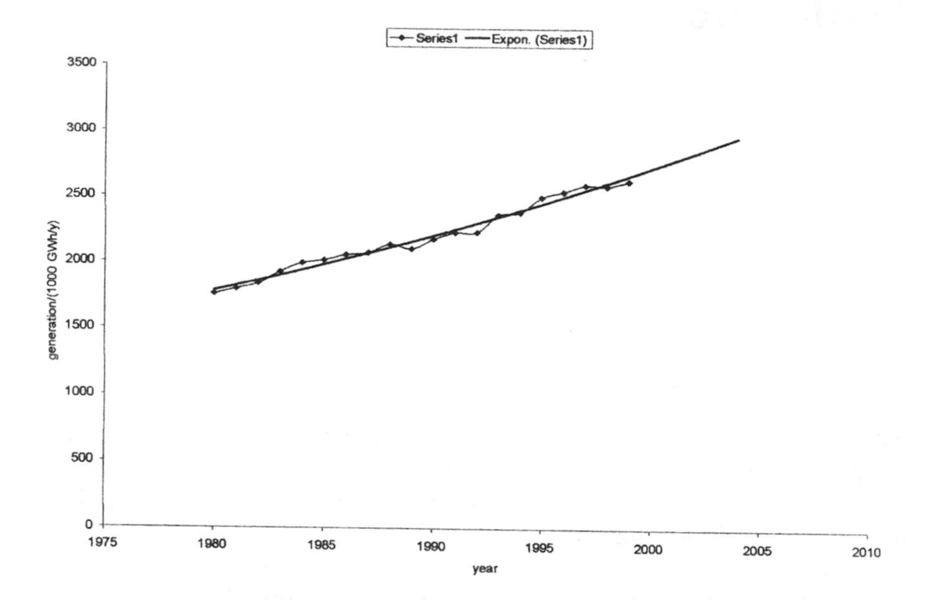


*Figure 8.1* World hydroelectricity generation (in TWh $y^{-1}$); mainly from large hydro. Extrapolated from the year of publication.

*Table 8.1* Hydro-power capacity by country/region

| A<br>country/region | B<br>technically exploitable potential/ TWh $y^{-1}$ | C<br>economically exploitable potential/ TWh $y^{-1}$ | D<br>actual generation/ TWh, 1999 | E<br>installed capacity/ GW, 1999 | F<br>under construction/ GW, 1999 | G<br>fraction harnessed = D/B/% |
|---|---|---|---|---|---|---|
| Africa | 1890 | n/a | 73 | 20 | 2 | 4% |
| North America | 1690 | n/a | 710 | 160 | 2 | 42% |
| Canada | 951 | 523 | 340 | 67 | 2 | 36% |
| USA | 530 | 376 | 320 | 80 | 0 | 60% |
| South America | 2800 | n/a | 500 | 106 | 16 | 18% |
| Brazil | 1490 | 810 | 285 | 57 | 11 | 19% |
| Asia | 4900 | n/a | 567 | 174 | 71 | 12% |
| China | 1920 | 1260 | 204 | 65 | 35 | 11% |
| India | 660 | n/a | 82 | 22 | 15 | 12% |
| Japan | 136 | 114 | 84 | 27 | 1 | 62% |
| Europe | 2700 | n/a | 735 | 214 | 9 | 27% |
| Austria | 60 | 56 | 41 | 11 | 0 | 68% |
| Norway | 200 | 180 | 121 | 27 | 0 | 61% |
| Russia | 1670 | 850 | 160 | 44 | 5 | 10% |
| UK | 6 | 1 | 5 | 1 | 0 | 88% |
| Middle East | 210 | n/a | 8 | 4 | 10 | 4% |
| Oceania | 240 | n/a | 42 | 13 | 0 | 18% |
| Australia | 35 | 30 | 17 | 8 | 0 | 49% |
| New Zealand | 77 | 40 | 23 | 5 | 0 | 30% |
| World | 14400 | n/a | 2630 | 692 | 110 | 18% |

Source: World Energy Council (2001)

of hydro-power in Table 8.1, being only about half the 'technical potential' assessed by summing (8.1) across the region.

Hydro installations and plants are long-lasting with routine maintenance, e.g. turbines for about fifty years and longer with minor refurbishment, dams and waterways for perhaps hundred years. Long turbine life is due to the continuous, steady operation without high temperature or other stress. Consequently established plant often produces electricity at low cost ($<4$ Eurocent/kWh) with consequent economic benefit.

Hydro turbines have a rapid response for power generation and so the power may be used to supply both base load and peak demand requirements on a grid supply. Power generation efficiencies may be as high as 90%. Turbines are of two types:

1 Reaction turbines, where the turbine is totally embedded in the fluid and powered from the pressure drop across the device.
2 Impulse turbines, where the flow hits the turbine as a jet in an open environment, with the power deriving from the kinetic energy of the flow.

Reaction turbine generators may be reversed, so water can be pumped to high levels for storage and subsequent generation at an overall efficiency of about 80%.

The main disadvantages of hydro-power are associated with effects other than the generating equipment, particularly for large systems. These include possible adverse environmental impact, effect on fish, silting of dams, corrosion of turbines in certain water conditions, social impact of displacement of people from the reservoir site, loss of potentially productive land (often balanced by the benefits of irrigation on other land) and relatively large capital costs compared with those of fossil power stations. For instance, there has been extensive international debate on the benefits and disadvantages of the Aswan Dam for Egypt and the Three Gorges project for China. All these issues are discussed further in Section 8.8.

This chapter considers certain fundamental aspects of hydro-power and does not attempt to be comprehensive in such a developed subject. In particular we have considered small-scale application, and we refer readers to the bibliography for comprehensive works at established engineering level. The fundamental equation (8.1) is often enough to give a guide to whether hydropower is worth considering at a particular location. If so, the methods of Section 8.3 give a more accurate assessment of the site potential. Sections 8.4 and 8.5 examine the mechanics of turbines, while Section 8.6 looks at other technical aspects of hydroelectric systems, Section 8.7 considers the hydraulic ram pump and Section 8.8 briefly reviews the social and environmental aspects of hydropower.

## 8.2 Principles

A volume per second, $Q$, of water falls down a slope. The density of the fluid is $\rho$. Thus the mass falling per unit time is $\rho Q$, and the rate of potential energy lost by the falling fluid is

$$P_0 = \rho Q g H \tag{8.1}$$

where $g$ *is* the acceleration due to gravity, $P_0$ is the energy change per second (a power measured in Watts) and $H$ is the vertical component of the water path.

The purpose of a hydro-power system is to convert this power to shaft power. Unlike some other power sources, there is no fundamental thermodynamic or dynamic reason why the output power of a hydro-system should be less than the input power $P_0$, apart from frictional losses which can be proportionately very small. The advantages of hydro-power can be inferred from (8.1). For a given site, $H$ is fixed and $Q$ can usually be held fairly constant by ensuring that the supply pipe is kept full. Hence the actual output is close to the design output, and it is not necessary to install a machine of capacity greater than normally required.

The main disadvantage of hydro-power is also clear from (8.1): the site must have sufficiently high $Q$ and $H$. In general this requires a rainfall $>\sim 40\,\mathrm{cm\,y^{-1}}$ dispersed through the year, a suitable catchment and, if possible, a water storage site. Where these are available, hydro-power is almost certainly the most suitable electricity generating source. However, considerable civil engineering (in the form of dams, pipework, etc.) is always required to direct the flow through the turbines. These civil works often cost more than the mechanical and electrical components. Note that the cost per unit power of turbines tends to increase with $Q$. Therefore for the same power output, systems with higher $H$ will be cheaper unless penstock costs become excessive.

## 8.3 Assessing the resource for small installations

Suppose we have a stream available, which may be useful for hydro-power. At first only approximate data, with an accuracy of about ±50%, are needed to estimate the power potential of the site. If this survey proves promising, then a detailed investigation will be necessary involving data, for instance rainfall, taken over several years. It is clear from (8.1) that to estimate the input power $P_0$ we have to measure the flow rate $Q$ and the available vertical fall $H$ (usually called the head, cf. Section 2.2). For example with $Q = 40\,\mathrm{L\,s^{-1}}$ and $H = 20\,\mathrm{m}$, the maximum power available at source is 8 kW. This might be very suitable for a household supply.

### 8.3.1 Measurement of head H

For nearly vertical falls, trigonometric methods (perhaps even using the lengths of shadows) are suitable; whereas for more gently sloping sites, the use of level and pole is straightforward. Note that the power input to the turbine depends not on the geometric (or total) head $H_t$ as measured this way, but on the available head $H_a$:

$$H_a = H_t - H_f \tag{8.2}$$

where $H_f$ allows for friction losses in the pipe and channels leading from the source to the turbine (see Section 2.6). By a suitable choice of pipework it is possible to keep $H_f <\sim H_t/3$, but according to (2.14) $H_f$ increases in proportion to the total length of pipe, so that the best sites for hydro-power have steep slopes.

### 8.3.2 Measurement of flow rate Q

The flow through the turbine produces the power, and this flow will usually be less than the flow in the stream. However, the flow in the stream varies with time, for example between drought and flood periods. For power generation we usually want to know the *minimum* (dry season) flow, since a turbine matched to this will produce power all the year without overcapacity of machinery. Such data are also necessary for environmental impact, e.g. maintaining a minimum flow for aquatic life. We also need to know the *maximum* flow and flood levels to avoid damage to installations.

The measurement of $Q$ is more difficult than the measurement of $H$. The method chosen will depend on the size and speed of the stream concerned. As in Section 2.2,

$$\text{flow rate } Q = (\text{volume passing in time } \Delta t)/\Delta t \tag{8.3}$$

$$= (\text{mean speed } \overline{u}) \times (\text{cross-sectional area } A) \tag{8.4}$$

$$= \int u \mathrm{d}A \tag{8.5}$$

where $u$ is the streamwise velocity (normal to the elemental area $\mathrm{d}A$). The methods that follow from each of these equations we call 'basic', 'refined' and 'sophisticated'. In addition, if water falls freely over a ledge or weir, then the height of the flow at the ledge relates to the flow rate. This provides a further method to measure flow rate.

1 *Basic method* (Figure 8.2(a)). The whole stream is either stopped by a dam or diverted into a containing volume. In either case it is possible to measure the flow rate from the volume trapped (8.3). This method

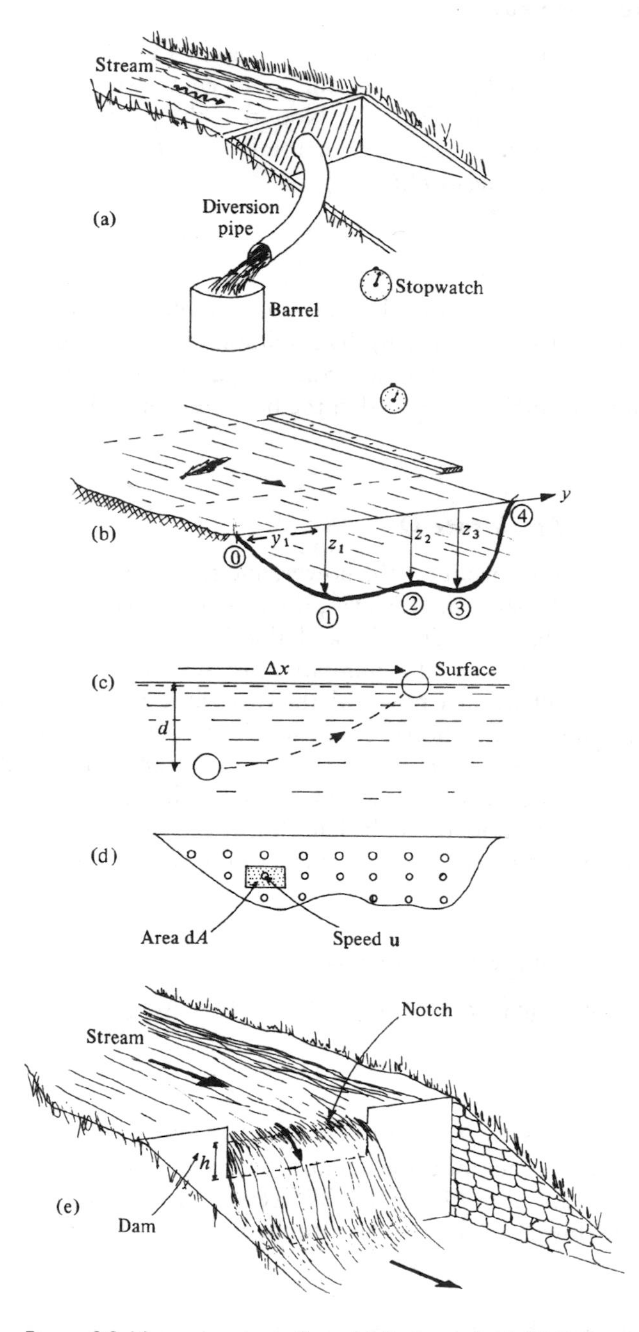


*Figure 8.2* Measuring water flow. (a) Basic method. (b) Refined method I. (c) Refined method II. (d) Sophisticated method. (e) Weir method.

makes no assumptions about the flow, is accurate and is ideal for small flows, such as those at a very small waterfall.

2 *Refined method I* (Figure 8.2(b)). Equation (8.4) defines the mean speed $\overline{u}$ of the flow. Since the flow speed is zero on the bottom of the stream (because of viscous friction), the mean speed will be slightly less than the speed $u_s$ on the top surface. For a rectangular cross-section, for example, it has been found that

$$\overline{u} \approx 0.8u_s$$

where $u_s$ can be measured by simply placing a float, e.g. a leaf, on the surface and measuring the time it takes to go a certain distance along the stream. For best results the measurement should be made where the stream is reasonably straight and of uniform cross-section.

The cross-sectional area $A$ can be estimated by measuring the depth at several points across the stream and integrating across the stream in the usual way

$$A \approx \frac{1}{2}y_1z_1 + \frac{1}{2}(y_2 - y_1)(z_1 + z_2) + \frac{1}{2}(y_3 - y_2)(z_2 + z_3) + \frac{1}{2}(y_4 - y_3)z_3$$

3 *Refined method II* (Figure 8.2(c)). A refinement which avoids the need for accurate timing can be useful on fast flowing streams. Here a float, e.g. a table tennis ball, is released from a standard depth below the surface. The time for it to rise to the surface is independent of its horizontal motion and can easily be calibrated in the laboratory. Measuring the horizontal distance required for the float to rise gives the speed in the usual way. Moreover what is measured *is* the mean speed (although averaged over depth rather than over cross section the difference is small).

4 *Sophisticated method* (Figure 8.1(d)). This is the most accurate method for large streams and is used by professional hydrologists. Essentially the forward speed $u$ is measured with a small flow meter at the points of a two-dimensional grid extending across the stream. The integral (8.5) is then evaluated by summation.

5 *Using a weir* (Figure 8.2(e)). If $Q$ is to be measured throughout the year for the same stream, measurement can be made by building a dam with a specially shaped calibration notch. Such a dam is called a weir. The height of flow through the notch gives a measure of the flow. The system is calibrated against a laboratory model having the same form of notch. The actual calibrations are tabulated in standard handbooks. Problem 8.2 shows how they are derived.

## 8.4 An impulse turbine

Impulse turbines are easier to understand than reaction turbines, so we shall consider a particular impulse turbine – the *Pelton wheel.*

### 8.4.1 Forces

The potential energy of the water in the reservoir is changed into kinetic energy of one or more jets. Each jet then hits a series of buckets or 'cups' placed on the perimeter of a vertical wheel, as shown in Figure 8.3. The resulting deflection of the fluid constitutes a change in momentum of the fluid. The cup has exerted a force on the fluid, and therefore the fluid has likewise exerted a force on the cup. This tangential force applied to the wheel causes it to rotate.

Figure 8.4(a) shows a jet, of density $\rho$ and volume flow rate $Q_j$, hitting a cup as seen in the 'laboratory' (i.e. earthbound) frame. The cup moves to the right with steady speed $u_c$ and the input jet speed is $u_j$. Figure 8.4(b) shows the frame of the cup with relative jet speed $(u_j - u_c)$; since the

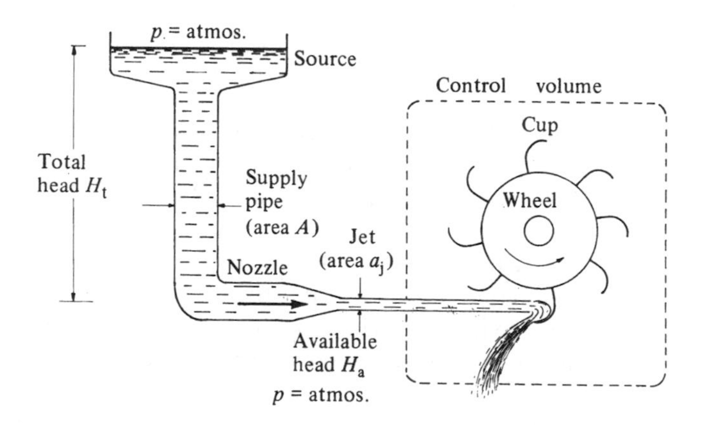


*Figure 8.3* Schematic diagram of a Pelton wheel impulse turbine.

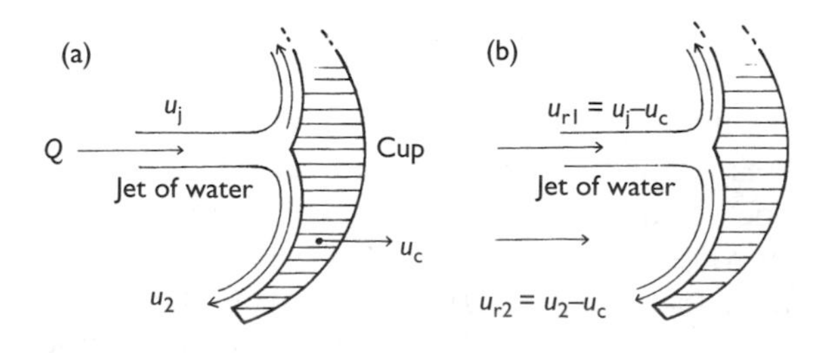


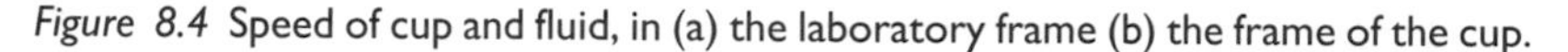

*Figure 8.4* Speed of cup and fluid, in (a) the laboratory frame (b) the frame of the cup.

polished cup is smooth, friction is negligible, and so the jet is deflected smoothly through almost 180° with no loss in speed; $u_2 = u_j$.

Thus in the frame of the cup, the change in momentum per unit time, and hence the force $F$ experienced by the cup, is

$$F = 2\rho Q_j(u_j - u_c) \tag{8.6}$$

(This force is in the direction of the jet.) The power $P_1$ transferred to the single cup is

$$P_j = Fu_c = 2\rho Q_j(u_j - u_c)u_c \tag{8.7}$$

where $Q_j$ is the flow through the jet. By differentiation with respect to $u_c$ this is a maximum for constant $u_j$ when

$$u_c/u_j = 0.5 \tag{8.8}$$

So substituting for $u_c$ in (8.7)

$$P_j = \frac{1}{2}\rho Q_j u_j^2 \tag{8.9}$$

i.e. the output power equals the input power, and this ideal turbine has 100% efficiency. For this ideal case, in the laboratory frame, the velocity of the water leaving the cup has zero component in the direction of the jet, i.e. $u_2 = 0$ in Figure 8.4(a). Therefore the water from the horizontal jet falls vertically from the cup.

The ideal efficiency can be 100% because the fluid impinges on the turbine in a constrained input flow, the jet, and can leave by a separate path; this contrasts with situations of extended flow, e.g. wind onto a wind turbine, where the extractable energy is significantly limited (see Section 9.3.1).

Although the ideal turbine efficiency is 100%, in practice values range from 50% for small units to 90% for accurately machined large commercial systems. The design of a practical Pelton wheel (Figure 8.5) aims for the ideal performance described. For instance, nozzles are adjusted so that the water jets hit the moving cups perpendicularly at the optimum relative speed for maximum momentum transfer. The ideal cannot be achieved in practice, because an incoming jet would be disturbed both by the reflected jet and by the next cup revolving into place. Pelton made several improvements in the turbines of his time (1860) to overcome these difficulties. Notches in the top of the cups gave the jets better access to the turbine cups. The shape of the cups incorporated a central splitter section so that the water jets were reflected away from the incoming water.

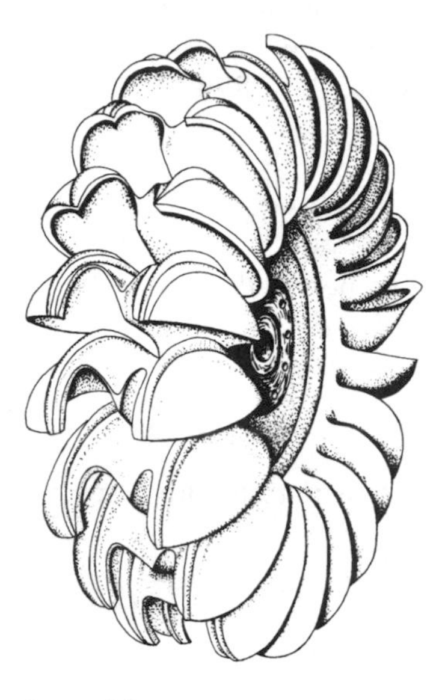

*Figure 8.5* Impulse turbine runner (Pelton type) with buckets cast integrally with the hub.

### *8.4.2 Jet velocity and nozzle size*

As indicated in Figure 8.3, the pressure is atmospheric both at the top of the supply pipe and at the jet. Therefore Bernoulli's theorem (2.3) implies that, in the absence of friction in the pipe, $u_j^2 = 2gH_t$. Pipe friction can be allowed for by replacing the total head $H_t$ by the available head $H_a$ defined by (8.2), so

$$u_j^2 = 2gH_a \tag{8.10}$$

In practice the size of the pipes is chosen so that $u_j$ is independent of the nozzle area. If there are $n$ nozzles, each of area $a$, then the total flow from all jets is

$$Q = nau_j = nQ_j \tag{8.11}$$

If the efficiency of transforming the water jet power into mechanical rotational power is $\eta_m$, then the mechanical power output $P_m$ from the turbine with $n$ jets is, from (8.8) and (8.9),

$$\begin{aligned} P_m &= \eta_{mn} P_j = \eta_{mn} \frac{1}{2} \rho Q u_j^2 = \eta_{mn} \frac{1}{2} \rho (auj) u_j^2 \\ &= \frac{1}{2} \eta_m n a \rho (2gH_a)^{3/2} \end{aligned} \tag{8.12}$$

This shows the importance of obtaining the maximum available head $H_a$ between turbine and reservoir. The output power is proportional to the total jet cross-sectional area $A = na$. However, $a$ is limited by the size of cup, so if $a$ is to be increased a larger turbine is needed. It is usually easier to increase the number of nozzles $n$ than to increase the overall size of the turbine, but the arrangement becomes unworkably complicated for $n \geq 4$. For small wheels, $n = 2$ is the most common.

Of course, the total flow $Q$ through the turbine cannot be more than the flow in the stream $Q_{stream}$. Using (8.10) and (8.11),

$$na_j \leq Q_{stream}/(2gH_a)^{1/2} \tag{8.13}$$

### 8.4.3 Angular velocity and turbine size

Suppose we have chosen the nozzle size and number in accordance with (8.11) and (8.12) to give the maximum power available. The nozzle size has fixed the size of the cups, but not the overall size of the wheel. The latter is determined by geometric constraints, and also by the required rotational speed. For electrical generation, the output variables, e.g. voltage, frequency and efficiency, depend on the angular speed of the generator. Most electric generators have greatest efficiency at large rotational frequency, commonly ~1500 rpm. To avoid complicated and lossy gearing, it is important that the turbine should also operate at such large frequency, and the Pelton wheel is particularly suitable in this respect.

If the wheel has radius $R$ and turns at angular velocity $\omega$, by (8.6) and (8.7),

$$P = FR\omega \tag{8.14}$$

Thus for a given output power, the larger the wheel the smaller its angular velocity. Since $u_c = R\omega$, and $u_c = 0.5u_j$ by (8.8), and using (8.10)

$$R = \frac{0.5(2gH_a)^{1/2}}{\omega} \tag{8.15}$$

The nozzles usually give circular cross-section jets of area $a$ and radius $r$. So $a = \pi r^2$ and from (8.12),

$$r^2 = \frac{P_m}{\eta_m \rho n \pi (gH_a)^{3/2} \sqrt{2}} \tag{8.16}$$

Combining (8.15) and (8.16), we find

$$r/R = 0.68(\eta_m n)^{-1/2} \mathcal{S} \tag{8.17}$$

where

$$\mathcal{S} = \frac{P_m^{1/2}\omega}{\rho^{1/2}(gH_a)^{5/4}} \tag{8.18}$$

$\mathcal{S}$ is a non-dimensional measure of the operating conditions, called the *shape number* of the turbine. Such non-dimensional factors are powerful functions in engineering, allowing 'scaling up' from smaller-scale laboratory measurements.

From (8.17), the mechanical efficiency $\eta_m$ at any instant is a function of (i) the fixed geometry of a particular Pelton wheel (measured by the non-dimensional parameters $r_j/R$ and $n$), and (ii) the non-dimensional 'shape number' $\mathcal{S}$ which characterises the operating conditions at that time.

Caution is needed, however, for instead of the *dimensionless* shape number $\mathcal{S}$ of (8.18), engineering texts usually use a *dimensioned* characteristic called 'specific speed', $N_s$, defined from the variables $P$, $\nu(=\omega/2\pi)$ and $H_a$:

$$N_s = \frac{P^{1/2}\nu}{H_a^{5/4}} \tag{8.19}$$

'Specific speed' does not include $g$ and $\rho$, since these are effectively constants. Consequently, $N_s$ has dimensions and units, and so, disturbingly, its numerical value depends on the particular units used. In practice, these units vary between USA (rpm, shaft Horsepower, ft) and Europe (rpm, metric Horsepower, m), with a standard version for SI units yet to become common.

Implicit in (8.17) is the relation (8.8) between the speed of the moving parts $u_c$ and the speed of the jet $u_j$. If the ratio $u_c/u_j$ is the same for two wheels of different sizes but the same shape, then the whole flow pattern is also the same for both. It follows that all non-dimensional measures of hydraulic performance, such as $\eta_m$ and $\mathcal{S}$, are the same for impulse turbines with the same ratio of $u_c/u_j$. Moreover, for a particular shape of Pelton wheel (specified here by $r_j/R$ and $n$), there is a particular combination of operating conditions (specified by $\mathcal{S}$) for maximum efficiency.

*Example 8.1*

Determine the dimensions of a single jet Pelton wheel to develop 160 kW under a head of (1) 81 m and (2) 5.0 m. What is the angular speed at which these wheels will perform best?

*Solution*

Assume that water is the working fluid. It is difficult to operate a wheel with $r_j > R/10$, since the cups would then be so large that they would interfere with each other's flow. Therefore assume $r_j = R/12$. Figure 8.7

(see Section 8.5) suggests that at the optimum operating conditions $\eta_m \approx 0.9$. From (8.17), the characteristic shape number for such a wheel is

$$\mathcal{S} = 0.11 \tag{8.20}$$

(1) From (8.18), the angular speed for best performance is

$$\omega_1 = \mathcal{S}\rho^{1/2}(gH_a)^{5/4}P^{-1/2}$$
$$= \frac{0.11(10^3\,\mathrm{kg\,m^{-3}})^{1/2}[(9.8\,\mathrm{m\,s^{-2}})(81\,\mathrm{m})]^{5/4}}{(16 \times 10^4\,\mathrm{W})^{1/2}}$$
$$= 36\,\mathrm{rad\,s^{-1}}$$

From (8.10),

$$u_j = (2gH_a)^{1/2} = 40\,\mathrm{m\,s^{-1}}$$

Therefore

$$R = \frac{1}{2}u_j/\omega = 0.55\,\mathrm{m}$$

(2) Similarly, with $H_a = 5\,\mathrm{m}$,

$$\omega_2 = \omega_1(5/81)^{5/4} = 1.1\,\mathrm{rad\,s^{-1}}$$
$$u_j = 10\,\mathrm{m\,s^{-1}}$$
$$R = 4.5\,\mathrm{m}$$

It can be seen from Example 8.1, by comparing cases (1) and (2), that Pelton wheels producing power from low heads should rotate slowly. Such wheels would be unwieldy and costly, especially because the size of framework and housing increases with the size of turbine. In practice therefore, Pelton wheels are used predominantly for high-head/small-flow installations.

## 8.5 Reaction turbines

It is clear from the fundamental formula (8.1) that to have the same power from a lower head, we have to maintain a greater flow $Q$ through the turbine. This can also be expressed in terms of the shape number $\mathcal{S}$ of (8.18). In order to maintain the same $\omega$ and $P$ with a lower $H$, we require a turbine with larger $\mathcal{S}$. One way of doing this is to increase the number of nozzles on a Pelton wheel; see (8.17) and Figure 8.6(a). However, the

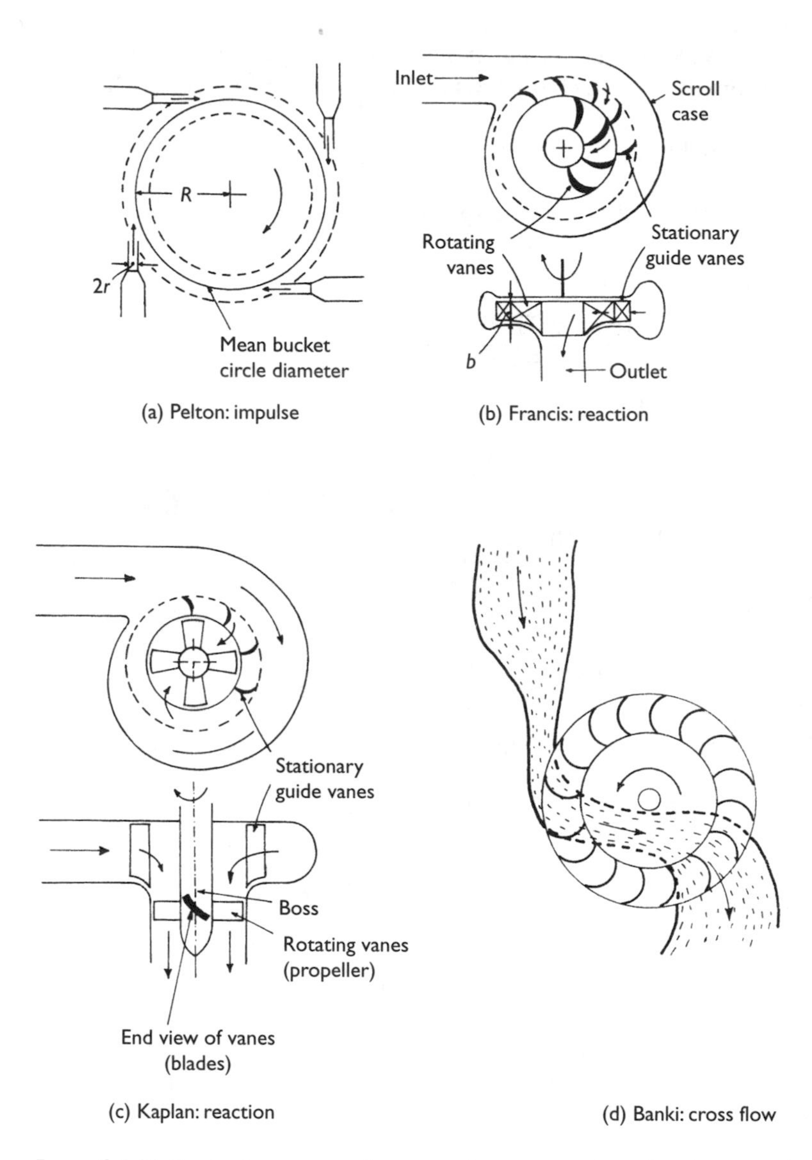


*Figure 8.6* Methods of increasing the power from a given size of machine, working at the same water pressure. (a) A four-jet Pelton wheel, the power of which is four times greater than that from a one-jet wheel of the same size and speed. (b) A reaction or radial flow turbine: e.g. Francis turbine; the jets supplying water to the rotor now exist all round the circumference as a slot and the water leaves the rotor axially. (c) A propeller turbine: e.g. Kaplan turbine; here large shape number, $\mathscr{S}$, is obtained if the jet is made the same size as the rotor and there is no radial flow over the rotor. (d) The Banki cross-flow turbine is an intermediate type in which the flow goes across an open, case-like, structure. Adapted from Francis (1974).

pipework becomes unduly complicated if $n > 4$, and the efficiency decreases because the many jets of water interfere with each other.

To have a larger flow through the turbine, it is necessary to make a significant change in the design. The entire periphery of the wheel is made into one large 'slot' jet which flows into the rotating wheel, as in Figure 8.6(b). Such turbines are called *reaction* machines' because the fluid pushes (or 'reacts') continuously against the blades. This contrasts with impulse machines, e.g. Pelton wheels, where the blades (cups) receive a series of impulses. For a reaction turbine, the wheel, called the runner, must be adapted so that the fluid enters radially perpendicular to the turbine axis, but leaves parallel to this axis. One design that accomplishes this is the *Francis* reaction turbine, shown in Figure 8.6(b). It can be seen that while the fluid is in the turbine it has a radial component of velocity in addition to the tangential velocity. This complicates analysis, which is, however, given in standard textbooks (Francis 1974; Massey and Ward-Smith 1998; Gulliver and Arndt 1991).

A larger water flow can be obtained by making the incoming water 'jet' almost as large in cross section as the wheel itself. This concept leads to a turbine in the form of a *propeller*, with the flow mainly along the axis of rotation; Figure 8.6(c). Note, however, that the flow is not exactly axial. Guide vanes on entry are used to give the fluid a whirl (rotary) component of velocity, like that in the Francis reaction turbine. It is the tangential momentum from this whirl component which is transferred to the propeller, thus making it rotate.

Since an axial flow machine is the most compact way of converting fluid power into mechanical power, it might be asked why designs of smaller $\mathcal{S}$ (such as the Pelton or Francis turbines) should ever be used rather than a reaction turbine. The main reason is the great pressure change in the fluid as it moves, sealed off from the outside air, through a reaction turbine. Bernoulli's equation (2.2) can be used to show that the smallest water pressure may even be less than the vapour pressure of water. If this happens, bubbles of water vapour will form within the fluid – a process called *cavitation*. Downstream from this, the water pressure might suddenly increase, so causing the bubble to collapse. The resulting force from the inrush of liquid water can cause considerable mechanical damage to nearby mechanical parts. These effects increase with flow speed and head, and so axial machines are restricted usually to sites with small values of $H$. Moreover, the performance of reaction turbines in general, and the propeller turbine in particular, is very sensitive to changes in flow rate. The efficiency drops off rapidly if the flow diminishes, because the slower flow no longer strikes the blade at the correct angle. It is possible to allow for this by automatically adjusting the blade angle, but this is complicated and expensive. Propeller turbines with automatically adjustable blade pitch were historically considered worthwhile only on large-scale installations, e.g. the

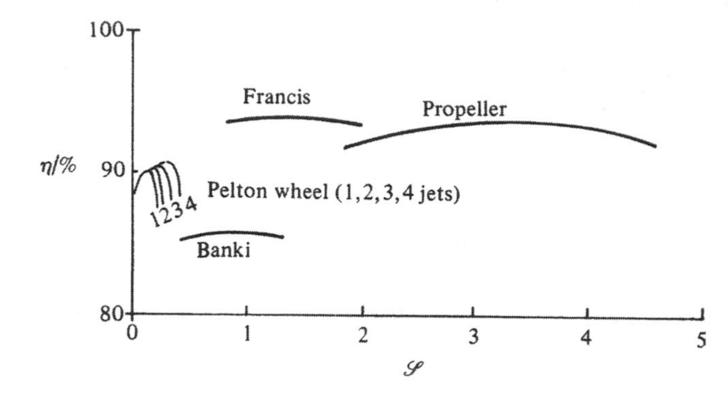


*Figure 8.7* Peak efficiencies of various turbines in relation to shape number. Adapted from Gulliver and Arndt (1991).

*Kaplan* turbine. However, smaller propeller turbines with adjustable blades are now available commercially for small-scale operation.

The operation of a Pelton wheel (and also the intermediate *Banki* cross-flow turbine; Figure 8.5(d)) is not so sensitive to flow conditions as a propeller turbine. The Banki turbine is particularly easy to construct with limited facilities (McGuigan 1978).

As a guide to choosing the appropriate turbine for given $Q$ and $H$, Figure 8.7 shows the range of shape number $\mathcal{S}$ over which it is possible to build an efficient turbine. In addition, for each type of turbine there will be a relationship between the shape number $\mathcal{S}$ (characterising the operating conditions under which the turbine performs best) and another non-dimensional parameter characterising the form of the turbine. One such parameter is the ratio $r/R$ of (8.17). These relationships may be established theoretically or experimentally, and are used to optimise design. Details are given in the recommended texts and engineering handbooks.

## 8.6 Hydroelectric systems

Most modern hydro-power systems are used to drive electric generators, although some special purpose mechanical systems are still useful, such as the hydraulic ram pump (Section 8.7). A complete hydroelectric system, such as that shown in Figure 8.8, must include the water source, the pipe (penstock), flow control, the turbine, the electric generator, fine control of the generator and wiring for electricity distribution (reticulation).

The dam ensures a steady supply of water to the system without fluctuations, and enables energy storage in the reservoir. It may also be used for purposes other than generating electricity, e.g. for roads or water supply. Small run-of-the-river systems from a moderately large and steady stream may require only a retaining wall of low height, i.e. enough to keep the penstock fully immersed, but this does not produce appreciable storage.

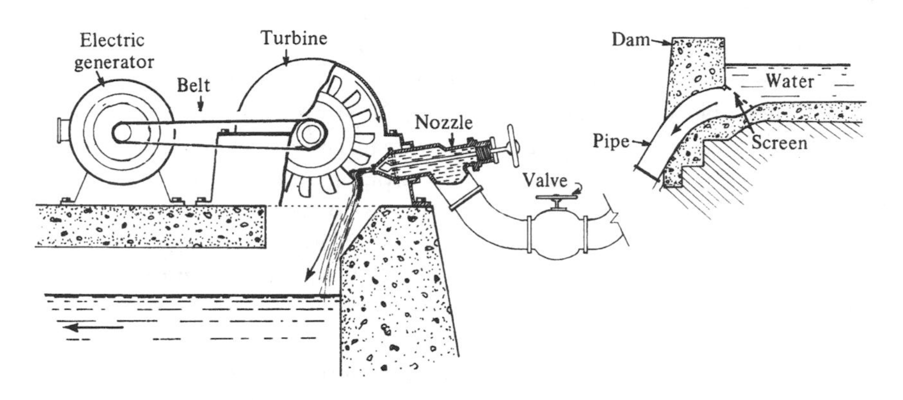


*Figure 8.8* Layout of a micro hydroelectric system using a Pelton wheel. Note that this diagram does not indicate the water head *H* required.

The supply pipe (penstock) is a major construction cost. It will be cheaper if it is thin walled, short and of small diameter $D$, but unfortunately it is seldom possible to meet these conditions. In particular the diameter cannot be decreased because of the head loss $H_f$(where $H_f \propto D^{-5}$; see Problem 2.6). Therefore if the pipe diameter is too small, almost all the power will be lost in the pipe. The greater cost of a bigger pipe has to be compared with the continued loss of power by using a small pipe. A common compromise is to make $H_f \leq 0.1\,H_t$.

The material of the pipe is required to be both smooth (to reduce friction losses) and strong (to withstand the static pressures, and the considerably larger dynamic 'water hammer' pressures that arise when the flow is turned suddenly on or off). For small installations, PVC plastic is suitable for the main length of the pipe, perhaps with a short steel section at the bottom to withstand the larger pressures there. A screen is needed at the top of the supply pipe to intercept the rubbish (e.g. leaves) before it blocks the pipe. This screen has to be regularly checked and cleared of debris. A settling chamber is essential to deposit suspended material before the water goes down the pipe.

The turbine speed is designed to be suitable for the electric generator, as explained in the previous sections, with generation usually at about 400 V AC. Large (megawatt) systems usually have a specially built generator running from the same shaft as the turbine, which minimises power losses between the turbine and the generator. Small systems (~10kW) generally use off-the-shelf generators or induction motors operated as generators, see Section 16.9.3(d). If the turbine speed is not large enough to match the generator, then gearing is used. A V-belt is a common gearing mechanism, which unfortunately may give power losses of 10–20% in very small systems.

### 8.6.1 Power regulation and control

Much of the material for this section is also considered in the discussions of electricity generation (Section 16.9), wind power (Section 9.8) and of renewable energy control (Section 1.5). With any hydro installation feeding electricity into a utility transmission grid, it is important that the voltage and frequency of the output match that of the rest of the grid. Although the primary generation is always at a relatively low voltage, the AC voltage can easily be increased by transformers, both to match the grid voltage and to minimise $I^2R$ losses in transmission. It is important that the voltage and frequency be controlled to maintain common grid standards and consumer requirements. This is done traditionally by mechanical feedback systems which control the flow through the turbine, so that it maintains constant frequency ('speed'). For example, with a Pelton wheel, a spear valve is made to move in and out of the nozzle (as indicated in Figure 8.8), thus regulating $Q$. For reaction turbines it is necessary to adjust the blade angles also. All such mechanical systems are relatively complicated and expensive, especially for smaller-scale application.

Stand-alone, autonomous, hydroelectric systems intended to supply electricity to a village or farm also require some regulation, but the devices they are required to run, such as lights and small electric motors (e.g. in refrigeration), can generally tolerate variations in voltage and frequency to $\pm 10\%$. Moreover the currents involved are easily switched by power electronic devices, such as thyristors. This gives the possibility of a much cheaper control than the conventional mechanical systems.

With an *electronic load control* system, major variations in output are accomplished by manually switching nozzles completely in or out, or by manually controlling the total flow through the turbine. Finer control is achieved by an electronic feedforward control which shares the output of the generator between the main loads (e.g. house lights) and a ballast (or 'off peak') heating circuit which can tolerate a varying or intermittent supply (see Figure 1.5c) . The generator thus always sees a constant total load (= main + ballast); therefore it can run at constant power output, and so too can the turbine from which the power comes. The flow through the turbine does not therefore have to be continually automatically adjusted, which greatly simplifies its construction. In one common type of system, the electronic control box is based on a thyristor, which responds to the difference between the actual and nominal voltage in the main load.

### 8.6.2 System efficiency

Even though the efficiency of each individual step is large, there is still a substantial energy loss in passing from the original power $P_0$ of the stream to the electrical output $P_e$ from the generator. Considering the successive energy transformations, we may obtain for small systems

$$\begin{aligned}\frac{P_e}{P_0} &= \frac{P_j}{P_0}\frac{P_m}{P_j}\frac{P_e}{P_m} \\ &= \frac{H_a}{H_t}\eta_m\eta_e \\ &\approx (0.8)(0.8)(0.8) \\ &\approx 0.5\end{aligned} \tag{8.21}$$

Further losses occur in the distribution and use of the electricity.

## 8.7 The hydraulic ram pump

This mechanical hydro-power device is well established for domestic and farm water pumping at remote sites, where there is a steady flow of water at a low level. The momentum of the stream flow is used to pump some of the water to a considerably higher level. For example, a stream falling 2 m below can be made to pump 10% of its flow to a height 12 m above. This is clearly a useful way of filling a header tank for piped water, especially in rural areas. It is an interesting device, perhaps surprising in its effectiveness and certainly intriguing for student practical study.

Figure 8.9 shows the general layout of a pumping system using a hydraulic ram. The water supply flows down a strong, inclined pipe called the 'drive pipe'. The potential energy $MgH$ of the supply water is first converted into kinetic energy and subsequently into potential energy $mgh$. The kinetic energy is obtained by a mass of water $M$ falling through a head $H$, and out through the impulse valve $V_1$. Operation is as follows:

1. The speed of flow increases under the influence of the supply head so a significant dynamic pressure term arises for the flow through the valve (see Bernoulli's equation (2.2)).
2. The static pressure acting on the underside of the impulse valve overcomes the weight of the valve, so it closes rapidly.
3. Consequently the water at the bottom becomes compressed by the force of the water still coming down the pipe.
4. The pressure in the supply pipe rises rapidly, forcing open the delivery valve $V_d$, and discharging a mass of water $m$ into the delivery section.
5. The air in the air chamber is compressed by the incoming water, but its compressibility cushions the pressure rise in the delivery pipe.

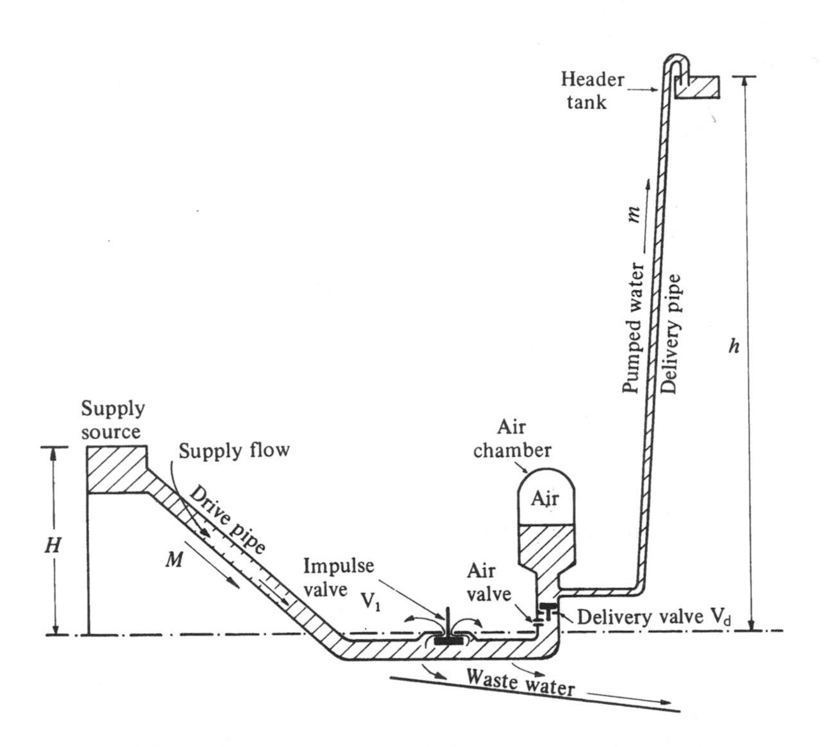


*Figure 8.9* General layout of a hydraulic ram pumping system. After Jeffrey *et al.* (1992).

6 The combined pressure of air and water forces the mass $m$ up the delivery pipe.
7 As soon as the momentum of the supply column is exhausted the delivery valve closes, and the water contained in the drive pipe recoils towards the supply.
8 This recoil removes the pressure acting on the underside of the impulse valve, which thereupon falls and again allows the escape of water.
9 Simultaneously, the recoil causes the small air charging valve to open, admitting a small amount of air into the impact chamber of the ram. This air is carried along with the water into the air chamber to compensate for that absorbed by the water.
10 The whole cycle repeats indefinitely at a rate which is usually set to be about 1 Hz.

A theory for calculating all these quantities is given by Krol (1951), which uses only one main empirical parameter, the drag coefficient of the impulse valve. The efficiency of the device over a period equals $mh/MH$. Very solid and reliable rams are available commercially. Their efficiency is about 60%.

It is also possible to build a ram (of slightly reduced performance) from commercial high pressure pipe fittings (Inverson 1978, Jeffrey *et al.* 1992).

## 8.8 Social and environmental aspects

Hydro-power is a mature technology in wide use in many countries of the world: it produces about 20% of the world's electric power. In at least twenty countries, including Brazil and Norway, hydro-power accounts for over 90% of the total electricity supply. Hydroelectric systems are long-lasting with relatively low maintenance requirements: many systems, both large and small, have been in continuous use for over fifty years and a few early installations still function after a hundred years. Their relatively large initial capital cost has been long since written off, so that the 'levelised' cost of power produced from them is much less than from non-renewable sources which require expenditure on fuel and more frequent replacement of plant. If the external costs are internalised (see Chapter 17), the non-renewable sources become even more expensive. For hydro plant with ample supply of water, the flow can be controlled to produce either base-load or rapidly peaking power as demanded; if the water is limited, then sale of electricity at peak demand is easy and most profitable. Nevertheless, the initial capital cost of hydro power is always relatively large, so it has been observed that 'all power producers wish they had invested in hydro-power twenty years ago, but unfortunately can't afford to do so now – and they said the same twenty years ago!'

The complications of hydro-power systems arise mostly from associated dams and reservoirs, particularly on the large-scale projects. Most rivers, including large rivers with continental-scale catchments, such as the Nile, the Zambesi and the Yangtze, have large seasonal flows making floods a major characteristic. Therefore most large dams (i.e. those >15m high) are built for more than one purpose, apart from the significant aim of electricity generation, e.g. water storage for potable supply and irrigation, controlling river flow and mitigating floods, road crossings, leisure activities and fisheries. The electricity provides power for industry and services, and hence economic development. For although social and economic development requires much more than just power and water, such projects appeal to politicians and financiers seeking a path to national development that is centralised and thus conceptually and administratively 'simple'. But as the World Commission on Dams (2000) pointed out, the enormous investments and widespread impacts of hydropower have made large dams, both those in place and those on the drawing board, one of the most hotly contested issues in sustainable development. Countering the benefits of large hydro referred to above are adverse impacts; examples are debt burden (dams are often the largest single investment project in a country!), cost over-runs, displacement and impoverishment of people, destruction of important ecosystems and fishery resources, and the inequitable sharing of costs and benefits. For example, over one million people were displaced by the construction of the Three Gorges dam in China, which has a planned capacity of over 17 000 MW; yet these displaced people may never consider they are, on balance, beneficiaries of the increased power capacity and industrialisation. Some dams have been built on notoriously silt-laden rivers, resulting in the depletion of reservoir volume predictable to all except the constructors and the proponents.

Hydro-power, like all renewable energy sources, mitigates emissions of the greenhouse gas $CO_2$ by displacing fossil fuel that would otherwise have been used. However, in some dam projects, in an effort to save construction time and cost, rotting vegetation (mostly trees) have been left in place as dam fills up, which results in significant emissions of methane, another greenhouse gas.

Such considerations have led to an almost complete cessation of dam-building in many industrialised countries, where the technically most attractive sites were developed decades ago. Indeed in the USA, dams have been decommissioned to allow increased 'environmental flow' through downstream ecosystems. However, in many countries, hydroelectric capacity has been increased by adding turbine generators to water supply reservoirs and, for older hydropower stations, installing additional turbines and/or replacing old turbines by more efficient or larger capacity modern plant. This has positive environmental impact, with no new negative impact, and is an example of using an otherwise 'wasted' flow of energy (cf. Section 1.4). Likewise, the installation of small 'run-of-river' hydroelectric systems, with only very small dams, is generally considered a positive development; the output of such systems in China is greater than the total hydro-power capacity of most other countries.

## Problems

8.1 Use an atlas to estimate the *hydro-potential of your country or state*, as follows:

a Call the place in question, X. What is the lowest altitude in X? What area of X lies more than 300 m above the lowest level? How much rain falls per year on this high part of X? What would be the potential energy per year given up by this mass of water if it all ran down to the lowest level? Express this in megawatts.

b Refine this power estimate by allowing for the following: (i) not all the rain that falls appears as surface run-off; (ii) not all the run-off appears in streams that are worth damming; (iii) if the descent is at too shallow a slope, piping difficulties limit the available head.

c If a hydroelectric station has in fact been installed at X, compare your answer with the installed capacity of X, and comment on any large differences.

8.2 The flow over a U-weir can be idealised into the form shown in Figure 8.10. In region 1, before the weir, the stream velocity $u_1$ is uniform with depth. In region 2, after the weir, the stream velocity increases with depth $h$ in the water.

a Use Bernoulli's theorem to show that for the streamline passing over the weir at a depth $h$ below the surface,

$$u_h = (2g)^{1/2}\left(h + u_1^2/2g\right)^{1/2}$$

*Hints*: Assume that $p_h$ in the water = atmospheric pressure, since this is the pressure above and below the water. Assume also that $u_1$ is small enough that $p_1$ is hydrostatic.

b Hence show that the discharge over the idealised weir is

$$Q_{th} = (8g/9)^{1/2} L H^{3/2}$$

c By experiment, the actual discharge is found to be

$$Q_{exp} = C_w Q_{th}$$

where $C_w \approx 0.6$. (The precise value of $C_w$ varies with $H/L'$ and $L/b$.) Explain why $C_w < 1$.

d Calculate $Q_{exp}$ for the case $L' = 0.3\,m$, $L = 1\,m$, $b = 4\,m$, $H = 0.2\,m$. Calculate also $u_1$ and justify the assumptions about $u_1$ used in (a) and (b).

8.3 Verify that $\mathcal{S}$ defined by (8.18) is dimensionless. What are the advantages of presenting performance data for turbines in dimensionless form?

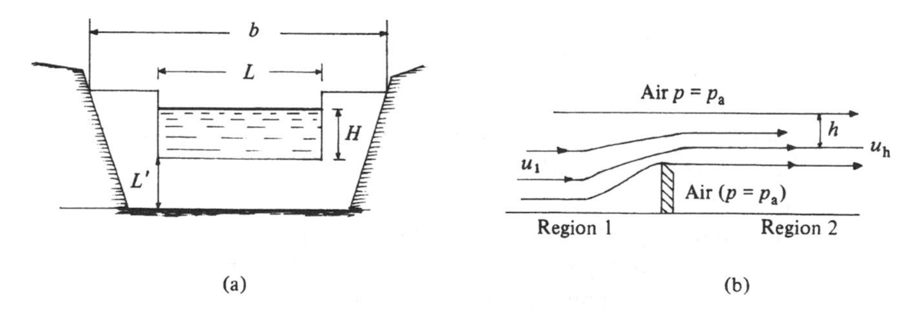


*Figure 8.10* A U-weir. (a) Front elevation. (b) Side elevation of idealised flow ($u_h$ is the speed of water over the weir where the pressure is $p_h$.)

8.4 A propeller turbine has shape number $\mathcal{S} = 4$ and produces 100 kW (mechanical) at a working head of 6 m . Its efficiency is about 70%. Calculate

a The flow rate
b The angular speed of the shaft
c The gear ratio required if the shaft is to drive a four-pole alternator to produce a steady 50 Hz.

8.5 A Pelton wheel cup is so shaped that the exit flow makes an angle $\theta$ with the incident jet, as seen in the cup frame. As in Figure 8.4, $u_c$ is the tangential velocity of the cup, measured in the laboratory frame. The energy lost by friction between the water and the cup can be measured by a loss coefficient $k$ such that

$$u_{r1}^2 = u_{r2}^2(1+k)$$

Show that the power transferred is

$$P = Q\rho u_c(u_j - u_c)\left[1 + \frac{\cos\theta}{\sqrt{(1+k)}}\right]$$

Derive the mechanical efficiency $\eta_m$. What is the reduction in efficiency from the ideal when $\theta = 7°, k = 0.1$? What is the angle of deflection seen in the laboratory frame?

8.6 A Pelton wheel is to be installed in a site with $H = 20\,\text{m}$, $Q_{min} = 0.05\,\text{m}^3\,\text{s}^{-1}$.

a Neglecting friction, find (i) the jet velocity (ii) the maximum power available (iii) the radius of the nozzles (assuming there are two nozzles).
b Assuming that the wheel has shape number

$$\mathcal{S} = \frac{\omega P_1^{1/2}}{\rho^{1/2}(gH)^{5/4}} = 0.1$$

where $P_1$ is the power per nozzle, find (iv) the number of cups (v) the diameter of the wheel (vi) the angular speed of the wheel in operation.
c If the main pipe (the penstock) had a length of 100 m, how would your answers to (a) and (b) be modified by fluid friction using: (vii) PVC pipe with a diameter of 15 cm? (viii) Common plastic hosepipe with a diameter of 5 cm? In each case determine the Reynolds number in the pipe.

# Bibliography

## *General articles and books on hydro-power*

Gulliver, J.S. and Arndt, R.E. (1991) *Hydropower Engineering Handbook*, McGraw Hill. {Professional level, but readable. Includes substantial chapters on preliminary studies, small dams, turbines, economic analysis. {Chapter on environment has strong slant towards US issues and regulations.}

Ramage, J. (2004, 2nd edn) Hydroelectricity, Chapter 5 of Boyle, G. (ed.) *Renewable Energy: Power for a Sustainable Future*, Oxford UP. {Non-technical survey, with many photos and illustrations.}

## *Mechanics of turbines*

Francis, J.R. (1974, 4th edn) *A Textbook of Fluid Mechanics*, Edward Arnold, London. {Has a clear chapter on hydraulic machinery, with thorough physics but not too much technical detail.}

Massey, B. and Ward-Smith, J. (1998, 7th edn) *Mechanics of Fluids*, Nelson Thomes, London. {Longer account of turbomachinery than Francis, but still at student level.}

Turton, R.K. (1984) *Principles of Turbomachinery*, E. and F.N. Spon, London. {Stresses technical details.}

## *Small scale (mini) hydropower* (~100 kW)

Cotillon, J. (1979) Micro-power: An old idea for a new problem, *Water Power and Dam Construction*, January. {Part of a special issue on mini-hydro.}

Francis, E.E. (1981) Small scale hydro electric developments in England and Wales, *Proc. Conf. Future Energy Concepts*, IEE, London.

Khennas, S. and Barnett, A. (2000) *Best Practices for Sustainable Development of Micro Hydro Power in Developing Countries*, ITDG, London. {Available on web at <www.microhydropower.net>; see also (much shorter) ITDG technical brief on micro-hydro power at < www.itdg.org>.}

Leckie, J., Masters, G., Whitehouse, H. and Young, L. (1976) *Other Homes and Garbage*, Sierra Club, San Francisco. {Another book from the 1970s, with a good discussion of energy principles; the chapter on hydro-power is particularly good on measurement techniques for small installations.}

McGuigan, D. (1978) *Harnessing Water Power for Home Energy*, Garden Way, Vermont {A useful guide for householders with technical aptitude in remote areas.}

Moniton, L., Le Nir, M. and Roux, J. (1984) *Micro Hydroelectric Power Stations*, Wiley. {Translation of a French book of 1981.}

Water Power and Dam Construction (1990) *Micro Hydro: Current Practice and Future Development*, Scottish Seminar – special issue of journal.

US Dept. of Energy (1988) *Small-scale Hydropower Systems*, NCIS Washington DC. {Non-technical account with many good line drawings.}

### *Hydraulic ram*

Inverson, A.R. (1978) *Hydraulic Ram Pump*, Volunteers in Technical Assistance, Maryland, USA, Technical Bulletin no. 32. Construction plans of the ram itself.

Jeffrey, T.D., Thomas, T.H., Smith, A.V., Glover, P.B., and Fountain, P.D. (1992) *Hydraulic Ram Pumps: A Guide to Ram Pump Water Supply Systems*, ITDG Publishing, UK. {See also ITDG Technical brief 'hydraulic ram pumps', online at <www.itdg.org> .}

Krol, J. (1951) The automatic hydraulic ram, *Proc. Inst. Mech. Eng.*, **165**, 53–65. {Mathematical theory and some supporting experiments. Clumsy writing makes the paper look harder than it is.}

### *Institutional and environmental issues*

International Energy Agency (1993) *Hydropower, Energy and the Environment* {Conference proceedings, but with useful overview. Focuses on implications of upgrades to existing facilities.}

World Commission on Dams (2000) *Dams and Development: A New Framework for Decision Making* (at www.dams.org). {The Commission was set up by the World Bank and the International Union for the Conservation of Nature to review the effectiveness of large dams in fostering economic and social development and to develop new criteria for assessing proposals for such dams.}

Moreira, J.R. and Poole, A.D. (1993) Hydropower and its constraints, in T. Johansson *et al.* (eds) *Renewable Energy: Sources for Fuels and Electricity*, Earthscan, London (pp. 71–119). {Good survey of global issues and potential, with focus on social and environmental constraints, and case studies from Brazil.}

### *Journals and websites*

*Water Power and Dam Construction*, monthly, Quadrant House, Sutton, UK. {General journal including production information, conference reports, articles, etc.}

<www.microhydropower.net> Portal with downloadable books and papers and an online discussion forum.}

World Energy Council (2001) *Survey of energy resources 2001* (chapter on hydropower), available on web at <www.worldenergy.org/wec-geis/publications/reports/ser/hydro/hydro.asp>. {Data on installed capacity and technical potential for numerous countries, compiled by utilities and energy agencies; publication covers other energy resources as well, including fossil and even OTEC.}

# Chapter 9

# Power from the wind

## 9.1 Introduction

The extraction of power from the wind with modern turbines and energy conversion systems is an established industry. Machines are manufactured with a capacity from tens of watts to several megawatts, and diameters of about 1 m to more than 100 m. Traditional mechanical-only machines have been further developed for water pumping, but the overriding commerce today is for electricity generation. Such 'wind turbine generators' have become accepted as 'mainstream generation' for utility grid networks in many countries with wind power potential, e.g. in Europe, the USA and parts of India and China; other countries are steadily increasing their wind power capacity. Smaller wind turbine generators are common for isolated and autonomous power production. The rapid growth of world wind turbine electricity generation capacity is shown in Figure 9.1. Since about 2002, much additional generation capacity is being installed at sea in offshore wind farms where the depth is moderate.

Later sections will show that in a wind of speed $u_0$ and density $\rho$, a turbine intercepting a cross-section $A$ of wind front will produce power to its rated maximum according to

$$P_T = \frac{1}{2} C_P A \rho u_0^3 \tag{9.1}$$

Here $C_p$ is an efficiency factor called 'the power coefficient'. Note that the power $P_T$ is proportional to $A$ and to the cube of wind speed $u_0$. Thus whereas doubling $A$ may produce twice the power, a doubling of wind speed produces eight times the power potential. The power coefficient $C_p$ also varies with wind speed for individual machines. Since wind speed distribution is skewed, at any one time speeds less than average are more likely than speeds greater than average. Therefore the optimum design size of rotor and generator at a particular site depends on the power requirement, either to maximise generated energy per year or to provide frequent power. Often the average annual power from a wind turbine approximates to the

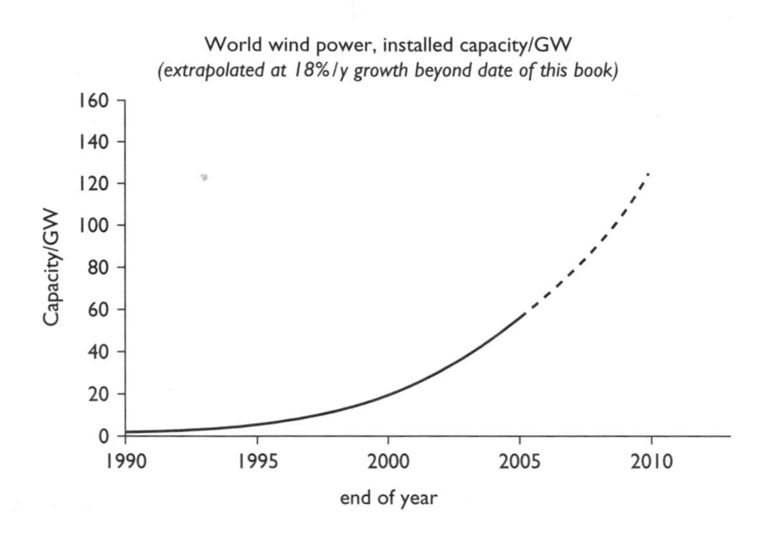


*Figure 9.1* World wind turbine power capacity/MW. Extrapolated from date of publication.

product of $C_p$, air density and the mean wind speed cubed: $\bar{P}_T \sim C_P A\rho(\bar{u}_0)^3$, see (9.74).

The structure comprising the rotor, its matched electricity generator and other equipment is sometimes called a *wind energy conversion system – WECS*, however it is increasingly common to use name *wind turbine* for the whole assembly, as in this edition. The maximum rated power capacity of a wind turbine is given for a specified 'rated' wind speed, commonly about $12\,\mathrm{m\,s^{-1}}$. At this speed, power production of about $0.3\,\mathrm{kW\,m^{-2}}$ of cross-section would be expected with power coefficients $C_p$ between 35 and 45%. The optimum rotation rate depends on the ratio of the blade tip speed to the wind speed, so small machines rotate rapidly and large machines slowly. Tables 9.1 and 9.2 give outline details of wind speeds and machine size. Machines would be expected to last for at least 20–25 years and cost about € 700–1000 ($US 850–1200) per kW rated capacity, ex-factory. When installed in windy locations and given some credit for not polluting, power production is competitive with the cheapest forms of other generation.

Wind power for mechanical purposes, milling and water pumping has been established for many hundreds of years. Wind electricity generators date from around 1890, with most early development from about 1930 to about 1955. At this time development almost ceased due to the availability of cheap oil, but interest reawakened and increased rapidly from about 1973. A few of the older machines kept operating for several tens of years, e.g. the Gedser 100 kW, 24-m diameter machine in Denmark, built in 1957.

*Table 9.1* Wind speed relationships based on the Beaufort scale

| *Beaufort number* | *Wind Speed range at 10 m height* | | | | *Description* | *Wind turbine effects* | *Power generation possibility for average speed in range at hub height* | *Observable effects* | |
|---|---|---|---|---|---|---|---|---|---|
| | $(ms^{-1})$ | $(km\,h^{-1})$ | $(mi\,h^{-1})$ | (knot) | | | | *Land* | *Sea* |
| 0 | 0.0 → 0.4 | 0.0 → 1.6 | 0.0 → 1 | 0.0 → 0.9 | Calm | None | – | Smoke rises vertically | Mirror smooth |
| 1 | 0.4 → 1.8 | 1.6 → 6 | 1 → 4 | 0.9 → 3.5 | Light | None | – | Smoke drifts but vanes unaffected | Small ripples |
| 2 | 1.8 → 3.6 | 6 → 13 | 4 → 8 | 3.5 → 7.0 | Light | None | Poor | Wind just felt across skin; leaves stir; vanes unaffected | Definite waves |
| 3 | 3.6 → 5.8 | 13 → 21 | 8 → 13 | 7.0 → 11 | Light | Start-up by turbines for light winds | Water pumping; minor electrical power | Leaves in movement; flags begin to extend | Occasional wave crest break, glassy appearance of whole sea |
| 4 | 5.8 → 8.5 | 21 → 31 | 13 → 19 | 11 → 17 | Moderate | | Useful electrical power production | Small branches move; dust raised; pages of books lifted | Larger waves, white crests common |
| 5 | 8.5 → 11 | 31 → 40 | 19 → 25 | 17 → 22 | Fresh | Useful power generation | Extremely good prospects for power | Small trees in leaf sway, wind noticeable for comment | White crests everywhere |
| 6 | 11 → 14 | 40 → 51 | 25 → 32 | 22 → 28 | Strong | Rated range at full capacity | Only for the strongest machines | Large branches sway, telephone lines whistle | Larger waves appear, foaming crests extensive |

*Table 9.1* (Continued)

| Beaufort number | Wind Speed range at 10 m height | | | | Description | Wind turbine effects | Power generation possibility for average speed in range at hub height | Observable effects | |
|---|---|---|---|---|---|---|---|---|---|
| | ($ms^{-1}$) | ($km\,h^{-1}$) | ($mi\,h^{-1}$) | (knot) | | | | Land | Sea |
| 7 | 14 ↓ 17 | 51 ↓ 63 | 32 ↓ 39 | 28 ↓ 34 | Strong | Full capacity reached | Life not worth living here | Whole trees in motion | Foam begins to break from crests in streaks |
| 8 | 17 ↓ 21 | 63 ↓ 76 | 39 ↓ 47 | 34 ↓ 41 | Gale | Shutdown or self-stalling initiated | | Twigs break off. Walking difficult | Dense streaks of blown foam |
| 9 | 21 ↓ 25 | 76 ↓ 88 | 47 ↓ 55 | 41 ↓ 48 | Gale | All machines shut down or stalled | | Slight structural damage, e.g. chimneys | Blown foam extensive |
| 10 | 25 ↓ 29 | 88 ↓ 103 | 55 ↓ 64 | 48 ↓ 56 | Strong gale | Design criteria against damage Machines shut down | | Trees uprooted. Much structural damage | Large waves with long breaking crests damage |
| 11 | 29 ↓ 34 | 103 ↓ 121 | 64 ↓ 75 | 56 ↓ 65 | Strong gale | Only strengthened machines would survive | | Widespread damage | |
| 12 | >34 | >121 | >75 | >65 | Hurricane | Serious damage likely unless pre-collapse | | Only occurs in tropical cyclones Countryside devastated. Disaster conditions. | Ships hidden in wave troughs. Air filled with spray |

$1\,m\,s^{-1} = 3.6\,km\,h^{-1} = 2.237\,mi\,h^{-1} = 1.943$ knot
$0.278\,m\,s^{-1} = 1\,km\,h^{-1} = 0.658\,mi\,h^{-1} = 0.540$ knot
$0.447\,m\,s^{-1} = 1.609\,km\,h^{-1} = 1\,mi\,h^{-1} = 0.869$ knot
$0.515\,m\,s^{-1} = 1.853\,km\,h^{-1} = 1.151\,mi\,h^{-1} = 1$ knot

Manufacturing growth since about 1980 has benefited much from the use of solid state electronics, composite materials and computer aided design.

A major design criterion is the need to protect the machine against damage in very strong winds, even though such gale-force winds are relatively infrequent. Wind forces tend to increase as the square of the wind speed. Since the 1-in-50-year gale speed will be five to ten times the average wind speed, considerable overdesign has to be incorporated for structural strength. Also wind speed fluctuates, so considerable fatigue damage can occur, especially related to the blades and drive train, from the frequent stress cycles of gravity loading (about $10^8$ cycles over twenty years of operation for a 20 m diameter, ~100 kW rated turbine, less for larger machines) and from fluctuations and turbulence in the wind. As machines are built to ever increasing size, the torque on the main shaft becomes a limiting factor.

Wind results from expansion and convection of air as solar radiation is absorbed on Earth. On a global scale these thermal effects combine with dynamic effects from the Earth's rotation to produce prevailing wind patterns (Figure 9.2). In addition to this general or synoptic behaviour of the atmosphere there is important local variation caused by geographical and environmental factors. Wind speeds increase with height, and the horizontal components are significantly greater than the vertical components. The latter are however important in causing gusts and short-term variations. The

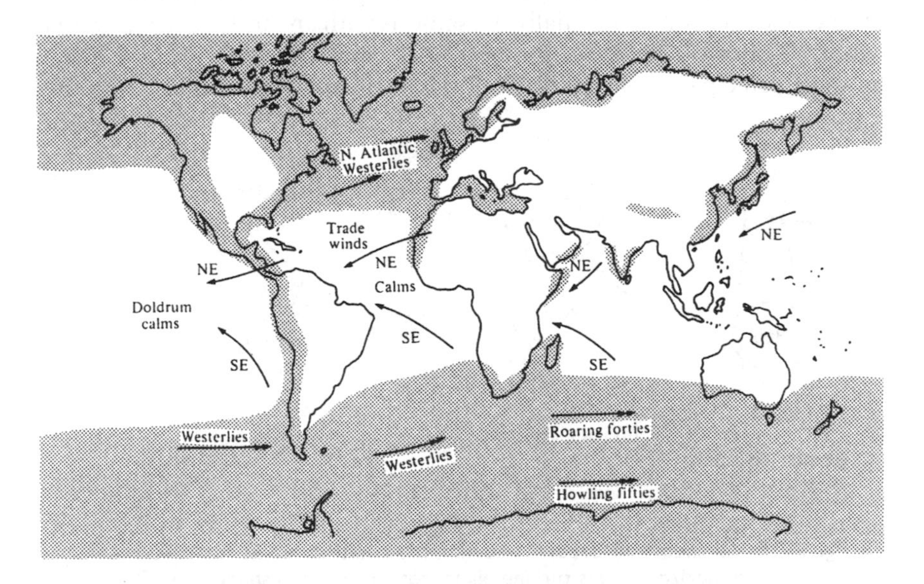


*Figure 9.2* Prevailing strong winds. The shaded areas indicate regions of wind attractive for wind power development, with average wind speed >5 m s$^{-1}$ and average generation ~33% of rated power. Note the importance of marine situations and upland sites.

kinetic energy stored in the winds is about $0.7 \times 10^{21}$ J, and this is dissipated by friction, mainly in the air, but also by contact with the ground and the sea. About 1% of absorbed solar radiation, 1200 TW ($1200 \times 10^{12}$ W), is dissipated in this way (refer to Figure 1.2).

The ultimate world use of wind power cannot be estimated in any meaningful way, since it is so dependent on the success and acceptance of machines and suitable energy end-use systems. However, without suggesting any major changes in electrical infrastructure, official estimates of wind power potential for the electrical supply of the United Kingdom are at least 25% of the total supply, a proportion now nearly attained in Denmark. With changes in the systems, e.g. significant load management and connection with hydro storage, significantly greater penetration would be possible. Autonomous wind power systems have great potential as substitutes for oil used in heating or for the generation of electricity from diesel engines. These systems are particularly applicable for remote and island communities.

## 9.2 Turbine types and terms

The names of different types of wind turbine depend on their constructional geometry, and the aerodynamics of the wind passing around the blades; also called aerofoils or airfoils. The basic aerodynamics is described in Chapter 2 (e.g. Figure 2.7), for, despite appearances, the relative motion of air with a turbine blade section is essentially the same as with an airplane wing section. Figure 9.3 shows a blade section of a horizontal axis wind turbine blade; the same principles apply to vertical-axis turbines. The section is rotating

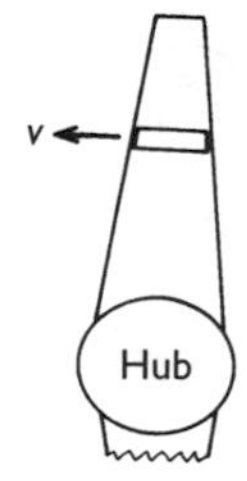


(a) Front view (section indicated)

(b) Perspective view

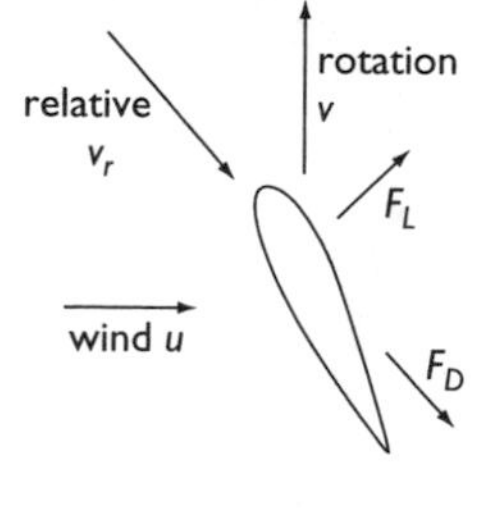


(c) Section profile

*Figure 9.3* Velocities and forces at a section of a rotating turbine blade. (a) Front view of horizontal axis turbine blade, rotating section speed $v$. (b) Perspective view, showing unperturbed wind speed $u$. (c) Section view from the blade tip, showing $v$ in the plane of rotation and distant wind speed $u$. The section 'sees' wind of relative speed $v_r$. Drag force $F_D$ is defined parallel to $v_r$. Lift force $F_L$ is defined perpendicular to $v_r$.

approximately perpendicular to the distant oncoming wind of speed $u$. Because of its own movement, the blade section experiences oncoming air at relative speed $v_r$. The comparison can be made with an airplane wing section by turning the page so Figure 9.3c has the relative air speed $v_r$ horizontal.

As the air is perturbed by the blade, a force acts which is resolved into two components. The main factors are:

1 The *drag force* $F_D$ is the component in line with the relative velocity $v_r$
2 The *lift force* $F_L$ is the component perpendicular to $F_D$. The use of the word 'lift' does not mean $F_L$ is necessarily upwards, and derives from the equivalent force on an airplane wing.
3 Rotational movement of the air occurs as the airstream flows off the blade. This may be apparent as distinct *vortices and eddies* (whirlpools of air) created near the surface. Vortex shedding occurs as these rotating masses of air break free from the surface and move away, still rotating, with this airstream.
4 The air is disturbed by the blade movement and by wind gusts, and the flow becomes erratic and perturbed. This *turbulence*, see Section 2.5, occurs before and after the rotating blades, so each individual blade may often be moving in the turbulence created by other blades.
5 The wind turbine presents a certain *solidity* to the airstream. This is the ratio of the total area of the blades at any one moment in the direction of the airstream to the swept area across the airstream. Thus, with identical blades, a four-bladed turbine presents twice the solidity of a two-bladed turbine.
6 The aerodynamic characteristics of the blades are crucial; roughness and protrusions should be avoided. Note that the predominantly 2-dimensional air flow over an airplane wing becomes 3 dimensional, and therefore more complex, for a rotating wind turbine blade.

The characteristics of a particular wind turbine are described by the answers to a number of questions; see Figure 9.4. The theoretical justification for these criteria will be given in later sections.

1 *Is the axis of rotation parallel or perpendicular to the airstream?* The former is a horizontal axis machine, the latter usually a vertical axis machine in a cross-wind configuration.
2 *Is the predominant force lift or drag?* Drag machines can have no part moving faster than the wind, but lift machines can have blade sections moving considerably faster than the wind speed. This is similar to a keeled sail boat which can sail faster than the wind.
3 *What is the solidity?* For many turbines this is described by giving the number of blades. Large solidity machines start easily with large initial

(a)

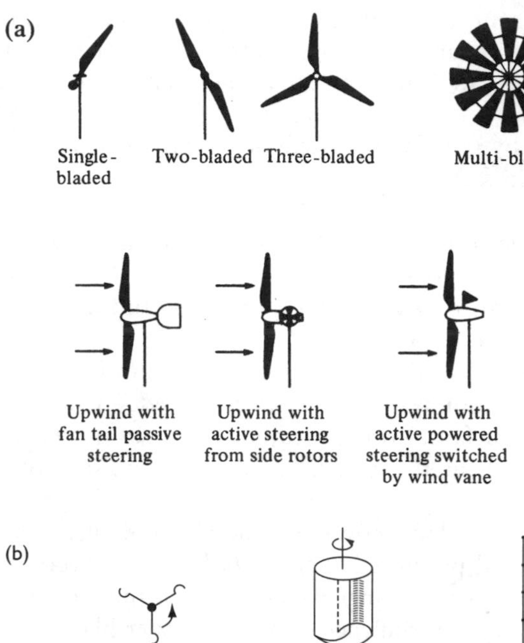


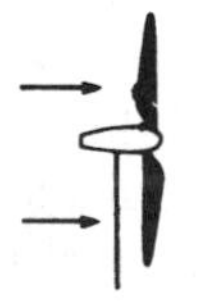


(b)

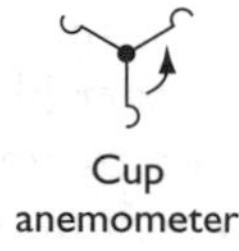


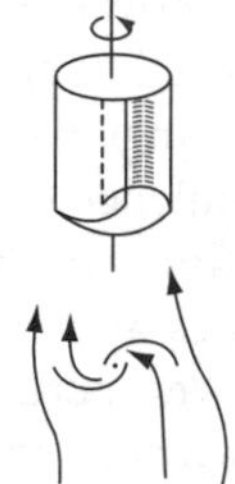


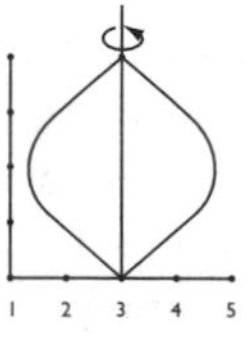


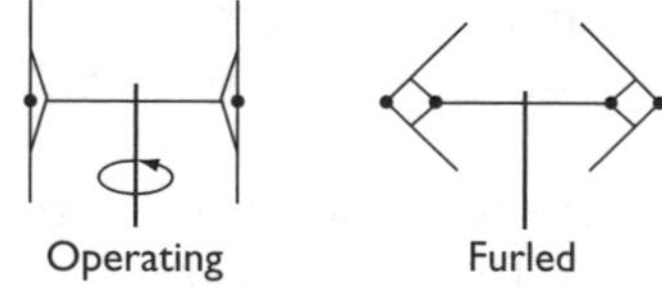

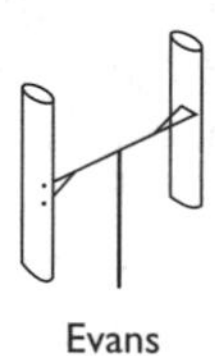


(c)

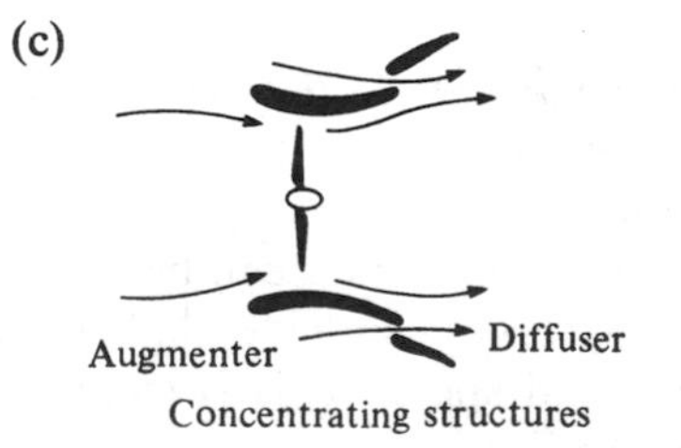


*Figure 9.4* Classification of wind machines and devices. (a) Horizontal axis. (b) Vertical axis. (c) Concentrators.

torque, but soon reach maximum power at small rotational frequency. Small solidity devices may require starting, but reach maximum power at faster rotational frequency. Thus large solidity machines are used for water pumping even in light winds. Small solidity turbines are used for electricity generation, since fast shaft rotational frequency is needed.

4 *What is the purpose of the turbine?* Historic grain windmills and water-pumping wind turbines produce mechanical power. The vast majority of modern wind turbines are for electricity generation; generally large for grid power and small for autonomous, stand-alone, power.

5 *Is the frequency of rotation controlled to be constant, or does it vary with wind speed?* A wind turbine whose generator is connected directly to a strong AC electrical grid will rotate at nearly constant frequency. However a turbine of variable frequency can be matched more efficiently to the varying wind speed than a constant frequency machine, but this requires an indirect connection through a power-electronic interface.

6 *Is the mechanical shaft directly coupled to its generator, or is connection through an intermediate step that acts as a smoothing device?* A decoupling of this kind filters out rapid torque fluctuations, and allows better matching of rotor to wind, and generator to load, than direct coupling. Partial decoupling of the turbine from the generator is a *soft coupling*. Since wind velocities fluctuate rapidly, see Section 9.6, the inertia of the wind turbine and the 'softness' of the rotor/generator coupling are used to prevent these fluctuations appearing in the electricity output. Similar effects occur if the blades are independently hinged against a spring, or hinged together (*teetered*), which smoothes forces and decreases mechanical stress.

A classification of wind machines and devices can now be given in association with Figure 9.4. This includes the main types, but numerous other designs and adaptations occur.

### 9.2.1 Horizontal-axis machines

We consider rotors with blades similar to airplane wings, sometimes misleadingly called 'propeller type'.

The dominant driving force is lift. Blades on the rotor may be in front (upwind) or behind (downwind) of the tower, see Figure 9.4(a). Wind veers frequently in a horizontal plane, and the rotor must turn in the horizontal plane (yaw) to follow the wind without oscillation. Upwind turbines need a tail or some other yawing mechanism, such as electric motor drives to maintain orientation. Downwind turbines are, in principle, self-orienting, but are more affected by the tower, which produces wind shadow and extra turbulence in the blade path. Perturbations of this kind cause cyclic stresses on the structure, additional noise and output fluctuations. Upwind and downwind machines of rotor diameter more than about 10 m use electric motors to control yaw.

Two- and three-bladed rotors are common for electricity generation. The three-bladed rotors operate more 'smoothly' and, generally, more quietly than two-bladed. Single-bladed rotors, with a counterweight, have been field tested at full scale, but the asymmetry produced too many difficulties for commercial prospects. Gearing and generators are usually at the top of the tower in a nacelle. Multi-blade rotors, having large starting torque in light winds, are used for water pumping and other low frequency mechanical power.

### 9.2.2 *Vertical-axis machines*

By turning about a vertical axis, a machine can accept wind from any direction without adjustment, whereas a horizontal-axis machine must yaw (i.e. turn in the horizontal plane to face the wind). An expectation for vertical-axis wind turbine generators is to have gear boxes and generators at ground level. Examples, from the smallest devices, are sketched in Figure 9.4(b):

1. *Cup anemometer*. This device rotates by drag force. The shape of the cups produces a nearly linear relationship between rotational frequency and wind speed, so measurement of the number of rotations per time period correlates to average wind speed over that period. The device is a standard anemometer for meteorological data.
2. *Savonius rotor* (*turbo machine*). There is a complicated motion of the wind through and around the two curved sheet airfoils. The driving force is principally drag. The construction is simple and inexpensive. The large solidity produces large starting torque, so Savonius rotors are mostly used for water pumping.
3. *Darrieus rotor*. This rotor has two or three thin curved blades with an airfoil section. The rotor shape is a catenary, with the aim of the rotating blades being only stressed along their length.
4. *Musgrove rotor*. The blades of this form of rotor are vertical for normal power generation, but tip or turn about a horizontal point for control or shutdown. There are several variations, see Figure 9.4, which are all designed to have the advantage of fail-safe shutdown in strong winds.
5. *Evans rotor*. The vertical blades change pitch about a vertical axis for control and failsafe shutdown.

For the Darrieus, Musgrove and Evans rotors, the driving wind forces are lift, with maximum turbine torque occurring when a blade moves twice per rotation across the wind. Uses are for electricity generation. The rotor is not usually self-starting. Therefore movement may be initiated with the electrical induction generator used as a motor.

A major advantage of vertical-axis machines is to eliminate gravity-induced stress/strain cycles (which occurs every rotation in the blades of horizontal axis turbines), so, in principle, vertical-axis blades may be very large. For small machines, gearing and generators may be directly coupled to the vertical main shaft at ground level. However, for larger vertical-axis machines, the high torque of the main shaft requires it to be short, so generators are not at ground level. Their principal disadvantages are (1) many vertical-axis machines have suffered from fatigue failures arising from the many natural resonances in the structure; (2) the rotational torque from the wind varies periodically within each cycle, and thus unwanted power periodicities appear at the output; and (3) guyed tower support is complex. As a result the great majority of working machines are horizontal axis, not vertical.

### 9.2.3 Concentrators

Turbines draw power from the intercepted wind, and it may be advantageous to funnel or concentrate wind into the turbine from outside the rotor section. Various systems have been developed or suggested for horizontal-axis propeller turbines, Figure 9.4.

1 *Blade tips* Various blade designs and adaptations are able to draw air into the rotor section, and hence harness power from a cross-section greater than the rotor area.
2 *Concentrating structures* Funnel shapes and deflectors fixed statically around the turbine draw the wind into the rotor. Concentrators are not yet generally used for commercial machines.

## 9.3 Linear momentum and basic theory

In this section we shall discuss important concepts for wind machines. Basic coefficients concerning power, thrust and torque will be defined. The analysis proceeds by considering the loss of linear momentum of the wind. More rigorous treatment will be outlined in later sections.

### 9.3.1 Energy extraction

In the unperturbed state (Figure 9.5) a column of wind upstream of the turbine, with cross-sectional area $A_1$ of the turbine disc, has kinetic energy passing per unit time of

$$P_0 = \frac{1}{2}(\rho A_1 u_0)\, u_0^2 = \frac{1}{2}\rho A_1 u_0^3 \tag{9.2}$$

Here $\rho$ is the air density and $u_0$ the unperturbed wind speed. This is the *power in the wind* at speed $u_0$.

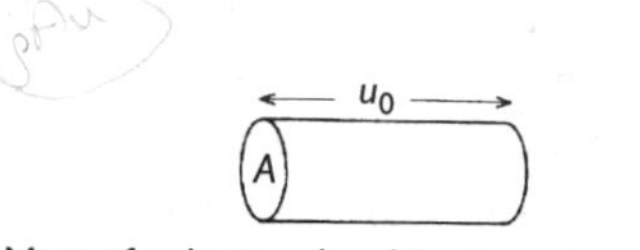


Mass of column $\rho A u_0$, kinetic energy $\frac{1}{2}(\rho A u_0)u_0^2$

*Figure 9.5* Power in wind.

Air density $\rho$ depends weakly on height and meteorological condition. Wind speed generally increases with height, is affected by local topography, and varies greatly with time. These effects are considered fully in Section 9.6, and for the present we consider $u_0$ and $\rho$ constant with time and over the area of the air column. Such incompressible flow has been treated in Chapter 2 on fluid mechanics. A typical value for $\rho$ is $1.2\,\text{kg}\,\text{m}^{-3}$ at sea level (Table B.1 in appendix), and useful power can be harnessed in moderate winds when $u_0 \sim 10\,\text{m}\,\text{s}^{-1}$ and $P_0 = 600\,\text{W}\,\text{m}^{-2}$. In gale force conditions, $u_0 \sim 25\,\text{m}\,\text{s}^{-1}$, so $P_0 \sim 10\,000\,\text{W}\,\text{m}^{-2}$; the cubic relationship of power and wind speed is strongly non-linear. Tables 9.1 and 9.2 give further details of meteorological wind conditions related to wind turbine size, and Section 9.7 considers wind properties related to power extraction.

*Table 9.2* Typical wind turbine generating characteristics at rated power in $12\,\text{m}\,\text{s}^{-1}$ wind speed. Data calculated assuming power coefficient $C_P = 30\%$, air density $\rho = 1.2\,\text{kg}\,\text{m}^{-1}$, tip-speed ratio $\lambda = 6$. Rated power $P_T = 1/2\rho[\pi D^2/4]u_0{}^3 C_P$. Hence $D = (2.02\,\text{m})\sqrt{(P/1\,\text{kW})}$, $T = (0.0436\,\text{s}\,\text{m}^{-1})D$

| class | *small* | | | | *intermediate* | | | | *large* | | |
|---|---|---|---|---|---|---|---|---|---|---|---|
| Rated power $P_T$ (kW) | 10 | 25 | 50 | 100 | 150 | 250 | 500 | 1000 | 2000 | 3000 | 4000 |
| Diameter $D$ (m) | 6.4 | 10 | 14 | 20 | 25 | 32 | 49 | 64 | 90 | 110 | 130 |
| Period $T$ (s) | 0.3 | 0.4 | 0.6 | 0.9 | 1.1 | 1.4 | 2.1 | 3.1 | 3.9 | 4.8 | 5.7 |

The theory proceeds by considering supposed constant velocity airstream lines passing through and by the turbine in laminar flow, Figure 9.6. The rotor is treated as an 'actuator disc', across which there is a change of

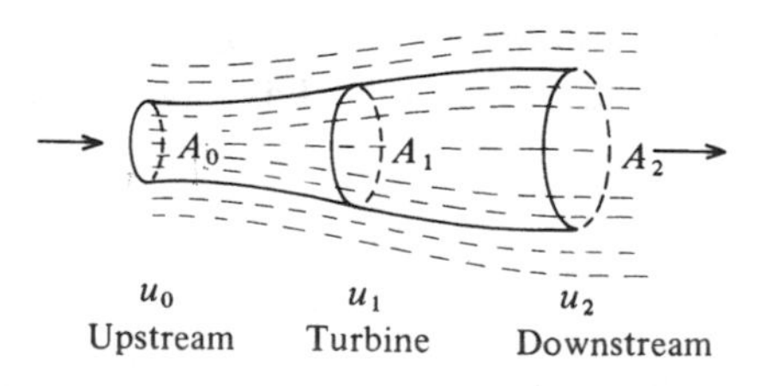


*Figure 9.6* Betz model of expanding airstream.

pressure as energy is extracted and a consequent decrease in the linear momentum of the wind. Perturbations to the smooth laminar flow are not considered here, although they undoubtedly occur because angular momentum is extracted and vortices in the air flow occur. Yet despite these severe limitations, the model is extremely useful.

Area $A_1$ is the rotor swept area, and areas $A_0$ and $A_2$ enclose the stream of constant air mass passing through $A_1$. $A_0$ is positioned in the oncoming wind front unaffected by the turbine, and $A_2$ at the position of minimum wind speed before the wind front reforms downwind. $A_0$ and $A_2$ can be located experimentally for wind speed determination. Such measurement at $A_1$ is not possible because of the rotating blades.

*Step 1: To determine $u_1$*. The force or thrust $F$ on the turbine is the reduction in momentum per unit time from the air mass flow rate $\dot{m}$

$$F = \dot{m}u_0 - \dot{m}u_2 \tag{9.3}$$

This force is applied by an assumed uniform air flow of speed $u_1$. The power extracted by the turbine is

$$P_T = Fu_1 = \dot{m}(u_0 - u_2)u_1 \tag{9.4}$$

The loss in energy per unit time by that airstream is the power extracted from the wind:

$$P_w = \tfrac{1}{2}\dot{m}({u_0}^2 - {u_2}^2) \tag{9.5}$$

Equating (9.4) and (9.5)

$$(u_0 - u_2)u_1 = \tfrac{1}{2}\left(u_0^2 - u_2^2\right) = \tfrac{1}{2}\left(u_0 - u_2\right)\left(u_0 + u_2\right) \tag{9.6}$$

Hence

$$u_1 = \tfrac{1}{2}(u_0 + u_2) \tag{9.7}$$

Thus according to this linear momentum theory, the air speed through the activator disc cannot be less than half the unperturbed wind speed.

*Step 2: knowing $u_1$, calculate the power extracted from the wind*. The mass of air flowing through the disc per unit time is given by

$$\dot{m} = \rho A_1 u_1 \tag{9.8}$$

So in (9.4),

$$P_T = \rho A_1 u_1^2(u_0 - u_2) \tag{9.9}$$

Now substitute for $u_2$ from (9.7)

$$P_T = \rho A_1 u_1^2[u_0 - (2u_1 - u_0)] = 2\rho A_1 u_1^2(u_0 - u_1) \tag{9.10}$$

The *interference factor* $a$ is the fractional wind speed decrease at the turbine. Thus

$$a = (u_0 - u_1)/u_0 \tag{9.11}$$

and

$$u_1 = (1 - a)u_0 \tag{9.12}$$

Using (9.7),

$$a = (u_0 - u_2)/(2u_0) \tag{9.13}$$

Other names for $a$ are the induction or the perturbation factor.
From (9.12), substituting for $u_1$ in (9.10),

$$\begin{aligned} P_T &= 2\rho A_1(1-a)^2 u_0^2[u_0 - (1-a)u_0] \\ &= [4a(1-a)^2](\frac{1}{2}\rho A_1 u_0^3) \end{aligned} \tag{9.14}$$

Comparing this with (9.1),

$$P_T = C_P P_0 \tag{9.15}$$

where $P_0$ is the power in the unperturbed wind, and $C_P$ is the fraction of power extracted, the *power coefficient*:

$$C_P = 4a(1 - a)^2 \tag{9.16}$$

Analysis could have proceeded in terms of the ratio $b = u_2/u_0$, sometimes also called an interference factor (see Problem 9.2).

The maximum value of $C_P$ occurs in the model when $a = 1/3$ (see Problem 9.1 and Figure 9.7):

$$C_{Pmax} = 16/27 = 0.59 \tag{9.17}$$

Note that the model predicts that, (i) when $a = 1/3$, then $u_1 = 3u_0/4$ and $u_2 = u_0/2$ and, (ii) when $a = 0.5$, $u_1 = u_0/2$ and $u_2 = 0$ (which is no longer meaningful for the turbine and indicates a change in mode of flow, as discussed later for Figure 9.13).

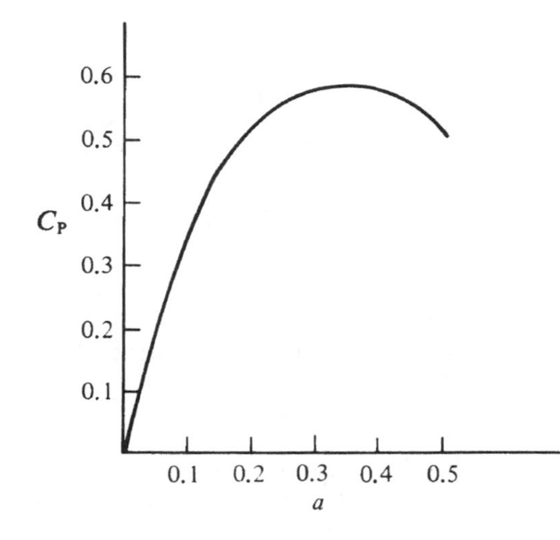


*Figure* 9.7 Power coefficient $C_p$ as a function of interference factor $a$. $C_p = 4a(1-a)^2$; $a = (u_0 - u_1)/u_0$; $C_{pmax} = 16/27 = 0.59$.

Note also that only about half the power in the wind is extracted because the air has to have kinetic energy to leave the turbine region. The criterion for maximum power extraction ($C_{pmax} = 16/27$) is called the *Betz criterion*, and may be applied to all turbines set in an extended fluid stream. Thus it applies to power extraction from tidal and river currents (see Chapter 13). With conventional hydro-power (Chapter 8) the water reaches the turbine from a pipe and is not in extended flow, so other criteria apply. In practical operation, a commercial wind turbine may have a maximum power coefficient of about 0.4, as discussed in Section 9.4. This may be described as having an efficiency relative to the Betz criterion of $0.4/0.59 = 68\%$.

The power coefficient $C_P$ is the efficiency of extracting power from the mass of air in the supposed stream tube passing through the actuator disc, area $A_1$. This incident air passes through area $A_0$ upstream of the turbine. The power extracted per unit area of $A_0$ upstream is greater than per unit area of $A_1$, since $A_0 < A_1$. It can be shown (see Problem 9.3) that the maximum power extraction per unit of $A_0$ is 8/9 of the power in the wind, and so the turbine has a maximum efficiency of 89% considered in this way. Effects of this sort are important for arrays of wind turbines in a wind farm array of turbines.

### 9.3.2 *Axial force on turbines*

It is important that a wind turbine and its tower not be blown over by strong winds. Therefore we need to estimate the forces involved. We shall

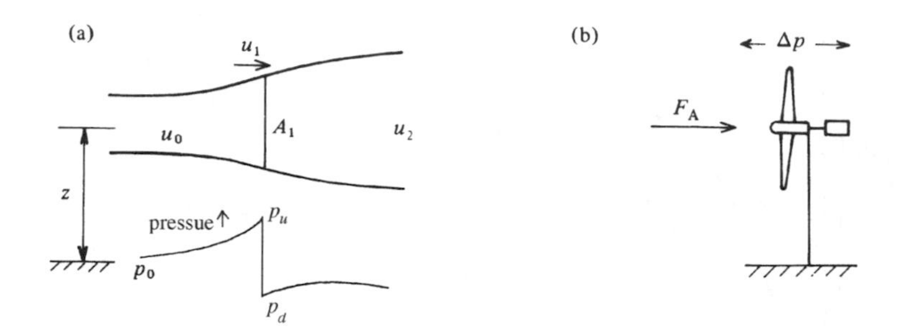


*Figure 9.8* Thrust on turbines. (a) Air flow speed $u$, pressure $p$, height $z$. (b) Axial thrust $F_A$, pressure difference $\Delta p$.

use Bernoulli's equation (2.2) to calculate the axial force, i.e. the thrust, on a wind turbine when treated as an actuator disc in streamlined flow, Figure 9.8. The effect of the turbine is to produce a pressure difference $\Delta p$ between the near upwind (subscript $u$) and near downwind (subscript $d$) parts of the flow.

From (2.2) supposing a disc with no energy extraction,

$$\frac{p_0}{\rho_0} + gz_1 + \frac{u_0^2}{2} = \frac{p_\mathrm{d}}{\rho_\mathrm{d}} + gz_\mathrm{d} + \frac{u_\mathrm{d}^2}{2} \tag{9.18}$$

The changes in $z$ and $\rho$ are negligible compared with the other terms, so if $\rho$ is the average air density then

$$\Delta p = (p_0 - p_\mathrm{d}) = (u_0^2 - u_\mathrm{d}^2)\rho/2 \tag{9.19}$$

$\Delta p$ is called the static pressure difference, and the terms in $u^2\rho/2$ are the dynamic pressures. The maximum value of static pressure difference occurs as $u_d$ approaches zero. So for a solid disc:

$$\Delta p_\mathrm{max} = \rho u_0^2/2 \tag{9.20}$$

and the maximum thrust on the disc is

$$F_{\mathrm{A\,max}} \approx \rho A_1 u_0^2/2 \tag{9.21}$$

On a horizontal axis machine the thrust is centred on the turbine axis and is called the axial thrust $F_\mathrm{A}$.

The thrust equals the rate of loss of momentum of the airstream:

$$F_A = \dot{m}(u_0 - u_2) \tag{9.22}$$

Using (9.8), (9.11) and (9.13),

$$\begin{aligned} F_A &= (\rho A_1 u_1)(2u_0 a) \\ &= \rho A_1 (1-a) u_0 (2u_0 A) \\ &= \frac{\rho A_1 u_0^2}{2} 4a(1-a) \end{aligned} \tag{9.23}$$

The term $\rho A_1 u_0^2/2$ would be the force given by this model for wind hitting a solid disc. The fraction of this force experienced by the actual turbine is the *axial force coefficient* $C_F$:

$$F_A = C_F \rho A_1 u_0^2/2$$

where

$$C_F = 4a(1-a) \tag{9.24}$$

and from (9.13)

$$a = (u_0 - u_1)/u_0 = (u_0 - u_2)/2u_0 \tag{9.25}$$

The maximum value of $C_F$ would be 1 when $a = 1/2$, equivalent to $u_2 = 0$ (i.e. the wind is stopped). Maximum power extraction by the Betz criterion occurs when $a = 1/3$, Figure 9.7 and (9.17), corresponding to $C_F = 8/9$.

In practice the maximum value of $C_F$ on a solid disc approaches 1.2 owing to edge effects. Nevertheless the linear momentum theory shows that the turbine appears to the wind as a near solid disc when extracting power. It is quite misleading to estimate the forces on a rotating wind turbine by picturing the wind passing unperturbed through the gaps between the blades. If the turbine is extracting power efficiently, these gaps are not apparent to the wind and extremely large thrust forces occur.

The term $A_1 u_0^2/2$ of (9.23) increases rapidly with increase in wind speed, and in practice wind turbines become unable to accept the thrust forces for wind speeds above about 15–20 m s$^{-1}$ unless evasive action is taken. The solutions to overcome this are (1) to turn (yaw) the turbine out of the wind, (2) to lessen power extraction and hence thrust by pitching the blades or extending spoil flaps, (3) to design fixed blades so they become extremely inefficient and self-stalling in large wind speed and (4) to stop the rotation by blade pitching and/or braking. Method (3) is perhaps the safest and cheapest, however, self-stalling blades may have a reduced power coefficient and may not give optimum power extraction in normal conditions nor smooth power control. Therefore method (2) is preferred for large commercial machines by blade pitching (not spoil flaps), since power performance can be optimised and controlled in strong winds, and the rotation stopped if necessary.

### 9.3.3 Torque

The previous calculation of axial thrust on a wind turbine provides an opportunity to introduce definitions for the torque causing rotational shaft power. At this stage no attempt is made to analyze angular momentum exchange between the air and the turbine. However, it is obvious that if the turbine turns one way the air must turn the other; full analysis must eventually consider the vortices of air circulating downwind of the turbine, see Section 9.5.

The maximum conceivable torque, $\Gamma_{max}$, on a turbine rotor would occur if the maximum thrust could somehow be applied in the plane of the turbine blades at the blade tip furthest from the axis. For a propeller turbine of radius $R$, this 'base-line' criteria would be

$$\Gamma_{max} = F_{max}R \tag{9.26}$$

Ignoring its direction for the moment, (9.21) suggests that the maximum thrust available to the turbine is

$$F_{max} = \rho A_1 u_0^2/2 \tag{9.27}$$

So

$$\Gamma_{max} = \rho A_1 u_0^2 R/2 \tag{9.28}$$

For a working machine producing an actual shaft torque $\Gamma$, the *torque coefficient* $C_\Gamma$ is defined by reference to the conceptual torque $\Gamma_{max}$

$$\Gamma = C_\Gamma \Gamma_{max} \tag{9.29}$$

In practice, for a commercial wind turbine in normal operation, $C_\Gamma <\sim 0.3$.

As will be discussed in Section 9.4, the tip-speed ratio $\lambda$ is defined as the ratio of the outer blade tip-speed $v_t$ to the unperturbed wind speed $u_0$:

$$\lambda = v_t/u_0 = R\omega/u_0 \tag{9.30}$$

where $R$ is the outer blade radius and $\omega$ is the rotational frequency.

From (9.28), substituting for $R$

$$\begin{aligned}\Gamma_{max} &= \rho A_1 u_0^2 (u_0\lambda)/2\omega \\ &= P_0\lambda/\omega\end{aligned} \tag{9.31}$$

where $P_0$ is the power in the wind from (9.2). Algebraic expressions for $\Gamma$ follow from this, see Problem 9.3(b).

The shaft power is the power derived from the turbine $P_T$, so

$$P_T = \Gamma\omega \tag{9.32}$$

Now from (9.15) $P_T = C_P P_0$. Equating the two expressions for $P_T$ and substituting for $\Gamma$ from (9.29) and (9.31) yields

$$C_P = \lambda C_\Gamma \tag{9.33}$$

So by this simplistic analysis, for the ideal turbine, $C_\Gamma$ is the slope of the $C_P$: $\lambda$ characteristic, Figure 9.12. In particular, the starting torque is the slope at the origin. However, it is important to realise that with a real rotor, it is not possible in practice to trace empirically the whole curve of $C_P$ vs $\lambda$.

Note that both $C_P$ and $C_\Gamma$ are strong functions of $\lambda$ and therefore not constant, unless the rotor has variable speed to maintain constant $\lambda$. By the Betz criterion (9.17) the maximum value of $C_P$ is 0.59, so in the 'ideal' case

$$C_{P,\max} \quad \text{at} \quad (C_\Gamma) = 0.59/\lambda \tag{9.34}$$

Figure 9.9 shows the torque characteristics of practical turbines. Large solidity turbines operate at small values of tip-speed ratio and have large starting torque. Conversely small solidity machines (e.g. with narrow two-bladed rotors) have small starting torque and perhaps may not be self-starting. At large values of $\lambda$, the torque coefficient, and hence the torque, drops to zero and the turbines 'freewheel'. Thus with all turbines there is a maximum rotational frequency in strong winds despite there being large and perhaps damaging axial thrust. Note also, that maximum torque

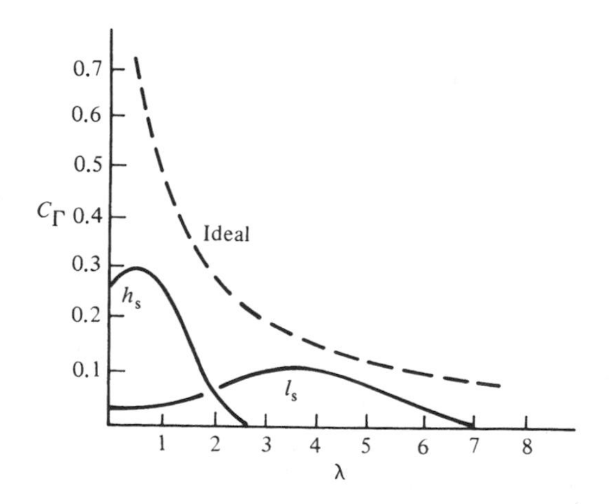


*Figure 9.9* Torque coefficient $C_\Gamma$ versus tip-speed ratio $\lambda$, sketched for large solidity $h_s$, small solidity $l_s$ and the 'ideal' criterion.

and maximum power extraction are not expected to occur at the same values of $\lambda$. The relationship of power coefficient $C_p$ to tip-speed ratio $\lambda$ is discussed in Section 9.4.

### 9.3.4 Drag machines

The ideal drag machine consists of a device with wind driven surfaces or flaps moving parallel to the undisturbed wind of speed $u_0$, Figure 9.10; compare the cup anemometer of Figure 9.4(b). The pressure difference across a stationary flap held perpendicular to the wind velocity is given by (9.20), if edge effects are neglected. For a flap of cross-section $A$ moving with a speed $v$, the maximum driving drag force is

$$F_{\max} = \rho A(u_0 - v)^2/2 \tag{9.35}$$

A dimensionless drag coefficient $C_D$ is used to describe devices departing from the ideal, so the drag force becomes

$$F_D = C_D \rho A(u_0 - v)^2/2 \tag{9.36}$$

The power transmitted to the flap is

$$P_D = F_D v = C_P \rho A(u_0 - v)^2 v/2 \tag{9.37}$$

This is a maximum with respect to $v$ when $v = u_0/3$, so

$$P_{D\max} = \frac{4}{27} C_D \frac{\rho A u_0^3}{2} \tag{9.38}$$

The power coefficient $C_P$ is defined from (9.15) as

$$P_{D\max} = C_P \rho A u_0^3/2$$

so

$$C_{P\max} = \frac{4}{27} C_D \tag{9.39}$$

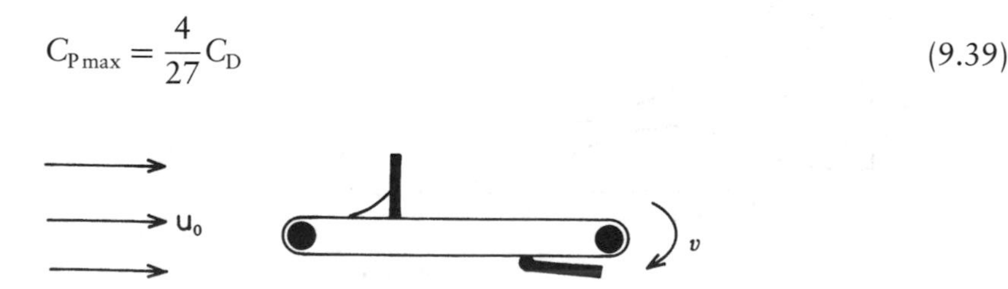


*Figure 9.10* Simplified drag machine with hinged flaps on a rotating belt.

Values of $C_D$ range from nearly zero for a pointed object, to a maximum of about 1.5 for a concave shape as used in standard anemometers. Thus maximum power coefficient for a drag machine is

$$C_{P\max} \approx \left(\frac{4}{27}\right)(1.5) = \frac{6}{27} = 22\% \tag{9.40}$$

This may be compared with the Betz criterion for an 'ideal' turbine of $C_p = 16/27 = 59\%$ (9.17). In Section 9.4 we show that lift force turbines may have power coefficients of 40% and more, and so it is possible for this Betz criterion to be approached.

Therefore drag-only devices have power extraction efficiencies of only about 33% that of lift force turbines for the same area of cross-section. Moreover returning drag flaps move against the wind, and power is reduced even more. Power extraction from drag machines may be increased slightly by incorporating more flaps or by arranging concentrated air flows. However, in practice a flap may easily meet the wakes of other flaps and power is reduced. The only way to improve drag machines is to incorporate lift forces, as happens in some forms of the Savonius rotor. Otherwise, drag machines are somewhat useless.

## 9.4 Dynamic matching

### *9.4.1 Optimal rotation rate; tip-speed ratio $\lambda$*

Wind power devices are placed in wide, extended, fluxes of air movement. The air that passes through a wind turbine cannot therefore be deflected into regions where there is no air already (unlike water onto a water turbine, Figure 8.3) and so there are distinctive limits to wind machine efficiency. Essentially the air must remain with sufficient energy to move away downwind of the turbine. The Betz criterion provides the accepted standard of 59% for the maximum extractable power, but the derivation of Section 9.3 tells us nothing about the dynamic rotational state of a turbine necessary to reach this criterion of maximum efficiency. This section explores this dynamic requirement with a qualitative analysis.

The non-dimensional characteristic for dynamic matching is the tip-speed-ratio, $\lambda$. We shall see in Section 9.5 that $\lambda$ is related to the angle, $\alpha$, at which the air is incident on the moving blade. The airfoil shape of the blades is designed for an optimum value of the angle, $\alpha_{opt}$, so the criterion for needing constant optimum tip-speed ratio $\lambda_0$ can be interpreted as the need to maintain $\alpha = \alpha_{opt}$ at all wind speeds. However, at this stage we give a qualitative analysis related to the practicalities of the fluid flow.

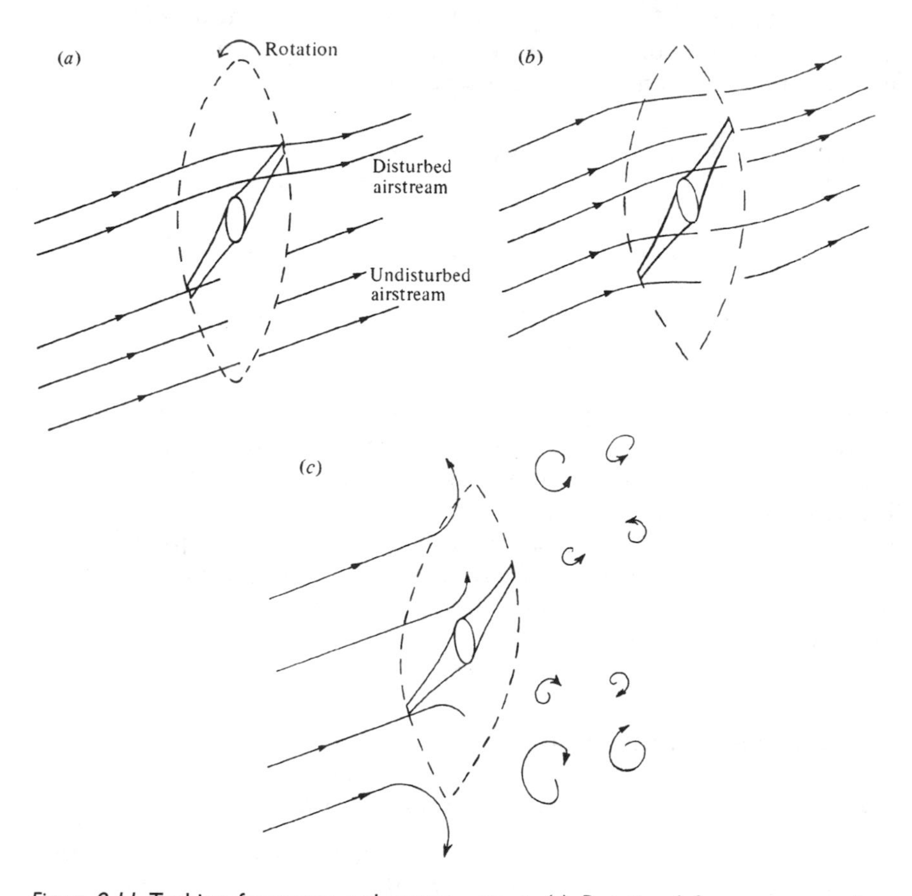


*Figure 9.11* Turbine frequency and power output. (a) Rotational frequency too slow: some wind passes unperturbed through the actuator disc. (b) Rotational frequency optimum: whole airstream affected, *d* is the 'length' of the wind strongly perturbed by the rotating blades. (c) Rotational frequency too fast: energy is dissipated in turbulent motion and vortex shedding.

Power extraction efficiency will decrease from an optimum, see Figure 9.11, if:

1 The blades are so close together, or rotating so rapidly, that a following blade moves into the turbulent air created by a preceding blade; or
2 The blades are so far apart or rotating so slowly that much of the air passes through the cross-section of the device without interfering with a blade.

It therefore becomes important to match the rotational frequency of the turbine to particular wind speeds so that the optimum efficiency is obtained.

Power extraction is a function of the time $t_b$ for one blade to move into the position previously occupied by the preceding blade, as compared with the time $t_w$ between the disturbed wind moving past that position and the normal airstream becoming re-established. $t_w$ varies with the size and shape of the blades and inversely as the wind speed.

For an $n$-bladed turbine rotating at angular velocity $\Omega$

$$t_b \approx \frac{2\pi}{n\Omega} \tag{9.41}$$

A disturbance at the turbine disc created by a blade into which the following blade moves will last for a time $t_w$, where

$$t_w \approx d/u_0 \tag{9.42}$$

Here $u_0$ is the speed of the oncoming wind and the equation defines a distance $d$ of the 'length' of the oncoming wind significantly perturbed by the rotating blades.

Maximum power extraction occurs when $t_w \approx t_b$ at the blade tips, where maximum incremental area is swept by the blades. From (9.41) and (9.42), therefore,

$$\frac{n\Omega}{u_0} \approx \frac{2\pi}{d} \tag{9.43}$$

If each side of this equation is multiplied by $R$, the blade-tip radius of rotation, then $R\Omega/u_0 \sim 2\pi R/(nd)$. Also if the tip-speed ratio $\lambda$ is defined as in (9.30) by

$$\lambda = \frac{\text{speed of tip}}{\text{speed of oncoming wind}} = \frac{R\Omega}{u_0} \tag{9.44}$$

then at optimum power extraction

$$\lambda_0 \approx \frac{2\pi}{n}\left(\frac{R}{d}\right) \tag{9.45}$$

Let $d = kR$, so in (9.45), the tip-speed ratio for maximum power extraction is

$$\lambda_0 \approx \frac{2\pi}{kn} \tag{9.46}$$

Practical results show that $k \sim 1/2$, so for an $n$-bladed turbine at optimum power extraction

$$\lambda_0 \approx 4\pi/n \tag{9.47}$$

For example, for a two-bladed turbine $C_{Pmax}$ occurs for $\lambda_0 \approx 4\pi/2 \approx 6$, and for a four-bladed turbine $C_{Pmax}$ occurs for $\lambda_0 \approx 4\pi/4 \approx 3$.

The preceding discussion is not rigorous, but it does describe the most important phenomena. With carefully designed airfoils, optimum tip-speed ratio $\lambda_0$ may be ~30% more than these values.

The Betz efficiency criterion of Section 9.3 takes no account of any dynamic effects. There are several approaches to such dynamic calculation (see the review by Shepherd). Glauert's criterion, describing the variation of power coefficient $C_P$ with tip-speed ratio $\lambda$, becomes of value. Both criteria and the relationship of $C_P$ and $\lambda$ for a variety of wind turbine types are shown in Figure 9.12. A further constraint on the design of high angular velocity turbines is that the tip-speed should not reach the speed of sound ($330\,\mathrm{m\,s^{-1}}$), so creating shock waves. This is possible for well-matched two-bladed turbines in strong winds of speed $\sim 50\,\mathrm{m\,s^{-1}}$.

Tip-speed ratio is probably the most important parameter of a wind turbine, since it relates to the angles of attack of the relative wind speed on the blade airfoil. It is a function of the three most important variables:

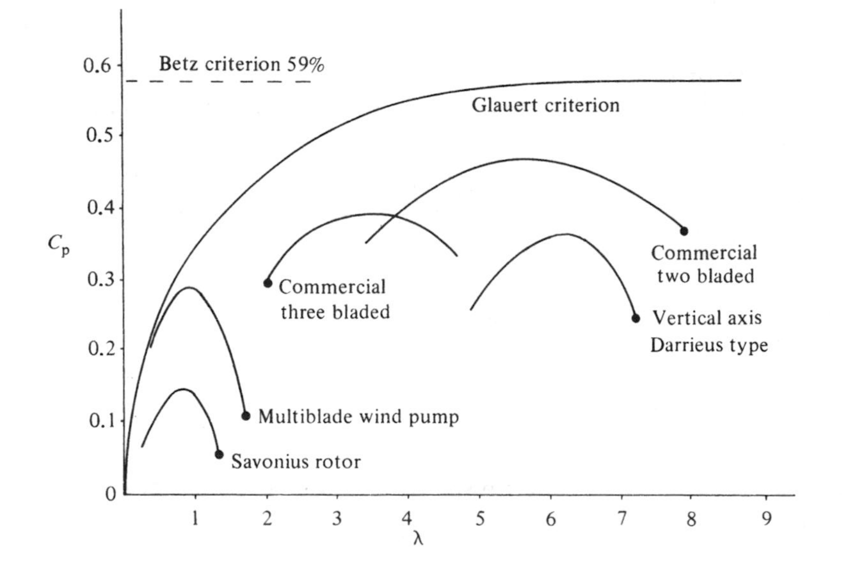


*Figure 9.12* Power coefficient $C_p$ as a function of tip-speed ratio $\lambda$ for a range of machine types.

blade swept radius, wind speed and rotor frequency. Being dimensionless, it becomes an essential scaling factor in design and analysis.

### 9.4.2 *Extensions of linear momentum theory*

Figure 9.7 is a graph of power coefficient $C_P$ against interference factor $a$ in the range $0 < a < 0.5$, as given by simple linear momentum theory. Thus from (9.16),

$$C_P = 4a(1-a)^2 \tag{9.48}$$

where, from (9.11),

$$a = 1 - (u_1/u_0) \tag{9.49}$$

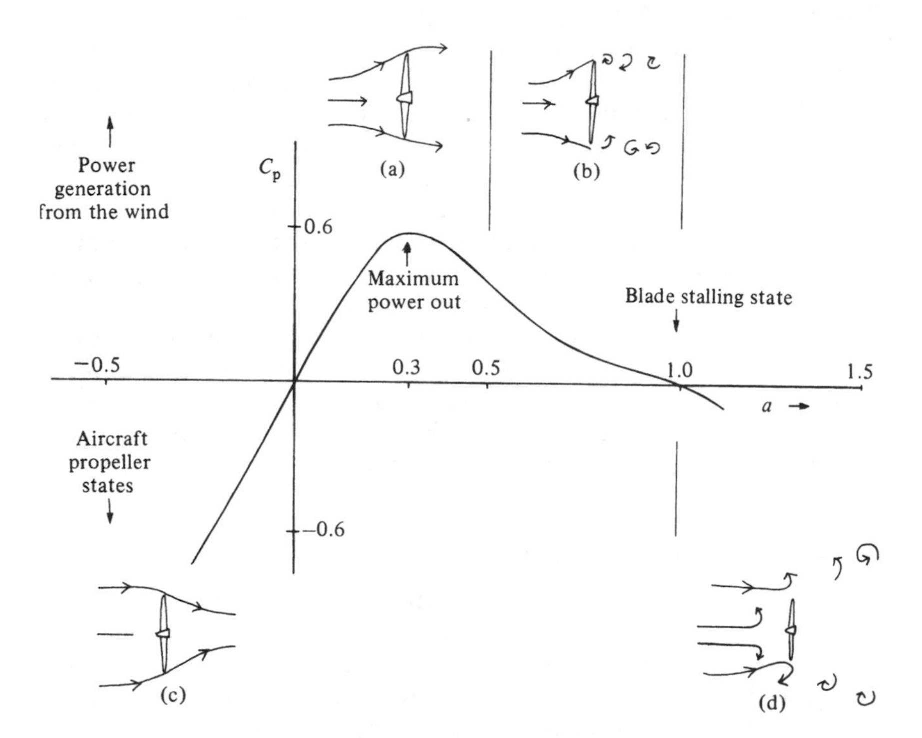


*Figure 9.13* Power coefficient $C_P$ versus interference factor $a = 1 - u_1/u_0$, as given by the linear momentum model. The results are related to practical experience of air motion and turbine/propeller states. (a) As Figure 9.6, normal energy abstraction by a wind turbine. (b) Turbulent wake reduces efficiency, as occurs with extreme wind speeds. (c) Normal airflow of an aircraft propeller; energy is added to the airstream. (d) Equivalent to aircraft propeller reverse-thrust for braking on landing.

Extensions to the simple theoretical model extend analysis into other regions of the interference factor, and link turbine driven performance with aircraft propeller characteristics. In Figure 9.13, the airstreams are sketched on the graph for specific regions that may be associated with actual air flow conditions:

1 $a < 0$, $C_P$ *negative*. This describes airplane propeller action where power is added to the flow to obtain forward thrust. In this way the propellers pull themselves into the incoming airstream and propel the airplane forward.
2 $0 < a < 0.5$, $C_P$ *positive and peaking*. At $a = 0$, $u_1 = u_0$ and $C_P = 0$; the turbine rotates freely in the wind and is not coupled to a load to perform work. As a load is applied, power is abstracted, so $C_P$ increases as $u_1$ decreases. Maximum power is removed from the airstream when $a = 1/3$ and $u_1 = 2u_o/3$, (9.17) and (9.12). At $a = 1/2$, the basic linear momentum theory models a solid disc, by predicting maximum thrust on the turbine (9.24) with axial force coefficient $C_F = 1$.
3 $0.5 < a < 1$, $C_P$ *decreasing to zero*. From (9.25), $a = (u_0 - u_2)/2u_0$. When $a = 0.5$, the model has $u_2 = 0$; i.e. the modelled wind exits perpendicular to the input flow. In practice it is possible to consider this region as equivalent to the onset of turbulent downwind air motion. It is equivalent to a turbine operating in extreme wind speeds when the power extraction efficiency decreases, owing to a mismatch of rotational frequency and wind speed. At $a = 1$, $C_P = 0$, the turbine is turning and causing extensive turbulence in the airstream, but no power is extracted. Real turbines may reach this state in a stall condition.
4 $a > 1$. This implies negative $u_1$ and is met when an airplane reverses thrust by changing blade pitch on landing. Intense vortex shedding occurs in the airstream as the air passes the propellers. In the airplane, additional energy is being added to the airstream and is apparent in the vortices, yet the total effect is a reverse thrust to increase braking.

## 9.5 Blade element theory

More advanced theory allows the calculation from basic principles of, for instance, the rotor power coefficient $C_P$ and hence the power production $P$ as a function of wind speed $u$. We give here only an outline of this theory, called 'blade element theory' or 'stream tube theory'. As the names suggest, the theory considers blade sections (elements) and the cylinders of the airstream (streamtubes) moving onto the rotor. Each blade element is associated with a standard aerofoil cross section. The lift and drag forces on most common aerofoil shapes have been measured as a function of speed and tabulated (notably by NASA and its predecessor agencies). Given these data and the pitch setting of the relative wind speed $v_r$ to the section, the forces that turn the rotor can be calculated by integration along each blade.

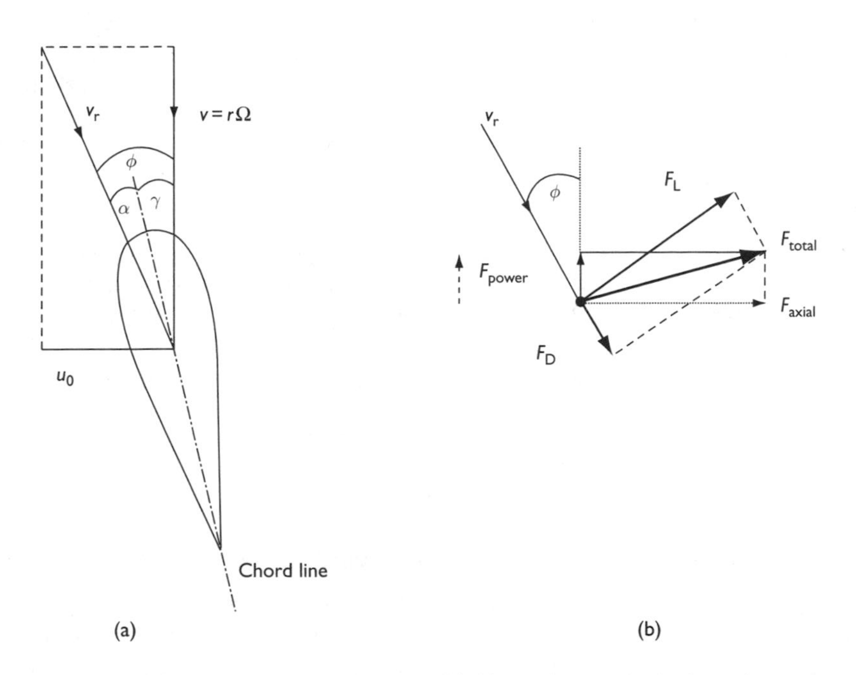


*Figure 9.14* Blade element parameters. (a) Upwind, unperturbed wind speed $u_0$. Rotating blade tip has speed $r\Omega$ ($r$: blade radius, $\Omega$: rotation rate). Relative speed of air and blade $v_r$. Angle of attack $\alpha$. Blade setting angle $\gamma$. (b) Same geometry, showing forces. The total force $F_{total}$ on the aerofoil section can be resolved into a drag force $F_D$ parallel to $v_r$ and a (usually larger) lift force $F_L$ perpendicular to $v_r$. $F_{total}$ is itself resolved into a force $F_{axial}$ along the axis of the turbine and a force $F_{power}$ in the plane of rotation of the blades. If $F_L > F_D$, $F_{power}$ should produce an accelerating torque on the rotor and hence wind turbine power production.

Such analysis of horizontal axis turbine performance models the forces from the oncoming airstream on each element of the rotating blades. Figure 9.14 shows the important initial parameters for a section of the blade whose orientation is defined by the direction of the chord (the line between the extremities of the leading and trailing edge of the blade section, which is effectively the 'zero lift line'). At radius $r$ from the axis, the blade element has a speed $r\Omega$ perpendicular to the unperturbed oncoming wind of speed $u_0$. The relative speed between this element and the moving air is $v_r$, with $v_r$ expected for electricity generation to be about five to ten times greater than $u_0$ at the blade tip. At this position along the blade, the blade setting angle is $\gamma$, the angle of attack of the relative wind speed is $\alpha$, with $\phi = \gamma + \alpha$. Note that cotan $\phi = R\Omega/u_0 = \lambda$, the tip-speed ratio. Best performance for the airfoil section occurs when the angle of attack $\alpha$ is maintained constant, i.e. in effect the tip speed is maintained constant at its optimum, hence the rotation speed should vary directly as the wind speed.

By definition, $F_D$ is the drag force parallel to $v_r$, and $F_L$ is the lift force perpendicular to $v_r$. If $F_{axial}$ is the increment of axial force, and $F_{power}$ that of tangential force (which produces rotor acceleration and power), then

$$F_{axial} = F_L \cos\phi + F_D \sin\phi \tag{9.50}$$

$$F_{power} = F_L \sin\phi + F_D \cos\phi \tag{9.51}$$

Stream tube theory considers a cylinder or tube of the oncoming wind incident on elements of the blades at radius $r$ from the axis, chord (width) $c = c(r)$, and incremental length $dr$. One such cylinder can be treated independently of others both upstream and downstream of the turbine disc. Advanced texts should be consulted for the further development, e.g. Burton *et al.*, (2000), Hansen (2000).

## 9.6 Characteristics of the wind

### 9.6.1 *Basic meteorological data and wind speed time series*

All countries have national meteorological services that record and publish weather related data, including wind speeds and directions. The methods are well established and co-ordinated within the World Meteorological Organisation in Geneva, with a main aim of providing continuous runs of data for many years. Consequently only the most basic data tend to be recorded at a few permanently staffed stations using robust and trusted equipment. Unfortunately for wind power prediction, measurements of wind speed tend to be measured only at the one standard height of 10 m, and at stations near to airports or towns where shielding from the wind might be a natural feature of the site. Therefore to predict wind power conditions at a specific site, standard meteorological wind data from the nearest station are only useful to provide first order estimates, but are not sufficient for detailed planning. Usually careful measurements around the nominated site are needed at several locations and heights for several months to a year. These detailed measurements can then be related to the standard meteorological data, and these provide a long-term base for comparison. In addition, information is held at specialist wind power data banks that are obtained from aircraft measurements, wind power installations and mathematical modelling, etc. Such organised and accessible information is increasingly available on the Internet. Wind power prediction models (e.g. the propriety WAsP models developed in Denmark) enable detailed wind power prediction for wind turbine prospective sites, even in hilly terrain.

Classification of wind speeds by meteorological offices is linked to the historical Beaufort scale, which itself relates to visual observations. Table 9.1 gives details together with the relationship between various units of wind speed.

A standard meteorological measurement of wind speed measures the 'length' or 'run' of the wind passing a 10 m high cup anemometer in 10 min. Such measurements may be taken hourly, but usually less frequently. Such data give little information about fluctuations in the speed and direction of the wind necessary for accurately predicting wind turbine performance. Continuously reading anemometers are better, but these too will have a finite response time. A typical continuous reading trace, Figure 9.15(a), shows the rapid and random fluctuations that occur. Transformation of such data into the frequency domain gives the range and importance of these variations, Figure 9.15(b).

The direction of the wind refers to the compass bearing *from* which the wind comes. Meteorological data are usually presented as a wind rose, Figure 9.16(a), showing the average speed of the wind within certain ranges of direction. It is also possible to show the distribution of speeds from these directions on a wind rose, Figure 9.16(b). Such information is of great importance when siting a wind machine in hilly country, near buildings, or in arrays of several machines where shielding could occur. Changes in wind direction may be called 'wind shift'; $0.5\,\text{rad}\,\text{s}^{-1}$ ($30°\,\text{s}^{-1}$) is a rapid change, e.g. in hilly terrain. Such changes may damage a wind turbine more than an extreme change in wind speed.

### 9.6.2 Variation with height

Wind speed varies considerably with height above ground; this is referred to as *wind shear*. A machine with a hub height of (say) 30 m above other obstacles will experience far stronger winds than a person at ground level. Figure 9.17 shows the form of wind speed variation with height $z$ in the near-to-ground boundary layer up to about 100 m. At $z = 0$ the air speed is always zero. Within the height of local obstructions wind speed increases erratically, and violent directional fluctuations can occur in strong winds. Above this erratic region, the height/wind speed profile is given by expressions of the form

$$z - d = z_0 \exp(u_z/V) \tag{9.52}$$

Hence

$$u_z = V \ln\left(\frac{z-d}{z_0}\right) \tag{9.53}$$

Here $d$ is the zero plane displacement with magnitude a little less than the height of local obstructions, the term $z_0$ is called the roughness length and $V$ is a characteristic speed. On Figure 9.17 the function is extrapolated to negative values of $u$ to show the form of the expression. Readers should

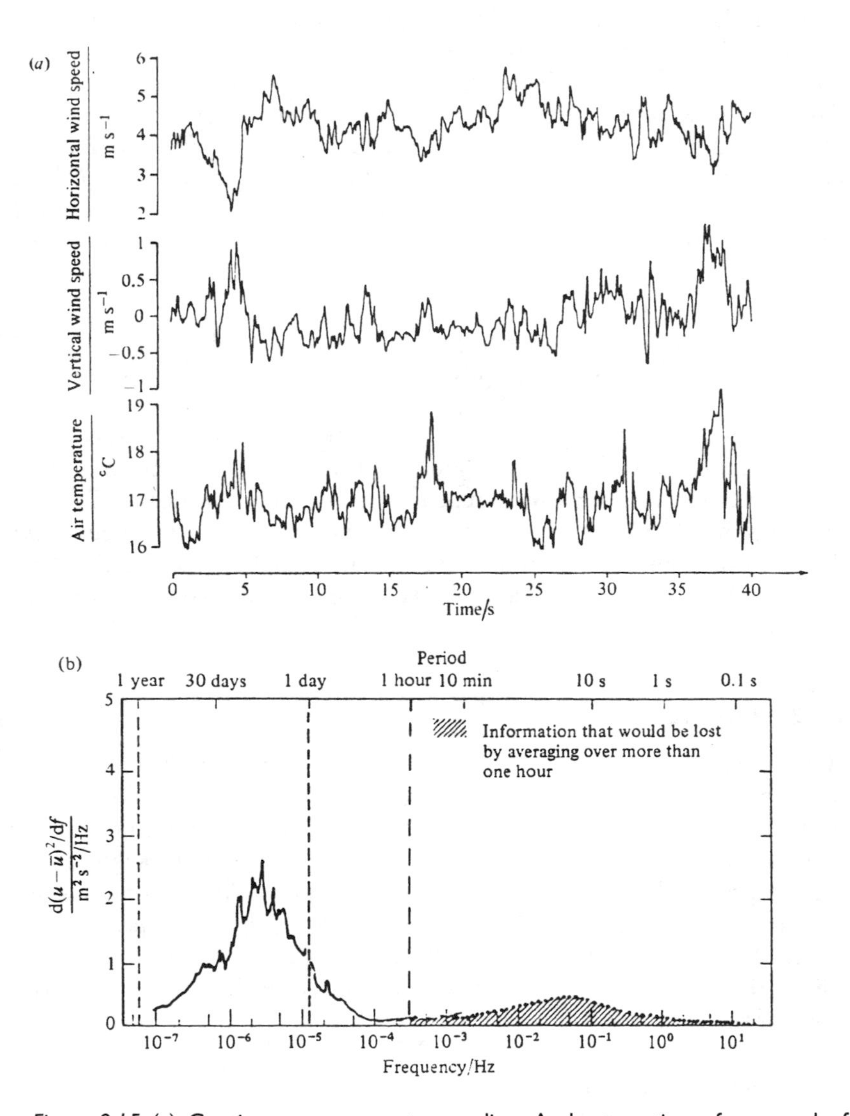


*Figure 9.15* (a) Continuous anemometer reading. A short section of a record of horizontal wind speed, vertical wind speed and temperature at a height of 2 m at the meteorological field, Reading University, UK. Note the positive correlations between vertical wind speed and temperature, and the negative correlations between horizontal and vertical wind speeds. (b) Frequency domain variance spectrum [after Petersen (1975)]. The graph is a transformation of many time series measurements in Denmark, which have been used to find the square of the standard deviation (the variance) of the wind speed $u$ from the mean speed $\bar{u}$. Thus the graph relates to the energy in wind speed fluctuations as a function of their frequency; it is sometimes called a 'Van der Hoven' spectrum.

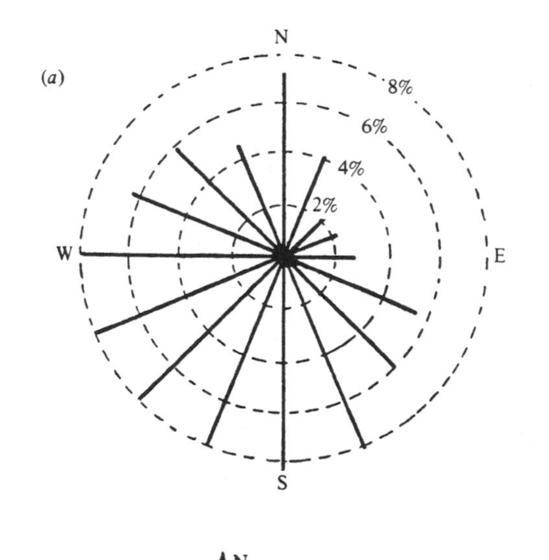


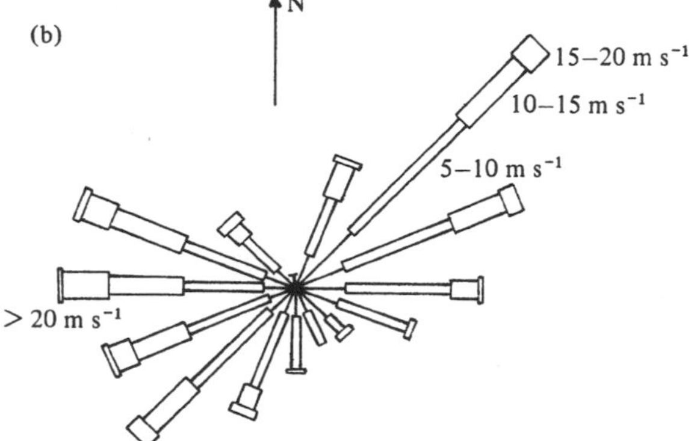


*Figure 9.16* Wind rose from accumulated data. (a) Direction. Station on the Scottish island of Tiree in the Outer Hebrides. The radial lines give percentages of the year during which the wind blows from each of 16 directions. The values are ten-year means and refer to an effective height above ground of 13 m. (b) Direction and distribution of speed. Malabar Hill on Lord Howe Island, New South Wales. The thicker sections represent the proportion of time the wind speed is between the specified values, within 16 directional sectors. After Bowden *et al.* (1983).

consult texts on meteorology and micrometeorology for correct detail and understanding of wind speed boundary layer profiles. However, the most important practical aspect is the need to place a turbine well above the height of local obstructions to ensure that the turbine rotor disc receives a strong uniform wind flux across its area without erratic fluctuations.

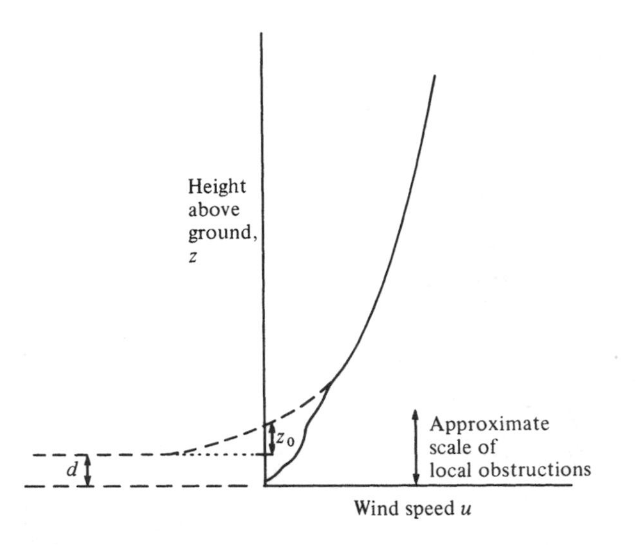


*Figure 9.17* Wind speed variation with height; 'wind shear', see (9.52).

The best sites for wind power are at the top of smooth dome-shaped hills that are positioned clear of other hills. In general the wind should be incident across water surfaces or smooth land for several hundred metres, i.e. there should be a good fetch. Most wind turbines operate at hub heights between 5 m (battery chargers) and 100 m (large, grid linked). However, it is common for standard meteorological wind speed measurements $u_s$ to be taken at a height of 10 m. An approximate expression often then used to determine the wind speed $u_z$ at height $z$ is

$$u_z = u_s \left(\frac{z}{10\,\mathrm{m}}\right)^{b'} \tag{9.54}$$

It is often stated that $b' = 1/7 = 0.14$ for open sites in 'non-hilly' country. Good sites should have small values of $b'$ to avoid changes in oncoming wind speed across the turbine disc, and large values of mean wind speed $\overline{u}$ to increase power extraction. Great care should be taken with this formula, especially for $z > 50\,\mathrm{m}$. Problem 9.12 shows that if (9.54) remains realistic, then extremely high towers ($>100\,\mathrm{m}$, say) are probably unwarranted unless the rotor diameter is very large also.

### 9.6.3 *Wind speed analysis, probability and prediction*

Implementation of wind power requires knowledge of future wind speed at the turbine sites. Such information is essential for the design of the machines and the energy systems, and for the economics. The seemingly random nature of wind and the site-specific characteristics makes such information challenging, yet much can be done from statistical analysis, from correlation of measurement time-series and from meteorology. The development of wind power has led to great sophistication in the associated analysis, especially involving data handling techniques and computer modelling. However, Example 9.1 and Table 9.3 illustrate the principles of such analysis, by showing how the power available from the wind at a particular site can be calculated from very basic measured data on the distribution of wind speed at that site. Commercial measurement techniques are more sophisticated, but the principles are the same.

*Example 9.1 Wind speed analysis for the island of North Ronaldsay, Orkney*

A ten-minute 'run of the wind' anemometer was installed at 10 m height on an open site near a proposed wind turbine. Five readings were recorded each day at 9 a.m., 12 noon, 3 p.m., 6 p.m. and 9 p.m., throughout the year. Table 9.3 gives a selection of the total data and analysis, with columns numbered as below.

*Table 9.3* Wind speed analysis for the example of North Ronaldsay. This is a selection of the full data of Barbour (1984), to show the method of calculation. Columns numbered as in Example 9.1

| 1 | 2 | 3 | 4 | 5 | 6 | 7 | 8 | 9 |
|---|---|---|---|---|---|---|---|---|
| $u'$ | $dN/du$ | $\Phi_u$ | $\Phi_{u\geq u'}$ | $\Phi_u u$ | $u^3$ | $\Phi_u u^3$ | $P_u$ | $P_u \Phi_u$ |
| $(m s^{-1})$ | $(m s^{-1})^{-1}$ | $(m s^{-1})^{-1}$ | | | $(m s^{-1})^3$ | $(m s^{-1})^2$ | $kW m^{-2}$ | $(W/m^2)/(m/s)$ |
| >26 | 1 | 0.000 | 0.000 | 0.000 | 17576 | 0.0 | 11.4 | 0.0 |
| 25 | 1 | 0.001 | 0.001 | 0.025 | 15625 | 15.6 | 10.2 | 10.2 |
| 24 | 1 | 0.001 | 0.002 | 0.024 | 13824 | 20.7 | 9.0 | 9.0 |
| 23 | 2 | 0.002 | 0.004 | 0.046 | 12167 | 18.3 | 7.9 | 15.8 |
| 22 | 4 | 0.002 | 0.006 | 0.044 | 10648 | 21.3 | 6.9 | 13.8 |
| . | . | . | . | . | . | . | . | . |
| . | . | . | . | . | . | . | . | . |
| 8 | 160 | 0.091 | 0.506 | 0.728 | 512 | 46.6 | 0.3 | 27.3 |
| 7 | 175 | 0.099 | 0.605 | 0.693 | 343 | 340 | 0.2 | 19.8 |
| 6 | 179 | 0.102 | 0.707 | 0.612 | 216 | 22.0 | 0.1 | 10.2 |
| 5 | 172 | 0.098 | 0.805 | 0.805 | 125 | 12.3 | 0.1 | 9.8 |
| 4 | 136 | 0.077 | 0.882 | 0.882 | 64 | 4.9 | 0.0 | 0.0 |
| . | . | . | . | . | . | . | . | . |
| . | . | . | . | . | . | . | . | . |
| 0 | 12 | 0.007 | | 0 | 0 | | 0 | 0 |
| Totals | 1763 | 1.000 | | 8.171 | | 1044.8 | | |
| Comment | | Peaks at $6.2 m s^{-1}$ | | $u_m = 8.2 m s^{-1}$ | | $(\overline{u^3})^{1/3} = 10.1 m s^{-1}$ | | |

1 Readings were classed within intervals of $\Delta u = 1\,\mathrm{m\,s^{-1}}$, i.e. 0.0–0.9; 1.0–1.9, etc. A total of $N = 1763$ readings were recorded, with 62 missing owing to random faults.
2 The number of occurrences of readings in each class was counted to give $\Delta N(u)/\Delta u$, with units of number per speed range ($dN/du$ in Table 9.3).
*Note:* $\Delta N(u)/\Delta u$ *is* a number per speed range, and so is called a frequency distribution of wind speed. Take care, however, to clarify the interval of the speed range $\Delta u$ (in this case 1 m/s , but often a larger interval).
3 $[\Delta N(u)/\Delta u]/N = \Phi_u$ is a normalized probability function, often called the *probability distribution of wind speed*. $\Phi_u$ is plotted against $u$ in Figure 9.18. The unit of $\Phi_u$ is the inverse of speed interval, in this case $(1\,\mathrm{m\,s})^{-1}$. $\Phi_u \Delta u$ is the probability that the wind speed is in the class defined by $u$ (i.e. $u$ to $u + \Delta u$). For one year $\Sigma \Phi_u \Delta u = 1$.
4 The cumulative total of the values of $\Phi_u \Delta u$ is tabulated to give the probability, $\Phi_{u>u'}$, of speeds greater than a particular speed $u'$. The units are number per speed range multiplied by speed ranges i.e. dimensionless. This function is plotted in Figure 9.19, and may be interpreted as the proportion of time in the year that $u$ exceeds $u'$.
5 The average or mean wind speed $u_{\mathrm{m}}$ is calculated from $u_{\mathrm{m}} \Sigma \Phi_u = \Sigma \Phi_u u$. The mean speed $u_{\mathrm{m}} = 8.2\,\mathrm{m\,s^{-1}}$ is indicated on Figure 9.18. Notice that $u_{\mathrm{m}}$ is greater than the most probable wind speed of $6.2\,\mathrm{m\,s^{-1}}$ on this distribution.

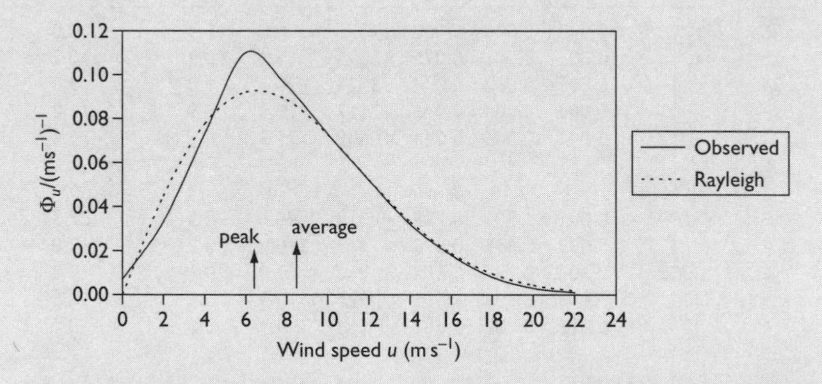


*Figure 9.18* Probability distribution of wind speed against wind speed. Data for North Ronaldsay from Barbour. (——) measured data (from Table 9.3) ; (- - -) Rayleigh distribution (9.69) fitted to match mean speed $\bar{u}$. Note that the average wind speed ($8.2\,\mathrm{m\,s^{-1}}$) exceeds the most probable wind speed ($6.2\,\mathrm{m\,s^{-1}}$); see Example 9.1 and (9.75).

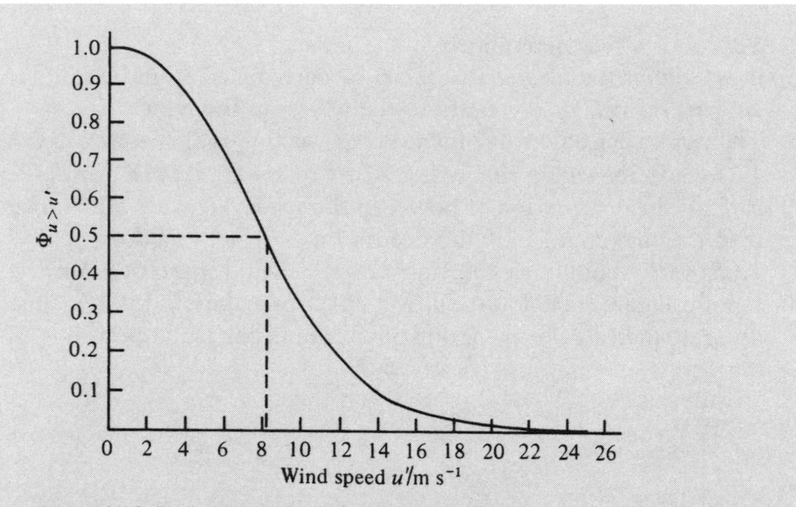


*Figure 9.19* Probability of wind speeds greater than a particular speed $u'$, for example of North Ronaldsay.

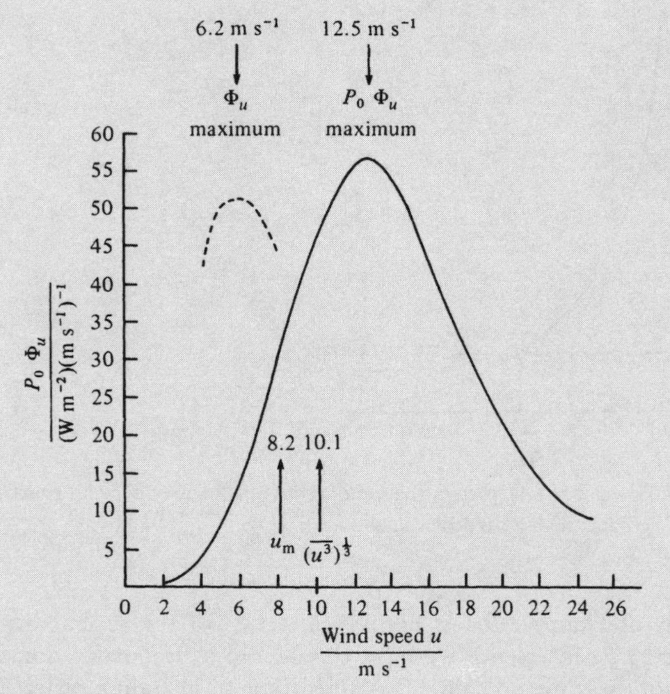


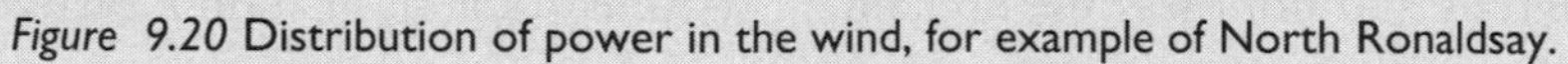

*Figure 9.20* Distribution of power in the wind, for example of North Ronaldsay.

6 Values of $u^3$ are determined.
7 $\Phi_u u^3$ allows the mean value $\overline{u^3}$ to be determined from $(\overline{u^3})\Sigma\Phi_u = \Sigma\Phi_u u^3$, see (9.71). $\overline{u^3}$ relates to the power in the wind.
8 The power per unit area of wind cross-section is $P_0 = \frac{1}{2}\rho u^3$. If $\rho = 1.3\,\mathrm{kg\,m^{-3}}$ then $P_u = Ku^3$ where $K = (0.65 \times 10^{-3})\,\mathrm{W\,m^{-2}(m/s)^{-3}}$.
9 $P_u\Phi_u$ is the distribution of power in the wind, Figure 9.20. Notice that the maximum of $P_u\Phi_u$ occurs on North Ronaldsay at $u = 12.5\,\mathrm{m\,s^{-1}}$, about twice the most probable wind speed of $6.2\,\mathrm{m\,s^{-1}}$.
10 Finally, Figure 9.21 plots the power unit area in the wind at $u'$ against $\Phi_{u>u'}$, to indicate the likelihood of obtaining particular power.

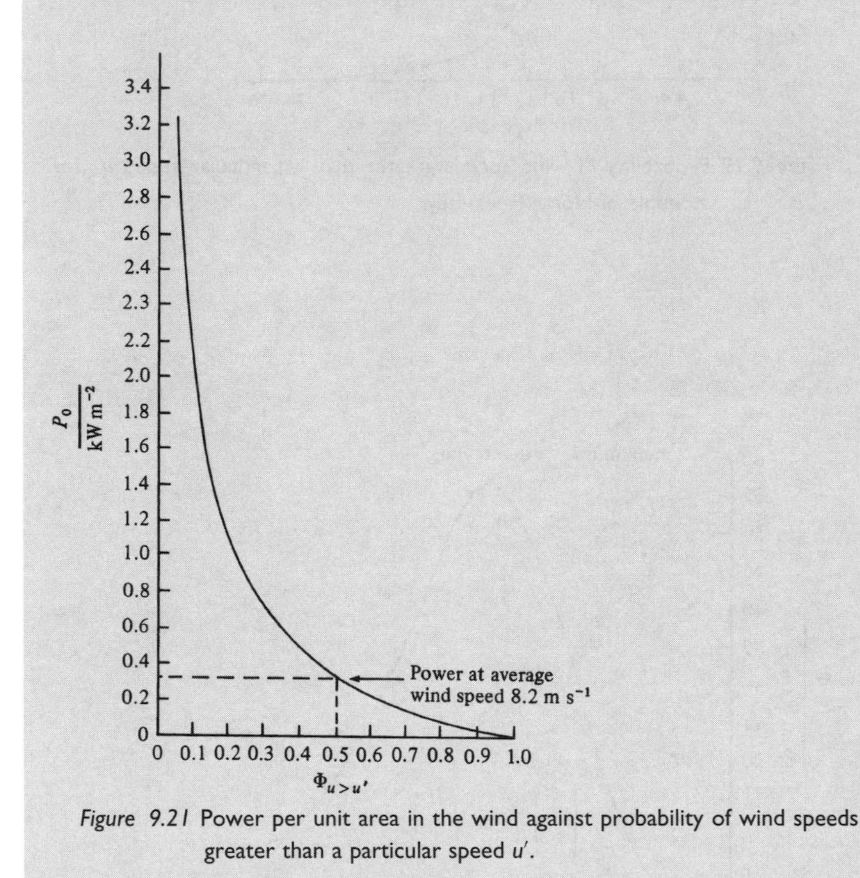


*Figure 9.21* Power per unit area in the wind against probability of wind speeds greater than a particular speed $u'$.

The analysis of Example 9.1 is entirely in terms of the probability of wind characteristics; in essence we have considered a 'frequency domain' analysis and not the 'time domain'. The time domain, including turbulence and gustiness, is considered in Section 9.6.5.

### 9.6.4 *Wind speed probability distributions: Weibull and Rayleigh*

The analysis of Example 9.1 depended solely on field data and repetitive numerical calculation. It would be extremely useful if the important function $\Phi_u$, the probability distribution of wind speed, could be given an algebraic form that accurately fitted the data. Two advantages follow: (1) fewer data need be measured, and (2) analytic calculation of wind machine performance could be attempted.

Using the symbols of the previous section,

$$\Phi_{u>u'} = \int_{u=u'}^{\infty} \Phi_u(u)\mathrm{d}u = 1 - \int_0^{u'} \Phi_u \,\mathrm{d}u \tag{9.55}$$

Therefore, by the principles of calculus,

$$\frac{\mathrm{d}\Phi_{u>u'}}{\mathrm{d}u'} = -\Phi_u \tag{9.56}$$

For sites without long periods of zero wind, i.e. the more promising sites for wind-power, usually with $\overline{u} > 5\,\mathrm{m\,s^{-1}}$, usually a two-parameter exponential function can be closely fitted to measure wind speed data. One such function, often used in wind speed analysis, is the *Weibull function* shown in Figure 9.22 obtained from

$$\Phi_{u>u'} = \exp\left[-\left(\frac{u'}{c}\right)^k\right] \tag{9.57}$$

so (Weibull):

$$\Phi_u = \frac{k}{c}\left(\frac{u}{c}\right)^{k-1} \exp\left[-\left(\frac{u}{c}\right)^k\right] \tag{9.58}$$

Figure 9.22 shows the form of $\Phi_{u>u'}$ and $\Phi_u$ for different values of $k$ around 2.0. Such curves often give very good fit to experimental data, with $k$ between 1.8 and 2.3 and $c$ near the mean wind speed for the site. See also Figure 9.18, which compares actual data for North Ronaldsay to a Rayleigh distribution, with $k = 2$. The dimensionless parameter $k$ is called the shape factor (see Figure 9.22(a) for the obvious reason), and $c$, unit $\mathrm{m\,s^{-1}}$, the scale factor. Note that when $\Phi_{u>u'} = 1/e$, $u'/c = 1$ and so $c$ can be obtained as equal to the wind speed measurement at that point.

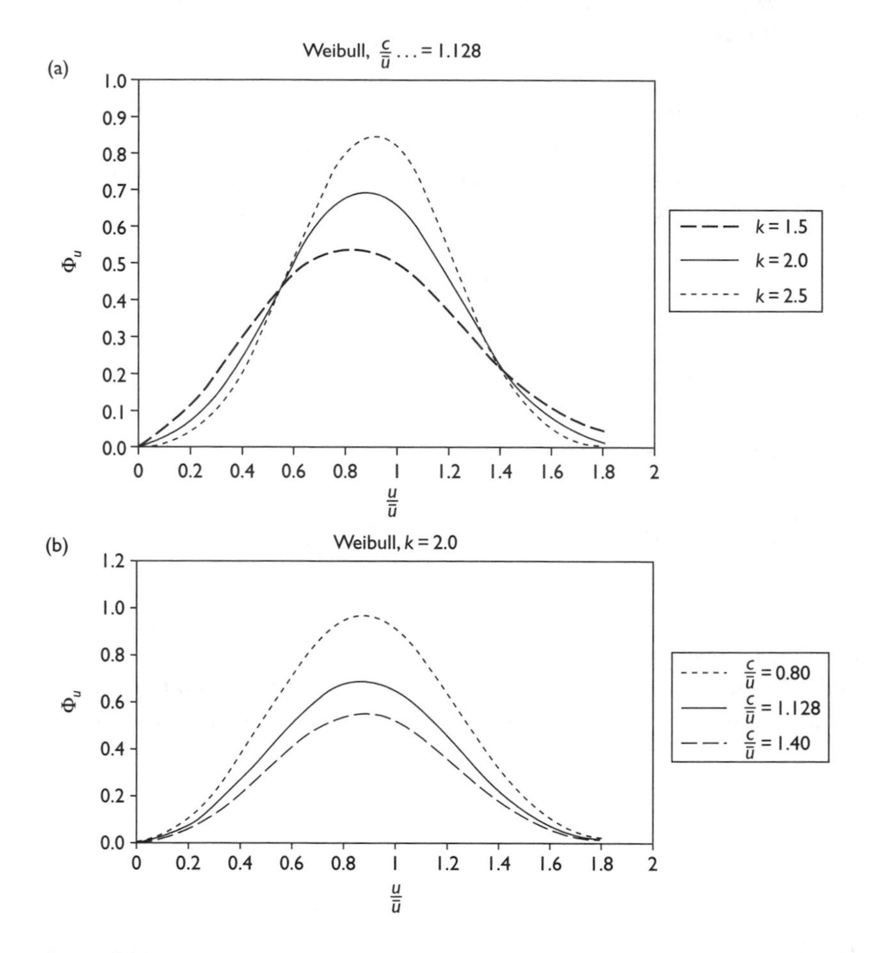


*Figure 9.22* Weibull distribution curves. (a) Varying $k$: curves of $\Phi_u$ for $c = 1.128\bar{u}$ and $k = 1.5, 2.0, 2.5$. Curve for $k = 2$, $c/\bar{u} = 1.128$ is the Rayleigh distribution. (b) Varying $c$: curves of $\Phi_u$ for $k = 2$ and $c/\bar{u} = 0.80, 1.128, 1.40$. As probability functions, areas under curves may be normalised to 1.0.

For many sites it is adequate to reduce (9.58) to the one-parameter *Rayleigh distribution* (also called the chi-squared distribution), by setting $k = 2$. So (Rayleigh)

$$\Phi_u = \frac{2u}{c^2} \exp\left[-\left(\frac{u}{c}\right)^2\right] \qquad (9.59)$$

For a Rayleigh distribution, $c = 2\bar{u}/\sqrt{\pi}$; see (9.68).

The mean wind speed is then

$$\bar{u} = \frac{\int_0^\infty \Phi_u u \, \mathrm{d}u}{\int_0^\infty \Phi_u \, \mathrm{d}u} \tag{9.60}$$

For the Weibull distribution of (9.58), this becomes

$$\bar{u} = \frac{\int_0^\infty u u^{k-1} \exp[-(u/c)^k] \mathrm{d}u}{\int_0^\infty u^{k-1} \exp[-(u/c)^k] \mathrm{d}u} \tag{9.61}$$

Let $(u/c)^k = v$, so $\mathrm{d}v = (k/c^k)u^{k-1}\,\mathrm{d}u$. Equation (9.61) becomes

$$\bar{u} = \frac{c \int_0^\infty v^{t/k} \exp[-v] \mathrm{d}v}{\int_0^\infty \exp[-v] \mathrm{d}v} \tag{9.62}$$

The denominator is unity, and the numerator is in the form of a standard integral called the factorial or gamma function, i.e.

$$\Gamma(z+1) = z! = \int_{v=0}^{\infty} v^z e^{-v} \, \mathrm{d}v \tag{9.63}$$

Note that the gamma function is unfortunately usually written, as here, as a function of $z+1$ and not $z$ (refer to Jeffreys and Jeffreys).

Thus

$$\bar{u} = c\Gamma(1+1/k) = c[(1/k)!] \tag{9.64}$$

Using the standard mathematics of the gamma function, the mean value of $u^n$ is calculated, where $n$ is an integer or fractional number, since in general for the Weibull function

$$\overline{u^n} = c^n \Gamma(1+n/k) \tag{9.65}$$

For instance the mean value of $u^3$ becomes

$$\overline{u^3} = c^3 \Gamma(1+3/k) \tag{9.66}$$

from which the mean power in the wind is obtained.

There are several methods to obtain values for $c$ and $k$ for any particular wind distribution (e.g. see Rohatgi and Nelson, or Justus *et al.*). Some examples are:

1 Fit the distribution to meteorological measurements. For instance if $\overline{u}$ and $\overline{u^3}$ are known, then (9.64) and (9.66) are simultaneous equations

with two unknowns. Modern data collection and online analysis methods enable mean values to be continuously accumulated without storing individual records, so $\overline{u}$ and $\overline{u^3}$ are easily measured.

2 Measure $\overline{u}$ and the standard deviation of $u$ about $\overline{u}$, to give $(\overline{u^2} - \overline{u}^2)$ and hence $\overline{u^2}$.

3 Plot the natural log of the natural log of $\Phi_{u>u'}$ in (9.57) against $\ln u$; the slope is $k$, and hence the intercept gives $c$.

The Rayleigh distribution is particularly useful for preliminary analysis of wind power potential across a large area, as often the only data that are available are maps showing interpolated curves of mean wind-speed – which is the only parameter needed to fit a Rayleigh distribution.

*Example 9.2 Rayleigh distribution analysis*
Show that for the Rayleigh distribution:

1 $\Phi_{u>u'} = \exp\left[-\frac{\pi}{4}\left(\frac{u'}{\bar{u}}\right)^2\right]$
2 $(\overline{u^3})^{1/3} = 1.24\bar{u}$
3 Power in the wind per unit area is $\bar{P}_0/A \approx \rho(\bar{u})^3$
4 $\Phi_u(\text{max})$ occurs at $u = (2/\pi)^{1/2}\bar{u} = 0.80\bar{u}$
5 $(\Phi_u u^3)(\text{max})$ occurs at $u = 2(2/\pi)^{1/2} = 1.60\bar{u}$

*Solution*
In (9.62) with $k = 2$,

$$\bar{u} = c\Gamma(1 + 1/2) = c[(1/2)!] \tag{9.67}$$

where by definition

$$(1/2)! = \int_0^\infty u^{1/2} e^{-u} \mathrm{d}u$$

By a standard integral

$$(1/2)! = \sqrt{\pi}/2$$

Hence in (9.67) for the Raleigh distribution,

$$c = 2\bar{u}/\sqrt{\pi} = 1.13\bar{u} \tag{9.68}$$

In (9.59), the Rayleigh distribution becomes

$$\Phi_u = \frac{\pi u}{2\bar{u}^2} \exp\left[-\frac{\pi}{4}\left(\frac{u}{\bar{u}}\right)^2\right] \tag{9.69}$$

and by (9.55)

$$\Phi_{u>u'} = \int_{u=u'}^{\infty} \Phi_u \, du = \exp\left[-\frac{\pi}{4}\left(\frac{u'}{\bar{u}}\right)^2\right] \tag{9.70}$$

Also

$$\overline{u^3} = \frac{\int_0^\infty \Phi_u u^3 du}{\int_0^\infty \Phi_u du} = \left[\frac{\pi}{2\bar{u}^2}\int_0^\infty u^4 \exp -\frac{\pi}{4}\left(\frac{u}{\bar{u}}\right)^2\right] du \tag{9.71}$$

By standard integrals of the gamma function this reduces to

$$\overline{u^3} = K(\bar{u})^3 \tag{9.72}$$

where $K$ is called the 'energy pattern factor'. For the Rayleigh distribution of (9.71), $K = (6/\pi) = 1.91$, see problem 9.5.
Hence

$$(\overline{u^3})^{1/3} = (1.91)^{1/3}\bar{u} = 1.24\bar{u} \tag{9.73}$$

A very useful relationship between mean wind speed and average annual power in the wind per unit area follows:

$$\frac{\overline{P_0}}{A} = \frac{1}{2}\rho\overline{u^3} \approx \rho\,(\bar{u})^3 \tag{9.74}$$

By differentiation to obtain the values of $u$ for maxima in $\Phi_u$ and $\Phi_u^3$, and again using the standard integral relationships of the gamma function (see Problem 9.5):

$$\Phi_u(\text{max}) \text{ occurs at } u = (2/\pi)^{1/2}\bar{u} = 0.80\bar{u} \tag{9.75}$$

and

$$(\Phi_u u^3)(\text{max}) \text{ occurs at } u = 2(2/\pi)^{1/2}\bar{u} = 1.60\bar{u} \tag{9.76}$$

*Example 9.3 Rayleigh distribution fitted to measured data*
Apply the results of Example 9.2 to the data from North Ronaldsay in Example 9.1.

*Solution*
For North Ronaldsay $\overline{u} = 8.2\,\mathrm{m\,s^{-1}}$. Therefore by (9.75), $\Phi_u(\max)$ is at $(0.80)(8.2\,\mathrm{m\,s^{-1}}) = 6.6\,\mathrm{m\,s^{-1}}$. The measured value from Figure 9.18 is $6.2\,\mathrm{m\,s^{-1}}$.

By (9.76), $(\Phi_u u^3)(\max)$ is at $(1.60)(8.2\,\mathrm{m\,s^{-1}}) = 13\,\mathrm{m\,s^{-1}}$. The measured value from Figure 9.20 is $12.5\,\mathrm{m\,s^{-1}}$.

By (9.73), $(\overline{u^3})^{1/3} = (1.24)(8.2\,\mathrm{m\,s^{-1}}) = 10.2\,\mathrm{m\,s^{-1}}$. The measured value from Figure 9.20 is $10.1\,\mathrm{m\,s^{-1}}$.
See also Figure 9.18.

### 9.6.5 Wind speed and direction variation with time

From Figure 9.15, note the importance of fluctuations $\sim$ 10 s. These not only contain significant energy, but lead to damaging stresses on wind machines. A measure of all such time variations is the *turbulence intensity*, equal to the standard deviation of the instantaneous wind speed divided by the mean value of the wind speed. Turbulence intensity is a useful measure over time intervals of a few minutes; values of around 0.1 imply a 'smooth' wind, as over the sea, and values of around 0.25 imply a very gusty, large turbulence, wind, as in mountainous locations. Turbulence can be expected to reduce with height above ground. There are similar expressions for the variation of wind direction with time, sometimes called 'wind shift'.

A wind turbine, especially medium to large size, will not respond quickly enough, or have the aerodynamic properties, to 'follow' rapid changes in wind speed and direction. Therefore energy in wind turbulence and shift may not be captured, but this is an advantage if fatigue damage is thereby lessened.

The more wind turbines and wind farms are dispersed on a grid, the less correlated are the short-term variations and the easier it is to accept increased capacity of wind power. The product of wind speed and the correlation time period is called the *coherence distance*. For short periods, say 10 s of turbulence, the coherence distance will be usually less than the 'length' of a wind farm, so such variations are averaged out. For periods of about 30 min, the correlation distance may be about 20 km; in which case wind farm output dispersed over distances of the order of 100 km will also not correlate, with variations in power not apparent over the whole grid. Only when the coherence distance becomes larger than the scale of the grid are fluctuations not smoothed out by diversity of the site locations.

## 9.7 Power extraction by a turbine

The fraction of power extracted from the wind by a turbine is, by (9.15), the power coefficient $C_P$ as discussed in Section 9.3. At any instant, $C_P$ *is* most dependent on the tip-speed ratio $\lambda$, unless the machine is controlled for other reasons (as happens below the cut-in wind speed and usually above the rated output). The strategy for matching a machine to a particular wind regime will range between the aims of (1) maximizing total energy production in the year (e.g. for fuel saving in a large electricity network), and (2) providing a minimum supply even in light winds (e.g. for water pumping to a cattle trough). In addition secondary equipment, such as generators or pumps, has to be coupled to the turbine, so its power matching response has to be linked to the turbine characteristic. The subject of power extraction is therefore complex, incorporating many factors, and in practice a range of strategies and types of system will be used according to different traditions and needs.

This section considers power extracted by the turbine, which will have a rated power capacity $P_R$ produced at the specified rated wind speed $u_R$. From Section 9.6.3, $\Phi_u$ *is* the probability per wind speed interval that the wind speed will be in the interval $u$ to $(u+du)$ (i.e. $\Phi_u\, du$ is the probability of wind speed between $u$ and $u+du$). Let $E$ be the total energy extracted by the turbine in the period $T$, and let $E_u$ be the energy extracted per interval of wind speed when the wind speed is $u$. Then

$$E = \int_{u=0}^{\infty} E_u du = \int_{u=0}^{\infty} (\Phi_u T)\, P_{T,u} du = \int_{u=0}^{\infty} A_1 \left[\frac{1}{2}\rho u_0^3 C_P(\Phi_u T)\right] du \quad (9.77)$$

where $A_1$ is the swept area of the turbine and $u$ the ambient wind speed ($u_0$ in Figure 9.6). The average power extracted if the air density is considered constant is

$$\frac{E}{T} = \overline{P}_T = \frac{\rho A_1}{2} \int_{u=0}^{\infty} \Phi_u u^3 C_P du \quad (9.78)$$

The *capacity factor* is the annual average power generated as a proportion of the turbine rated power. So, in principle:

$$\text{Capacity factor} = (\rho A_1/2) \int_{u=0}^{\infty} [\Phi_u u^3 C_p\, du]/[(C_p \rho A_1/2) u_R{}^3]$$

This integral or summation cannot be evaluated until the dependence of $C_P$ on the upstream wind speed $u = u_0$ is established.
It is usually considered that there are four distinct wind speed regions of operation, see. Figure 9.23:

1 *$u_0$ less than cut-in speed $u_{ci}$:*

$$E_u = 0 \textit{ for } u_0 < u_{ci} \quad (9.79)$$

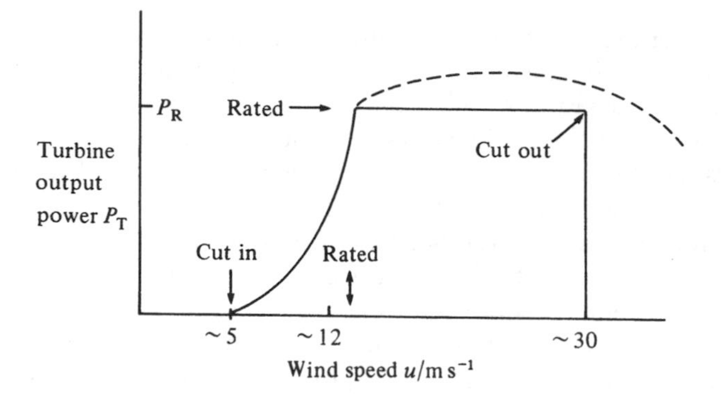


*Figure 9.23* Wind turbine operating regions and power performance
_______standard characteristics; requiring exact blade pitch control
_ _ _ _ _actual operating characteristics of many machines, including stall regulation.

2 *$u_0$ greater than rated speed $u_R$, but less than cut-out speed $u_{co}$:*

$$E_u = \Phi_{u>u_R} P_R T \tag{9.80}$$

where $P_R$ is the rated power output (i.e. in this range the turbine is producing constant power $P_R$).

3 *$u_0$ greater than cut-out speed $u_{co}$:*

$$E_u = 0 \textit{ for } u_0 > u_{co} \tag{9.81}$$

However, in practice many machines do not cut out in high wind speeds because of stall regulation, but continue to operate at greatly reduced efficiency at reasonably large power.

4 *$u$ between $u_{ci}$ and $u_R$*

The turbine power output $P_T$ increases with $u$ in a way that depends on the operating conditions and type of machine. For many machines

$$P_T \approx a u_0^3 - b P_R \tag{9.82}$$

where $a$ and $b$ are constants.

At cut-in, $P_T = 0$; so

$$u_{ci}^3 = b P_R / a$$

At rated power $P_T = P_R$; so

$$u_R^3 = (1+b) P_R / a$$

Thus

$$(u_{ci}/u_R)^3 = b/(1+b) \tag{9.83}$$

Hence $a$ and $b$ can be determined in terms of $u_{ci}$, $u_R$ and $P_R$.

In practice, turbines will often be operating in the region between cut-in and rated output, and it is wasteful of energy potential if the machine is unduly limited at large wind speeds. There are two extreme theoretical conditions of operation, see Figure 9.24:

1 *Variable rotor speed for constant tip-speed ratio $\lambda$, hence constant $C_P$.* This is the most efficient mode of operation and captures the most energy, see Problem 9.13 (and its answer) for details of calculating the energy capture. Variable speed turbines usually cut in at wind speeds less than for constant speed turbines, which also increases energy capture.
2 *Constant (fixed) turbine rotational frequency, hence varying $C_P$.* Although less efficient than variable speed turbines, the use of standard induction generators allows easy grid connection (the small frequency slip of induction generators is not significant, so the machines are described as 'constant speed').

From Figure 9.24, it can be seen that $C_P$ can be obtained as a function of unperturbed wind speed $u_0$, and the turbine power calculated by numerical methods. By operating at constant frequency there is a loss of possible energy extraction. This may be particularly serious if there is a mismatch of optimum performance at larger wind speeds.

In practice, a measured (or estimated) operating power curve of a wind turbine is usually supplied by the maker, in the form of a curve like Figure 9.23 or as a data table, and the term 'capacity factor' is used for the ratio of actual annual average generated power at a site, divided by the generator name-plate rated power.

## 9.8 Electricity generation

### 9.8.1 *Basics*

See Section 16.9 for a basic description of the various types of electrical generator and of electricity networks or 'grids'.

Electricity is an excellent energy vector to transmit the captured mechanical power of a wind turbine. Generation is usually ~95% efficient, and transmission losses should be less than 10%. There are already many designs of wind/electricity systems including a wide range of generators. Future

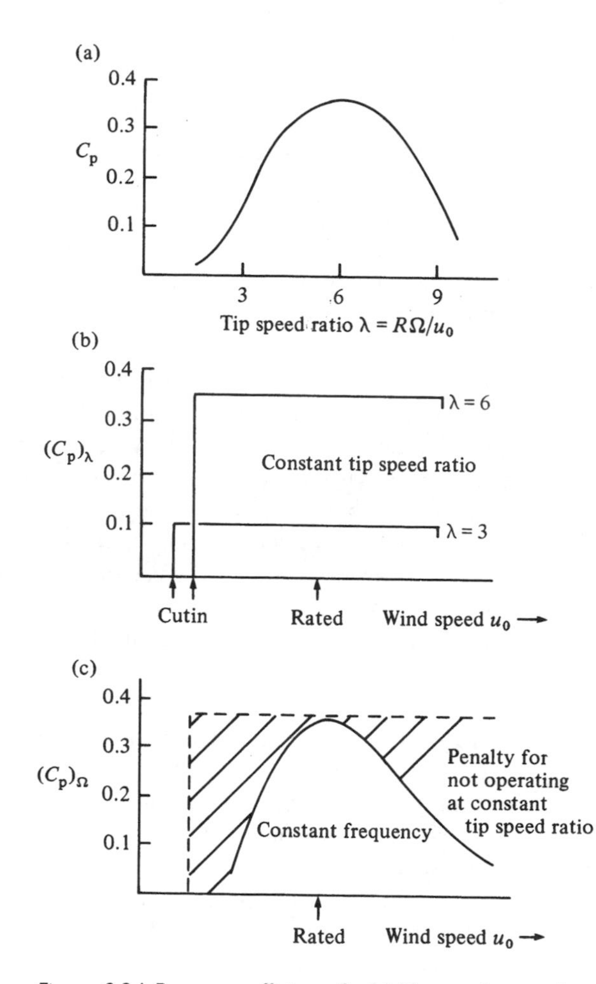


*Figure 9.24* Power coefficient $C_p$: (a) Versus tip-speed ratio. (b) Versus wind speed at constant tip-speed ratio and so variable rotor speed. (c) Versus wind speed at constant turbine frequency, compared with variable speed at tip-speed ratio of 6.

development will certainly produce new and improved designs of generators and control systems as wind-generated power becomes an electrical engineering speciality.

Grid connected turbines and wind farms dispatch power to be integrated with other forms of generation, e.g. thermal power stations. Consumers use power at nearly constant voltage and frequency, as controlled by the grid operators for the power transmission system. However, power from the wind varies significantly and randomly. If the power from wind is no more than about 20% of the total power at one time, then the variations are usually acceptable within the ever changing conditions of the consumer loads.

For stand-alone applications, the frequency and voltage of transmission need not be so standardised, since end-use requirements vary. Heating in particular can accept wide variations in frequency and voltage.

In all applications it will be necessary to match carefully the machine characteristics to the local wind regime. Obviously extended periods of zero or light wind will limit wind-power applications. In particular, sites with an average wind speed less than $5\,\mathrm{m\,s^{-1}}$ usually have unacceptably long periods at which generation would not occur, although water pumping into water storage may still be feasible. Usually, if the annual average wind speed at 10 m height is $5\,\mathrm{m\,s^{-1}}$ or more, electricity generation may be contemplated.

The distinctive features of wind/electricity generating systems are:

1 Wind turbine efficiency is greatest if rotational frequency varies to maintain constant tip-speed ratio, yet electricity generation is most efficient at constant or nearly constant frequency.
2 Mechanical control of a turbine by blade pitch or other mechanical control at powers less than rated increases complexity and expense. An alternative method, usually cheaper and more efficient but seldom done, is to vary the electrical load on the turbine to control the rotational frequency.
3 The optimum rotational frequency of a turbine (its 'speed') in a particular wind speed decreases with increase in radius in order to maintain constant tip-speed ratio. Thus only small ($\sim$2 m radius) turbines can be coupled directly to conventional four or six pole-pair generators. Larger machines require a gearbox to increase the generator drive frequency or special multipole generators. Gearboxes are relatively expensive and heavy; they require maintenance and can be noisy.
4 The rotor can be decoupled from the load, with the advantage of allowing the rotor to be optimised to the wind. Some developments experimented with a mechanical accumulator (e.g. a weight lifted by hydraulic pressure), but predominantly an electrical method is used. For autonomous systems, chemical batteries provide both this decoupling and longer-term energy storage. For grid connected systems, generated AC electricity may be rectified to DC and then inverted to grid frequency AC; very short term, but useful, 'rotor inertia' energy storage occurs, which smoothes wind turbulence. Even the provision of a 'soft coupling' using teetered blades, shock absorbers or other mechanisms is useful to reduce electrical spikes and mechanical strain.
5 There are always periods without wind. Thus wind turbines must be linked to energy storage or parallel generating systems, e.g. through the utility power grid, so supplies are to be maintained.

Often, the best wind-power sites are in remote rural, island or marine areas. Energy requirements in such places are distinctive, and almost certainly will not require the intense electrical power of large industrial complexes. In such cases, explicit end-use requirements for controlled electricity (e.g. 240 V/50 Hz or 110 V/60 Hz for lighting, machines and electronics) are likely to be only 5–10% of the total energy requirement for transport, cooking and heat. As wind power experience increases, further developments can be expected so the wind provides affordable energy for heat and transport, in addition to standard electrical uses. Such novel developments occurred first in some remote area power systems, e.g. the Fair Isle system described in Example 9.4 (Section 9.8.4). Moreover, rural grid systems are likely to be 'weak', since they carry relatively low-voltage supplies (e.g. 33 kV) over relatively long distances with complicated inductive and resistive power loss problems. Interfacing a large wind turbine in weak grids is challenging, but certainly not impossible with modern power electronics; indeed the wind power can be used to strengthen the grid supply, for instance by controlling reactive power and voltage.

### 9.8.2 *Classification of electricity systems using wind power*

There are three classes of wind turbine electricity system, depending on the relative size of the wind-turbine generator, $P_T$, and other electricity generators connected in parallel with it, $P_G$ (Table 9.4).

*Class A: wind turbine capacity dominant,* $P_T \geq 5P_G$
Usually this will be a single autonomous stand-alone machine without any form of grid linking. Other generators are not expected. For electricity supply, a battery is necessary to stabilize the voltage and store the electricity. For remote communication sites, household lighting, marine lights etc.,

*Table 9.4* Classes of wind turbine electricity systems

| *Class* | *A* | *B* | *C* |
|---|---|---|---|
| $P_T$: wind turbine generator<br>$P_G$: other generator Capacity | $P_T \gg P_G$ | $P_T \sim P_G$ | $P_T \ll P_G$ |
| Example | Autonomous | Wind/diesel | Grid embedded |
| Control modes | (a) Blade pitch<br><br>(b) Load matching | (a) Wind or diesel separately<br>(b) Wind and diesel together | (a) Direct induction generator<br>(b) To DC then AC<br>(c) Increased slip induction generator |

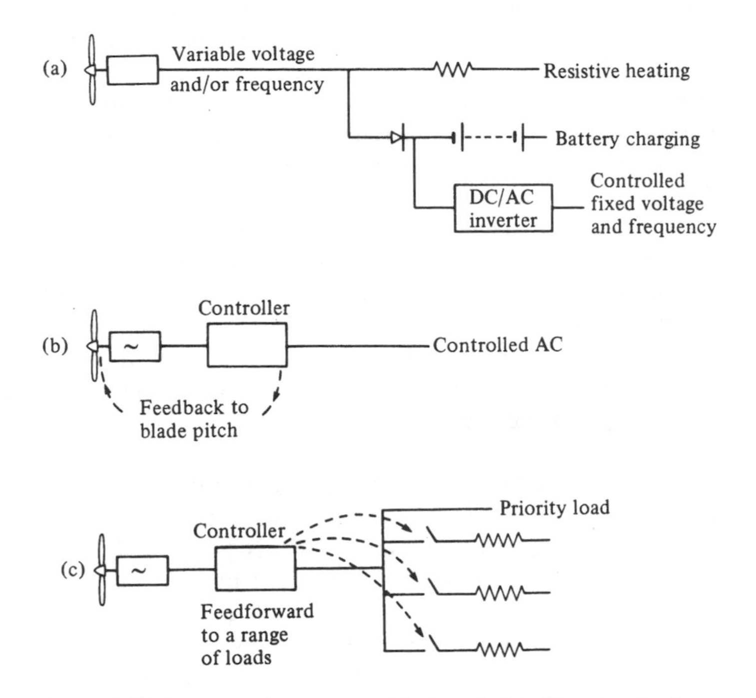


*Figure 9.25* Some supply options with the wind turbine the dominant supply.

$P_T \leq 2\,\text{kW}$. For full household supplies, including heat, $P_T \sim 10\,\text{kW}$. Wind turbines of large capacity are likely to have standby generation of class B.

Control options have been discussed in Section 1.5.4 and are of extreme importance for efficient cost-effective systems, Figure 9.25. One choice is to have very little control so the output is of variable voltage (and, if AC, frequency) for use for direct resistive heating or for rectified power, Figure 9.25(a). There are many situations where such a supply will be useful. The relatively small amount of power that usually has to be controlled at, say, 240 V/50 Hz or 110 V/60 Hz can be obtained from batteries by inverters. Thus the high quality controlled electricity is obtained 'piggy-back' on the supply of less quality, and can be costed only against the marginal extras of battery and inverter.

However, it may be preferred to have the electricity at controlled frequency. There are two extreme options for this:

1 *Mechanical control of the turbine blades.* As the wind changes speed, the pitch of the blades or blade tips is adjusted to control the frequency of turbine rotation, Figure 9.25(b). The disadvantages are that power in the wind is 'spilt' and therefore wasted (see Section 1.5) and the control method may be expensive and unreliable.

2 *Load control*. As the wind changes speed, the electrical load is changed by rapid switching, so the turbine frequency is controlled, Figure 9.25(c). This method makes greater use of the power in the wind by optimising tip-speed ratio $\lambda$. Moreover local control by modern electronic methods is cheaper and more reliable than control of mechanical components exposed in adverse environments.

Permanent magnet multipole generators are common for small machines. DC systems can be smoothed and the energy stored in batteries. AC systems may have synchronous generators producing uncontrolled output for heat, or controlled output by mechanical or load control. AC induction generators can be self-excited with a capacitor bank to earth, or may operate with an idling synchronous generator as a compensator (see Section 16.9.3 for further discussion of generator types).

*Class B: wind turbine capacity ≈ other generator capacity*, $P_T \approx P_G$
This is a common feature of remote area, small grid systems. We shall first assume that the 'other generator' of capacity $P_G$ is powered by a diesel engine, perhaps fuelled by biodiesel. The principal purpose of the wind turbine is then to be a saver of the fuel. The diesel generator will be the only supply at windless periods and will perhaps augment the wind turbine at periods of weak wind. There are two extreme modes of operation:

1 *Single-mode electricity supply distribution*. With a single set of distribution cables (usually a three-phase supply that takes single phase to domestic dwellings), the system must operate in a single mode at fixed voltage for 240 V or 110 V related use, Figure 9.26(a). A 24 h maintained

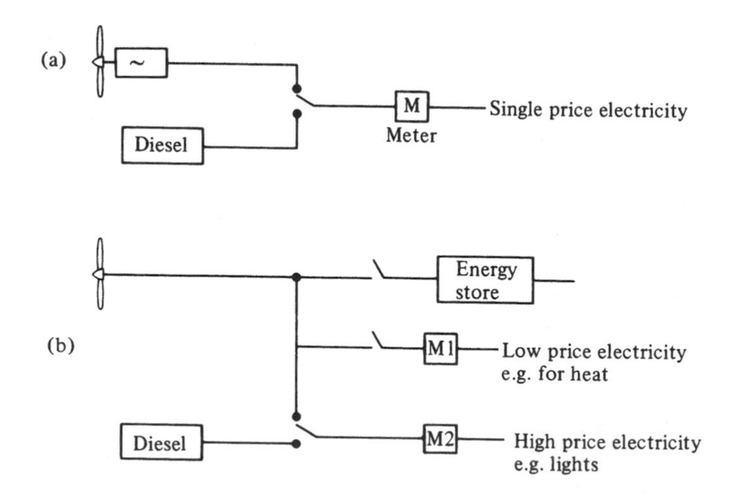


*Figure 9.26* Wind/diesel supply modes. (a) Single mode. (b) Multiple mode.

supply without load management control will still depend heavily (at least 50% usually) on diesel generation since wind is often not available. The diesel is either kept running continuously (frequently on light load, even when the wind power is available) or switched off when the wind power is sufficient. In practice a large amount (sometimes over 70%) of wind-generated power has to be dumped into outside resistor banks owing to the mismatch of supply and demand in windy conditions.

2 *Multiple-mode distribution.* The aim is to use all wind-generated power by offering cheap electricity for many uses in windy conditions, Figure 9.26(b). As the wind speed decreases, the cheaper serviced loads are automatically switched off to decrease the demand, and vice versa. The same system can be used to control the rotation of the wind turbine. When no wind power is available, only the loads on the expensive supply are enabled for supply by the diesel generator. The pragmatic economic advantage of successful multiple-mode operation is that the full capital value of the wind machine is used at all times, and since the initial power in the wind is free, the maximum benefit is obtained. It is also advantageous in using less fuel with the abatement of pollution and noise.

*Class C: grid linked, wind turbine embedded in a large system,* $P_T \leq 0.2P_G$

This is the most common arrangement for large (~3 MW), medium (~250 kW) and small (~50 kW) machines where a public utility or other large capacity grid is available. In recent years, institutional factors have led to the bulk of new wind-power capacity being in 'wind farms', in which a number (10–1000) of turbines in a group are all feeding into the grid (see next section and Chapter 17). For smaller systems, the owner of the machine may use the wind power directly and sell (export) any excess to the grid, with electricity purchased (imported) from the grid at periods of weak or no wind (Figure 9.27).

The cheapest type of generator is an induction generator connected directly to the grid. The turbine has to operate at nearly constant frequency,

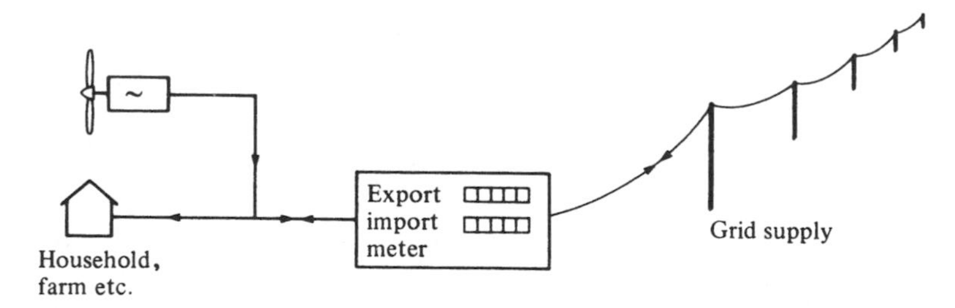


*Figure 9.27* Grid linked wind turbine slaved in a large system.

within a maximum slip usually less than 5% ahead of the mains-related frequency; this is usually called 'fixed speed'. In weak wind, there is an automatic cut-out to prevent motoring. The disadvantage of a directly coupled induction generator is that the turbine frequency cannot change sufficiently to maintain even approximate constant tip-speed ratio.

However, there are several ways in which the system can be made to produce electricity at fairly constant frequency while allowing variable turbine frequency. They include: (1) multiple (usually two) combination windings in an induction generator to connect more pole pairs in weak winds for smaller rotational frequency; (2) some intermediate scale machines use two generators in the same nacelle, say 5 kW and 22 kW, for automatic connected to a two-speed gearbox in light and strong winds; (3) using a variable speed generator and rectifying its output to direct current and then producing the prescribed alternating current mains frequency with an inverter; and (4) increasing the effective slip on an induction generator by active change of the current and phase in the generator's rotor, e.g. in a doubly-fed induction generator; this requires external electrical connection to the rotor winding via slip rings and brushes.

### 9.8.3 Electricity generation for utility grids: wind farms

Commercial wind turbines are a 'mainstream' form of power generation into grid distribution and transmission networks. Machines of megawatt capacity have operated successfully for many years. Multiple numbers of machines arrayed on 'wind farms' (typically with 10–100 turbines) make convenient and manageable units. Grouping machines in this way allows savings (10–20%) in the construction costs (e.g. getting specialised cranes etc. on site), grid connection (fewer step-up transformers required), management and maintenance. Wind farms are most likely in countries where there is (1) a commitment to sustainable, low-carbon, energy supplies, (2) previous strong dependence on brown energy and (3) open areas with an average wind speed at 10 m height $>6\,\mathrm{m\,s^{-1}}$.

Since wind speeds are usually larger offshore than onshore, except in mountainous regions, it can be beneficial to locate wind farms up to several kilometre offshore. This approach is particularly attractive in marine countries where potential onshore sites are limited by dense population, visual intrusion, and failure to gain planning permissions.

Because the output from wind turbine power is less predictable than that from a conventional system (fossil, nuclear or hydro), adding a wind turbine rated at 1 MW capacity to a grid is not equivalent to adding 1 MW capacity from a 'brown' source. In general the site-dependent annual capacity factor of a wind turbine is 20–35%, whereas a thermal power station is about 70–90%. Yet, not all 'brown' sources are equivalent, e.g. nuclear power is suitable only for base load, whereas gas turbines are best for rapid response to peak demands. Energy economists describe this in terms of *capacity credit*, the power rating of conventional plant that is displaced by the installation of wind power or other renewable energy. Theoretical studies by numerous utilities have predicted that 1000 MW (rated) wind power has a capacity credit of 250–400 MW (Milborrow 2001). The larger figure corresponds to sites of larger average wind speeds, since these have more extended periods of wind. If the wind power comes from a diversity of sites, there is less chance of them all having reduced output at the same time, and so the predicted capacity credit is larger – in some cases arguably close to 100%. A related opinion is that wind power requires the construction of back-up power stations. Therefore a utility network always has to have reserve generating capacity available for all forms of generation, especially since large power stations fail at times, and so to the authors' knowledge no additional back-up power has yet been needed or constructed due to extra wind power capacity.

### 9.8.4 Individual machines and integrated systems

Most wind turbine capacity is associated with commercial wind farms for grid connected power, and therefore large machines (~3 MW) are the most common now. However small machines of capacity between about 50 W and1 kW are common for boats, holiday caravans and houses, small power public service (e.g. rural bus shelters) and small meteorological and other measurement sites. Slightly larger, but still 'small' are 5–100 kW wind turbines installed for household, farm and institutional use. Cost-effective operation is most likely in locations where other energy supplies are expensive (e.g. oil) and grid electricity not available. However, where there is a grid and if excess electricity can be sold to a utility grid at a price of at least half the buying price, grid connection is no discouragement for wind power projects.

The principles of renewable energy supply, developed in Chapter 1, indicate that the renewables technology has to operate within quite different constraints than have fossil fuel and nuclear sources. The dispersed and highly fluctuating nature of wind attracts radically different approaches than those used for steady intensive sources. In particular, there is scope for adaptation of the end-uses of the wind-generated power so that the load responds to the changing supply, and energy storage is incorporated, see Chapter 16. The multi-mode system at Fair Isle (Example 9.4 and Figure 1.6) illustrates what can be achieved by taking an integrated whole-system approach, covering both supply and use of energy. Such an approach is possible, but very uncommon, on much larger systems.

*Example 9.4 Fair Isle multimode wind power system*

Fair Isle is an isolated Scottish island in the North Sea between mainland Shetland and Orkney. The population of 70 is well established and progressive within the limits of the harsh yet beautiful environment. Previously the people depended on coal and oil for heat, petroleum for vehicles, and diesel for electricity generation. The electricity co-operative installed first a 50 kW rated-capacity wind turbine that operates in the persistent winds, of average speed 8–9 $m s^{-1}$. The control system was mentioned in Chapter 1, see Figure 1.6. Lighting and power outlets receive electricity metered at larger price, and a reduced price controlled-supply is available (wind permitting) for hard-wired comfort heat and water heating, see Section 1.5.4. At the frequent periods of excessive wind power, further heat is available, e.g. for growing food in a glasshouse or for a small swimming pool. An electric vehicle was charged from the system to include transport as an end-use. Despite the strong winds, the total generating capacity is small for the population served, and acceptable standards are only possible because the houses are well insulated, careful energy strategies are maintained and sophisticated reliable control systems are incorporated.

## 9.9 Mechanical power

Historically the mechanical energy in the wind has been harnessed predominantly for transport with sailing ships, for milling grain and for pumping water. These uses still continue and may increase again in the future. This section briefly discusses those systems, bearing in mind that electricity can be an intermediate energy vector for such mechanical uses.

### 9.9.1 *Sea transport*

The old square rigged sailing ships operated by drag forces and were inefficient. Modern racing yachts, with a subsurface keel harnessing lift forces, are much more efficient and can sail faster than the wind. Some developments to modern cargo ships have used fixed sails set by mechanical drives.

### 9.9.2 *Grain milling*

The traditional windmill (commonly described as a Dutch windmill) has been eclipsed by engine- or electrical-driven machines. It is unlikely that the intermittent nature of wind over land will ever be again suitable for commercial milling in direct mechanical systems.

### *9.9.3 Water pumping*

Pumped water can be stored in tanks and reservoirs or absorbed in the ground. This capacitor-like property gives smoothing to the intermittent wind source, and makes wind-powered pumping economic. Farm scale pumps to about 10 kW maximum power are common in many countries including Argentina, Australia and the United States. The water is used mostly for cattle, irrigation or drainage. Continuity of supply is important, so large solidity multi-blade turbines are suitable, having large initial torque in weak winds. The small rotational speed is not a handicap for direct mechanical action. The traditional cylinder pump with a fixed action, Figure 9.28, is simple and reliable. At best, however, the delivered power is proportional to turbine rotational frequency ($P' \propto \Omega$), whereas at constant tip-speed ratio the power at the turbine is proportional to $\Omega^3 (P_T \propto \Omega^3)$: Therefore the efficiency $P'/P_T$ drops as $1/\Omega^2$. Improved pumps that match the wind turbine characteristics and maintain simplicity of operation are important for more efficient water pumping. Since water is usually available at low locations, and wind increases with height, it is often sensible to have an electricity-generating wind turbine placed on a hill operating an electric pump placed at the nearby water supply.

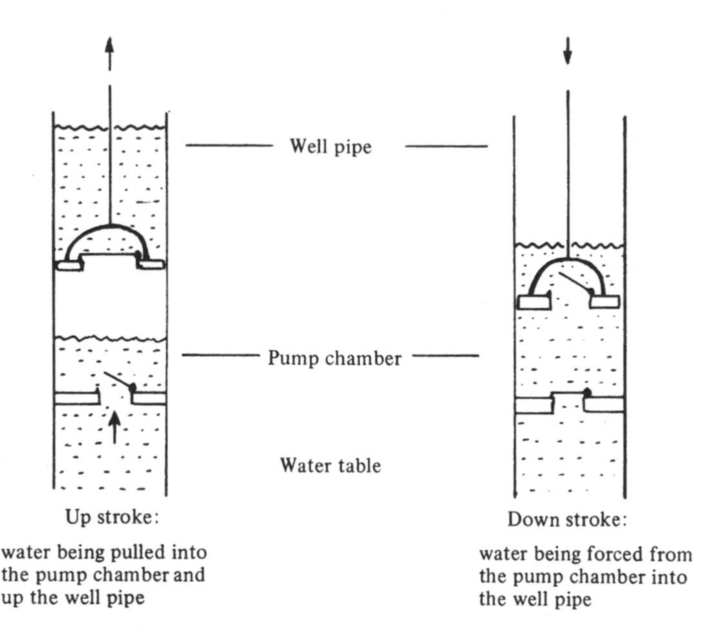


*Figure 9.28* Positive displacement water pump. The shaft would be connected to the rotating crankshaft of the wind turbine.

### 9.9.4 Heat production

The direct dissipation of the mechanical power from a wind turbine (e.g. by paddlewheel systems) produces heat with 100% efficiency. However, electrical generators are so common and efficient that it is certain that electricity will be favoured as the intermediate energy vector to electrically powered heating.

## 9.10 Social and environmental considerations

Some of the potentially best locations for wind-power are in areas of natural beauty, such as coastlines, high ground and mountain passes. Proposals to use such locations usually attract opposition with arguments of loss of visual amenity, irritating acoustic noise and bird strikes. Similar objections have been raised to wind farms on farmland. Manufacturers of modern wind turbines responded by having architects influence the shape of towers and nacelles, by making the machines (especially the gear boxes) much quieter and by employing ecologists to advise on sites with least adverse and most advantageous impact on animals and plants. Some early wind farms in western USA were located in mountain passes on bird migration paths of birds and consequently bird-kill became a concern. However in most locations, birds can and do fly around the turbines without hazard. In general, since normal ecological and farming processes can continue underneath the spread of the rotor wherever it is, as in Figure 9.29, there is no environmental impact, other than on human opinion and perhaps on certain types of bird. The impacts on access for low flying military planes, on radar generally and on TV and communication channels have also to be considered.

There is a definite danger that wind power development will be pushed only by those with technical understanding. This is a definite mistake, since there should be appreciation of ecology, aesthetics, cultural heritage and public perceptions. These 'other' aspects are well reviewed in Pasqualetti *et al.* (2002) *Wind Power in View*; readers are strongly recommended to consider the lateral perceptions of this study.

In most countries, wind farm developers have to obtain local planning permission before installing a wind farm. Consequently the process of preparing for an application has become comprehensive and professional. Simulation software is used to give a dynamic visual impression of the wind farm from all viewpoints, in-depth bird and other ecological studies are funded, acoustic noise is predicted in the vicinity, effect on roads is studied, local benefit may be publicly offered (e.g. cheaper electricity supplies, donations to schools) and many other issues considered. If an application is refused, then appeals may be made. All these procedures are necessary, but time-consuming and expensive.

Yet the final outcome is that national and world wind power capacities are increasing, carbon and other emissions are being abated, the technology

*Figure 9.29* A wind farm in Minnesota, USA, with agricultural activity continuing underneath. [Photo by Warren Gretz, courtesy of [US] National Renewable Energy Laboratory.]

is improving and many adverse impacts, per unit of generated output, are decreasing.

## Problems

9.1 From (9.16) the fraction of power extracted from the wind is the power coefficient $C_P = 4a(1-a)^2$. By differentiating with respect to $a$, show that the maximum value of $C_P$ is 16/27 when $a = 1/3$.

9.2 The calculation of power coefficient $C_P$ by linear momentum theory (Section 9.3.1) can proceed in terms of $b = u_2/u_0$. Show that (a) $C_P = (1-b^2)(1+b)/2$, (b) $C_P$ is a maximum at 16/27 when $b = 1/3$, (c) $a = (1-b)/2$ where $a = (u_0 - u_1)/u_0$, and (d) the drag coefficient $= (1-b^2)$.

9.3 a By considering the ratio of the areas $A_0$ and $A_1$ of Figure 9.6, show that the optimum power extraction (according to linear momentum theory) per unit of area $A_0$ is 8/9 of the incident power in the wind.

b Prove that the torque produced by a wind turbine rotor of radius $R$ can be expressed as $\Gamma = (\pi/2)C_P R^3 u_0^2/\lambda$

9.4 The flow of air in the wind will be turbulent if Reynolds number $\mathcal{R} \geq 2000$ (see Section 2.5). Calculate the maximum wind speed for laminar flow around an obstruction of dimension 1.0 m. Is laminar flow realistic for wind turbines?

9.5 For a wind speed pattern following a Rayleigh distribution, prove that:

a The most probable wind speed is $0.80\overline{u}$
b The most probable power in the wind occurs at a wind speed of $1.60\overline{u}$.
c $\overline{u^3} = \frac{6}{\pi}(\bar{u})^3$

where $\overline{u^3}$ is the mean of $u^3$, and $\overline{u}$ is the mean of $u$, and so $(\overline{u^3})^{1/3} = 1.24\overline{u}$

9.6 Compare the solutions for Problem 9.5 with the wind speed factors indicated on Figures 9.18 and 9.20 for North Ronaldsay, and comment on how well the relationships between these factors is explained by the wind having a Rayleigh distribution.

9.7 A number of designs of wind turbine pass the output wind from one set of blades immediately on to a second identical set (e.g. two contrary rotating blades set on the same horizontal axis). By considering two actuator disks in series, and using linear momentum theory, show that the combined maximum power coefficient $C_P$ equals 0.64.
*Note*: This is only slightly larger than the maximum of $16/27 = 0.59$ for a single pass of the wind through one set of blades. Thus in a tandem horizontal axis machine of identical blade sets, and indeed in a vertical axis turbine, little extra power is gained by the airstream passing a second set of blades at such close proximity.

9.8 a A wind turbine maintains a tip-speed ratio of 8 at all wind speeds. At which wind speed will the blade tip exceed the speed of sound?

a A large wind turbine has a blade diameter of 100 m. At what rotor speed (i.e. frequency) will the tip-speed exceed the speed of sound?

9.9 a Calculate the possible maximum axial thrust per unit area of rotor for a wind turbine in a $20\,\mathrm{m\,s^{-1}}$ wind.

b The Danish standard for axial thrust design is $300\,\mathrm{N\,m^{-2}}$ of rotor area. What is the minimum possible wind speed that this corresponds to?

9.10 From Figure 9.14a and equation (9.12), prove that at maximum power extraction, tip-speed ratio $\lambda \sim 1.5 \cot\phi$. Hence maximum power extraction in varying wind speed relates to maintaining $\lambda$ constant, as aerodynamic design also requires.

9.11 A wind turbine rated at 600 kW has a cut-in speed of $5\,\mathrm{m\,s^{-1}}$, a rated speed of $20\,\mathrm{m\,s^{-1}}$ and a cut-out speed of $22\,\mathrm{m\,s^{-1}}$. Its power output as a function of wind speed at hub height is summarised in the following table. Its hub height is 45 m.

| Speed/($\mathrm{m\,s^{-1}}$) | 0 | 2.0 | 4.0 | 6.0 | 8.0 | 10.0 |
|---|---|---|---|---|---|---|
| Power output/kW | 0 | 0 | 0 | 80 | 220 | 360 |
| Speed/($\mathrm{m\,s^{-1}}$) | 12.0 | 14.0 | 16.0 | 18.0 | 20.0 | 22.0 |
| Power output/kW | 500 | 550 | 580 | 590 | 600 | 0 |

Calculate approximately the likely annual power output, and hence its capacity factor at:

a an extremely windy site where the wind follows a Rayleigh distribution with mean speed $8.2\,\mathrm{m\,s^{-1}}$, measured at a height of 10 m (i.e. conditions like North Ronaldsay, Section 9.6.3),

b at a potentially attractive site where the mean wind speed at 10 m is $6\,\mathrm{m\,s^{-1}}$.

9.12 According to (9.54) the wind speed $u_z$ at height $z(>10\,\mathrm{m})$ is approximately proportional to $z^{0.14}$, whereas the power density in the wind varies as $u_z{}^3$. By plotting $u_z{}^3$ against $z$ show that for $z > 100\,\mathrm{m}$ the variation of power density with height is relatively small. It follows that is not worthwhile to have very high towers (i.e. >100 m or so) for small wind turbines. How might the argument be different for large wind turbines?

9.13 Consider a turbine which maintains constant tip-speed ratio (and hence constant $C_p$ for output power $P_T >$ rated power $P_R$). If its cut-out speed $u_{co}$ is large ($\gg$ rated speed $u_R$), and the wind follows a Rayleigh distribution, show that the mean output power can be expressed as

$$\overline{P}_T = \frac{\rho C_P A_1}{2}\frac{6}{\pi}(\overline{u})^3 + P_R \exp\left[-\frac{\pi}{4}\left(\frac{u_R}{\overline{u}}\right)^2\right]$$

Evaluate this expression for some typical cases, e.g. $\overline{u} = 8\,\mathrm{m\,s^{-1}}$, $u_R = 5\,\mathrm{m\,s^{-1}}$, $P_R = 600\,\mathrm{kW}$, $A = 800\,\mathrm{m^2}$.

## Bibliography

### General

Burton, T., Sharpe, D., Jenkins, N. and Bossanyi, E. (2001) *Wind Energy Handbook*, Wiley. {This is the wind turbine 'bible', with advanced fundamental theory and professional experience of designing, manufacturing and implementing wind power.}

Eggleston, D.M. and Stoddard, F.S. (1987) *Wind Turbine Engineering Design*, Nostrand Reinhold. {A foundation text for engineers.}

Freris, L.L. (1990) (ed.) *Wind Energy Conversion Systems*. Prentice Hall, London. {Postgraduate level chapters by leading experts.}

Gipe, P. (1995) *Wind Power*, James and James, London. {An update of other similar publications and editions, with thorough and personal analysis of wind power development, especially in the USA; a bias to the independent owner.}

Golding, E.W. (1976) *The Generation of Electricity by Wind Power*, reprinted with additional material by E. and F.N. Spon, London. {The classic text that became a guide for much modern work.}

Hansen, M.O. (2000) *Aerodynamics of Wind Turbines*, James & James, London. {Clearly presented but advanced text from an experienced lecturer; moves from fundamental aeronautics to blade element theory, with physical explanations of further intricacies and applications.}

Milborrow, D. (2001) Wind energy review, in *Solar Energy : The State of the Art*, (ed.) J. Gordon, International Solar Energy Society and James & James , London.

Pasqualetti, M.J., Gipe, P. and Righter, R.W. (2002) *Wind Power in View*, Academic Press with Reed Elsevier, San Diego; Academic Press, London. {An edited set of chapters, mostly by experts other than engineers, concerning the visual and other non-engineering impacts of wind power installations. Important insights into personal aesthetics and cultural heritage.}

Shepherd, D.G. (1978) Wind power, in Amer, P. (ed.) *Advances in Energy Systems and Technology*, vol. 1, Academic Press, New York. {An excellent summary of theory and application at research level. Includes summaries of extension of linear momentum theory and of aerodynamic analysis.}

Strategies Unlimited (1988) *The Potential for Wind Energy in Developing Countries*, Windbooks Inc., Vermont, USA.

van Est, R. (1999) *Winds of Change*, International Books, Utrecht (in English). {A comparative study of the politics and institutional factors of wind energy development on California and Denmark.}

Walker, J.F. and Jenkins, N. (1997) *Wind Energy Technology*, John Wiley & Sons. {An introductory text with analytical and quantitative explanations.}

### Specifically referenced

Barbour, D. (1984) Energy study of the island of North Ronaldsay, Orkney, MSc thesis, University of Strathclyde.

Bowden, G.J., Barker, P.R., Shestopal, V.O. and Twidell, J.W. (1983) 'The Weibull distribution function and wind power statistics', *Wind Engineering*, 7, 85–98.

Jeffreys, H. and Jeffreys, B. (1966) *Methods of Mathematical Physics*, Cambridge University Press. Carefully presented text of advanced maths for engineers, etc.

Justus, C.G., Hargreaves, W.R., Mikherl, A.S. and Graves, D. (1977) 'Methods for estimating wind speed frequency distribution', *J. Ap. Meteorology*, **17**, 673–678.

Kaimal, J.C. and Finnigan, J.J. (1994) *Atmospheric Boundary Layer Flows*, Oxford University Press, UK. {Fundamental analysis and explanation by leading experts.}

Panofsky, H.A. and Dutton J.A., (1984) *Atmospheric Turbulence, Models and Methods for Engineering Applications*, Wiley, New York. {Useful analysis and background for wind turbine generation.}

Petersen, E.L. (1975) On the Kinetic Energy Spectrum of Atmospheric Motions in the Planetary Boundary Layer, Report 285 of the Wind Test site, Riso, Denmark.

Rohatgi, J.S. and Nelson, V. (1994) *Wind Characteristics, An Analysis for the Generation of Power*, Burgess Publishing, Edina M.A., USA.

World Meteorological Organisation (1981) *Meteorological Aspects of the Utilization of Wind as an Energy Source*, Technical Note no. 175, Geneva, Switzerland.

## Journals and websites

*Wind directions*. (6 issues per year), published by the European Wind Energy Association, 26 rue de Trône, Brussels.

*Wind Power Monthly*. (12 issues per year), published by Windpower Monthly News Magazine A/S, P.O. Box 100, Vrinners Hoved, 8420 Knebel, Denmark; ISSN 0109–7318. In English; world wide news and articles.

*Wind Statistics* (4 issues per year), published by Forlaget Vistoft, through WindStata Newslatter, P.O. Box 100, Vrinners Hoved, 8420 Knebel, Denmark; ISSN 0903–5648.

*Wind Engineering*. (6 issues per year), published by Multi-Science Publishing Co., 5 Wates Way, Brentwood UK; UK ISSN 0309–524X. Academic and research journal.

*Wind Energy* (4 issues per year), published by Wiley InterScience. ISSN 1095–4244. Academic and research journal.

*The World of Wind Atlases – Wind Atlases of the World* <http://www.windatlas.dk/>. Web site for wind atlas methodology and wind atlases of the world; established and maintained by members of the Wind Energy Department at Risø National Laboratory, Roskilde, Denmark.

Chapter 10

# The photosynthetic process

## 10.1 Introduction

Photosynthesis is the making (synthesis) of organic structures and chemical energy stores by the action of solar radiation (photo). It is by far the most important renewable energy process, because living organisms are made from material fixed by photosynthesis, and our activities rely on oxygen in which the solar energy is mostly stored. For instance, human metabolism continuously releases about 150 W per person from food. Thus, both the materials and the energy for all life are made available in gases circulating in the Earth's atmosphere, namely carbon dioxide and oxygen. Sadly, despite photosynthesis being a physically induced process and the driving function of natural engineering, the subject is missing from most physics and engineering texts. This chapter tries to rectify this omission by describing a cheap process that provides abundant stored energy – an engineer's dream, but a natural phenomenon.

The continuous photosynthetic output flux on the Earth is about $0.9 \times 10^{14}$ W (i.e. about 15 kW per person; the power output of 100 000 large nuclear power stations). This chapter discovers how the process occurs within molecules and cells, and how eventually it may be utilised at these levels. Energy supply from plant and animal materials, biomass, is discussed in Chapter 11. Solar radiation incident on green plants and other photosynthetic organisms relates to two main effects: (1) temperature control for chemical reactions to proceed, especially in leaves, and (2) photo excitation of electrons for the production of oxygen and carbon structural material. It is so important to maintain leaf temperature in the correct range that some solar radiation is reflected or transmitted, rather than absorbed (this is why leaves are seldom black). The energy processes in photosynthesis depend on the photons (energy packets) of the solar radiation, labelled '$h\nu$', where $h$ is Planck's constant and $\nu$ is the frequency of the radiation. The organic material produced is mainly carbohydrate, with carbon in a medium position of oxidation and reduction (e.g. glucose, $C_6H_{12}O_6$). If this (dry) material is burnt in oxygen, the heat released is about 16 MJ kg$^{-1}$ (4.8 eV per carbon atom, 460 kJ per mole of carbon). The fixation of one carbon atom from atmospheric $CO_2$ to carbohydrate proceeds by a series of stages in green plants, including algae:

1 Reactions in light, in which photons produce protons from $H_2O$, with $O_2$ as an important by-product, and electrons are excited in two stages to produce strong reducing chemicals.
2 Reactions not requiring light (called dark reactions), in which these reducing chemicals reduce $CO_2$ to carbohydrates, proteins and fats.

Combining both the light and the dark reactions gives an overall reaction, neglecting many intermediate steps:

$$CO_2 + 2H_2\dot{O} \xrightarrow{\text{light}} \dot{O}_2 + [CH_2O] + H_2O \qquad (10.1)$$

where the products have about 4.8 eV per carbon atom more enthalpy (energy production potential) than the initial material because of the absorption of at least eight photons. Here $[CH_2O]$ represents a basic unit of carbohydrate, so the reaction for sucrose production is

$$12CO_2 + 24H_2\dot{O} \xrightarrow{\text{light}} 12\dot{O}_2 + C_{12}H_{22}O_{11} + 13H_2O \qquad (10.2)$$

In these equations, the oxygen atoms initially in $CO_2$ and $H_2\dot{O}$ are distinguished, the latter being shown with a dots over the O.

Most studies of photosynthesis depend on biochemical analysis considering the many complex chemical processes involved. This chapter, however, will emphasise the physical processes, and will relate to the branch of spectroscopy called photophysics. There will also be interesting similarities and comparisons with photovoltaic devices (Chapter 7). We shall proceed in three stages:

1 The trophic level (Figure 10.1)
2 The plant level (Figure 10.2)
3 The molecular level (Figure 10.3): this is a complex system, which will be studied in Section 10.6.

There is extensive variety in all aspects of photosynthesis, from the scale of plants down to molecular level. It must not be assumed that any one system is as straightforward as described in this chapter, which concentrates on the general physical principles. However, the end result is that energy from the Sun is stored in stable chemicals for later use – a principle goal of renewable energy technology, yet happening all around us.

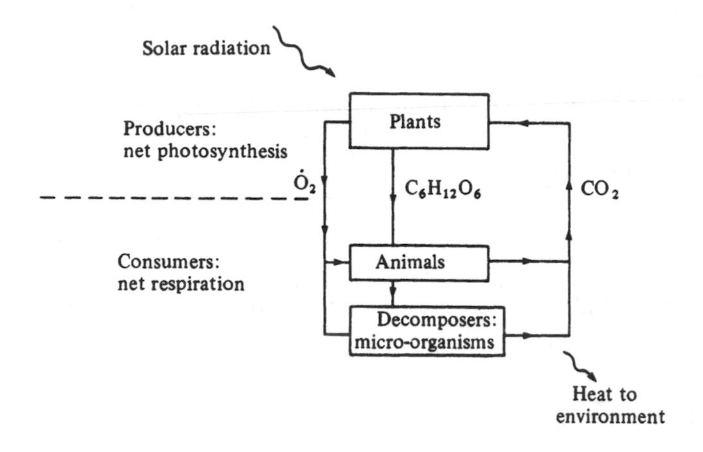


*Figure 10.1* Trophic level global photosynthesis, also requiring water. Fluxes: energy, $10^{14}$ W; carbon, $10^{11}$ t/y; $CO_2$, $4 \times 10^{11}$ t/y; oxygen, $3 \times 10^{11}$ t/y; water (as reactant) $3 \times 10^{11}$ t/y. Atmospheric concentrations: oxygen, 21%; $CO_2$, 0.030% by volume (pre-industrial, in 1850) but increasing due to anthropogenic activity (was 0.037% by year 2000, now increasing at a proportional rate of $\sim$ 0.4%/y).

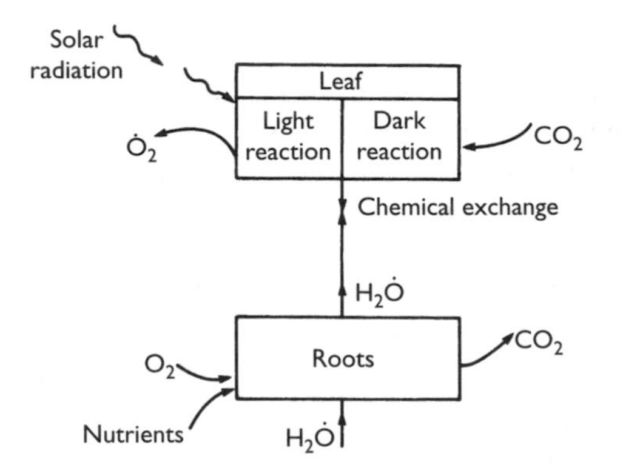


*Figure 10.2* Plant level photosynthesis.

## 10.2 Trophic level photosynthesis

Animals exist by obtaining energy and materials directly or indirectly from plants. This is called the trophic (feeding) system. Figure 10.1 is an extremely simplified diagram to emphasise the essential processes of natural ecology. We should remember, however, that the box labelled 'animals' might also include the human fossil fuel–based activities of industry, transport, heating, etc.

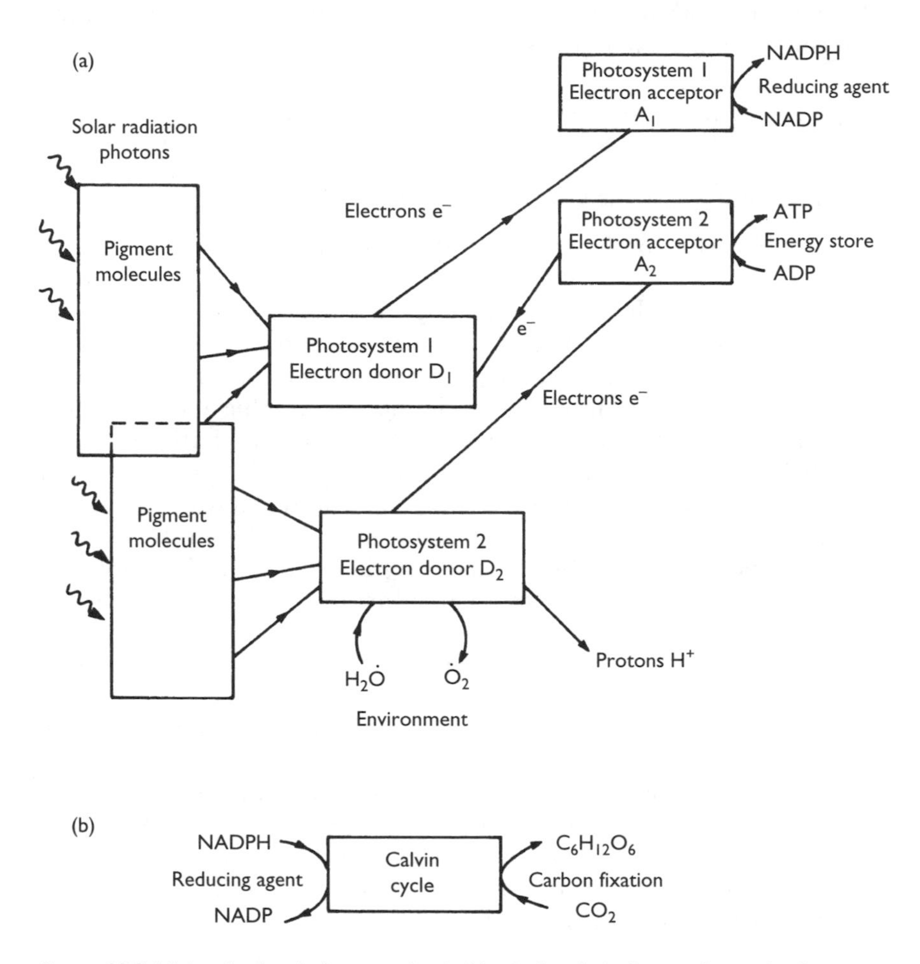


*Figure 10.3* Molecular level photosynthesis. Vertical scale indicates the excitation energy of the electron. (a) Light reaction, indicating the flow of energy and materials in the two interacting photosystems of green plants. (b) Dark reaction, using the reducing agent produced from the light reaction of photosystem I.

During photosynthesis $CO_2$ and $H_2O$ are absorbed to form carbohydrates, proteins and fats. The generalised symbol $[CH_2O]$ is used to indicate the basic building block for these products. $CO_2$ is released during respiration of both plants and animals, and by the combustion of biological material. This simplified explanation is satisfactory for energy studies, but neglects the essential roles of nitrogen, nutrients and environmental parameters in the processes.

The net energy absorbed from solar radiation during photosynthesis can be measured from combustion, since

$$\Delta H + CO_2 + 2H_2O \underset{\text{combustion}}{\overset{\text{photosynthesis}}{\rightleftharpoons}} [CH_2O] + O_2 + H_2O \tag{10.3}$$

$$\Delta H = 460\,\text{kJ per mole C} = 4.8\,\text{eV per atom C}$$
$$\approx 16\,\text{MJ}\,\text{kg}^{-1}\text{of dry carbohydrate material}$$

Here $\Delta H$ is the enthalpy change of the combustion process, equal to the energy absorbed from the photons of solar radiation in photosynthesis, less the energy of respiration during growth and losses during precursor reactions. $\Delta H$ may be considered as the heat of combustion. Note that combustion requires temperatures of ~400 °C, whereas respiration proceeds by catalytic enzyme reactions at ~20 °C. The uptake of $CO_2$ by a plant leaf is a function of many factors, especially temperature, $CO_2$ concentration and the intensity and wavelength distributions of light (Figure 10.4).

Photosynthesis can occur by reducing $CO_2$ in reactions with compounds other than water. In general these reactions are of the form

$$CO_2 + 2H_2X \rightarrow [CH_2O] + X_2 + H_2O \tag{10.4}$$

For example, X = S, relating to certain photosynthetic bacteria that grow in the absence of oxygen by such mechanisms, as was the dominant process on Earth before the present 'oxygen-rich' atmosphere was formed.

The efficiency of photosynthesis $\eta$ is defined for a wide range of circumstances. It is the ratio of the net enthalpy gain of the biomass per unit area ($H/A$) to the incident solar energy per unit area ($E/A$), during the particular biomass growth over some specified period:

$$\eta = \frac{H/A}{E/A} \tag{10.5}$$

Here $A$ may range from the surface area of the Earth (including deserts) to the land area of a forest, the area of a field of grain, and the exposed or total surface area of a leaf. Periods range from several years to minutes, and conditions may be natural or laboratory controlled. It is particularly important with crops to determine whether quoted growth refers to just the growing season or a whole year. Table 10.1 gives values of $\eta$ for different conditions.

The quantities involved in a trophic level description of photosynthesis can be appreciated from the following example. Healthy green leaves in sunlight produce about 3 litres of $O_2$ per hour per kg of leaf (wet basis). This is an energy flow of 16 W, and would be obtained from an exposed

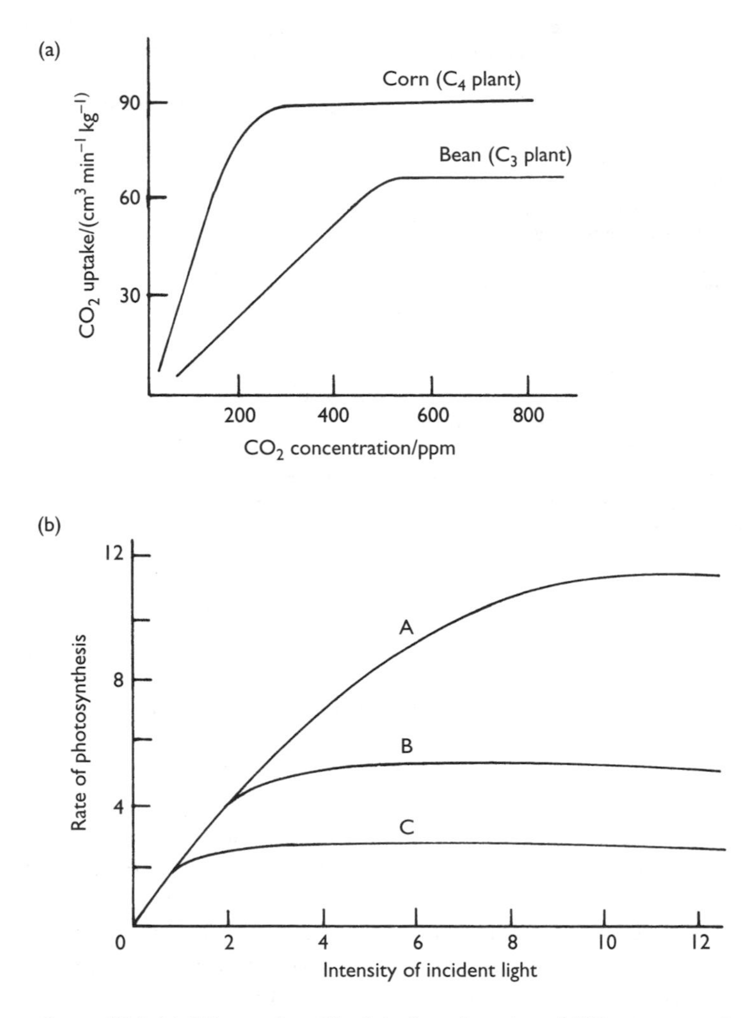


*Figure 10.4* (a) $CO_2$ uptake of fresh leaf as a function of $CO_2$ concentration. (b) Effect of external factors on rate of photosynthesis. Effect of light intensity at (A) 25 °C and 0.4% $CO_2$ (B) 15 °C and 0.4% $CO_2$ (C) 25 °C and 0.01% $CO_2$. All units arbitrary. Adapted from Hall and Rao (1999).

leaf area of about 1 $m^2$. A person metabolizes at about 100 W (resting), 200 W (active). Thus each person obtains metabolic energy for 24 h from reaction with oxygen derived from about 15–30 $m^2$ of leaf area. Thus in temperate regions, one person's annual bodily oxygen intake is provided by approximately one large tree. In the tropics such a tree would provide metabolic energy for about three people. Industrial, transport and domestic fuel consumption require far more oxygen per person, e.g. about 100 trees

*Table 10.1* Approximate photosynthetic efficiency for a range of circumstances. Reported data vary widely for many different circumstances

| *Conditions* | *Photosynthetic efficiency, (%): Approximate guide* |
|---|---|
| *Whole plant (net photosynthesis)* | |
| Whole earth: 1 year average (radiation incident beneath the atmosphere on to all land and sea) | 0.1 |
| Forest: annual general average | 0.5 |
| Grassland: annual (tropical, average; temperate, well managed) | 1 |
| Cereal crop close planted, good farming, growing season only temperate or tropical | 3 |
| Continuing crop: e.g. cassava | 2 |
| Laboratory conditions: enhanced $CO_2$, temperature and lighting optimised, ample water and nutrients | 5 |
| *Initial photosynthetic process (i.e. not including plant respiration)* | |
| Filtered light, controlled conditions etc. Theoretical maxima: primary photosynthetic: exciton process only | 36 |
| with the reaction centres | 20 |
| with carbohydrate formation | 10 |

per person in the USA, about 60 in Europe, about 20 in much of the developing world.

Morowitz (1969) argued that consideration of the bond energies in photosynthesis implies that the absorbed solar energy is predominantly stored in the oxygen molecules and not the carbon compounds. The oxygen gas is able to move freely in the atmosphere, and eventually disperses this store of energy evenly over the Earth's surface. In a similar manner carbon, in $CO_2$, is also dispersed. The coupled reaction of oxygen with carbon compounds has to occur before the energy can be released. Nevertheless, the usual assumption that food, biomass and fossil fuels contain the stored energy is acceptable as a basis for calculation. However, appreciating that plants and trees provide our energy and carbon-based materials through easily dispersed gases in the atmosphere provides an insight into the mechanisms of sustainable ecology and the need to conserve trees.

## 10.3 Photosynthesis at the plant level

### *10.3.1 Absorption of light*

Solar radiation incident on a leaf is reflected, absorbed and transmitted. Part of the absorbed radiation ($<5\%$) provides the energy stored in photosynthesis and produces oxygen; the remainder is absorbed as sensible heat producing a temperature increase, or as latent heat for water evaporation (kinetic and potential energy changes are negligible). Oxygen production is a function of the wavelength of the radiation and may be plotted as an *action spectrum*, as in Figure 10.5(b).

Figure 10.5 shows typical absorption and action spectra for a leaf. Note that photosynthesis, as measured by oxygen production, occurs across

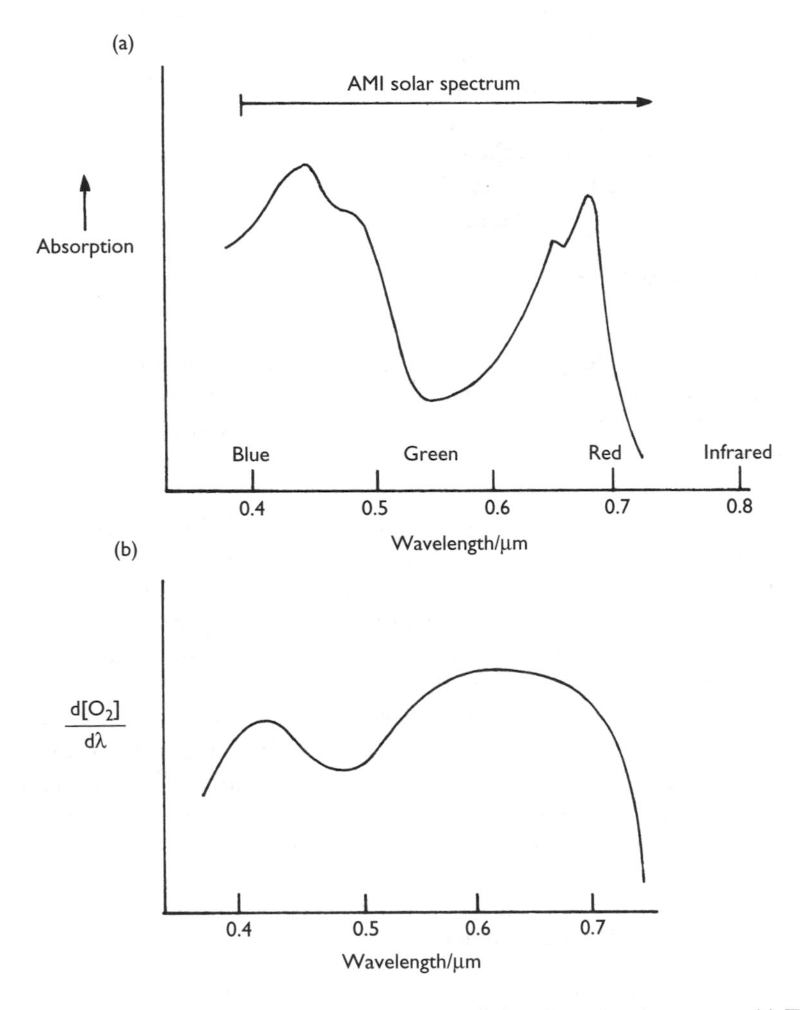


*Figure 10.5* Absorption and action spectra of plant leaves and pigments. (a) Typical spectral absorption spectrum of a green leaf *in vivo*. (b) Action spectrum of a typical green plant. $d[O_2]/d\lambda$ is the spectral distribution of the rate of oxygen production per unit area per unit radiation intensity. (c) Absorption spectra of important pigments when separated in laboratory conditions *in vitro*. The graphs give the form of the spectra, where 'absorption' is the proportion of monochromatic solar irradiance absorbed.

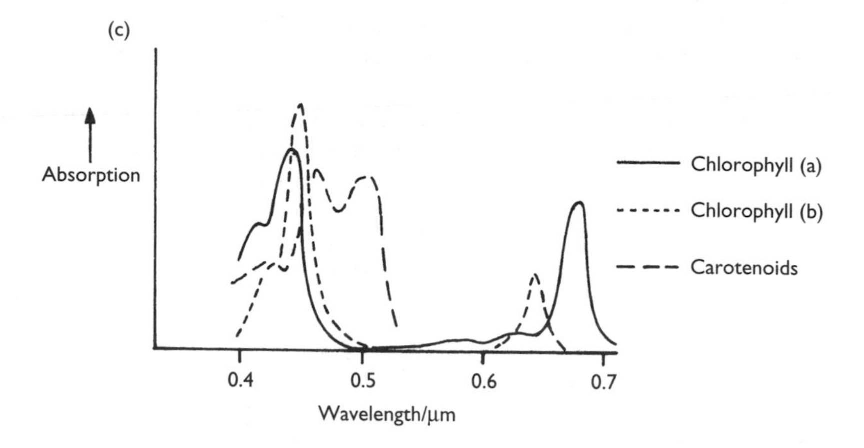


*Figure 10.5* (Continued).

nearly the whole visible spectrum. Absorption is usually most marked in the blue and red spectral regions; hence the green colour of most leaves. Spectroscopic techniques of great precision and variety are used to investigate photosynthesis, but two basic experiments are fundamental:

1 The absorption spectrum from the pigments of a living leaf is different from the sum of the individual absorption spectra of the same pigments separated by chemical methods. Thus within the cooperative structure of an assembly of pigments *in vivo*, the absorption spectrum of any one pigment molecule is changed because it is no longer isolated and there are cooperative interactions with other pigments.
2 For green plants, photosynthesis increases if there is absorption in one part of the spectrum, e.g. 700 nm, and also simultaneously at another wavelength, e.g. 650 nm. This is a further cooperative effect, called the *Emerson effect*, giving substantial evidence that photosynthesis in green plants occurs in two systems that act together in series.

### *10.3.2 Structure of plant leaves*

A general structure and scale of green leaves is given in Figure 10.6. In practice there is extensive variety and complexity in all these aspects down to molecular scale, as described in texts on plant physiology. In outline:

1 Photosynthesis occurs in plant material, usually green leaves and algae, which we consider here. It also occurs in some simple organisms (e.g. purple bacteria) without the associated $O_2$-producing system.

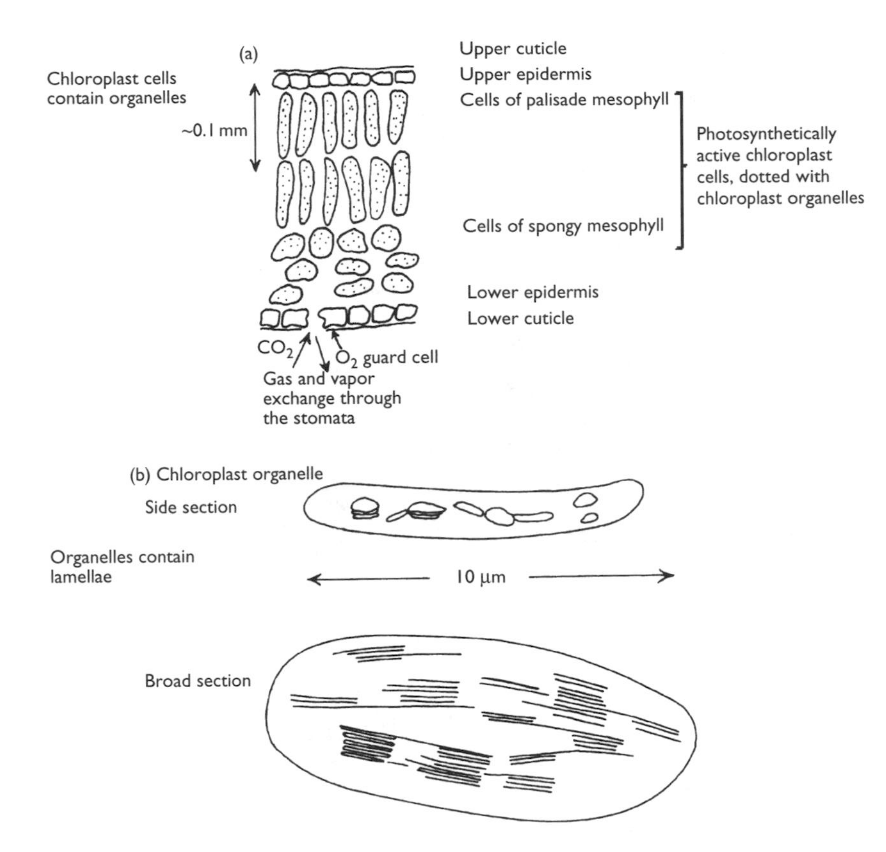


*Figure 10.6* Structure and scale of plant leaves. (a) Section of a typical leaf of a broad-leafed plant. Photosynthetically active green cells are shown dotted with chloroplast organelles. Approximate scale only. Actual cells press together more closely than shown, i.e. do not have gaps as big as indicated in the figure for clarity. (b) Section through chloroplast organelle. The thylakoid internal membranes are shown in the liquid stroma. Certain regions have stacked thylakoid membranes (the grana) which are connected by unstacked stroma lamellae membrane. (c) Perspective of the stacked and unstacked thylakoid membrane structure. Stacked grana linked by bridges of the stroma lamellae, all within the liquid stroma of the chloroplast organelle. Approximate scale only. (d) Thylakoid membrane shown fractured by freezing techniques along natural lines of weakness. Four distinct faces and surfaces appear: outer, PS (protoplasmic surface); inner of outer, PF (protoplasmic face); inner of inner, EF (ectoplasmic face); and outer of inner, ES (ectoplasmic surface). Stacked (subscripts) and unstacked (u) regions are identified. The inner surfaces show distinct bumps with electron microscopy. These bumps seem to be associated with ATP and NADPH production. Loose protein structures associated with $CO_2$ assimilation are on the outer PS surface. Structures associated with $O_2$ and proton production are on the outer inner surface $ES_s$. (Note: *In vitro* there is a single bilayer membrane, as indicated in (e).)

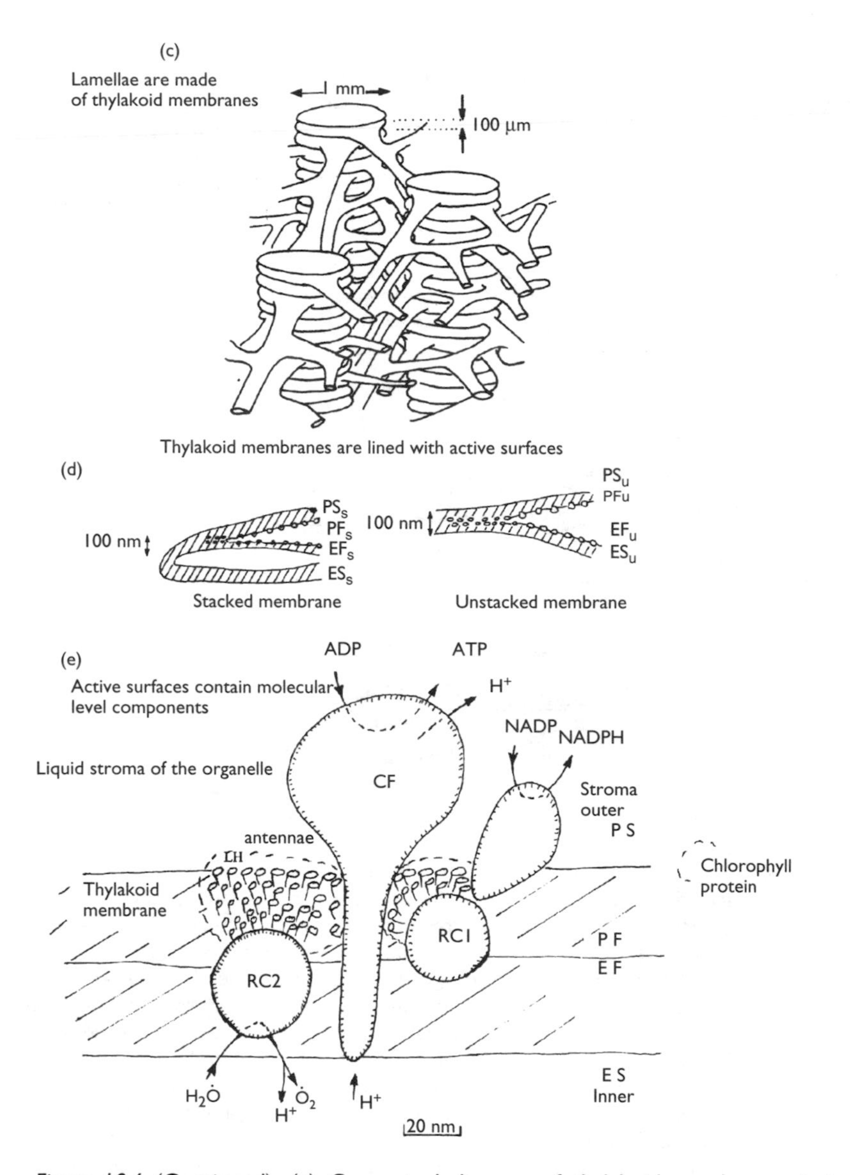


*Figure 10.6* (Continued). (e) Conceptual diagram of thylakoid membrane of the grana stacks, with the liquid stroma beyond the outer surface. LH light harvesting system of pigment molecules about 5 nm in length. About 200–400 molecules per reaction centre. RC reaction centres spanning the thylakoids membrane, about 20 nm diameter containing protein molecule complexes of about 50 000 molecular weight, 1 and 2 indicate the photosystems. CF: coupling factor. Very large complexes producing ATP and allowing protons to 'pump' through the membrane.

2 The active cells of green plants, e.g. palisade and spongy mesophyll cells of ~0.5 mm in length, have membranes permeable to gases and water (Figure 10.6(a)). These cells contain distinct intracellular bodies, organelles, also having membranes.
3 The photosynthetically active ellipsoidal shaped organelles are called chloroplasts. These are ~10 μm long, and contain liquid, the stroma, and membrane structure, the lamellae; see Figure 10.6(b).
4 The lamellae have a layered structure that is in general either stacked or open (unstacked) as shown in Figure 10.6(c). The stacks are called grana, and the unstacked membranes are stroma lamellae. The open structure is linked to the stacks and maintains the enclosed pockets within the chloroplast.
5 The lamellae are made of thylakoids. These are like flattened balloons having a double membrane structure with four different surfaces (outer top, inner top, inner lower, outer lower) as in Figure 10.6(d). The structure divides the chloroplast so the fluid may be different each side of the thylakoid lamella. This division is not easily discerned in two-dimensional sections of chloroplasts.
6 The thylakoid membrane contains the components of the photosynthetic light reaction as in Figure 10.6(e). These include the pigment molecules, mostly chlorophyll, which absorb photons in a structural array like a telecommunication antenna. This array is called the *antenna*. The pigment molecules act cooperatively to channel the absorbed energy 'packet' (called an exciton) to central reaction centres. The light trapping and energy channelling system is called the *light harvesting* (LH) system. There are about 30 pigment molecules associated with each reaction centre.
7 The reaction centres contain the final pigment molecules of the LH systems in chemical contact with large molecular weight enzyme molecules. At the reaction centre the energy from the LH system enables oxidation/reduction reactions to occur in complex catalytic sequences.
8 There are two types of reaction centre in green plants that may partly, but not entirely, share LH systems. The centres are of photosystem 1 and photosystem 2 (PS1, PS2).
9 At PS2 reaction centres, protons and electrons are separated from water as $O_2$ is produced, and some excess energy is used to form energy storage molecules ATP (adenosine triphosphate).
10 At PS1 reaction centres, a strong reducing agent NADPH (reduced nicotinamide adenine dinucleotide phosphate) is produced.
11 NADPH is able to initiate reactions to fix $CO_2$ outside the thylakoid membrane in the outer liquid of the stroma. These reactions can occur in dark or light, since the production of NADPH has separated the $CO_2$ uptake from the immediate absorption of light. The reactions are called the *Calvin cycle dark reactions*.

12 Protons, formed at oxygen and ATP production, are held by the thylakoid membrane within the inner regions of the lamellae. During ATP formation the protons move down an electrochemical gradient through the membrane to maintain the cycle of elements in the entire process.
13 In the Calvin cycle, $CO_2$ from solution is fixed into carbohydrate structures, and also protein and fat formation can be initiated.
14 *Photosynthetic bacteria* are prokaryotic cells (without internal nuclear membranes) and the photosynthetically active pigments are located in the membrane of the cell itself. Only one photosystem, PS1, operates and oxygen is not produced.

## 10.4 Thermodynamic considerations

In this section we shall first consider photosynthesis as an aspect of thermodynamics. The implications are important to guide strategy for renewable energy research and to give basic understanding.

Consider an ideal Carnot heat engine operated from solar energy (see Section 6.9) and producing work at efficiency $\eta$ with a heat sink at ambient temperature, say 27 °C(300 K). The heat supply is solar radiation. With a flat plate absorbing collector, the maximum source temperature is about 200 °C(473 K) and the maximum Carnot efficiency is $(473 - 300)/473 = 37\%$. For a power tower with radiation concentration on to the collector, the maximum Carnot efficiency might be $(773 - 300)/773 = 61\%$. If the Sun's outer temperature could be used as the source, the maximum Carnot efficiency would be $(5900 - 300)/5900 = 95\%$. Thus from an engineering point of view, there is much interest in seeking to link processes to the largest temperature available to us, namely the Sun's temperature.

The only connection between the Earth and the Sun is via solar radiation, so a radiation absorbing process is needed. If the absorption is on a black collector, the process is temperature limited by the melting point of the collector material. However, it is possible to absorb the radiation by a photon process into the electron states of a material without immediately increasing the bulk temperature. Such a process occurs in photovoltaic power generation (see Chapter 7). Let us compare the two processes, namely thermal and photon excitation. Figure 10.7 represents a material that can exist in two electronic states, normal and excited. The difference between these states is solely the different electronic configuration; the core or 'lattice' of the material remains unaffected.

In Figure 10.7(a) the excited state can only be reached by heating the whole material, and the proportion of excited states $N_e$ to normal states $N_n$ is calculated as for intrinsic semiconductors:

$$N_e/N_n = \exp(-\Delta E/kT) \tag{10.6}$$

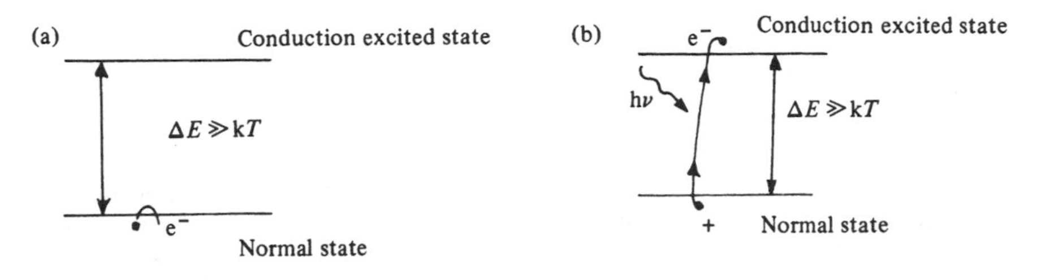


*Figure 10.7* Electron excitation by (a) heat and (b) photon absorption. The vertical scale indicates the excitation energy of the electron.

We shall be considering pigment molecules where $\Delta E \sim 2\,\text{eV}$, and $T < 373\,\text{K} = 100\,°\text{C}$, since the cellular material is water based. Thus $N_e/N_n \sim 10^{-27}$. Even at the Sun's temperature of 5900 K, $N_e/N_n = 0.02$. It is concluded that thermal excitation does not produce many excited states!

However, in Figure 10.7(b) the excited electronic state is formed by electromagnetic absorption of a photon of energy $h\nu \geq \Delta E$. This process does not immediately add energy to the remaining 'lattice', which remains at the same temperature. The population of the excited state depends on the rate of absorption of photons and coupling of the excited electronic states to the 'lattice'. Ideally, the population limit is $N_e = N_n$, when the radiation is transforming equal numbers of states back and forth and the electronic temperature is effectively infinite. This ideal is not reached in practice, but the theory of the process explains how $10^{10}$ more electronic excited states can be formed by electronic-state photon absorption, than by thermal excitation.

The thermodynamic analysis is not complete until the energy has performed a function. In photosynthesis the solar energy is transformed into excited states by photon absorption, and the energy is eventually stored in chemical products. There is no production of 'work' in the normal mechanical engineering sense, but work has been done in the production of organic material structures and of chemical stores of energy.

The chemical changes occurring in photosynthesis are in some ways similar to energy state changes in semiconductor physics. In chemistry the changes occur by reduction and oxidation. The *reduction level* ($R$) is the number of oxygen molecules per carbon atom needed to transform the material to $CO_2$ and $H_2O$. For carbon compounds of the general form $C_cH_hO_o$, the reduction level is

$$R = \frac{c + 0.25h - 0.5o}{c} \tag{10.7}$$

The energy to form these compounds from $CO_2$ and $H_2O$ per unit reduction level $R$ is about 460 kJ per mole carbon.

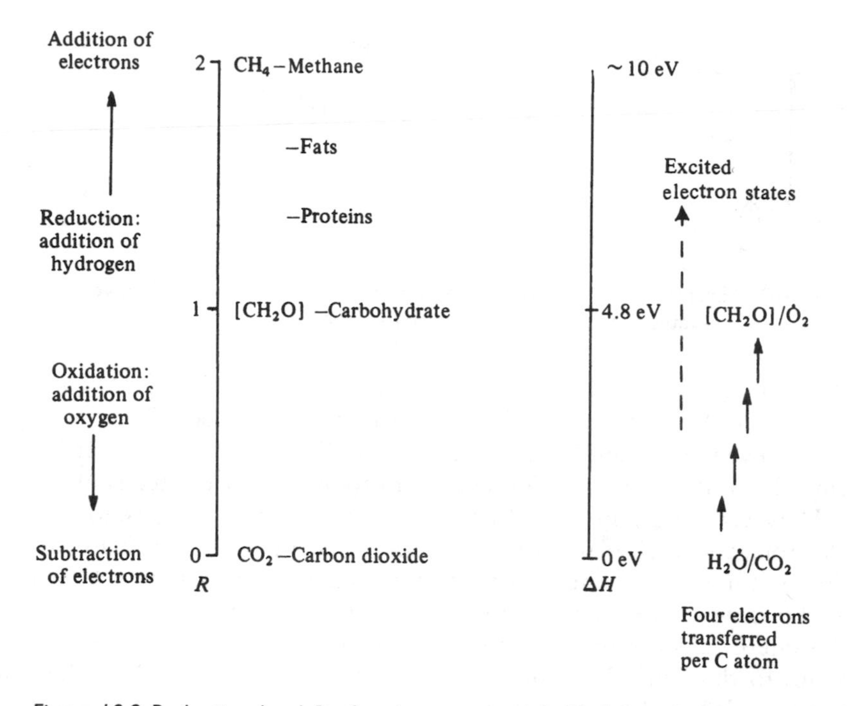


*Figure 10.8* Reduction level *R* of carbon compounds. Enthalpy change per carbon atom, $\Delta H$, of chemical couples referred to $CO_2/H_2O$.

The relationship of reduction level to the energy states involved in photosynthesis is shown in Figure 10.8. Photosynthesis is essentially the reduction of $CO_2$ in the presence of $H_2O$ to carbohydrate and oxygen. In the process four electrons have to be removed from four molecules of water, Figure 10.9. The full process will be explained in Section 10.6.

$$4H_2\dot{O} \xrightarrow{h\nu} 2H_2\dot{O} + \dot{O}_2 + 4H^+ + 4e^-$$

## 10.5 Photophysics

The physical aspects of photosynthesis involve the absorption of photons of light by electrons within pigment molecules. These molecules absorb the energy to form excited states. When the molecules are isolated, the energy is usually re-emitted as fluorescent radiation and heat. However, when the pigments are bound in chloroplast structures, the majority of the energy is transferred cooperatively to reaction centres for chemical reductions, with the excess coming out as heat, and there is little fluorescence.

The isolated properties are explained by the Franck-Condon diagram (Figure 10.10). This portrays the ground and excited energy states of the

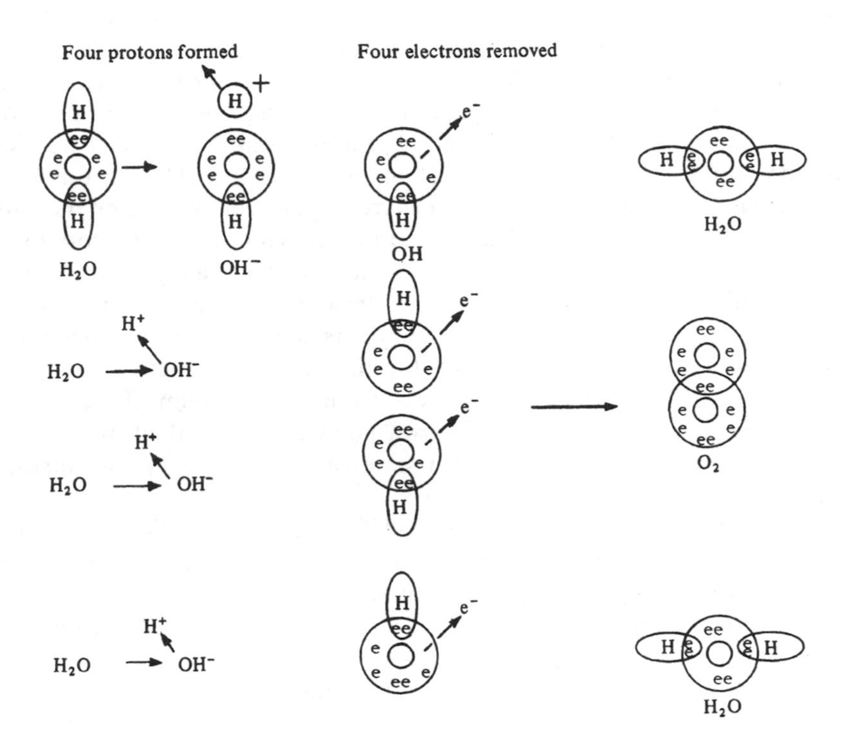


*Figure 10.9* Reduction of water to oxygen and protons at reaction centre of PS2 as four electrons are removed:

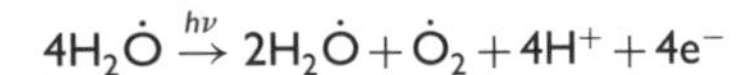

$$4H_2\dot{O} \xrightarrow{h\nu} 2H_2\dot{O} + \dot{O}_2 + 4H^+ + 4e^-$$

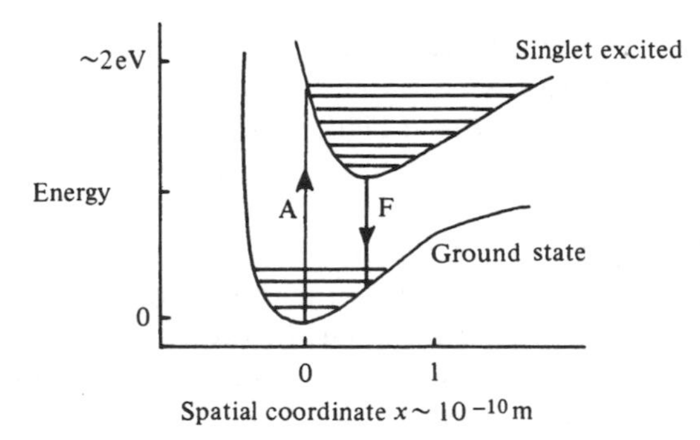


*Figure 10.10* Franck-Condon diagram illustrating Stokes shift in energy between the absorbed photon A and the fluorescent photon F. The spatial co-ordinate *x* indicates the change in position or size between the excited system and its ground state.

molecule as a function of the relative position of its atoms. This relative position is measured by some spatial coordinate, such as the distance $x$ between two particular neighbouring atoms. Note that the minima in energy occur at different values of $x$ owing to molecular changes in size or position after excitation. A photon in radiation, travelling at $3 \times 10^8\,\mathrm{m\,s^{-1}}$, passes the molecule of dimension $\sim 10^{-9}\,\mathrm{m}$ in time $\sim 10^{-18}\,\mathrm{s}$. During this time electromagnetic interaction with the electronic state can occur, and the photon energy of $\sim 2\,\mathrm{eV}$ is absorbed (A). However, vibrational and rotational motions are occurring in the molecule, with thermal energy $kT \sim 0.03\,\mathrm{eV}$ and period $\sim 10^{-13}\,\mathrm{s}$. These states are indicated by horizontal lines on the diagram as the molecule oscillates about its minimum energy positions. Absorption (A) takes place too fast for the molecule structure to adjust, and so the excited state is formed away from the minimum. If the excited electron is paired with another electron (as will be probable), the excited state will be expected to be a singlet state (spin $= \frac{1}{2} - \frac{1}{2} = 0$) with lifetime $\sim 10^{-8}\,\mathrm{s}$.

During this time of $10^{-8}\,\mathrm{s}$, there are $\sim 10^5$ molecular vibrations and so the excited state relaxes to the minimum of excited energy by thermal exchange to the surroundings. After this, one of the following two main processes occurs with the release or transfer of the remaining excitation energy.

1. The molecule is close to other similar molecules, and the absorbed energy (called an exciton) is passed to these by resonant transfer linked with the thermal motion during the $10^{-8}\,\mathrm{s}$ lifetime. This is the dominant process for pigment molecules *in vivo*.
2. After $\sim 10^{-8}\,\mathrm{s}$, fluorescent emission (F) may occur as the molecule returns to the ground state. The wavelength of fluorescence is longer than the absorbed light, as described by the *Stokes shift*. Alternatively the electron may change orientation in the excited state, by magnetic interaction with the nucleus, to form a triplet state (spin $= \frac{1}{2} + \frac{1}{2} = 1$). The lifetime of triplet states is long ($\sim 10^{-3}\,\mathrm{s}$) and again loss of energy occurs, by phosphorescence (P: Figure 10.11) or by resonant transfer.

*Resonant transfer* can occur between molecules when they are close ($\sim 5 \times 10^{-10}\,\mathrm{m}$), and when the fluorescence radiation of the transferring molecule overlaps with the absorption band of the neighbour (Figure 10.12). In these conditions, the excited electronic state energy (the exciton) may transfer without radiation to the next molecule. Separate energy level diagrams of the form of Figure 10.13(a) may describe this or, when molecules are very close, by a graded band gap diagram like Figure 10.13(b). In either description there is a spatial transfer of energy down a potential gradient through the assembly of molecules. The process is similar to conduction band electron movement in graded gap photovoltaic cells; see Section 7.7.2(5). However,

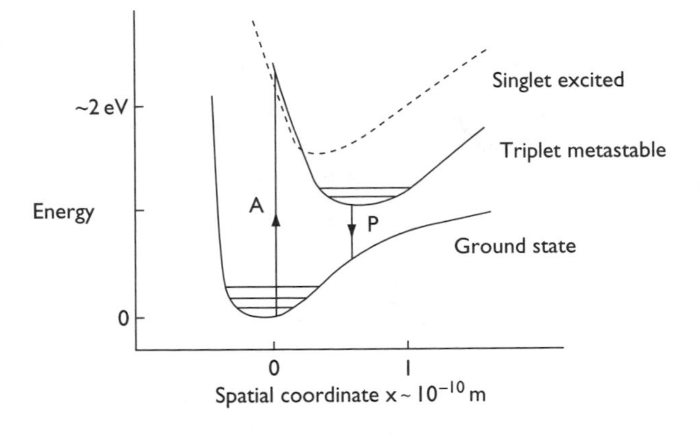


*Figure 10.11* Triplet state and phosphorescent photon P. However, the excited and ground states may overlap, so return to ground state may occur without emitted radiation.

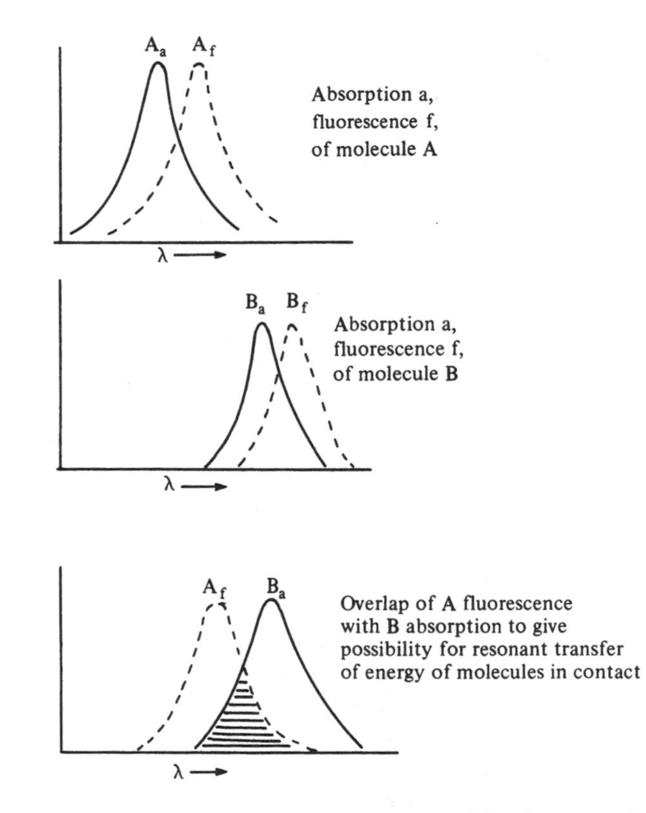


*Figure 10.12* Resonant transfer of energy. The abscissa is the wavelength of the fluorescence and absorption bands of separated molecules. The ordinate indicates the intensity of the bands.

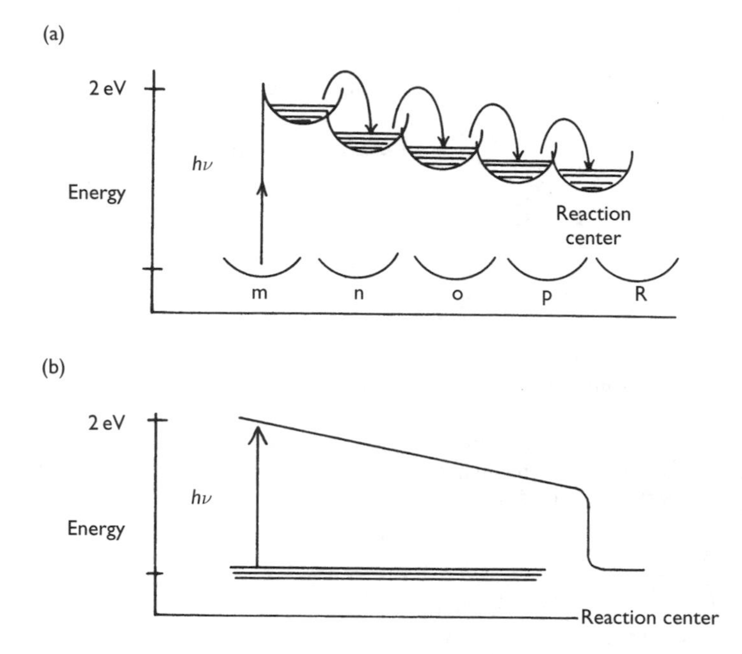


*Figure 10.13* Transfer of energy by pigment molecules of the LH system to the particular reaction centre. (a) Spatial position of LH pigment molecules (m, n, o, p) transferring energy to a reaction centre R. (b) Graded band gap model: continuous electronic structure of LH pigment molecules acting as a continuous 'super molecule'.

in photosynthesis, energy is transferred as whole molecules slightly adjust their position and structure during electronic excitation and relaxation, and not just by the transport of a free electron.

There is nevertheless a most significant difference between electron transport in photovoltaic semiconductors and energy transport in pigment molecules. In photovoltaics the structural material is manufactured with graduated dopant properties across the cell. Each element of material has a precise dopant level and must remain at the suitable location. If the photovoltaic cell is broken up, each piece keeps its distinguishing characteristic. However, in the photosynthetic LH system, it is the *cooperative structure* of all the pigments that gives each pigment the necessary electronic structure required for its precise location. It does not matter where a pigment molecule finds itself; it will always be given the correct properties to fit into the LH array, suitable for its position. So when the 'array' is broken up, each pigment reverts to its isolated properties. This accounts for the difference between *in vivo* and *in vitro* properties of pigment molecules, during absorption and fluorescence.

## 10.6 Molecular level photosynthesis

The previous sections have outlined the whole photosynthetic process. In this section we shall consider some of the physical details, although we stress that many of these are not fully understood and are the subject of active research. Note that we purposely do not include biological and chemical details, since these may detract from physical understanding and also such detail is well described elsewhere, e.g. Wrigglesworth, 1997.

### *10.6.1 Light harvesting and the reaction centre*

The photosynthetically active light is absorbed in pigment 'antenna' molecules, of which the most common is chlorophyll type (a). Other pigments occur of similar molecular structure and shape. A chlorophyll molecule consists of a hydrophobic (water fearing) tail and hydrophilic (water loving) ring. The tail is a long chain hydrocarbon (phytyl residue $C_{20}H_{39}$), about four times longer than the 'diameter' of the ring. It seems that most of the antenna chlorophyll molecules are embedded in chlorophyll pigment proteins, which in turn are embedded in the thylakoid membrane. The tail is presumed to be fixed in the thylakoid membrane's outer surface, with the ring in the water-based stroma of the chloroplast (Figure 10.6(e)). Light photons may be absorbed by the molecules, which are close enough to (1), cooperatively interact to produce wideband *in vivo* absorption, and (2), to channel (harvest) excited state energy excitons down a potential gradient by resonant transfer to the reaction centre. Assemblies of about 200–400 pigment molecules occur as 'cores', each containing one reaction centre. Assemblies with less or more pigment molecules per reaction centre occur in some species of plant, algae or photosynthetic bacteria. In general, systems adapted to weak light have more pigment molecules per core.

Photons may be absorbed at the outer or inner LH molecules, and the resulting energy exciton passes to the reaction centre along paths of least resistance. This consists of the final pigment molecule and some very large specialised molecules (molecular weight ~50 000) in which the 'chemistry' occurs. Most evidence for this general structure of parallel LH paths to a central centre has come from optical absorption, fluorescence and action spectra linked with light flashes of varying interval and intensity. Two types of core exist, one for each of photosystems 1 and 2. It is possible for different cores to share the same LH arrays.

### *10.6.2 The light reactions*

The physical processes of photosystems 1 and 2 (PS1 and PS2) are similar (Figure 10.14). The light harvesting (LH) system channels excitons from the initial quanta to the reaction centre, where the last molecule of chlorophyll

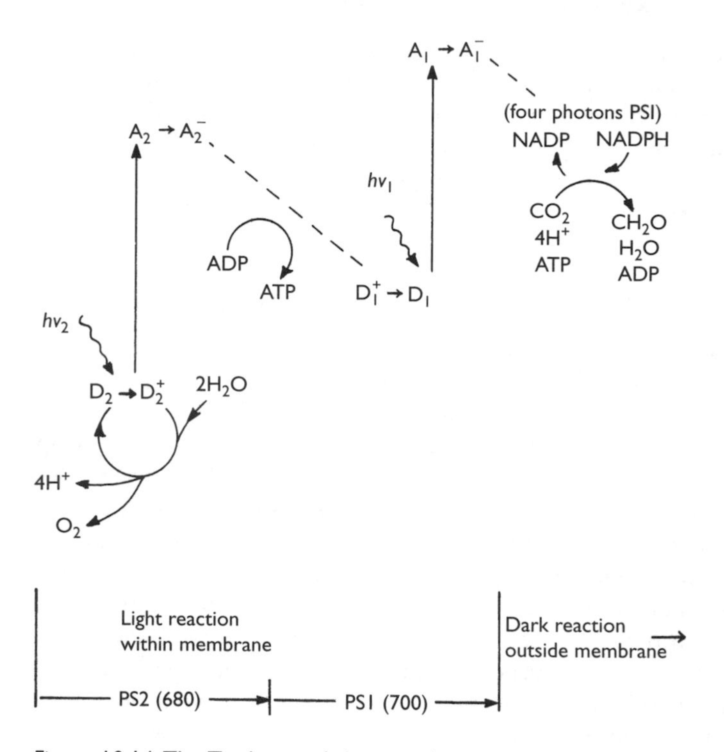


*Figure 10.14* The Z scheme of photosynthesis at the reaction centres: eight photons per C fixed. The vertical scale indicates the excitation energy of the electron.

in the system is used to identify the particular centre. This is called P680 for PS2, and P700 for PS1, since the respective molecules absorb distinctively at $680 \pm 10\,\text{nm}$ (1.82 eV) and $700 \pm 10\,\text{nm}$ (1.77 eV) respectively. Note that both these molecules absorb at the red end of the spectrum indicating their position at the lower potential energy end of the LH system.

In the reaction centre there are both donor and acceptor molecules. (These names are used in the sense of semiconductor physics. However, beware that the letter p may be used for pigment donor molecules, in the *opposite* sense of p-type semiconductors.) The channelled exciton from each absorbed photon lifts an electron from the particular donor (D) to the acceptor (A). Chemical reactions then occur with the $D^+$ and $A^-$ radicals so formed. These reactions are cyclic, so D and A are reformed by the transfer of another electron to D, and the excited electron from A. Two sets of chemical reactions occur, one set as $D^+$ returns to D, and the other set as $A^-$ returns to A. These chemical reactions are complex and not yet fully determined for the many variations of detail that occur. However, the physical processes are clear. Note especially how the complete process utilises multiple photon absorption and electron excitation steps per molecule of $O_2$ released and $CO_2$ absorbed.

For green plants there are five distinct steps of the Z scheme as illustrated in Figure 10.14 and also Figure 10.3:

1 *$D_2^+ \rightarrow D_2$.* $D_2^+$ is formed by electron excitation of PS2. It is a strong oxidant (attracting electrons) with a redox potential of +0.9 V. There is evidence that $D_2^+$ is a chlorophyll (a) molecule, or molecular pair, in the non-aqueous environment of the thylakoid membrane. The oxidising action of $D_2^+$ is to extract an electron from $H_2O$; with the ultimate result that one molecule of $O_2$ is formed by four $D_2^+$ reactions, see Section 10.4 and Figure 10.9. The $O_2$ molecules diffuse through the various membranes and out of the plant cells. The protons, $H^+$, remain in the thylakoid membrane double layer.

2 *Photon absorption in PS2.* The donor $D_2$ releases an electron that is lifted to acceptor $A_2$, forming $A_2^-$. $A_2$ has been given the label Q by research workers and may be a plastoquinone. (Note that Mitchell's 'Q-cycle' is involved in the movement of protons from PS1 to PS2.) $A_2^-$ is a mild reductant of redox potential about −0.1 V, i.e. the potential energy stored by photon excitation when an electron is transferred from $A_2$ to $D_2$ changes from 1.0 to 1.1 eV. $D_2$ absorbs at 680 ± 10 nm.

3 *Link between PS2 and PS1; ATP production.* After $A_2^-$ formation, a series of reactions occur to transfer the electron down an energy potential to $D_1^+$ at redox potential +0.4 V. The energy difference is used to form the energy storage molecule ATP (adenosine triphosphate) from ADP (adenosine diphosphate) at 0.34 eV per molecule. Minimum ATP production seems to be three ATP per four operations of PS2 and PS1, as required for the Calvin cycle (Section 10.6.3):

$$ADP + P_i \rightarrow ATP + H_2O \tag{10.8}$$

where $P_i$ denotes inorganic phosphate.

4 *Photon absorption in PS1.* Further photon absorption through the LH system to the reaction centre of PS1 results in the formation of $D_1^+$ and $A_1^-$. $D_1$ appears to be a further chlorophyll (a) monomer or pair (absorbing at 700 ± 10 nm). $A_1$ is a ferredoxin protein.

5 *NADPH production.* The electron raised in energy potential to $A_1$ then passes to form NADPH from $NADP^+$. NADP stands for nicotinamide adenine dinucleotide phosphate. Its reduced form is NADPH, and its oxidised form is $NADP^+$, sometimes confusingly written NADP.

$$NADP^+ + 2e^- + H^+ \rightarrow NADPH \tag{10.9}$$

Thus each reduction of $NADP^+$ requires two electrons, from two operations of PS1.

### *10.6.3 Dark reactions (Calvin cycle)*

The formation of NADPH by PS1 supplies the main reducing agent required to fix $CO_2$ into carbohydrate. Subsequent reactions occur by ordinary 'thermal' chemistry, not photochemistry, and so may occur in the dark *after* a period of light absorption, as well as *during* periods of light absorption. The reactions – the Calvin cycle – occur in the stroma outside the thylakoid structures, but within the chloroplast. The reactions may be followed using the radioisotope $^{14}C$. In photosynthetic bacteria, PS1 is the only photosystem, and $CO_2$ fixation occurs with no $O_2$ production (since there is no PS2). In green plants, chemical products from all three sets of reactions associated with PS2 and PS1 ($D_2^+$, $A_2^-$ and $D_1^+$) are used in the dark reaction. Thus the main initial inputs for the Calvin cycle are NADPH, ATP and $H^+$ from the photosystems, and $CO_2$ and $H_2O$ from the environment. A complex of intermediate chemicals and numerous catalytic enzymes is required within the cycle (e.g. the enzyme Rubisco, i.e. ribulose biphosphate carboxylaseoxygenase). The protons are made available from within the thylakoid membrane by the various 'pumping' mechanisms. In the chemical equations below, $P_i$ indicates inorganic phosphate.

Assimilation of $CO_2$ is described by:

$$3CO_2 + 9ATP + 6NADPH + 5H^+ \rightarrow C_3H_5O_3P + 9ADP + 8P_i + 6NADP^+ + 3H_2O + e^- \qquad (10.10)$$

while the overall Calvin cycle reaction can be summarised as

$$CO_2 + 2NADPH + 3ATP + 2H^+ \rightarrow [CH_2O] + 2NADP^+ + 3ADP + 3P_i + H_2O \qquad (10.11)$$

Thus two NADPH are required per C fixed from reduced $CO_2$. Each NADPH requires two electrons from two operations of PS1 to be reduced again to NADP, cf. (10.9). Thus the Calvin cycle is powered by four photon absorptions in PS1.

The first product of the Calvin cycle is a three-carbon ($C_3$) compound in most plants, as in (10.10). Certain tropical plants (e.g. sugarcane, maize and sorghum) have a preliminary chemical cycle involving a $C_4$ compound before the Calvin cycle. These $C_4$ plants have two different types of photosynthetic cells that function cooperatively in the plant. In moderate to strong light intensity ($\sim 0.5\,kW\,m^{-2}$) and elevated temperatures in the leaves ($\sim 40\,°C$), the C fixation and hence biomass production of $C_4$ plants may be twice that of $C_3$ plants.

### *10.6.4 Number of photons per carbon fixed*

The main requirement for light absorption is that individual photons can be absorbed, and the energy stored for sufficient time to be used in later chemical reactions or further photon excitation. Thus each photosystem is triggered by single photons. If the molecules of one system are in operation and so 'saturated', then it seems that the exciton can be passed to others. A minimum of four operations of PS2 are needed to produce one molecule of $O_2$, i.e. four electrons have to be lifted off $H_2O$, see Figure 10.9. Four other photons are needed to produce the NADPH for $CO_2$ reduction. So in green plants with coupled PS2 and PS1, *at least eight photons are needed to fix one C atom* as carbohydrate. In practice it seems that more photons are needed, either because an effective chemical saturation or loss occurs, or because further ATP is required. Thus most plants probably operate at about ten photons per C fixed in optimum conditions.

### *10.6.5 Energy states*

In the simplified reaction (10.1), the energy difference of C in $[CH_2O]$ and $CO_2$ is 4.8 eV. This averages 1.2 eV each cycle, per four cycles of PS2 and PS1 together. The 1.2 eV may be accounted for at the reaction centres by the sum of:

| | |
|---|---|
| −0.2 eV | $O_2$ formation by $D_2^+$ |
| +1.1 eV | absorption by PS2 |
| −0.5 eV | link of PS2 and PS1, ATP formation |
| +1.0 eV | absorption by PS1 |
| −0.2 eV | NADPH formation |

The signs of these energies are used in the sense of electron excitation of physics. The chemical redox potentials have opposite sign, but the essential meaning is the same. The zero energy is with separated $CO_2$ and $H_2O$ before solutions are formed and the reactions begin.

### *10.6.6 Efficiency of photosynthesis*

This may be defined in various ways. The minimum photon energy input at the outside antenna pigment molecules (i.e. *not* at the reaction centre) may be given as four photons of 1.77 eV (PS2 absorption for $D_2$ at 700 nm) and four of 1.82 eV (PS1 absorption for $D_1$ at 680 nm), totalling 14.4 eV. The actual excitations $D_2$ to $A_2$, and $D_1$ to $A_1$, are about 1.1 eV each. So four operations of each require 8.8 eV. The outputs may be considered to be four electrons lifted from $H_2O$ to NADP over redox potential 1.15 eV (4.60 eV) plus three ATP molecules at 0.34 eV each (1.02 eV) to give a total output of 5.6 eV. The output may also be considered as one $O_2$ molecule, and one C atom fixed in carbohydrate, requiring 4.8 eV.

A reasonable maximum efficiency from light absorption to final product can be taken as 4.8/14.4 = 33%. The larger proportions of 5.6/14.4, 5.6/8.8 and 4.8/8.8 may sometimes be considered.

In discussing photon interactions, the unit of the *einstein* is often used. One einstein is Avogadro's number of photons of the same frequency, i.e. one mole of identical or similar photons.

The solar spectrum consists of many photons with quantised energy too small to be photosynthetically active ($\lambda > 700$ nm, $h\nu < 1.8$ eV), and photons of greater energy than the minimum necessary ($h\nu > 1.8$ eV). The situation is very similar to that with photovoltaic cells (see Figure 7.12 and Section 7.4), so that only about 50% of the energy of the photons photosynthetically absorbed is used to operate PS2 and PS1. This effect would reduce the maximum efficiency to about 16%. However, leaves are not black and there is considerable reflection and transmission, reducing the maximum efficiency to about 12%, considering only the solar radiation incident at the photosystems. Efficiencies near to this have been obtained in controlled laboratory conditions. Considering solar irradiation on land generally, which will include parts other than leaves, such large efficiencies are not reached in even the best agriculture, or in natural conditions (Table 10.1).

## 10.7 Applied photosynthesis

Technology continually advances from fundamental studies in science. The same process will follow the eventual full understanding of photosynthesis in its many varied details. This section considers some energy-related applications, both current and potential.

### *10.7.1 Plant physiology and biomass*

Plants are being selected and bred to be better suited to their environment. Examples have already resulted from understanding why $C_4$ plants are more productive in stronger light intensity than $C_3$ plants, see Section 10.6. As biomass energy becomes more important (see Chapter 11), plants are being selected and developed to optimise fuel supplies rather than just their fruit, grain or similar part product. For instance, propagation from clones of best plants and application of genetic engineering have increased photosynthetic efficiency for biomass production. Considerable research concerns the functioning of the Rubisco enzyme, with a view to eventually 'designing' a form of Rubisco which allows increased carboxylation at the expense of the side reactions which now occur naturally, notably oxygenation. It is conceivable that artificial carbon-based structures, materials and food could be manufactured from synthetically controlled forms of photosynthesis.

### 10.7.2 Hydrogen production

The free protons formed during the operation of the photosynthetic cycles may, in some instances, be emitted as hydrogen gas. The $H^+$ ions are reduced (have an electron added) from the acceptor $A_1^-$ of PS1. This reaction is aided by certain natural enzymes (hydrogenases) and does occur naturally, e.g. with the bacteria *bacteriorhodopsin* and in human metabolism associated with digestion deficiencies. In general, however, hydrogenases are inoperative in the presence of $O_2$ and, if $H_2$ is emitted, the concentration is extremely small. Nevertheless, the prospect of producing considerable amounts of hydrogen from chemical reactions activated by sunlight is potentially very important commercially and warrants continuing research.

### 10.7.3 Photochemical electricity production

The driving function of photosynthesis is the photon-induced molecular excitations of PS2 and PS1. These involve electronic excitations within molecular structure and charge separations, and are not exactly comparable with electron–hole separation across the band gap photovoltaic devices. Nevertheless these molecular excitons have sufficient energy to drive an external electric circuit, and research and development continues for photochemical power devices. A number of related devices are available, including the dye-sensitive cell (Figure 7.23). The advantage over conventional photovoltaics might be the manufacture of the base material by liquid chemical methods in bulk and the manufacture of single devices that both store electricity in sunlight and provide electricity as a battery. Such developments from photochemistry may lead to radical innovation.

## Problems

10.1 Calculate very approximately how many trees are needed (i) to produce the oxygen used for your own metabolism, and (ii) to maintain the per capita total fuel consumption of your country. Compare this with the approximate number of trees per person in your country.

10.2 The heat of combustion of sucrose $C_{12}H_{22}O_{11}$ is $5646\,kJ\,mol^{-1}$. Calculate using the Avogadro constant, the energy per atom of carbon in units of eV.

## Bibliography

### *Undergraduate-level books*

Hall, D.O. and Rao, K.K. (1999) *Photosynthesis* (6th edn) Edward Arnold, London. {A short and stimulating introduction with more physical bias than many others.}

Lawlor, D.W. (2001, 3rd edn) *Photosynthesis: Molecular, Physiological and Environmental Processes*, BIOS Scientific Publications, Oxford. {Concise text for biology undergraduates.}

Monteith, J. and Unsworth, K. (1990, 2nd edn) *Principles of Environmental Physics*, Edward Arnold, London. {Considers the physical interaction of plant and animal life with the environment. Chemical aspects are not considered. Of background relevance to photosynthesis.}

Morowitz, H.J. (1969) *Energy Flow in Biology: Biological Organisation as a Problem in Thermal Physics*, Academic Press, New York. {A stimulating book with lateral thinking for considering physical processes and constraints.}

Wrigglesworth, J. (1997) *Energy and Life*, Taylor & Francis, UK. {The biochemistry of metabolism and photosynthesis, clearly presented.}

### *Journals and websites*

Photosynthesis is an extremely active area of scientific research, with most of that research (>1000 papers per year) reported in specialist scientific journals such as *Photosynthesis Research, Annual Review of Plant Physiology and Plant Molecular Biology, Nature* and *Photochemistry and Photobiology*. Much of this work is then distilled into graduate-level monographs.

Chapter 11

# Biomass and biofuels

## 11.1 Introduction

The material of plants and animals, including their wastes and residues, is called *biomass*. It is organic, carbon-based, material that reacts with oxygen in combustion and natural metabolic processes to release heat. Such heat, especially if at temperatures >400 °C, may be used to generate work and electricity. The initial material may be transformed by chemical and biological processes to produce *biofuels*, i.e. biomass processed into a more convenient form, particularly liquid fuels for transport. Examples of biofuels include methane gas, liquid ethanol, methyl esters, oils and solid charcoal. The term *bioenergy* is sometimes used to cover biomass and biofuels together.

The initial energy of the biomass-oxygen system is captured from solar radiation in photosynthesis, as described in Chapter 10. When released in combustion the biofuel energy is dissipated, but the elements of the material should be available for recycling in natural ecological or agricultural processes, as described in Chapter 1 and Figure 11.1. Thus the use of industrial biofuels, when linked carefully to natural ecological cycles, may be non-polluting and sustainable. Such systems are called *agro-industries*, of which the most established are the sugarcane and forest products industries; however, there are increasing examples of commercial products for energy and materials made from crops as a means of both diversifying and integrating agriculture.

The dry matter mass of biological material cycling in the biosphere is about $250 \times 10^9\,\mathrm{t\,y^{-1}}$ incorporating about $100 \times 10^9\,\mathrm{t\,y^{-1}}$ of carbon. The associated energy bound in photosynthesis is $2 \times 10^{21}\,\mathrm{J\,y^{-1}} (= 0.7 \times 10^{14}\,\mathrm{W})$. Of this, about 0.5% by weight is biomass as crops for human food. Biomass production varies with local conditions, and is about twice as great per unit surface area on land than at sea.

Biomass provides about 13% of mankind's energy consumption, including much for domestic use in developing countries but also significant amounts in mature economies; this percentage is comparable to that of fossil gas. The domestic use of biofuel as wood, dung and plant residues

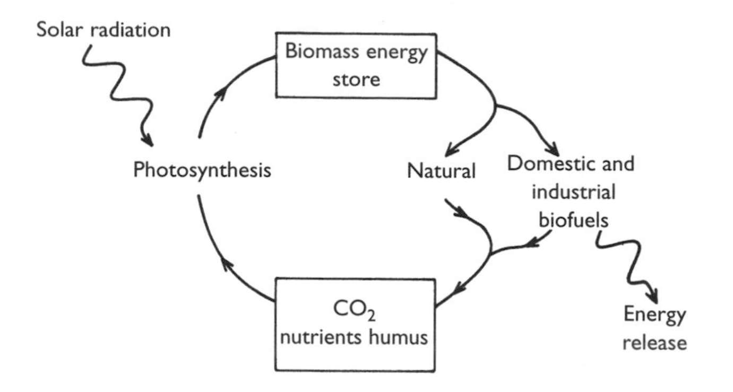


*Figure 11.1* Natural and managed biomass systems.

for cooking is of prime importance for about 50% of the world's population. The industrial use of biomass energy is currently comparatively small for most countries, except in a few sugarcane-producing countries where crop residues (bagasse) burnt for process heat may be as much as 40% of national commercial supply. Nevertheless, in some industrialised countries, the increasing use of biomass and wastes for heat and electricity generation is becoming significant, e.g. USA (about 2% of all electricity at 11 $GW_e$ capacity); Germany (at 0.5 $GW_e$ capacity) and in several countries for co-firing with coal.

If biomass is to be considered renewable, growth must at least keep pace with use. It is disastrous for local ecology and global climate control that firewood consumption and forest clearing is significantly outpacing tree growth in ever increasing areas of the world.

The carbon in biomass is obtained from $CO_2$ in the atmosphere via photosynthesis, and not from fossil sources. When biomass is burnt or digested, the emitted $CO_2$ is recycled into the atmosphere, so not adding to atmospheric $CO_2$ concentration over the lifetime of the biomass growth. Energy from biomass is therefore 'carbon neutral'. This contrasts with the use of fossil fuels, from which extra $CO_2$ is added to the atmosphere. The use of biomass in place of fossil fuels leaves the fossil fuel underground and harmless; the use of biomass 'abates' the extra $CO_2$ otherwise emitted. Thus use of renewable biofuels, on a large scale, is an important component of most medium- to long-term policies for reducing greenhouse gas emissions.

The *energy storage* of sunshine as biomass and biofuels is of fundamental importance. All of the many processes described in this chapter have the aim of producing convenient fuels, at economical prices, for a full range of end-uses, including liquid fuel for transport. The heat energy available in combustion, equivalent in practice to the enthalpy or the net energy density, ranges from about 8 $MJ\,kg^{-1}$ (undried 'green' wood) and

$15\,MJ\,kg^{-1}$ (dry wood), to about $40\,MJ\,kg^{-1}$ (fats and oils) and $56\,MJ\,kg^{-1}$ (methane). Biomass is, however, mostly carbohydrate material with a heat of combustion of about $20\,MJ\,kg^{-1}$ dry matter; refer to Table B.6, Appendix B for detail.

The success of biomass systems is regulated by principles that are often not appreciated:

1 Every biomass activity produces a wide range of products and services. For instance where sugar is made from cane, many commercial products can be obtained from the otherwise waste molasses and fibre. If the fibre is burnt, then any excess process heat can be used to generate electricity. Washings and ash can be returned to the soil as fertilizer.
2 Some high-value fuel products may require more low-value energy to manufacture than they produce, e.g. ethanol from starch crops, hydrogen. Despite the energy ratio being $>1$, such an energy deficiency need not be an economic handicap provided that process energy can be available cheaply by consuming otherwise waste material, e .g. straw, crop fibre, forest trimmings.
3 The full economic benefit of agro-industries is likely to be widespread and yet difficult to assess. One of many possible benefits is an increase in local 'cash flow' by trade and employment.
4 Biofuel production is only likely to be economic if the production process uses materials *already concentrated*, probably as a by-product and so available at low cost or as extra income for the treatment and removal of waste. Thus there has to be a supply of biomass already passing near the proposed place of production, just as hydro-power depends on a natural flow of water already concentrated by a catchment. Examples are the wastes from animal enclosures, offcuts and trimmings from sawmills, municipal sewage, husks and shells from coconuts and straw from cereal grains. It is extremely important to identify and quantify these flows of biomass in a national or local economy *before* specifying likely biomass developments. If no such concentrated biomass already exists as a previously established system, then the cost of biomass collection is usually too great and too complex for economic development. Some short-rotation crops may be grown primarily for energy production as part of intensive agriculture; however, within the widespread practice of agricultural subsidies it is difficult to evaluate fundamental cost-effectiveness.
5 The main dangers of extensive biomass fuel use are deforestation, soil erosion and the displacement of food crops by fuel crops.
6 Biofuels are organic materials, so there is always the alternative of using these materials as *chemical feedstock* or *structural materials*. For instance, palm oil is an important component of soaps; many plastic and

pharmaceutical goods are made from natural products; and much building board is made from plant fibres constructed as composite materials.

7 Poorly controlled biomass processing or combustion can certainly produce unwanted pollution, especially from relatively low temperature combustion, wet fuels and lack of oxygen supply to the combustion regions. Modern biomass processes require considerable care and expertise.

8 The use of sustainable biofuels in place of fossil fuels abates the emission of fossil-$CO_2$ and so reduces the forcing of climate change. Recognition of this is a key aspect of climate change policies.

Section 11.2 sets out a classification of biofuels, while the sections following it consider the different types one by one. The concluding section brings together the social, economic and environmental considerations that are vital if biofuels are to contribute positively, and not negatively, to sustainable development.

## 11.2 Biofuel classification

Biomass is largely composed of organic material and water. However, significant quantities of soil, shell or other extraneous material may be present in commercial supplies. It is essential that biomass is clearly assessed as either wet or dry matter mass, and the exact moisture content should be given.

If $m$ is the total mass of the material as it is and $m_0$ is the mass when completely dried, the dry basis moisture content is $w = (m - m_0)/m_0$ and the wet basis moisture content is $w' = (m - m_0)/m$. The moisture content is in the form of extracellular and intracellular water, and so drying processes may be necessary, see Section 6.3. When harvested, the wet basis moisture content of plants is commonly 50%, and may be as large as 90% in aquatic algae including seaweed (kelps). The material is considered 'dry' when it reaches long-term equilibrium with the environment, usually at about 10–15% water content by mass.

Carbon-based fuels may be classified by their reduction level, Section 10.4. When biomass is converted to $CO_2$ and $H_2O$, the energy made available is about 460 kJ per mole of carbon (38 MJ per kg of carbon; ~16 MJ per kg of dry biomass), per unit of reduction level $R$. This is not an exact quantity because of other energy changes. Thus sugars ($R = 1$) have a heat of combustion of about 450 kJ per 12 g of carbon content. Fully reduced material, e.g. methane $CH_4$($R = 2$), has a heat of combustion of about 890 kJ per 12 g of carbon (i.e. per 16 g of methane).

The presence of moisture in biomass fuel usually leads to a significant loss in useful thermal output because (i) evaporation uses 2.3 MJ kg$^{-1}$ of water and (ii) the subsequently reduced combustion temperature increases smoke

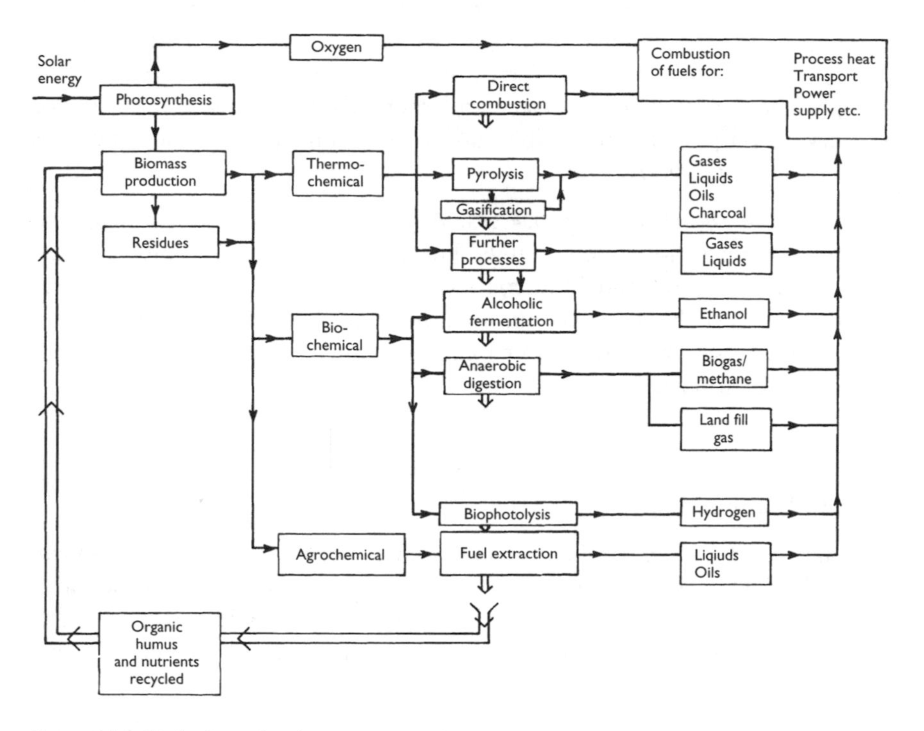


*Figure 11.2* Biofuel production processes.

and air pollution. With *condensing boilers*, much of such latent heat can be recovered by condensing water vapour in the emission and so pre-heating incoming cold water. Nevertheless, the task of clean combustion remains.

The density of biomass, and the bulk density of stacked fibrous biomass, is important. In general three to four times the volume of dry biological material has to be accumulated to provide the same energy as coal. Thus transport and fuel handling become difficult and expensive, especially if the biofuels are not utilised at source.

We have identified three classifications and nine general types of biomass energy process for fuller discussion in the later sections. These are as follows (Figure 11.2).

### 11.2.1 Thermochemical, heat

1 Direct *combustion* for immediate heat. Dry homogeneous input is preferred.

2 *Pyrolysis*. Biomass is heated either in the absence of air or by the partial combustion of some of the biomass in a restricted air or oxygen supply.

The products are extremely varied, consisting of gases, vapours, liquids and oils, and solid char and ash. The output depends on temperature, type of input material and treatment process. In some processes the presence of water is necessary and therefore the material need not be dry. If output of combustible gas is the main product, the process is called *gasification*.

3 Other *thermochemical processes*. A wide range of pre-treatment and process operations are possible. These normally involve sophisticated chemical control and industrial scale of manufacture; methanol production is such a process, e.g. for liquid fuel. Of particular importance are processes that break down cellulose and starches into sugars, for subsequent fermentation.

### 11.2.2 Biochemical

4 *Aerobic digestion*. In the presence of air, microbial aerobic metabolism of biomass generates heat with the emission of $CO_2$, but not methane. This process is of great significance for the biological carbon cycle, e.g. decay of forest litter, but is not used significantly for commercial bioenergy.

5 *Anaerobic digestion*. In the absence of free oxygen, certain micro-organisms can obtain their own energy supply by reacting with carbon compounds of medium reduction level (see Section 10.4) to produce both $CO_2$ and fully reduced carbon as $CH_4$. The process (the oldest biological 'decay' mechanism) may also be called 'fermentation', but is usually called 'digestion' because of the similar process that occurs in the digestive tracts of ruminant animals. The evolved mix of $CO_2$, $CH_4$ and trace gases is called *biogas* as a general term, but may be named *sewage gas* or *landfill-gas* as appropriate.

6 *Alcoholic fermentation*. Ethanol is a volatile liquid fuel that may be used in place of refined petroleum. It is manufactured by the action of micro-organisms and is therefore a fermentation process. Conventional fermentation has sugars as feedstock.

7 *Biophotolysis*. Photolysis is the splitting of water into hydrogen and oxygen by the action of light. Recombination occurs when hydrogen is burnt or exploded as a fuel in air. Certain biological organisms produce, or can be made to produce, hydrogen in biophotolysis. Similar results can be obtained chemically, without living organisms, under laboratory conditions. Commercial exploitation of these effects has not yet occurred, see Section 10.7.2.

### *11.2.3 Agrochemical*

8 *Fuel extraction.* Occasionally, liquid or solid fuels may be obtained directly from living or freshly cut plants. The materials are called exudates and are obtained by cutting into (tapping) the stems or trunks of the living plants or by crushing freshly harvested material. A well-known similar process is the production of natural rubber latex. Related plants to the rubber plant *Herea*, such as species of *Euphorbia*, produce hydrocarbons of less molecular weight than rubber, which may be used as petroleum substitutes and turpentine.

9 *Biodiesel and esterification.* Concentrated vegetable oils from plants may be used directly as fuel in diesel engines; indeed Rudolph Diesel designed his original 1892 engine to run on a variety of fuels, including natural plant oils. However, difficulties arise with direct use of plant oil due to the high viscosity and combustion deposits as compared with standard diesel-fuel mineral oil, especially at low ambient temperature ($\leq\sim 5\,°C$). Both difficulties are overcome by converting the vegetable oil to the corresponding ester, which is arguably a fuel better suited to diesel engines than conventional (petroleum-based) diesel oil.

## 11.3 Biomass production for energy farming

This section links the discussion of photosynthesis in Chapter 10 to the production of crops. Of particular importance are the efficiencies of photosynthesis and biomass production, Sections 10.2 and 10.6.6.

### *11.3.1 Energy farming*

We use this term in the very broadest sense to mean the production of fuels or energy as a main or subsidiary product of agriculture (fields), silviculture (forests), aquaculture (fresh and sea water), and also of industrial or social activities that produce organic waste residues, e.g. food processing, urban refuse. Table 11.1 gives some examples from an almost endless range of opportunities. The main purpose of the activity may be to produce energy (as with wood lots), but more commonly it is found best to integrate the energy and biofuel production with crop or other biomass material products.

An outstanding and established example of energy farming is the sugarcane industry (Figure 11.3). The process depends upon the combustion of the crushed cane residue (bagasse) for powering the mill and factory operations. With efficient machinery there should be excess energy for the production and sale of by-products, e.g. molasses, chemicals, animal feed, ethanol, fibre board and electricity. Commonly the ethanol becomes a component of transport fuel and the excess electricity is sold to the local grid.

*Table 11.1* Biomass supply and conversion examples

| *Biomass source or fuel* | | *Biofuel produced* | *Conversion technology* | *Approx. conversion efficiency/%* | *Energy required in conversion: (n) necessary, (o) optional* | *Approx. range of energy from biofuel/MJ* |
|---|---|---|---|---|---|---|
| Forest logging | | (Heat) | Combustion | 70 | Drying (o) | 16–20 $(kg\ wood)^{-1}$ |
| Wood from timber mill residues | either (a) | (Heat) | Combustion | 70 | Drying (o) | 16–20 $(kg\ wood)^{-1}$ |
| | | Gas | | | | *40 $(kg\ gas)^{-1}$ |
| Wood from fuel lot cropping | or (b) | Oil | Pyrolysis | 85 | Drying (o) | 40 $(kg\ oil)^{-1}$ |
| | | Char | | | | 20 $(kg\ char)^{-1}$ |
| Grain crops | | Straw | Combustion | 70 | Drying (o) | 14–16 $(kg\ dry\ straw)^{-1}$ |
| Sugarcane pressed juice | | Ethanol | Fermentation | 80 | Heat (n) | 3–6 $(kg\ fresh\ cane)^{-1}$ |
| Sugarcane pressed residue | | Bagasse | Combustion | 65 | Drying (o) | 5–8 $(kg\ fresh\ cane)^{-1}$ |
| Sugarcane total | | – | – | – | – | 8–14 $(kg\ fresh\ cane)^{-1}$ |
| Animal wastes (tropical) | | Biogas | Anaerobic digestion | 50 | – | 4–8 $(kg\ dry\ input)^{-1}$ |
| Animal wastes (temperate) | | Biogas | Anaerobic digestion | 50 | Heat (o) | †2–4 $(kg\ dry\ input)^{-1}$ |
| Sewage gas | | Biogas | Anaerobic digestion | 50 | – | 2–4 $(kg\ dry\ input)^{-1}$ |
| Landfill gas (from MSW††) | | Biogas | Anaerobic digestion | 40 | – | 2–4 $(kg\ dry\ compostable)^{-1}$ |
| Urban refuse (MSW)†† | | (Heat) | Combustion | 50 | – | 5–16 $(kg\ dry\ input)^{-1}$ |

Notes
* Nitrogen removed.
† This value is net, having deducted the biogas fed back to heat the boiler.
†† Municipal Solid Waste.

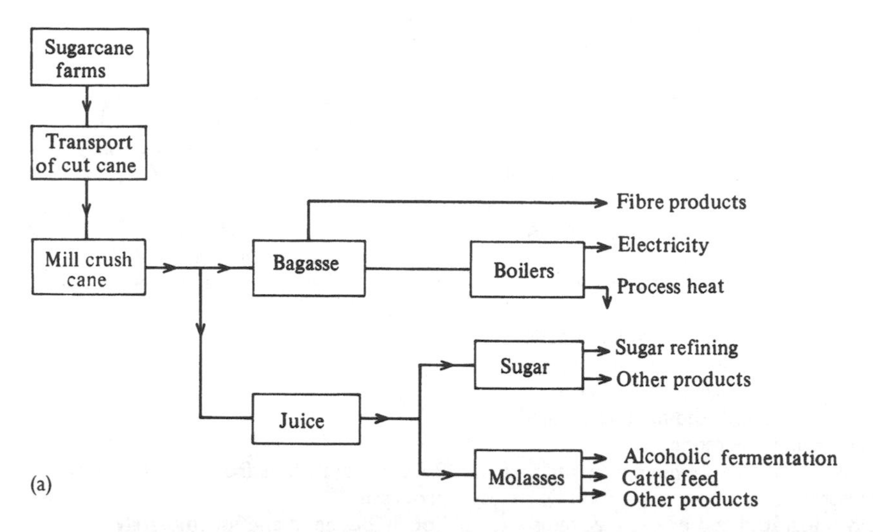

(b)

*Figure 11.3* (a) Sugar cane agro-industry; process flow diagram. Bagasse is plant fibre residue: molasses is sugar-rich residue. (b) A 30 $MW_e$ co-generation plant in Queensland (Australia) fuelled by bagasse and wood waste from nearby industry. Sugar cane is growing in the foreground; wood waste and bagasse are stockpiled in the open. [Photo by courtesy of Stanwell Corporation.]

*Table 11.2* Advantages and dangers of energy farming

| *Advantages* | *Dangers and difficulties* |
|---|---|
| Large potential supply | May lead to soil infertility and erosion |
| Variety of crops | May compete with food production |
| Variety of uses (including transport fuel and electricity generation) | |
| Efficient use of by-products, residues, wastes | Bulky biomass material handicaps transport to the processing factory |
| Link with established agriculture and forestry | May encourage genetic engineering of uncontrollable organisms |
| Encourages integrated farming practice | |
| Establishes agro-industry that may include full range of technical processes, with the need for skilled and trained personnel | |
| Environmental improvement by utilising wastes | Pollutant emissions from poorly controlled processes |
| Fully integrated and efficient systems need have little water and air pollution (e.g. sulphur content low) | Poorly designed and incompletely integrated systems may pollute water and air |
| Encourages rural development | Large-scale agro-industry may be socially disruptive |
| Diversifies the economy with respect to product, location and employee skill | |
| Greatest potential is in tropical countries, frequently of the Third World | Foreign capital may not be in sympathy with local or national benefit |

The variety of opportunities for energy farming has distinct advantages and disadvantages (Table 11.2). A major disadvantage is that energy crops may substitute for necessary food production. For example, the grain farms of the United States grow about 10% of the world's cereal crops, and export about one-third of this. A sudden change to producing biofuels, e.g. ethanol from corn, on a large scale would therefore decrease world food supplies before alternatives could be established. A second major danger is that intensive energy farming would be a further pressure towards soil infertility and erosion. The obvious strategy to avoid these excesses is (a) to always grow plants that can supply both human foods (e.g. grain) and energy (e.g. straw), (b) to decrease dramatically the feeding of animals from crops; and (c) to use all resources more efficiently.

### *11.3.2 Geographical distribution*

Perhaps the greatest potential for energy farming occurs in tropical countries, especially those with adequate rainfall and soil condition. Table 11.3 gives estimates of the potential bioenergy production of various regions of the world.

*Table 11.3* Potential bioenergy by region ($EJ\,y^{-1} = 10^{18}\,J\,y^{-1} = 32\,GW$)

| *Region* | *A: Recoverable residues* | | | | *B: Potential Biomass* | *(A+B)/ (national* |
|---|---|---|---|---|---|---|
| | *Crops* | *Forests and* | *Dung* | *Total* | *plantations* | *energy use)* |
| see note: | [a] | *woodland* [b] | [c] | | [d] | |
| *Industrialised* | | | | | | |
| US+Canada | 1.7 | 3.8 | 0.4 | 5.9 | 19 | 0.3 |
| Europe | 1.3 | 2 | 0.5 | 3.8 | 6 | 0.1 |
| Aust.+NZ | 0.3 | 0.2 | 0.2 | 0.6 | 10 | 2.8 |
| Former USSR | 0.9 | 2 | 0.4 | 3.3 | 25 | 0.5 |
| *Developing* | | | | | | |
| Latin America | 2.4 | 1.2 | 0.9 | 4.5 | 27 | 1.8 |
| Africa | 0.7 | 1.2 | 0.7 | 2.6 | 28 | 3.3 |
| China | 1.9 | 0.9 | 0.6 | 3.4 | 9 | 0.5 |
| other Asia | 3.2 | 2.2 | 1.4 | 6.8 | 18 | 0.9 |
| *world* | 12.5 | 13.6 | 5.2 | 31.2 | 142 | 0.5 |

*Source*: After Hall *et al.* (1993), based on country estimates by Biomass Uses Network.
Notes:
[a] 25% of residues from cereals, vegetables and sugar cane.
[b] 75% of mill wastes +25% of forestry residues.
[c] 12% of dung from farm animals.
[d] 8 dry tonnes per hectare per year on 10% of land now in forest or cropland or pasture.

The assumptions used are generally optimistic about what is 'recoverable', how much land is available for plantations and the possible biomass yields on that land (see notes on Table 11.3). Thus, present fuelwood use appears to be only about half of the 'forest' residue potential, although assessment is difficult, since most use is non-commercial. Moreover, it would be negligent not to leave significant amounts of rotting wood for ecological sustainability. Dung from cattle is often more valuable as fertiliser than as fuel and only where animals are penned is it easily collected for biogas production. Nevertheless, only by establishing extensive sustainable biomass plantations with the associated post-harvest mechanisms and markets can biomass become a significant proportion of commercial energy supply.

### *11.3.3 Crop yield*

It is not possible to predict crop yields without detailed knowledge of meteorological conditions, soil type, farming practice, fertiliser use, irrigation, etc. Comparison between different crops is made even more difficult by differences in growing seasons and harvesting methods. Some arable crops are planted annually, e.g. cereal grains, and may be cropped more than once, e.g. grasses. Others are planted every few years and harvested annually, e.g. sugarcane, or may grow for long periods before harvesting, e.g. more than ten months for some varieties of cassava. Trees may grow for many years and be totally harvested (timber logging); other tree crops may grow from the continuing roots and be harvested as coppice every few years, e.g. willow, hazel and some eucalyptus. Table 11.4 is a summary of data to estimate the maximum biofuel potential of crops in terms of heat of combustion and continuing energy supply. The data for aquatic crops assume abundant nutrients. Grasses are assumed to have frequent harvesting in the growing season. We emphasise the considerable uncertainty of such data and the rule that such generalisations should never be applied for actual developments without site-specific testing.

### 11.3.4 Energy and greenhouse gas analysis

Crop growth requires two forms of energy: (1) solar irradiance and (2) energy expended in labour, fuel for tractors, and manufacturing machines and fertiliser, etc. The total of the second form of energy is the gross energy requirement (GER), also called the *embedded energy*, and is the total of all forms of energy other than incident solar energy sequestered (used up) in producing the crop. It is best to explain the technique of energy analysis by an example. Note, however, that such energy analysis neglects the thermodynamic 'quality' of the energy involved, i.e. it neglects the proportion of the energy that can be transformed to work; such analysis involves the thermodynamic functions of Free Energy and Exergy and will not be pursued here.

The GER analysis of Table 11.5 lists the energy sequestered for all the market inputs to make ethanol from various crop substrates; solar energy is not considered in the analysis here as an input, since it arrives as a free input. The total GER is given per kg of ethanol (row 7 of the table). Note that in most instances the energy obtainable from the final product (ethanol at $30\,\text{MJ}\,\text{kg}^{-1}$) is less than the GER, which gives the impression of a useless fuel manufacturing process, with negative net energy production (row 8).

However, the greatest amount of energy in production is associated with process heat and factory machines (rows 3, 4 and 5). Frequently all, or a large part, of this energy is available at very low cost within the factory from the combustion or pyrolysis of otherwise waste material (bagasse from sugarcane, trimmings and other waste from timber, part of the straw from cereal crops). Thus such energy supplies can be treated as low cost gains rather like solar radiation. Row 10 gives the net energy production (the 'gain') without the components of rows 3, 4 and 5, showing the dramatic change produced if otherwise waste material is used as a free (i.e. no cost) energy gain.

Another important parameter is the energy ratio (ER), being the ratio of the heat of combustion (strictly the enthalpy) of the crop to the Gross Energy Requirement. Published values much depend on what is included in GER, how the analysis is performed and what crop yields are assumed. For

*Table 11.4* Maximum practical biomass yields. Total plant mass, not just the grain; 'R' indicates the mass is coppiced, with the roots remaining in the soil. The data are from a variety of sources and summarised by the authors. Accuracy of no more than ±25% is claimed. The majority of plants and crops yield much less than these maxima, with yields much dependent on soil, climate, fertilisers and farming practice

| *Crop (Assume one crop per year unless indicated otherwise)* | *(a)* | *Biomass yield* ($t\,ha^{-1}\,y^{-1}$) *Wet basis* | *Dry basis* | *Energy density* ($MJ\,(kg\ dry)^{-1}$) | *Energy from dried yield* ($GJ\,ha^{-1}\,y^{-1}$) |
|---|---|---|---|---|---|
| *Natural* | | | | | |
| Grassland | | 7 | 3 | | |
| Forest, temperate | C3 | 14 | 7 | 18 | 130 |
| Forest, tropical | C3 | 22 | 11 | 18 | 200 |
| *Forage* | | | | | |
| Sorghum (3crops) | R, $C_4$ | 200 | 50 | 17 | 850 |
| Sudangrass (6 crops) | R, $C_4$ | 160 | 40 | 15 | 600 |
| Alfalfa | $C_3$ | 40 | 25 | | |
| Rye grass, temperate | $C_3$ | 30 | 20 | | |
| Napier grass | $C_4$ | 120 | 80 | | |
| *Food* | | | | | |
| Cassava (60% tubers) | | 50 | 25 | | |
| Maize (corn) (35% grain) | $C_4$ | 30 | 25 | 18 | 77[(b)] |
| Wheat (35% grain) | $C_3$ | 30 | 22 | | |
| Rice (60% grain) | $C_3$ | 20 | | | |
| Sugarbeet | $C_3$ | 45 | | | |
| Sugarcane | R, $C_4$ | 100 | 30 | 18 | 150[(b)] |
| Soya beans | $C_3$ | | | | 20[(c)] |
| Rapeseed | $C_3$ | | | | 30[(c)] |
| *Plantation* | | | | | |
| Oil palm | R, $C_3$ | 50 | 40 | | |
| *Combustion energy* | | | | | |
| Eucalyptus | R, $C_3$ | 55 | 20 | 19 | 380 |
| Sycamore | R, $C_3$ | 20 | 10 | 19 | 190 |
| Populus | R, $C_3$ | 18 | 29 | 19 | 380 |
| Willow (salix) | R, $C_3$ | 25 | 15 | 19 | 350 |
| Miscanthus | R, $C_4$ | 21 | 18 | 18 | 330 |
| Water hyacinth | $C_3$ | 300 | 36 | 19 | 680 |
| Kelp (macro-algae) | $C_3$ | 250 | 54 | 21 | 1100 |
| Algae (micro-algae) | $C_3$ | 230 | 45 | 23 | 1000 |
| *Tree exudates* | | | | | |
| Good output | | 1 | 1 | 40 | 40 |

Notes

(a) $C_3$, $C_4$: photosynthesis type (see Section 10.6).
R: harvested above the root (coppiced).

(b) As ethanol.

(c) As biodiesel.

*Table 11.5* Ethanol production: energy analysis from various crop substrates. Data refer to the gross energy requirement for the crop input and each component of manufacture: unit $MJ\,kg^{-1}$ of anhydrous ethanol produced. The heat of combustion of the output ethanol is $H_o = 30\,MJ\,kg^{-1}$. Rows 1 to 7 are from Slesser and Lewis (1979). Rows 8, 9 and 10 show the importance of using free waste products in such processes, which, as with solar radiation, need not be accounted to the process as are purchased products. Many of these processes have improved since 1980 – see text

| | *Sugarcane* | *Cassava* | *Timber (enzyme hydrolysis)* | *Timber (acid hydrolysis)* | *Straw* |
|---|---|---|---|---|---|
| (1) Substrate | 7.3 | 19.2 | 12.7 | 20.0 | 4.4 |
| (2) Chemicals | 0.6 | 0.9 | 4.7 | 6.4 | 4.7 |
| (3) Water pumping | 0.3 | 0.4 | 0.8 | 0.3 | 0.8 |
| (4) Electricity | 7.0 | 10.5 | 176 | 7.8 | 167 |
| (5) Fuel oil | 8.0 | 29 | 42 | 62 | 42 |
| (6) Machinery and buildings | 0.5 | 1.2 | 3.3 | 0.6 | 3.3 |
| (7) Total (1)–(6) ($MJ\,kg^{-1}$) | 24 | 61 | 239 | 98 | 222 |
| (8) Net energy: [$= H_o - (7)$] | +8 | −31 | −209 | −68 | −192 |
| *If inputs (3), (4), (5) from waste:* | | | | | |
| (9) Total [(1) + (2) + (6)] | 8.4 | 21 | 21 | 27 | 12 |
| (10) Net energy [$= H_o - (9)$] | +21 | +9 | +9 | +3 | +18 |
| (11) Energy ratio[$= H_o/(9)$] | 3.6 | 1.4 | 1.4 | 1.1 | 2.5 |

instance, post-2000 data from process and cultivation improvements for bioethanol from sugar cane in Brazil, beyond those assumed in Table 11.5, give an increased value of ER of about 6. However, data for ethanol production from corn (maize) in the USA gives an ER of only about 1.3 because no use is made of the crop residues (Keshgi *et al.* 2000). Improved processes for ethanol from lignocellulose may have ER $> 4$, although this has yet to be confirmed in commercial practice (see Section 11.7.1(4)). Energy ratio is a useful indicator for food and energy crops since the values can relate to the farming practice, e.g. 'organic' or 'intensive'. Best practice in energy crop production may give ER of about 18 (hybrid poplar), 16 (sorghum) and 11 (switchgrass) (Hall *et al.* 1993).

Other concerns regarding the energy characteristics of fuels are:

a Why are inputs for non-sustainable sources (fossil and nuclear fuels) usually neglected, but included for sustainable, renewable, resources?
b Why treat all energy inputs and outputs purely in terms of Enthalpy (which neglects 'quality'), and not as the more thermodynamically comprehensive parameters of Free Energy or Exergy (which enumerate 'quality')? Good quality energy supplies, e.g. electricity and methane, are much more valuable than those of poor quality energy, e.g. brown coal, wet wood or warm water.

In practice the energy analysis and subsequent economic analysis of biomass agro-industries is much more complicated than this simplistic approach. However, the crucial factor remains that use of cheap biomass residues for process heat and electricity production can be of overriding importance.

Energy analysis is a useful tool in assessing energy-consuming and energy-producing systems, since it emphasises the technical aspects and choices of the processes. For instance, from Table 11.5 it would seem obvious that ethanol production from sugarcane is most reasonable. Nevertheless, the final choice must involve other factors, such as the market for non-fossil fuel supplies, the value of the alternative products and by-products, and government polices and incentives.

By enumerating the $CO_2$ emissions associated with the energy use in each step of a process, energy analysis can readily be turned into analysis of the associated greenhouse gas emissions. Such analyses distinguish 'automatically' between (i) the use of biomass residues for process energy (zero net $CO_2$ emissions, since the residues would have decayed naturally anyway) and (ii) the use of fossil fuel energy inputs (see Keshgi *et al.* 2000).

## 11.4 Direct combustion for heat

Biomass is burnt to provide heat for cooking, comfort heat (space heat), crop drying, factory processes and raising steam for electricity production and transport. Traditional use of biomass combustion includes (a) cooking with firewood, with the latter supplying about 10–20% of global energy use (a proportion extremely difficult to assess) and (b) commercial and industrial use for heat and power, e.g. for sugarcane milling, tea or copra drying, oil palm processing and paper making. Efficiency and minimum pollution is aided by having dry fuel and controlled, high temperature combustion. Table B.6 gives the heat of combustion for a range of energy crops, residues, derivative fuels and organic products, assuming dry material. Such data are important for the industrial use of biomass fuel.

### 11.4.1 Domestic cooking and heating

A significant proportion of the world's population depends on fuelwood or other biomass for cooking, heating and other domestic uses. Average daily consumption of fuel is about 0.5–1 kg of dry biomass per person, i.e. $10\text{–}20\,\text{MJ}\,\text{d}^{-1} \approx 150\,\text{W}$. Multiplied by, say, $2 \times 10^9$ people, this represents energy usage at the very substantial rate of 300 GW. Most domestic fuelwood use, but certainly not all, is in developing countries, with the majority not included in commercial energy statistics. Here we assume the fuel has dried thoroughly, since this is an essential first step for biomass combustion, see Sections 6.4 and 11.4.2; using wet or damp fuel should be avoided.

An average consumption of 150 W 'continuous', solely for cooking, may seem surprisingly large. Such a large consumption arises from the widespread use of inefficient cooking methods, the most common of which is still an open fire. This 'device' has a thermal efficiency of only about 5%. That is, only about 5% of the heat that could be released by complete combustion of the wood reaches the interior of the cooking pot. The rest is lost by incomplete combustion of the wood, by wind and light breezes carrying heat away from the fire, and by radiation losses, etc. resulting from the mismatch of fire and pot size. Considerable energy is also wasted in evaporation from uncovered pots and from wet fuel. Smoke (i.e. unburnt carbon and tars) from a fire is evidence of incomplete combustion, and there may be little control over the rate at which wood is burnt. Moreover, the smoke is a health hazard to the cook unless there is an efficient extraction chimney. However, a reason for allowing internal smoke may be to deter vermin and pests from the roof, and to cure ('smoke') dried food. Efficiently burnt dry wood, in which the initially produced unburnt gases and tars burn in a secondary reaction, emits only $CO_2$ and $H_2O$ with fully combusted ash.

Cooking efficiency and facilities can be improved by

1 Using dry fuel.
2 Introducing alternative foods and cooking methods, e.g. steam cookers.
3 Decreasing heat losses using enclosed burners or stoves, and well-fitting pots with lids.
4 Facilitating the secondary combustion of unburnt flue gases.
5 Introducing stove controls that are robust and easy to use.
6 Explanation, training and management.

With these improvements, the best cooking stoves using fuelwood and natural air circulation can place more than 20% of the combustion energy into the cooking pots. Designs using forced and actively controlled ventilation, say with an electric fan, can be more than 80% efficient. There are many scientifically based programs to improve cooking stoves, yet full market acceptability is not always reached, especially if cultural and gender factors are not considered adequately. A major difficulty of 'efficient' stoves may be the difficulty of obtaining rapid heat.

The combustion of firewood is a complex and varying process. Much depends on the type of wood and its moisture content. Initial combustion releases CO, which itself should burn in surplus air. At temperatures greater than 370 °C, calcium oxalate in the wood breaks down with the release of some oxygen, so improving combustion and reducing particulate and combustible emissions. Good design ensures that high temperature combustion is restricted to a 'white-hot' small volume and that pyrolytic gases are themselves burnt in a secondary combustion region where further air enters.

If space heating is needed, then the seemingly wasted heat from cooking becomes useful. If space heat alone is needed, then closed stoves with controlled combustion can be 80–90% efficient.

A parallel method for reducing domestic fuelwood demand is to encourage alternative renewable energy supplies, such as biogas (methane with $CO_2$), see Section 11.6 fuel from crop wastes and small-scale hydro-power, Section 8.5. The need for such improvements is overwhelming when forests are dwindling and deserts increasing.

Figure 11.4 shows two types of wood burning stoves, designed to make better use of wood as a cooking fuel. Both designs are cheap enough to allow widespread use in developing countries. More expensive stoves (often called ranges) for both cooking and water-heating are luxury items in many kitchens of northern European and North American homes, where some designs allow wood burning.

In the stove shown in Figure 11.4(a), the fire is completely enclosed in the firebox at the left. The iron (dark coloured) door is removed only when fuel is inserted. Air enters through a hole of adjustable size beneath the door (fully shut in the photo). Thus the rate of combustion can be closely controlled to match the type of cooking being done. Hot gases from the fire are led through a narrow channel underneath the cooking pots, which are sized to fit closely in holes on the top. At this stage air can enter through further channels for secondary combustion. The fully burnt gases and vapours pass to the outside environment through the chimney at the far end of the stove; this prevents pollution in the cooking place and encourages air flow.

The stove shown in Figure 11.4(b) is simpler and cheaper, but has less control and less flexibility. Nevertheless its small mass means that it is transportable and little heat is used in heating the stove as distinct from the pot, which is an advantage for quick cooking. Air reaches the fuel from below, through a grate. Since the fire is contained and the heat is channelled towards the pot, the efficiency is high. This stove is well suited for use with charcoal as a fuel, since charcoal burns cleanly without smoke.

Many of these remarks on cooking apply also to the use of biofuels for comfort (space) heat in buildings. It is important to have a controlled fire with good secondary combustion. In some systems air for combustion is introduced directly to the stove from outside the building. This decreases air circulation and heat losses in the room. Some very sophisticated and efficient wood-burners for heating are in wide use, especially in some wood-rich industrialised countries (e.g. Norway, Canada and New Zealand) and countries encouraging markets in fuelwood chips and pellets from otherwise waste biomass residues (e.g. Northern Europe). For the latter, sophisticated

(a)

(b)

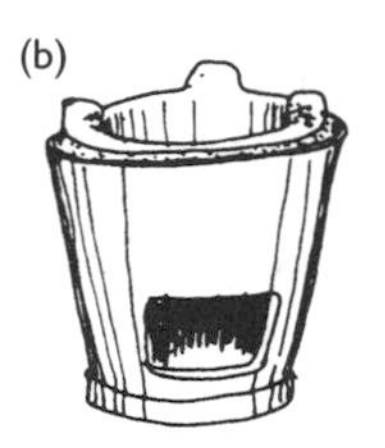

(c)

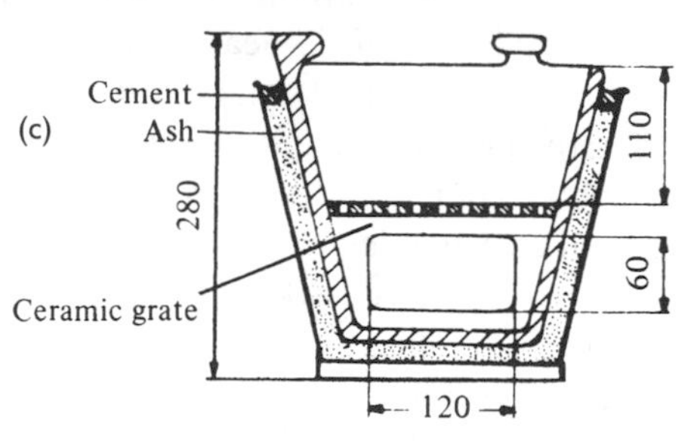


*Figure 11.4* Improved efficiency cooking stoves. (a) A large stove designed by the Fiji Ministry of Energy. It is a modification of the Indian (Hyderabad) *chula*, and is constructed mainly from concrete mouldings. Its operation is described in the text. (b) The 'Thai bucket' stove (sketch). (c) Vertical section through same (units millimetres).

stoves, usually with automatic input of fuel, are manufactured to ensure ease of use, fuel efficiency and minimum pollution.

### *11.4.2 Crop drying*

The drying of crops (e.g. fruit, copra, cocoa, coffee, tea), for storage and subsequent sale, is commonly accomplished by burning wood and the crop residues, or by using the waste heat from electricity generation. The material to be dried may be placed directly in the flue exhaust gases, but there is a danger of fire and contamination of food products. More commonly air is heated in a gas/air heat exchanger before passing through the crop. Drying theory is discussed in Section 6.4.

Combustion of residues for crop drying is a rational use of biofuel, since the fuel is close to where it is needed. Combustion in an efficient furnace yields a stream of hot clean exhaust gas ($CO_2 + H_2O +$ excess air) at about 1000 °C, which can be diluted with cold air to the required temperature. In almost all cases, the flow of biomass exceeds that required for crop drying, leaving an excess of the residues for other purposes, such as producing industrial steam.

### *11.4.3 Process heat and electricity*

Steam process heat is commonly obtained for factories by burning wood or other biomass residues in boilers, perhaps operating with fluidized beds. It is physically sensible to use the steam first to generate electricity before the heat degrades to a lower useful temperature. The efficiency of electricity generation from the biomass may be only about 20–25% due to low temperature combustion, so 75–80% of the energy remains as process heat and a useful final temperature is maintained. Frequently the optimum operation of such processes treats electricity as a by-product of process heat generation, with excess electricity being sold to the local electricity supply agency, as in modern sugarcane mills (Figure 11.3(b)).

Perhaps the easiest way to use energy crops and biomass residues is co-firing in coal-burning power stations. The combustion method is adapted for the known mixture of coal and biomass. Such substitution (abatement) of coal may be one of the most effective ways for biomass to reduce greenhouse gas emissions.

### *11.4.4 Wood resource*

Wood is a renewable energy resource only if it is grown as fast as it is consumed. Moreover there are ecological imperatives for the preservation of natural woodland and forests. The world's wood resource is consumed not just for firewood, but for sawn timber, paper making and other industrial uses. In addition, much forest is cleared for agriculture with its timber just burnt as 'waste'. Approximate estimates of present and potential biomass energy resources are given in Table 11.3, including forests and woodland, but the accuracy in most countries is no better than a factor of two, partly because of different definitions of exactly what are 'forests' and 'woodlands'.

In many countries, firewood consumption exceeds replacement growth, so that fuelwood is a depleting resource. In India, for example, present consumption of fuelwood is estimated to be around $200\,\mathrm{Mt\,y^{-1}}$, of which only about $20\,\mathrm{Mt\,y^{-1}}$ constitutes sustainable availability from forests. About $100\,\mathrm{Mt\,y^{-1}}$ is derived from non-forest sources such as from village woodlots, trees or shrubs on the edge of fields and roads, etc. The balance represents non-sustainable extraction from forests plus miscellaneous gathering of woody material. Moreover, the populations of firewood-using countries are increasing at 2–3% per year, thus tending to increase the demand for cheap cooking fuels. Fuelwood collection for household consumption, usually a task for women and children, is becoming more burdensome as fuelwood becomes scarcer. The proportion of rural women affected by fuelwood scarcity is around 60% in Africa, 80% in Asia and 40% in Latin America. Moreover, gathering firewood may require 1–5 hours per day. Alleviating these difficulties requires both intensive reforestation and a switch to more efficient cooking methods.

Regeneration may occur in natural forest or in man-made plantations (which usually grow faster and are to be encouraged). Plantations grown especially for energy supply need different management (silviculture) techniques than plantations grown primarily for timber (Sims 2002). Combustible wood need not be grown in long straight lengths, and can therefore be harvested much more often (at 3–5 years rather than ~30 years). Coppicing (i.e. leaving the roots in the ground and cropping only the branches) is successful with many tree species; it reduces (costly) labour for planting and weeding, and also reduces soil erosion compared with repeated replanting.

## 11.5 Pyrolysis (destructive distillation)

Pyrolysis is a general term for all processes whereby organic material is heated or partially combusted to produce secondary fuels and chemical products. The input may be wood, biomass residues, municipal waste or indeed coal. The products are gases, condensed vapours as liquids, tars and oils, and solid residue as char (charcoal) and ash. Traditional *charcoal making* is pyrolysis with the vapours and gases not collected. *Gasification* is pyrolysis adapted to produce a maximum amount of secondary fuel gases.

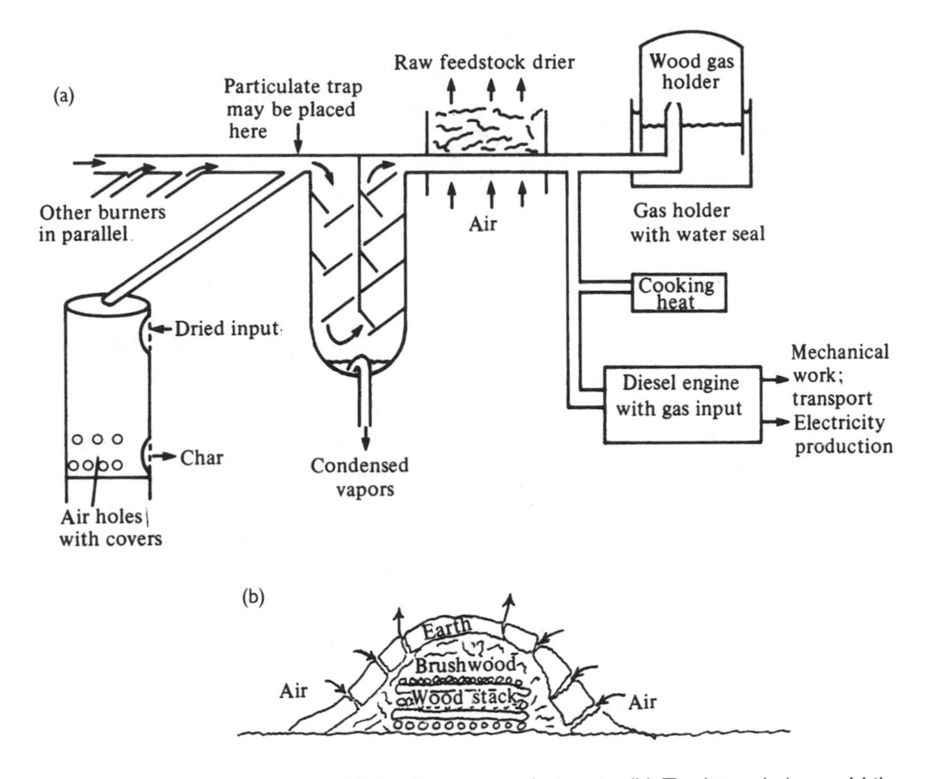


*Figure 11.5* Pyrolysis systems. (a) Small-scale pyrolysis unit. (b) Traditional charcoal kiln.

Various pyrolysis units are shown in Figure 11.5. Vertical top-loading devices are usually considered the best. The fuel products are more convenient, clean and transportable than the original biomass. The chemical products are important as chemical feedstock for further processes or as directly marketable goods. Partial combustion devices, which are designed to maximise the amount of combustible gas rather than char or volatiles, are usually called *gasifiers*. The process is essentially pyrolysis, but may not be described as such.

*Efficiency* is measured as the heat of combustion of the secondary fuels produced divided by the heat of combustion of the input biomass as used. Large efficiencies of 80–90% can be reached. For instance gasifiers from wood can produce 80% of the initial energy in the form of combustible gas (predominantly $H_2$ and CO – *producer gas*), suitable for operation in converted petroleum–fuelled engines. In this way the overall efficiency of electricity generation (say 80% of 30% = 24%) could be greater than that obtained with a steam boiler. Such gasifiers are potentially useful for small-scale power generation (<150 kW).

The chemical processes in pyrolysis are much related to similar distillations of coal to produce synthetic gases, tars, oils and coke. For instance the large-scale use of piped town gas ($H_2+CO$) in Europe, before the change to fossil 'natural' gas (mainly $CH_4$), was possible from the reaction of water on heated coal with reduced air supply:

$$H_2O+C \rightarrow H_2+CO$$
$$C+O_2 \rightarrow CO_2; \qquad CO_2+C \rightarrow 2CO \tag{11.1}$$

The following is given as a summary of the wide range of conditions and products of pyrolysis. The input material needs to be graded to remove excessive non-combustible material (e.g. soil, metal), dried if necessary (usually completely dry material is avoided with gasifiers, unlike boilers), chopped or shredded, and then stored for use. The air/fuel ratio during combustion is a critical parameter affecting both the temperature and the type of product. Pyrolysis units are most easily operated at temperatures less than 600 °C. Increased temperatures of 600–1000 °C need more sophistication, but more hydrogen will be produced in the gas. At less than 600 °C there are generally four stages in the distillation process:

1 ~100–120 °C. The input material dries with moisture passing up through the bed.
2 ~275 °C. The output gases are mainly $N_2$, CO and $CO_2$; acetic acid and methanol distil off.
3 ~280–350 °C. Exothermic reactions occur, driving off complex mixtures of chemicals (ketones, aldehydes, phenols, esters), $CO_2$, CO, $CH_4$, $C_2H_6$ and $H_2$. Certain catalysts, e.g. $ZnCl_2$, enable these reactions to occur at smaller temperature.
4 >350 °C. All volatiles are driven off, a larger proportion of $H_2$ is formed with CO, and carbon remains as charcoal with ash residues.

The condensed liquids, called tars and pyroligneous acid, may be separated and treated to give identifiable chemical products (e.g. methanol, $CH_3OH$, a liquid fuel). Table 11.6 gives examples and further detail.

The secondary fuels from pyrolysis have less total energy of combustion than the original biomass, but are far more convenient to use. Some of the products have significantly greater energy density (e.g. $CH_4$ at 55 MJ kg$^{-1}$) than the average input. Convenience includes easier handling and transport, piped delivery as gas, better control of combustion, greater variety of end-use devices and less air pollution at point of use. The following comments consider the solid, liquid and gaseous products respectively.

*Table 11.6* Pyrolysis yields from dry wood

| *Approximate yields per 1000 kg (tonne) dry wood* | |
|---|---|
| Charcoal | ~300 kg |
| Gas (combustion 10.4 MJ $m^{-3}$) | ~140 $m^3$ (NTP) |
| Methyl alcohol | ~14 litre |
| Acetic acid | ~53 litre |
| Esters | ~8 litre |
| Acetone | ~3 litre |
| Wood oil and light tar | ~76 litre |
| Creosote oil | ~12 litre |
| Pitch | ~30 kg |

### 11.5.1 *Solid charcoal (mass yield 25–35% maximum)*

Modern charcoal retorts operating at about 600 °C produce 25–35% of the dry matter biomass as charcoal. Traditional earthen kilns usually give yields closer to 10%, since there is less control. Charcoal is 75–85% carbon, unless great care is taken to improve quality (as for chemical grade charcoal), and the heat of combustion is about 30 MJ $kg^{-1}$. Thus if charcoal alone is produced from wood, between 15 and 50% of the original chemical energy of combustion remains. Charcoal is useful as a clean controllable fuel. Chemical grade charcoal has many uses in laboratory and industrial chemical processes. Charcoal is superior to coal products for making high quality steel.

### 11.5.2 *Liquids (condensed vapours, mass yield ~30% maximum)*

These divide between (1) a sticky phenolic tar (creosote) and (2) an aqueous liquid, pyroligneous acid, of mainly acetic acid, methanol (maximum 2%) and acetone. The liquids may be either separated or used together as a crude, potentially polluting and carcinogenic fuel with a heat of combustion of about 22 MJ $kg^{-1}$. The maximum yield corresponds to about 400 litres of combustible liquid per tonne of dry biomass. The liquids may be better used as a source of chemicals, but this requires larger-scale and sophisticated operation.

### 11.5.3 *Gases (mass yield ~80% maximum in gasifiers)*

The mixed gas output with nitrogen is known as *wood gas*, *synthesis gas*, *producer gas* or *water gas*, and has a heat of combustion in air of 5–10 MJ $kg^{-1}$ (4–8 MJ $m^{-3}$ at STP). It may be used directly in diesel cycle or spark ignition engines with adjustment of the fuel injector or carburettor, but extreme care has to be taken to avoid intake of ash and condensable vapours. The gas is mainly $N_2$, $H_2$ and CO, with perhaps small amounts of $CH_4$ and $CO_2$. The gas may be stored in gas holders near atmospheric pressure, but is not conveniently compressed. A much cleaner and more uniform gas may be obtained by gasification of wet charcoal rather than wood, since the majority of the tars from the original wood have already been removed.

## 11.6 Further thermochemical processes

In the previous sections, biomass has been used directly after preliminary sorting and cutting for combustion or pyrolysis. However, the biomass may be treated chemically (1) to produce material suitable for alcoholic fermentation, Section 11.7, or (2) to produce secondary or improved fuels.

Consider the following few important examples from the great number of possibilities.

### *11.6.1 Hydrogen reduction*

Dispersed, shredded or digested biomass, e.g. manure, is heated in hydrogen to about 600 °C under pressure of about 50 atmospheres. Combustible gases, mostly methane and ethane, are produced, which may be burnt to give about $6\,\mathrm{MJ\,kg^{-1}}$ of initial dry material.

### *11.6.2 Hydrogenation with CO and steam*

The process is as above, but heating is within an enclosure with CO and steam to about 400 °C and 50 atmospheres. A synthetic oil is extracted from the resulting products that may be used as a fuel. A catalyst is needed to produce reactions of the following form:

$$\begin{aligned} CO + H_2O &\rightarrow CO_2 + H_2 \\ C_n(H_2O)_n + (n+1)H_2 &\rightarrow nH_2O + H(CH_2)_nH \end{aligned} \tag{11.2}$$

where the latter reaction implies the conversion of carbohydrate material to hydrocarbon oils. The energy conversion efficiency is about 65%.

### *11.6.3 Acid and enzyme hydrolysis*

Cellulose is the major constituent (30–50%) of plant dry biomass and is very resistant to hydrolysis and hence to fermentation by micro-organisms (see Section 11.7.1(4)). Conversion to sugars, which can be fermented, is possible by heating in sulphuric acid or by the action of enzymes (cellulases) of certain micro-organisms. The products may also be used as cattle feed.

### 11.6.4 Methanol liquid fuel

Methanol, a toxic liquid, is made from the catalytic reaction of $H_2$ and CO at 330 °C and at 150 atmosphere pressure:

$$2H_2 + CO \rightarrow CH_3OH \qquad (11.3)$$

The gases are the components of synthesis gas, Section 11.5.3, and may be obtained from gasification of biomass. Methanol may be used as a liquid fuel in petroleum spark-ignition engines with an energy density of 23 MJ kg$^{-1}$. It is also used as an 'anti-knock' fuel additive to enhance the octane rating, and is potentially a major fuel for fuel cells, Section 16.6.

## 11.7 Alcoholic fermentation

### 11.7.1 Alcohol production methods

*Ethanol*, $C_2H_5OH$, is produced naturally by certain micro-organisms from sugars under acidic conditions, i.e. pH 4 to 5. This alcoholic fermentation process is used worldwide to produce alcoholic drinks. The most common micro-organism, the yeast *Saccharomyces cerevisiae*, is poisoned by $C_2H_5OH$ concentration greater than 10%, and so stronger concentrations up to 95% are produced by distilling and fractionating (Figure 11.6). When distilled, the remaining constant boiling point mixture is 95% ethanol and 5% water. Anhydrous ethanol is produced commercially with azeotropic

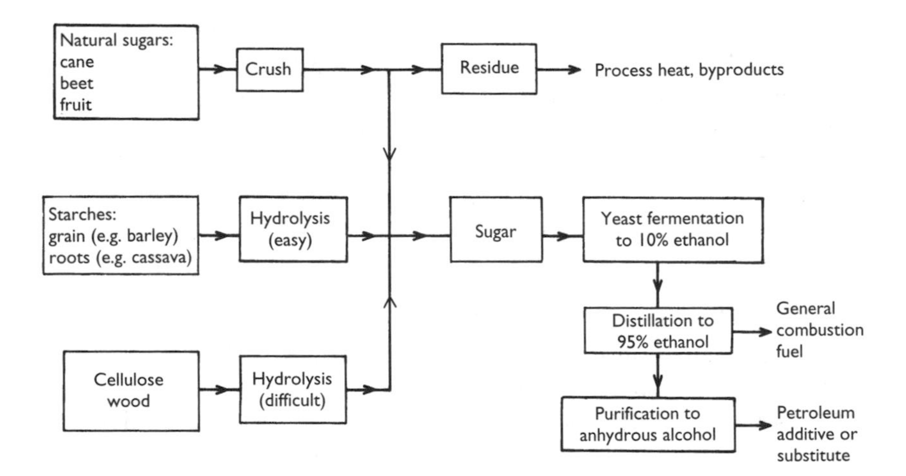


*Figure 11.6* Ethanol production.

removal of water by co-distillation with solvents such as benzene. Only about 0.5% of the energy potential of the sugars is lost during fermentation, but significant amounts of process heat are required for the concentration and separation processes, see Table 11.5. This process heat may be provided from the combustion or gasification of otherwise waste biomass and from waste heat recovery.

The sugars may be obtained by the following routes, listed in order of increasing difficulty.

1 *Directly from sugarcane.* Usually commercial sucrose is removed from the cane juices, and the remaining molasses used for the alcohol production process (Figure 11.3(a)). These molasses themselves have about 55% sugar content. But if the molasses have little commercial value, then ethanol production from molasses has favourable commercial possibilities, especially if the cane residue (bagasse) is available to provide process heat. In this case the major reaction is the conversion of sucrose to ethanol:

$$C_{12}H_{22}O_{11} + H_2O \xrightarrow{\text{yeast}} 4C_2H_5OH + 4CO_2 \tag{11.4}$$

In practice the yield is limited by other reactions and the increase in mass of yeast. Commercial yields are about 80% of those predicted by (11.4). The fermentation reactions for other sugars, e.g. glucose, $C_6H_{12}O_6$, are very similar.

2 From *sugar beet.* Sugar beet is a mid-latitude root crop for obtaining major supplies of sugar. The sugar can be fermented, but obtaining process heat from the crop residues is, in practice, not as straightforward as with cane sugar, so ethanol production is more expensive.

3 From *starch crops.* Starch crops, e.g. grain and cassava, can be hydrolyzed to sugars. Starch is the main energy storage carbohydrate of plants, and is composed of two large molecular weight components, amylose and amylopectin. These relatively large molecules are essentially linear, but have branched chains of glucose molecules linked by distinctive carbon bonds. These links can be broken by enzymes from malts associated with specific crops, e.g. barley or corn, or by enzymes from certain moulds (fungi). Such methods are common in whisky distilleries, corn syrup manufacture and ethanol production from cassava roots. The links can also be broken by acid treatment at pH 1.5 and at 2 atmospheres pressure, but yields are small and the process more expensive than enzyme alternatives. An important by-product of the enzyme process is the residue used for cattle feed or soil conditioning.

4 From *cellulose.* Cellulose comprises about 40% of all biomass dry matter. Apart from its combustion as part of wood, cellulose is potentially a primary material for ethanol production on a large scale. It has

a polymer structure of linked glucose molecules, and forms the main mechanical-structure component of the woody parts of plants. These links are considerably more resistant to breakdown into sugars under hydrolysis than the equivalent links in starch. Cellulose is found in close association with lignin in plants, which discourages hydrolysis to sugars. Acid hydrolysis is possible as with starch, but the process is expensive and energy intensive. Hydrolysis is less expensive, and less energy input is needed if enzymes of natural, wood-rotting fungi are used, but the process is slow. Prototype commercial processes have used pulped wood or, more preferably, old newspaper as input. The initial physical breakdown of woody material is a difficult and expensive stage, usually requiring much electricity for the rolling and hammering machines. Substantial R&D in the USA and Scandinavia led to processes with improved yields and potentially cheaper production, key features of which are acid-catalysed hydrolysis of hemicellulose, more effective enzymes to breakdown cellulose, and genetically engineered bacteria that ferment all biomass sugars (including 5-carbon sugars, which resist standard yeasts) to ethanol with high yields. Although not yet generally applied commercially, these processes may allow ethanol from biomass to compete commercially with fossil petroleum (Wyman 1999).

### 11.7.2 Ethanol fuel use

Liquid fuels are of great importance because of their ease of handling and controllable combustion in engines. Anhydrous ethanol is a liquid between −117 and +78 °C, with a flash point of 130 °C and an ignition temperature of 423 °C, and so has the characteristics of a commercial liquid fuel, being used as a direct substitute or additive for petrol (gasoline), and is used in three ways:

1. As 95% (hydrous) ethanol, used directly in modified and dedicated spark-ignition engines;
2. Mixed with the fossil petroleum in dry conditions to produce *gasohol*, as used in unmodified spark-ignition engines, perhaps retuned;
3. as an emulsion with diesel fuel for diesel compression engines (this may be called *diesohol*, but is not common).

In general such *bioethanol* fuel has the proportion of ethanol indicated as EX, where X is the percentage of ethanol, e.g. E10 has 10% ethanol and 90% fossil petroleum. Gasohol for unmodified engines is usually between E10 and E15, and larger proportions of ethanol require engine modification to some extent. Note that water does not mix with petrol, and so water is often present as an undissolved sludge in the bottom of petroleum vehicle fuel tanks, as in cars, without causing difficulty. If gasohol is added to such a tank, the fuel will be spoilt and may not be suitable for an unmodified engine. With the ethanol mostly from sugarcane, gasohol is now standard in Brazil, e.g. as E22, and in countries of southern Africa. In USA, gasohol is also common, but with the ethanol likely to be from corn grain.

The ethanol additive has antiknock properties and is preferable to the more common tetraethyl lead, which produces serious air pollution. The excellent combustion properties of ethanol enable an engine to produce up to 20% more power with ethanol than with petroleum. The mass density and calorific value of ethanol are both less than those of petroleum, so the energy *per unit volume* of ethanol ($24\,GJ\,m^{-3}$) is 40% less than for petroleum ($39\,GJ\,m^{-3}$), see Table B.6. However, the better combustion properties of ethanol almost compensate when measured as volume per unit distance, e.g. litre/100 km. Fuel consumption by volume in similar cars using petrol, gasohol or pure ethanol is in the ratio 1:1:1.2, i.e. pure ethanol is only 20% inferior by this criteria. We note, however, that the custom of measuring liquid fuel consumption per unit volume is deceptive, since measurement per unit mass relates better to the enthalpy of the fuel.

Production costs of ethanol fuels depend greatly on local conditions and demand relates to the prices paid for alternative products. Government policy and taxation are extremely important in determining the market sales and hence the scale of production, see Section 11.11.

### 11.7.3 Ethanol production from crops

Table 11.7 gives outline data of ethanol production and crop yield. Commercial operations relate to many other factors, including energy analysis (see Section 11.3.4 and Table 11.5) and economic analysis. We emphasise that the use of otherwise waste biomass residues for electricity production and factory process heat is crucial in these analyses. As a benchmark, global production of ethanol for fuel was over 20 billion litres in 2003 – a year of considerable political uncertainty affecting the international price of fossil oil.

*Table 11.7* Approximate yields of *ethanol* from various crops, based on average yields in Brazil (except for corn, which is based on US yields). Two crops a year are possible in some areas. Actual yields depend greatly on agricultural practice, soil and weather

| | *Litres of ethanol per tonne of crop* | *Litres of ethanol per hectare per year* |
|---|---|---|
| Sugarcane | 70 | 5200 |
| Cassava | 180 | 2160 |
| Sweet sorghum | 86 | 3010 |
| Sweet potato | 125 | 1875 |
| Corn (maize) | 370 | 2800 |
| Wood | 160 | 3200 |

## 11.8 Anaerobic digestion for biogas

### 11.8.1 Introduction

Decaying biomass and animal wastes are broken down naturally to elementary nutrients and soil humus by decomposer organisms, fungi and bacteria. The processes are favoured by wet, warm and dark conditions. The final stages are accomplished by many different species of bacteria classified as either aerobic or anaerobic.

*Aerobic* bacteria are favoured in the presence of oxygen with the biomass carbon being fully oxidised to $CO_2$. This composting process releases some heat slowly and locally, but is not a useful process for energy supply. To be aerobic, air has to permeate, so a loose 'heap' of biomass is essential. Domestic composting is greatly helped by having layers of rumpled newspaper and cardboard, which allows air pockets and introduces beneficial carbon from the carbohydrate material. Such aerobic digestion has minimal emission of methane, $CH_4$, which, per additional molecule, is about eight times more potent as a greenhouse gas than $CO_2$, see Section 4.6.2.

In closed conditions, with no oxygen available from the environment, *anaerobic* bacteria exist by breaking down carbohydrate material. The carbon may be ultimately divided between fully oxidised $CO_2$ and fully reduced $CH_4$, see (10.8). Nutrients such as soluble nitrogen compounds remain available in solution, so providing excellent fertilizer and humus. Being accomplished by micro-organisms, the reactions are all classed as fermentations, but in anaerobic conditions the term 'digestion' is preferred.

It is emphasised that both aerobic and anaerobic decompositions are a fundamental processes of natural ecology that affect all biomass irrespective of human involvement. As with all other forms of renewable energy, we are able to interface with the natural process and channel energy and resources for our economy. The decomposed waste should then be released for natural ecological processes to continue.

*Biogas* is the $CH_4/CO_2$ gaseous mix evolved from digesters, including waste and sewage pits; to utilise this gas, the digesters are constructed and controlled to favour methane production and extraction (Figure 11.7). The energy available from the combustion of biogas is between 60 and 90% of the dry matter heat of combustion of the input material. However, the gas is obtainable from slurries of up to 95% water, so in practice the biogas energy is often available where none would otherwise have been obtained. Another, perhaps dominant, benefit is that the digested effluent forms significantly less of a health hazard than the input material. Note, however, that *not* all parasites and pathogens are destroyed in the digestion.

The economics and general benefit of biogas are always most favourable when the digester is placed in a flow of waste material already present. Examples are sewage systems, piggery washings, cattle shed slurries, abattoir wastes, food processing residues, sewage and municipal refuse landfill

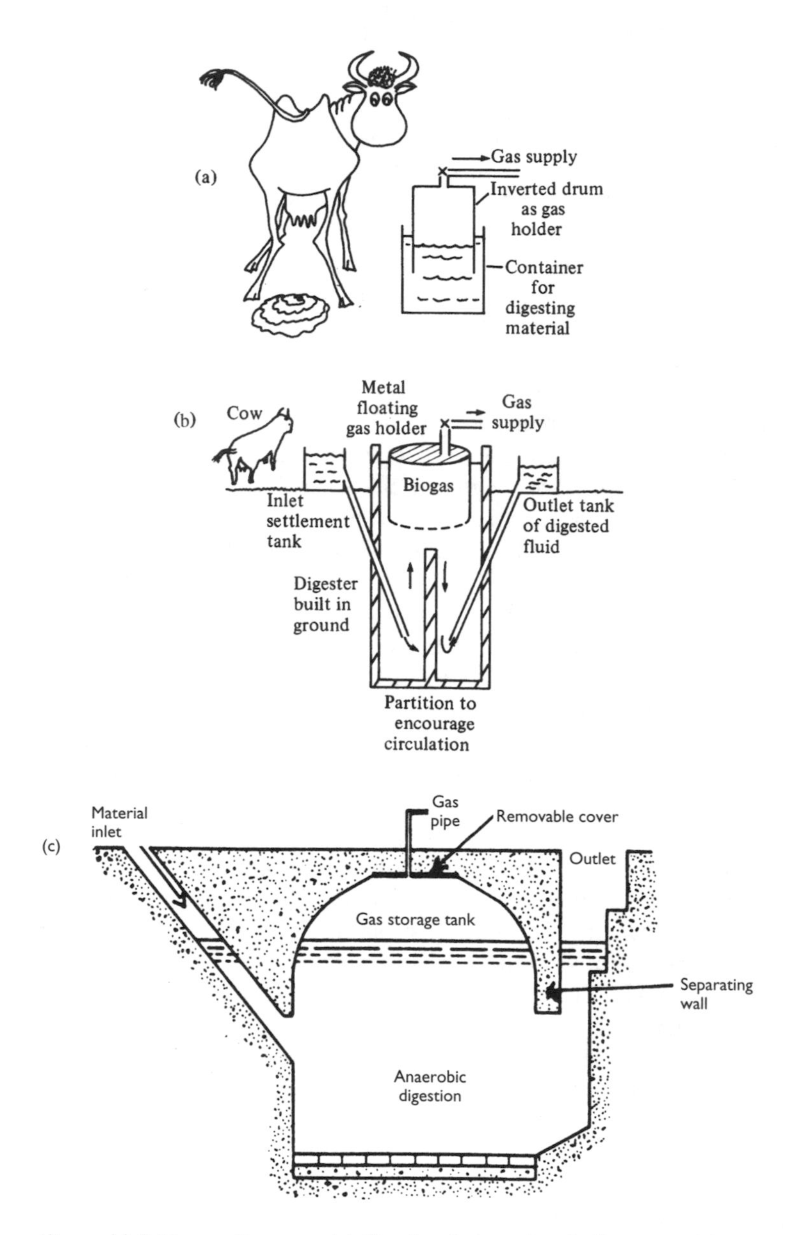


*Figure 11.7* Biogas digesters. (a) Simple oil drum batch digester, with cow to scale (b) Indian 'gobar gas' digester. (c) Chinese 'dome' for small-scale use [adapted from Van Burren]. (d) Accelerated rate farm digester with heating, for use in middle latitudes [adapted from Meynell]. (e) A large system near Aalborg in Denmark, which produces 10 000 $m^3$ of biogas per day. The biogas store is in the foreground, with the digester tanks visible behind it. The lorry (truck) in front of the white shed indicates the scale. [Photo by courtesy of NIRAS als].

(d)

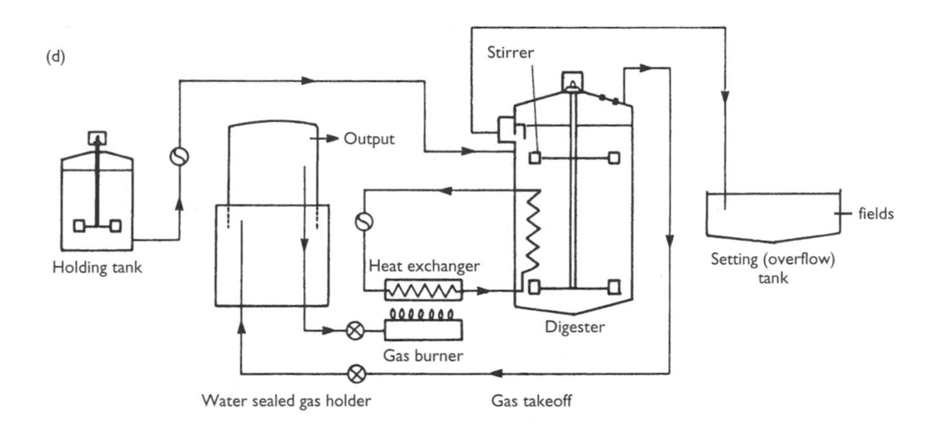


(e)

*Figure 11.7* (Continued).

dumps. The economic benefits are that input material does not have to be specially collected, administrative supervision is present, waste disposal is improved, and uses are likely to be available for the biogas and nutrient-rich effluent. However, in high and middle latitudes, tank digesters have to be heated for fast digestion (especially in the winter); usually such heat would come from burning the output gas, hence reducing net yield significantly. Slow digestion does not require such heating. Obviously obtaining biogas from, say, urban landfill waste is a different engineering task than from cattle slurries. Nevertheless the biochemistry is similar. Most of the following refers to tank digesters, but principles apply to other biogas systems.

Biogas generation is suitable for small- to large-scale operation. Several million household-scale systems have been installed in developing countries, especially in China and India, with the gas used for cooking and lighting. However, successful long-term operation requires (a) trained maintenance and repair technicians, (b) the users to perceive benefits and (c) alternative fuels, e.g. kerosene, not to be subsidised.

Biogas systems may be particularly attractive as part of *integrated farming*, where the aim is to emulate a full ecological cycle on a single farm. Thus plant and animal wastes are digested with the collection of the biogas as a fuel, with the effluent passing for further aerobic digestion in open tanks, before dispersal. The biogas is used for lighting, machines, vehicles, generators, and domestic and process heat. Algae may be grown on the open air tanks and removed for cattle feed. From the aerobic digestion, the treated effluent passes through reed beds, perhaps then to fish tanks and duck ponds before finally being passed to the fields as fertilizer. Success for such schemes depends ultimately on total integrated design, good standards of construction, and the enthusiasm and commitment of the operator, not least for the regular maintenance required.

### 11.8.2 Basic processes and energetics

The general equation for anaerobic digestion is

$$\begin{aligned} C_xH_yO_z + (x - y/4 - z/2)H_2O \\ \rightarrow (x/2 - y/8 + z/4)CO_2 + (x/2 + y/8 - z/4)CH_4 \end{aligned} \tag{11.5}$$

For cellulose this becomes

$$(C_6H_{10}O_5)_n + nH_2O \rightarrow 3nCO_2 + 3nCH_4 \tag{11.6}$$

Some organic material, e.g. lignin, and all inorganic inclusions do not digest in the process. These add to the bulk of the material, form a scum and can easily clog the system. In general 95% of the mass of the material is water.

The reactions are slightly exothermic, with typical heat of reaction being about $1.5\,\mathrm{MJ\,kg^{-1}}$ dry digestible material, equal to about 250 kJ per mole of $C_6H_{10}O_5$. This is not sufficient to significantly affect the temperature of the bulk material, but does indicate that most enthalpy of reaction is passed to the product gas.

If the input material had been dried and burnt, the heat of combustion would have been about $16\,\mathrm{MJ\,kg^{-1}}$. Only about 10% of the potential heat of combustion need be lost in the digestion process. This is 90% conversion efficiency. Moreover very wet input has been processed to give the convenient and controllable gaseous fuel, whereas drying 95% aqueous input would have required much energy (about $40\,\mathrm{MJ\,kg^{-1}}$ of solid input). In practice, digestion is seldom allowed to go to completion because of the long time involved, so 60% conversion is common. Gas yield is about 0.2 to $0.4\,\mathrm{m^3}$ per kg of dry digestible input at STP, with throughput of about 5 kg dry digestible solid per cubic metre of liquid.

It is generally considered that three ranges of temperature favour particular types of bacteria. Digestion at higher temperature proceeds more rapidly than at lower temperature, with gas yield rates doubling at about every 5 °C of increase. The temperature ranges are (1) psicrophilic, about 20 °C, (2) mesophilic, about 35 °C, and (3) thermophilic, about 55 °C. In tropical countries, unheated digesters are likely to be at average ground temperature between 20 and 30 °C. Consequently the digestion is psicrophilic, with retention times being at least 14 days. In colder climates, the psicrophilic process is significantly slower, so it may be decided to heat the digesters, probably by using part of the biogas output; a temperature of about 35 °C is likely to be chosen for mesophilic digestion. Few digesters operate at 55 °C unless the purpose is to digest material rather than produce excess biogas. In general, the greater is the temperature, the faster is the process time.

The *biochemical processes* occur in three stages, each facilitated by distinct sets of anaerobic bacteria:

1. Insoluble biodegradable materials, e.g. cellulose, polysaccharides and fats, are broken down to soluble carbohydrates and fatty acids (hydrogenesis). This occurs in about a day at 25 °C in an active digester.
2. Acid forming bacteria produce mainly acetic and propionic acid (acidogenesis). This stage likewise takes about one day at 25 °C.
3. Methane forming bacteria slowly, in about 14 days at 25 °C, complete the digestion to a maximum $\sim 70\%\,CH_4$ and minimum $\sim 30\%\,CO_2$ with trace amounts of $H_2$ and perhaps $H_2S$ (methanogenesis). $H_2$ may play an essential role, and indeed some bacteria, e.g. *Clostridium*, are distinctive in producing $H_2$ as the final product.

The methane forming bacteria are sensitive to pH, and conditions should be mildly acidic (pH 6.6–7.0) but not more acidic than pH 6.2. Nitrogen should be present at 10% by mass of dry input and phosphorus at 2%. A golden rule for successful digester operation is to *maintain constant conditions* of temperature and suitable input material. As a result a suitable population of bacteria is able to become established to suit these conditions.

### 11.8.3 Digester sizing

The energy available from a biogas digester is given by:

$$E = \eta H_b V_b \tag{11.7}$$

where $\eta$ is the combustion efficiency of burners, boilers, etc. (~60%). $H_b$ is the heat of combustion per unit volume biogas ($20\,\text{MJ}\,\text{m}^{-3}$ at 10 cm water gauge pressure, 0.01 atmosphere) and $V_b$ is the volume of biogas. Note that some of the heat of combustion of the methane goes to heating the $CO_2$ present in the biogas, and is therefore unavailable for other purposes. The net effect is to decrease the efficiency.

Alternatively:

$$E = \eta H_m f_m V_b \tag{11.8}$$

where $H_m$ is the heat of combustion of methane ($56\,\text{MJ}\,\text{kg}^{-1}$, $28\,\text{MJ}\,\text{m}^{-3}$ at STP) and $f_m$ is the fraction of methane in the biogas. As from the digester, $f_m$ should be between 0.5 and 0.7, but it is not difficult to pass the gas through a counterflow of water to dissolve the $CO_2$ and increase $f_m$ to nearly 1.0.

The volume of biogas is given by

$$V_b = cm_0 \tag{11.9}$$

where $c$ is the biogas yield per unit dry mass of whole input ($0.2$–$0.4\,\text{m}^3\,\text{kg}^{-1}$) and $m_0$ is the mass of dry input.

The volume of fluid in the digester is given by

$$V_f = m_0/\rho_m \tag{11.10}$$

where $\rho_m$ is the density of dry matter in the fluid ($\sim 50\,\text{kg}\,\text{m}^{-3}$).

The volume of the digester is given by

$$V_d = \dot{V}_f t_r \tag{11.11}$$

where $\dot{V}_f$ is the flow rate of the digester fluid and $t_r$ is the retention time in the digester (~8–20 days).

Typical parameters for animal waste are given in Table 11.8.

*Table 11.8* Typical manure output from farm animals; note the large proportion of liquid in the manure that favours biogas production rather than drying and combustion

| *Animal* | *Total wet manure per animal per day/kg* | *Of which, total solids per kg* | *Moisture mass content per wet mass* |
|---|---|---|---|
| Dairy cow (~500 kg) | 35 | 4.5 | 87% |
| Beef steer (~300 kg) | 25 | 3.2 | 87% |
| Fattening pig (~60 kg) | 3.3 | 0.3 | 91% |
| Laying hen | 0.12 | 0.03 | 75% |

*Example 11.1*
Calculate (1) the volume of a biogas digester suitable for the output of 6000 pigs, (2) the power available from the digester, assuming a retention time of 20 days and a burner efficiency of 0.6.

*Solution*
Mass of solids in waste is approximately

$$m_0 = (0.3\,\mathrm{kg\,d^{-1}})(6000) = 1800\,\mathrm{kg\,d^{-1}} \tag{11.12}$$

From (11.10), fluid volume is

$$\dot{V}_f = \frac{(1800\,\mathrm{kg\,d^{-1}})}{(50\,\mathrm{kg\,m^{-3}})} = 36\,\mathrm{m^3\,d^{-1}} \tag{11.13}$$

In (11.11), digester volume is

$$V_d = (36\,\mathrm{m^3\,d^{-1}})(20\,\mathrm{d}) = 720\,\mathrm{m^3} \tag{11.14}$$

From (11.9), volume of biogas is

$$V_b = (0.24\,\mathrm{m^3\,kg^{-1}})(1800\,\mathrm{kg\,d^{-1}}) = 430\,\mathrm{m^3\,d^{-1}} \tag{11.15}$$

So, from (11.7), energy output is

$$\begin{aligned} E &= (0.6)(20\,\mathrm{MJ\,m^{-3}})(430\,\mathrm{m^3\,d^{-1}}) \\ &= 5200\,\mathrm{MJ\,d^{-1}} = 1400\,\mathrm{kWh\,d^{-1}} \\ &= 60\,\mathrm{kW}(\text{continuous, thermal}) \end{aligned} \tag{11.16}$$

If continuously converted to electricity, this would yield about 15 $\mathrm{kW_e}$ of electricity from a biogas-fired generator set at 25% overall efficiency.

### *11.8.4 Working digesters*

Figure 11.7 shows a progression of biogas digesters, from the elementary to the sophisticated, allowing principles to be explained.

a *Household batch unit for the tropics.* This is the simplest method, comprising an upturned metal cylinder in another larger tank, e.g. a 200-litre oil drum with the top removed. The biogas is trapped in the top cylinder to be piped to the household for cooking and lighting. The tank has to be filled for each batch with fresh animal manure, seeded if possible with anaerobic bacteria from a previous batch. Systems like this are messy and usually do not last for more than a brief period of enthusiasm. Such batch treatment does not give a constant yield, so a continuous process is preferable.

b *Indian gobar gas system.*Gobar means cow dung, and is the word used for the sun-dried cow pats used in tropical countries and previously in Europe for cooking fuel. The diagram shows the principles of the original method. Material is placed in the inlet settlement tank to separate out non-digestible straw and inclusions. The flow moves slowly through the buried brick tank in about 14–30 days to the outlet, from which nutrient-rich fertilizer is obtained. Gas pressure of ~10 cm water column is maintained by the heavy metal gas holder, which is the most expensive item of the original design. The holder is lifted regularly (approximately six-monthly intervals) so that any thick scum at the top of the fluid can be removed. Daily inspection of pipes, etc. and regular maintenance are essential. Lack of maintenance is the predominant reason for the failure of biogas digesters generally.

c *Chinese digester.* This is a recommended design in the Republic of China for households and village communes, where several million have been installed. The main feature of the design is the permanent concrete top which enables pressurised gas to be obtained. This top is much cheaper than the heavy metal floating gas holder of (b) in older Indian systems. The flow moves slowly through the buried brick tank in about 14–30 days to the outlet, from which nutrient-rich fertilizer is obtained. As the gas evolves, its volume replaces digester fluid and the pressure increases. Frequent (~daily) inspection of pipes, etc. and regular maintenance are essential to avoid clogging by non-digestible material.

d *Industrial design.* The diagram shows a design for commercial operation in mid-latitudes for accelerated digestion under fully controlled conditions. The digester tank is usually heated to at least 35 °C. A main purpose of such a system is likely to be the treatment of the otherwise unacceptable waste material, with biogas being an additional benefit.

e *Case study.* The photograph shows an industrial-scale biogas digestion system operating in Denmark. It processes some 120 tonnes per day of slurry (manure) from nearby farms plus about 35 tonnes per day of other organic matter including slaughterhouse wastes, fatty industrial wastes. The raw slurry comes in 20 $m^3$ tanker lorries from the farms, with the digested product returned in the same lorries to the farms for use as fertiliser. The mixed biomass is fed into two 950 $m^3$ steel digester tanks by computer-controlled pumps and valves. The temperature in the digesters is about 53 °C for thermophilic digestion, as this gives optimum process efficiency. About 80% of the 10 000 $m^3 d^{-1}$ of biogas is passed through a pressurised water 'scrubber' to remove $CO_2$ and traces of $H_2S$, compressed and transferred to a nearby district combined heat and power (CHP) plant; most of the rest is used on-site in another CHP unit which provides the necessary heat for the digesters. The payback time for such a system, based on the sale of gas and electricity, is about 10 years.

## 11.9 Wastes and residues

Wastes and residues from human activity and economic production are a form of 'indirect' renewable energy, since they are unstoppable flows of energy potential in our environment. Wastes and residues arise from (a) primary economic activity, e.g. forestry, timber mills, harvested crops, abattoirs and food processing; and (b) urban, municipal and domestic refuse, including sewage. The energy generation potential from such wastes is primarily from the biomass content. However, there is usually a significant proportion of combustible waste from mineral sources, e.g. most plastics; however, such combustion requires regulation to reduce unacceptable emissions. A key factor regarding wastes and refuse is that they are usually available at points of concentration, where they easily become an environmental hazard. Dealing with this 'problem' becomes a necessity. The wastes operator will therefore be paid for the material and so be subsidised in later energy production.

Major wastes are (i) municipal solid waste (MSW), (ii) landfill and (iii) sewage. MSW is the wastes removed by municipal authorities from domestic and industrial sources; it usually contains significant amounts of metal, glass and plastic (i.e. non-biomass) material. Recycling of most plastics, metal, glass and other materials should occur before landfill or combustion. Nevertheless, non-biomass materials usually remain in significant amounts. MSW is loose, solid material of variable composition, available directly for combustion and pyrolysis. If the composition is acceptable, it may be pressurised and extruded as '*refuse derived fuel, RDF*', usually available as dried pellets of about 5 cm dimension for combustion in domestic-scale boilers.

'*Landfill*' is waste, usually MSW, deposited in large pits. A large proportion of MSW is biological material which, once enclosed in landfill, decays anaerobically. The process is slower than in most biogas digesters because of the reduced ground temperature, but when stabilised after many months the gas composition is similar; see Section 11.8. If not collected, the gas leaks slowly into the atmosphere, along with various smellier gases such as $H_2S$, so causing unpleasant environmental pollution. Therefore, the landfill site should be constructed and capped, e.g. with clay, so the gas can be collected when the pit is full, e.g. by an array of perforated pipes laid horizontally as the landfill is completed or drilled vertically into the buried refuse of an existing site. Regulations in several countries require capture of at least 40% of the methane from landfill, in order to reduce greenhouse gas emissions. Even without monetary credit for the greenhouse benefits, it is usually profitable to capture and utilise landfill gas if there is an industrial facility nearby which can use the fuel for direct combustion in boilers or engines for process heat and electricity generation. Faced with limited land for landfills, many municipalities have reduced the amount of landfill per household by obliging households to separate much of the biological material that previously went to landfill, e.g. garden clippings and food scraps; once collected separately, this is made into garden and horticultural compost by chopping and aerobic digestion, as described in Section 11.2.2(4).

## 11.10 Vegetable oils and biodiesel

Vegetable oils are extracted from biomass on a substantial scale for use in soap-making, other chemical processes and, in more refined form, for cooking.

Categories of suitable materials are:

1 Seeds: e.g. sunflower, rape, soya beans; ~50% by dry mass of oil.
2 Nuts: e.g. oil palm, coconut copra; ~50% by dry mass of oil, e.g. the Philippines' annual production of coconut oil is $\sim 10^6\,\mathrm{t\,y^{-1}}$.
3 Fruits: e.g. world olive production $\sim 2\,\mathrm{Mt\,y^{-1}}$.
4 Leaves: e.g. eucalyptus has ~25% by wet mass of oil.
5 Tapped exudates: e.g. rubber latex; jojoba, *Simmondsia chinensis* tree oil.
6 By-products of harvested biomass, for example oils and solvents to 15% of the plant dry mass, e.g. turpentine, rosin, oleoresins from pine trees, oil from *Euphorbia*.

Concentrated vegetable oils may be used directly as fuel in diesel engines, but difficulties arise from the high viscosity and from the combustion deposits, as compared with conventional (fossil) petroleum-based diesel oil, especially at low ambient temperature ($\leq 5\,^\circ\mathrm{C}$). Both difficulties are overcome by reacting the extracted vegetable oil with ethanol or methanol to form the equivalent ester. Such esters, called *biodiesel*, have technical characteristics as fuels that are better suited to diesel-engines than petroleum-based diesel oil. The reaction yields the equivalent ester and glycerine (also called 'glycerol'). The process usually uses KOH as a catalyst. Glycerol is also a useful and saleable product.

The esterification process is straightforward for those with basic chemical knowledge, and, with due regard for safety, can be undertaken as a small batch process (see website 'journeytoforever' amongst others). Continuous commercial production obviously needs more sophistication, and uses whatever oil is most readily and cheaply available in the country concerned, e.g. rapeseed oil in Europe (called 'canola' in some other countries) and soya oil in USA. Biodiesel can also be made from waste (used) cooking oil and from animal fat (tallow). The use of waste cooking oil as the raw material is attractive in both environmental and cost terms, especially on a small scale; the cost of collection is an issue on a larger scale.

When some governments removed institutional barriers to the production and sale of biodiesel, world commerce grew dramatically from near-zero in 1995 to over 1.5 million tons by 2003. Examples are Austria and Germany, whose federal and state governments established financial and taxation policy-instruments to encourage biodiesel production and use, either as 100% biodiesel or blended with petroleum-based diesel. Although the production cost of biodiesel substantially exceeds that of conventional diesel fossil fuel, such governments justified the policy in terms of the 'external' benefits for agriculture and for the environment, e.g. absence of sulphur emissions. Similar considerations apply to many other biofuels, notably bio-ethanol, see Sections 11.7 and 11.11.

The energy density of biodiesel as an ester varies with composition and is typically about $38\,\mathrm{MJ\,kg^{-1}}$, which is greater than for the raw oil and near to petroleum-based diesel fuel at about $46\,\mathrm{MJ\,kg^{-1}}$. Nevertheless, in practice fuel consumption per unit volume of a diesel-engine vehicle running on biodiesel is little different from that of fossil diesel. Quality standards have been established for the compatibility of biodiesel with most vehicles. A minor benefit of using biodiesel is that the exhaust smell is reminiscent of cooking, e.g. of popcorn.

Energy analysis of biodiesel produced from soya oil and methanol in the general economy indicates that the production of 1 MJ of the fuel may use about 0.3 MJ of fossil fuel input. The production of methanol from (fossil) natural gas accounts for nearly half of the 0.3 MJ, so the analysis would be even more favourable if the methanol (or ethanol) came from biomass.

## 11.11 Social and environmental aspects

### *11.11.1 Bioenergy in relation to agriculture and forestry*

The use and production of biomass for energy are intimately connected with wider policies and practices for agriculture and forestry. An overriding consideration is that such use and production should be ecologically sustainable, i.e. that the resource be used in a renewable manner, with (re-)growth keeping pace with use. Moreover, for ethical reasons, it is vital that biomass production for energy is not at the expense of growing enough food to feed people.

Nevertheless, in the European Union and the USA, a major issue in agriculture is over-production of food, as encouraged by agricultural subsidies. Such subsidies increase general taxation and the consequent surpluses of agricultural products distort world trade to the disadvantage of developing countries. As a partial response to such concerns, the European Union introduced financial incentives for its farmers to set aside land from food production, and either to maintain it unproductively or for biomass for energy. Such policies retain the social benefits of an economically active rural population while also bringing the environmental benefits, described below, of substituting biofuels for fossil fuels.

Utilising waste biomass increases the productivity of agriculture and forestry. This is especially so for the acceptable disposal of otherwise undesirable outputs, e.g. biodigestion of manure from intensive piggeries, so the integrated system brings both economic and environmental benefits. As emphasised in Section 11.1, successful biofuel production utilises already concentrated flows of biomass, such as offcuts and sawdust from sawmilling, straw from crops, manure from penned animals and sewage from municipal works. Biofuel processes that depend first upon transporting and then upon concentrating diffuse biomass resources are unlikely to be viable.

Energy developments utilising local crops and established skills are most likely to be socially acceptable. Thus the form of biomass most likely to be viable as an energy source will vary from region to region. Moreover, as with any crop, sustainable agriculture and forestry is required, for instance extensive monocultures are vulnerable to disease and pests and unfriendly to native fauna. Note, too, that greenhouse gas benefits only occur when the biomass is used to replace fossil fuel use, so leaving the abated fossil fuel underground.

### *11.11.2 Food versus fuel*

Production of liquid biofuels has been based historically on biomass from grain, sugar and oil crops, all of which are essential food crops, generally grown on the best agricultural land available. Despite crop production surpluses in the USA and Europe, the increasing worldwide demand for food indicates that these crops will not be diverted significantly from food to energy. Therefore, biofuel production as a major contribution to world energy supplies requires other feedstock and land than for food and other strategies. For instance, there is a need for cheaper, more energy-efficient processes for producing ethanol from widely available lignocellulosic materials, e.g. corn-stalks, straw and wood, especially sawdust and other woody residues, rather than from food-related crops.

### *11.11.3 Greenhouse gas impacts: bioenergy and carbon sinks*

When a plant grows, carbon is extracted from the air as $CO_2$ is absorbed in photosynthesis, so becoming 'locked into' carbohydrate material both above and below ground. Significant amounts of $CO_2$ are released in plant metabolism, but the net carbon flow is *into* the plant. Carbon concentrations in the soil may also increase 'indirectly' from organic matter formed from plant detritus in fallen leaves and branches. Such removal of the greenhouse gas $CO_2$ from the atmosphere is called a 'carbon sink'. Consequently a dedicated programme to increase plant growth will offset temporarily an increase in atmospheric $CO_2$ from burning fossil fuels. However, all plants die and the vast majority of all such direct and indirect carbon eventually returns to the atmosphere, so joining a natural cycle which neither depletes nor increases atmospheric $CO_2$ concentrations. Only if the plant material is burnt to replace (abate) specific use of fossil fuel will there be a long-term benefit, by preventing that fossil carbon from otherwise reaching the atmosphere. It follows that such abated fossil carbon should always stay beneath the ground and never be extracted.

The Kyoto Protocol is an international treaty which aims to mitigate climate change induced by greenhouse gas emissions; amongst other things it encourages countries to plant new forests. However, as explained above, such carbon sinks are only temporary because when the forest is harvested all its above-ground carbon will be returned relatively quickly to the atmosphere: within months if used for paper, years if used for construction. Thus only a *new* and thereafter continuing forest can be a carbon sink, albeit 'once only'. However, a replanted forest cannot be so considered, unless the biomass of the previous forest was used to abate fossil fuel. Figure 11.8 illustrates such effects. Consequently, the effectiveness of forest plantations to act as carbon sinks is limited by the availability of 'unused' land, which, from a global perspective, is itself limited by the increasing need for land for food.

Biomass can be both a carbon sink and a substitute for fossil fuels, in the many ways described in this chapter. Used as bioenergy, biomass has an additional value as a carbon offset because the fossil fuel which it displaces remains underground. Its value in this way has not been fully recognised in the Kyoto Protocol. The carbon offset of $CO_2$ emissions by using bioenergy is sustainable (at least in principle), since each annual crop replaces the same amount of would-be fossil fuel emissions, regardless of previous or subsequent harvests – unlike the sink effect, which is negated by each successive harvest (see Figure 11.8). As indicated in Figure 11.8, coppicing can give a similar effect.

### *11.11.4 Internal and external costs of biofuels for transport*

Within most national economies, both bio-ethanol and biodiesel can be expected to be more expensive to produce commercially than refined fossil fuels. This is not surprising, since no one has paid for the initial growth of the fossil deposits and given the maturity of the petroleum industry and the political importance placed on it. However, the great majority of

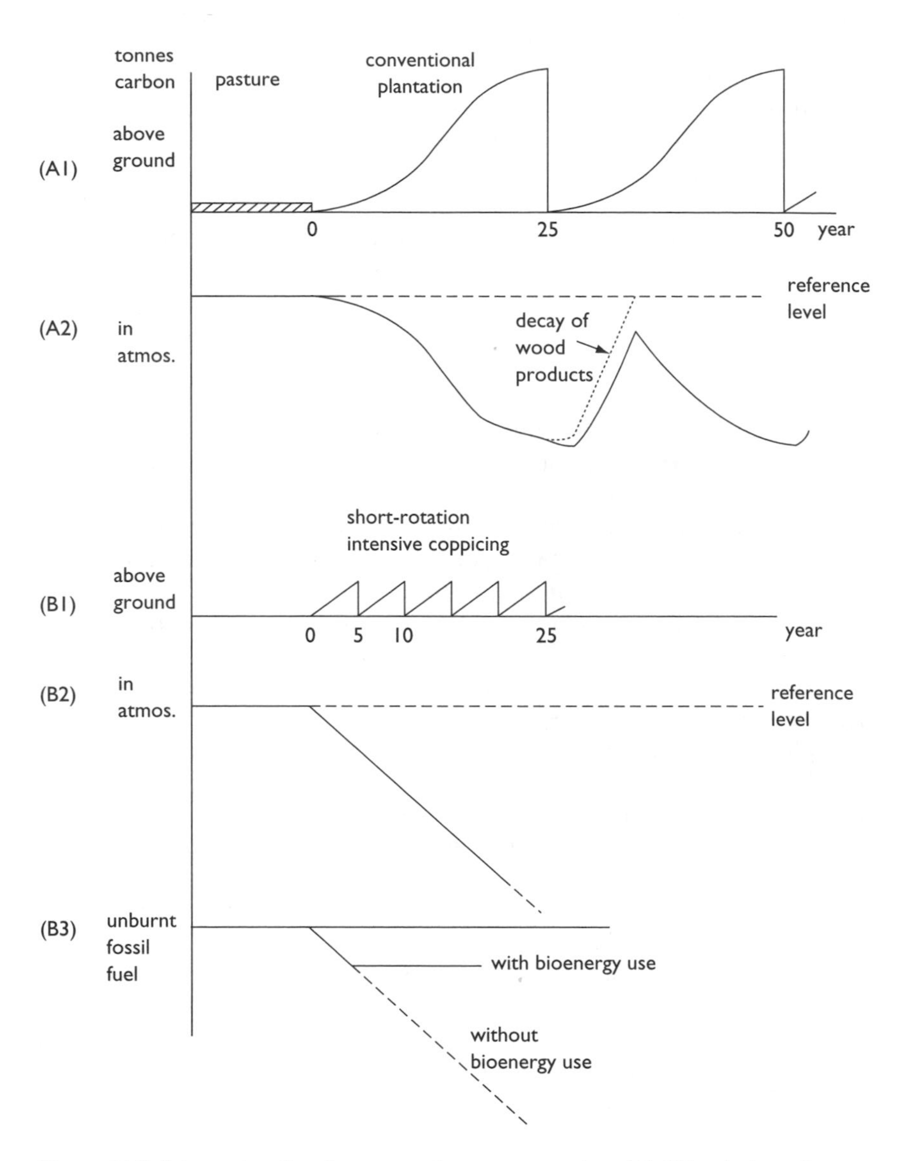


*Figure 11.8* Schematic of carbon mass for two scenarios: (A) Wood plantation grown and harvested for wood products (assumed for this diagram to be of short lifetime and then discarded to landfill or waste incineration); (B) Biofuel from coppiced trees, used to abate fossil fuel. For both scenarios, 'carbon in the atmosphere' is the additional carbon present from previously burning fossil fuels. For simplicity, no other plant growth is included.

governments tax automotive petroleum fuels, e.g. the UK with ~400% total taxation. Such taxation both raises revenue and discourages unnecessary driving to reduce pollution, road congestion and, usually, imports costing foreign exchange. Governments may therefore encourage use of biofuels with a smaller tax on biofuels than on fossil petroleum. If the proportion of biofuel in the total fuel mix is relatively small, the effect on government taxation revenue may not be great, but the infrastructure for the biofuels will have been initiated. Alternatively, government can mandate that all transport fuel sold must contain a certain percentage of biofuel, as Brazil since the 1970s; in this case consumers pay the extra cost. Neither measure is politically feasible unless motor manufacturers, as most now do, uphold warranties using the 'new' fuel. Subsidies to the agricultural producers of biofuels, as in the EC as part of the general agricultural subsidies, are another policy tool.

The Brazilian ethanol programme is the most famous example of such large-scale support for biofuels. It was established in the 1970s to reduce the country's dependence on imported oil and to help stabilise sugar production in the context of cyclic variation of world prices. The programme both increased employment in the sugar industry and generated several hundred thousand new jobs in processing and manufacturing. It is claimed that the investment cost per job in the ethanol industry was less than for other sectors at the time. The programme certainly led to economies of scale and technological development which reduced the production cost of ethanol from sugar.

Environmentally, substituting biofuel for fossil petroleum reduces greenhouse gas emissions, provided the biofuel comes from a suitable process; see Table 11.5 and its discussion. Moreover, biofuel combustion under properly controlled conditions is usually more complete than for fossil petroleum, and unhealthy emissions of particulates and $SO_2$ are less. This is because

*Figure 11.8* (Continued). (A1) Conventional plantation for wood products. Carbon in plants above ground (25 y rotation). (A2) Carbon in atmosphere (as $CO_2$). As the trees grow, carbon in the atmosphere decreases, but when the wood products decay (shown here during ~10 y), the bound carbon returns to the atmosphere. Therefore the total atmospheric carbon returns to its original concentration, less any permanent residue of leaves and roots in the soil.
(B1) Carbon above ground in a short-rotation intensive plantation, coppiced every 5 years for biofuel.
(B2) Carbon in atmosphere (as $CO_2$) as a result of (B1), assuming the biomass is used to abate fossil fuel as an energy source.
(B3) Coal remaining unburnt (i.e. in the ground).
For all graphs similarly, the ordinate indicates amount of carbon; the abscissa indicates years. For further discussion, see text.

all biofuels have a larger proportion of oxygen and a smaller proportion of sulphur in their chemical composition than fossil petroleum hydrocarbons.

Therefore, with the correct technology, biofuels not only abate fossil fuels but they reduce urban air pollution. Greenhouse gases and unhealthy air pollutants impose real costs on the community, e.g. increased damage from severe weather, flood prevention costs, medical costs. Unless such costs are internalised specifically as a levy in the price of the fossil fuel, they are 'external costs'; see Section 17.3 and Box 17.1.

### *11.11.5 Other chemical impacts*

Every country has regulations concerning the allowed and the forbidden emissions of gases, vapours, liquids and solids. This is a huge and complex subject within environmental studies.

The most vital aspect for the optimum combustion of any fuel is to control temperature and input of oxygen, usually as air. The aim with biomass and biofuel combustion, as with all fuels, is to have emissions with minimum particulates (unburnt and partially burnt material), with fully oxidised carbon to $CO_2$ and not CO or $CH_4$, and with minimum oxides of nitrogen which usually result from excessive temperature of the air. Therefore, in practice, the combustion should be confined to a relatively small space at almost white hot temperature; this volume has to be fed with air and fresh fuel. In addition, only fully burnt ash should remain (at best this is a fine powder that moves almost as a liquid). Useful heat is extracted by radiation from the combustion and by conduction from the flue gases through a heat exchanger, usually to water. Combustion of biofuels in engines, including turbines, has similar basic requirements, but occurs with much greater sophistication. Such combustion is possible according to different circumstances, e.g.

- With firewood. Position the wood so the fire is contained within two or three burning surfaces, e.g. at the tips of three logs (the classic '3-stone fire', or in the lengthwise space between three parallel logs.
- With wood chips or pellets. Feed the fuel by conveyor or slope from a hopper to a relatively small combustion zone, onto which compressed air is blown and from which the ash falls.
- With general timber and forest waste. Feed the fuel as above, but probably with a moving or shaking grate.
- With liquid and gaseous biofuels. The combustion should be controlled in boilers and engines as with liquid and gaseous fossil fuels, but with different air flow and fuel/air mixing requirements

Combustion of contaminated biomass, e.g. when mixed with plastics etc. in municipal solid waste or under less controlled conditions (most notably cooking over an open fire in a confined space), has considerable adverse environmental impact, unless great care is taken. Over a million deaths per year of women and children in developing countries have been attributed to kitchen smoke, 'the killer in the kitchen'. Improving domestic air quality is a major motivation for the improved cook-stoves of Section 11.4.1. On an industrial scale, particulates can be removed by improved combustion, filters, cyclones and flue condensation, which also recovers the latent heat of the condensate and increases efficiency. Nitrogen oxides, $NO_x$, formed as part of combustion in air can be alleviated by controlling the temperature of combustion. Straw from cereal crops may contain relatively large concentrations of potassium and chlorine, which can cause corrosion in boiler grates; this can be reduced by having rotating grates to prevent a solid mass of ash forming. Nevertheless, the ash from the complete combustion of any biomass is always a valued fertilizer, especially for the phosphate content.

Although the natural carbon cycle of plant growth fully renews the carbon in a crop or plantation, there may be a net loss of nitrogen and possibly other nutrients when the biomass is burnt or otherwise processed. That is, nitrogen is not returned sufficiently to the soil 'automatically' and has to be put back as a chemical input, possibly in the form of manure or by rotation with nitrogen-fixing crops such as beans, clover or *leucaena*.

### *11.11.6 Bioenergy in relation to the energy system*

Biomass is a major part of the world energy system now, although mainly in the form of inefficiently used firewood in rural areas, especially where cooking is over an open fire. A more sustainable energy system for the world will necessarily have to involve this widely distributed and versatile resource, but used in more efficient and more modern ways. For example, the 'renewables intensive' scenario of Johansson *et al.* (1993) has over 25% of world fuel use coming from biomass, drawing on plantations on presently uncropped land but of an area comparable to that currently cropped.

## Problems

11.1 A farmer with 50 pigs proposed to use biogas generated from their wastes to power the farm's motor car.

- a Discuss the feasibility of doing this. You should calculate both the energy content of gas and the energy used in compressing the gas to a usable volume, and compare these with the energy required to run the car.
- b Briefly comment on what other benefits (if any) might be gained by installing a digester.

You may assume that a 100 kg pig excretes about 0.5 kg of volatile solid (VS) per day (plus 6 kg water), and that 1 kg of VS yields 0.4 $m^3$ of biogas at STP.

11.2 Studies show that the major energy consumption in Fijian villages is wood which is used for cooking on open fires. Typical consumption of wood is 1 kg per person per day.

a Estimate the heat energy required to boil a two-litre pot full of water. Assuming this to be the cooking requirement of each person, compare this with the heat content of the wood, and thus estimate the thermal efficiency of the open fire.

b How much timber has to be felled each year to cook for a village of 200 people?

Assuming systematic replanting, what area of crop must the village therefore set aside for fuel use if it is not to make a net deforestation? *Hint*: refer to Table 11.4.

c Comment on the realism of the assumptions made, and revise your estimates accordingly.

11.3 a A butyl rubber bag of total volume 3.0 $m^3$ is used as a biogas digester. Each day it is fed an input of 0.20 $m^3$ of slurry, of which 4.0 kg is volatile solids, and a corresponding volume of digested slurry is removed. (This input corresponds roughly to the waste from 20 pigs.)

Assuming that a typical reaction in the digestion process is

$$C_{12}H_{22}O_{11} + H_2O \overset{\text{bacteria}}{\rightarrow} 6CH_4 + 6CO_2$$

and that the reaction takes 7 days to complete, calculate (i) the volume of gas (ii) the heat obtainable by combustion of this gas for each day of operation of the digester (iii) how much kerosene would have the same calorific value as one day's biogas?

b The reaction rate in the digester can be nearly doubled by raising the temperature of the slurry from 28 (ambient) to 35 °C. (i) What would be the advantage of doing this? (ii) How much heat per day would be needed to achieve this? (iii) What proportion of this could be contributed by the heat evolved in the digestion reaction?

11.4 a Write down a *balanced* chemical equation for the conversion of sucrose ($C_{12}H_{22}O_{11}$) to ethanol ($C_2H_5OH$). Use this to calculate how much ethanol could be produced in theory from one tonne of sugar. What do you think would be a realistic yield?

b Fiji is a small country in the South Pacific, whose main export crop is sugar. Fiji produces 300 000 t/y of sugar, and imports 300 000 t of fossil petroleum fuel. If all this sugar were converted to ethanol, what proportion of petroleum imports could it replace?

11.5 Consider a pile of green wood chips at 60% moisture content (wet basis) and weighing 1 t. What is the oven-dry mass of biomass in the pile?

The biomass has a heat of combustion of 16 MJ per oven-dry kg. This is the 'gross calorific value' corresponding to the heat output in a reaction of the type

$$[CH_2O] + O_2 \rightarrow CO_2(\text{gas}) + H_2O(\text{liq})$$

The net calorific value (or 'lower heating value') is the heat evolved when the final water is gaseous; in practice this is the maximum thermal energy available for use when biomass is burnt.

i The pile is left to dry to 50% moisture content (wet basis), when it looks much the same but has less water in it.
ii The pile is left to dry for a few more weeks, and reaches 20% m.c. (w.b.), at which stage it has shrunk a little in volume and much in mass.

For each situation calculate the total mass of the pile, the net heat energy available from burning the pile, and its net calorific value per wet kg.

11.6 Figure 11.8 shows *total* (cumulative) carbon above ground in a plantation and in the atmosphere as trees grow and are harvested. For the same situation sketch the corresponding curves for *annual* emissions of $CO_2$ by sources and removals by sinks. (These are the terms in which greenhouse gas inventories are compiled for purposes of the UN Framework Convention on Climate Change and its Kyoto Protocol.)

## Bibliography

### Overviews

Chum, H.L. and Overend, R.P. (2003) Biomass and Bioenergy in the United States, *Advances in Solar Energy*, **15**, 83–148. {Comprehensive review of commercial and near-commercial technologies, and supporting policies and R&D, with emphasis on large- and medium-scale.}

Johansson, T.B, Kelly, H., Reddy, A.K.N., Williams, R.H. and Burnham, L. (eds) (1993) *Renewable Energy; sources for fuels and electricity*, Earthscan (London) and Island Press (Washington D.C.). {Much respected, detailed, text and report of renewable energy for commercial and developmental supplies of electricity and fuels (i.e. not for heat as such). Includes 8 chapters on biomass or biofuels.}

Klass, D.L. (1998) *Biomass for Renewable Energy, Heat and Chemicals*, Academic Press. {An extremely comprehensive and reliable text. Based on chemical principles, but aware of all appropriate disciplines, including economics. Uses S.I. units.}

Lewis, C.W. (1983) *Biological Fuels*, Edward Arnold, London. {Excellent basic text.}

Sims, R.E. (2002) *The Brilliance of Bioenergy in Business and in Practice*, James & James, London. {Illuminating text with emphasis on modern industrial production and applications; includes numerous illustrated case studies of power systems, including with biogas.}

Slesser, M. and Lewis, C. (1979) *Biological Energy Resources*, E. and F.N. Spon, London. {Recommended for microbiological information and energy analysis of processes.}

Sørensen, B. (2000, 2nd edn) *Renewable Energy*, Academic Press. {Pages 471–506 review the many paths of conversion of biological material for energy purposes.}

### Resource estimates

Hall, D.O., Rosillo-Calle, F., Williams, R.H., Woods, J. (1993) Biomass for energy: supply prospects, in Johansson *et al.* (1993), pp. 593–651.

World Energy Council (2001) *World Energy Resources* (chapters on wood and other biomass), on web at www.worldenergy.org.

### Direct combustion, especially wood fuel

De Lepeleire, G., Prasad, K.K., Verhaart, P. and Visser, P. (1981) *A Woodstove Compendium*, Eindhoven University, Holland. {Gives principles of woodburning, and technical descriptions of many stoves designed for domestic cooking in developing countries.}

Dutt, G.S. and Ravindranath, N. (1993) Bioenergy: direct applications in cooking, in Johansson *et al.* (1993), pp. 653–697. {Emphasis on developing countries, appliance efficiency, and resource implications.}

Kammen, D.M. (1995) Cookstoves for the developing world, *Scientific American*, **273**, 64–67.

Wahlund, B., Yan, J. and Westermark, M. (2004) Increasing biomass utilisation in energy systems: A comparative study of $CO_2$ reduction and cost for different bioenergy processing options, *Biomass and Bioenergy*, **26**, 531–544. {Focuses on wood in Sweden; concludes pelletisation for coal substitution is the best option in that case.}

### Biogas

Chynoweth, D.P., Owens, J.M. and Legrand, R. (2001) Renewable methane from anaerobic digestion of biomass, *Renewable Energy*, **22**, 1–8. {Advocates anaerobic digestion as the principal pathway to use of energy crops.}

Lettinga, G. and Van Haandel, A. (1993) *Anaerobic Digestion for Energy Production and Environmental Protection*, in Johansson *et al.* {Survey with emphasis on industrial scale.}

Meynell, P.J. (1976) *Methane – Planning a Digester*, Prism Press, Dorchester, UK. {An old but still useful short and practical book with basic technical and biochemical explanations.}

Van Buren, A. (1979) *A Chinese Biogas Manual*, Intermediate Technology Publications, London. {A stimulating and useful handbook, based on the considerable experience of small-scale digesters in rural China. Reprinted several times.}

CADDET (2000) *Danish Biogas Plant with Separate Line for Organic Household Waste*, CADDET Technical Brochure No.125, on web at <www.caddet.org>. {CADDET is a database of 'demonstrated technology' projects on renewable energy and energy efficiency; it has case studies of all the main RE technologies.}

### *Biofuels*

IEA Bioenergy (2004) *Biofuels for Transport*, overview report by Task Group 39 (ed. D Stevens), International Energy Agency, Paris. {On web at <www.ieabioenergy.com>; a good summary of current status.}

Keshgi, H.S., Prince, R.C. and Marland, G. (2000) The potential of biomass fuels in the context of global climate change: Focus on transportation fuels, *Annual Review of Energy and the Environment*, **25**, 199–244. {Wide background, plus energy analysis of bioethanol in USA and Brazil.}

Moreira, J.R. and Goldemberg, J. (1999) The alcohol program, *Energy Policy*, 27, 229–245. {The economics and energy balance of the Brazilian bioethanol program. See also Keshgi *et al.* (2000) and the relevant chapter in Johansson *et al.* (1993).}

Twidell, J.W. and Pinney, A.A.(1985) The quality and exergy of energy systems, using conventional and renewable resources, *Sun-at-Work in Britain*, L.F. Jesch (ed.), UK-ISES. {Comments on energy analysis.}

Wyman, C.E. (1999) Biomass ethanol: Technical progress, opportunities, and commercial challenges, *Annual Review of Energy and the Environment*, **24**, 189–226 {Emphasises potential of new technology to produce ethanol from cellulose, e.g. 'waste' from food crops.}

### *Journals and websites*

*Biomass and Bioenergy*, monthly, Elsevier. {Covers a wide range of basic science and applications.}

website, http://journeytoforever.org. {A guide to do-it-yourself-biodiesel.}

www.biodiesel.org. {Basics of biodiesel technology, and lots of links onwards to news and developments; emphasis on USA.}

www.ieabioenergy.com. {Reports of international collaborative research on technology and policy.}

www.itdg.org, *Boiling point*. {Stoves and domestic energy.}

Chapter 12

# Wave power

## 12.1 Introduction

Very large energy fluxes can occur in deep water sea waves. The power in the wave is proportional to the square of the amplitude and to the period of the motion. Therefore the long period (~10 s), large amplitude (~2 m) waves have considerable interest for power generation, with energy fluxes commonly averaging between 50 and 70 kW m$^{-1}$ width of oncoming wave.

The possibility of generating electrical power from these deep water waves has been recognised for many years, and there are countless ideas for machines to extract the power. For example, a wave power system was used in California in 1909 for harbour lighting. Modern interest has revived, particularly in Japan, the UK, Scandinavia and India, so research and development has progressed to commercial construction for meaningful power extraction. Very small scale autonomous systems are used for marine warning lights on buoys and much larger devices for grid power generation. The provision of power for marine desalination is an obvious attraction. As with all renewable energy supplies, the scale of operation has to be determined, and present trends support moderate power generation at about 100 kW–1 MW from modular devices each capturing energy from about 5 to 25 m of wavefront. Initial designs are for operation at shore-line or near to shore to give access and to lessen, hopefully, storm damage.

It is important to appreciate the many difficulties facing wave power developments. These will be analysed in later sections, but may be summarised here:

1 Wave patterns are irregular in amplitude, phase and direction. It is difficult to design devices to extract power efficiently over the wide range of variables.
2 There is always some probability of extreme gales or hurricanes producing waves of freak intensity. The structure of the power devices must be able to withstand this. Commonly the 50 year peak wave is

10 times the height of the average wave. Thus the structures have to withstand ~100 times the power intensity to which they are normally matched. Allowing for this is expensive and will probably reduce normal efficiency of power extraction.

3 Peak power is generally available in deep water waves from open-sea swells produced from long fetches of prevailing wind, e.g. beyond the Western Islands of Scotland (in one of the most tempestuous areas of the North Atlantic) and in regions of the Pacific Ocean. The difficulties of constructing power devices for these types of wave regimes, of maintaining and fixing or mooring them in position, and of transmitting power to land, are fearsome. Therefore more protected and accessible areas near to shore are most commonly used.
4 Wave periods are commonly ~5–10 s (frequency ~0.1 Hz). It is extremely difficult to couple this irregular slow motion to electrical generators requiring ~500 times greater frequency.
5 So many types of device may be suggested for wave power extraction that the task of selecting a particular method is made complicated and somewhat arbitrary.
6 The large power requirement of industrial areas makes it tempting to seek for equivalent wave energy supplies. Consequently plans may be scaled up so only large schemes are contemplated in the most demanding wave regimes. Smaller sites of far less power potential, but more reasonable economics and security, may be ignored.
7 The development and application of wave power has occurred with spasmodic and changing government interest, largely without the benefit of market incentives. Wave power needs the same learning curve of steadily enlarging application from small beginnings that has occurred with wind power.

The distinctive advantages of wave power are the large energy fluxes available and the predictability of wave conditions over periods of days. Waves are created by wind, and effectively store the energy for transmission over great distances. For instance, large waves appearing off Europe will have been initiated in stormy weather in the mid-Atlantic or as far as the Caribbean.

The following sections aim to give a general basis for understanding wave energy devices. First we outline the theory of deep water waves and calculate the energy fluxes available in single frequency waves. Then we review the patterns of sea waves that actually occur. Finally we describe attempts being made to construct devices that efficiently match variable natural conditions. With the complex theory of water waves we have sacrificed mathematical rigor for, we hope, physical clarity, since satisfactory theoretical treatments exist elsewhere.

## 12.2 Wave motion

Most wave energy devices are designed to extract energy from deep water waves. This is the most common form of wave, found when the mean depth of the sea bed $D$ is more than about half the wavelength ($\lambda$). For example, an average sea wave for power generation may be expected to have a wavelength of $\sim$100 m and amplitude of $\sim$3 m, and to behave as a deep water wave at depths of sea bed greater than $\sim$30 m. Figure 12.1(a) illustrates the motion of water particles in a deep water wave. The circular particle motion has an amplitude that decreases exponentially with depth and becomes negligible for $D > \lambda/2$. In shallower water, Figure 12.1 (b), the motion becomes elliptical and water movement occurs against the sea bottom, producing energy dissipation.

The properties of deep water waves are distinctive, and may be summarised as follows:

1 The surface waves are sets of unbroken sine waves of irregular wavelength, phase and direction.
2 The motion of any particle of water is circular. Whereas the surface form of the wave shows a definite progression, the water particles themselves have no net progression.
3 Water on the surface remains on the surface.
4 The amplitudes of the water particle motions decrease exponentially with depth. At a depth of $\lambda/2\pi$ below the mean surface position, the amplitude is reduced to $1/e$ of the surface amplitude ($e = 2.72$, base of natural logarithms). At depths of $\lambda/2$ the motion is negligible, being less than 5% of the surface motion.
5 The amplitude $a$ of the surface wave is essentially independent of the wavelength $\lambda$, velocity $c$ or period $T$ of the wave, and depends on the history of the wind regimes above the surface. It is rare for the amplitude to exceed one-tenth of the wavelength, however.
6 A wave will break into white water when the slope of the surface is about 1 in 7, and hence dissipate energy potential.

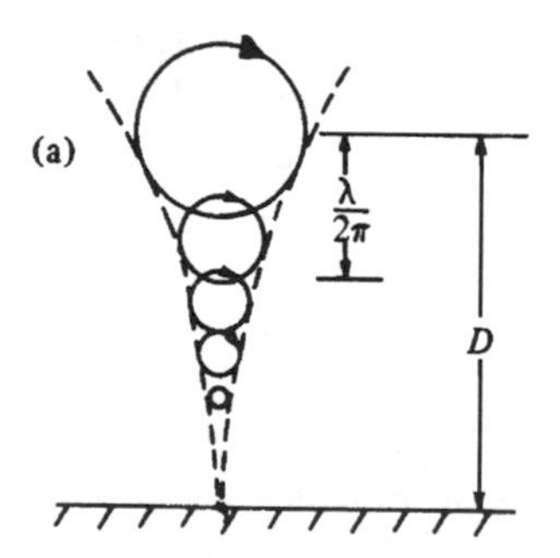


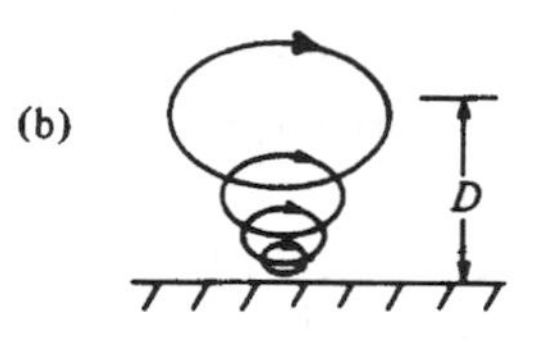


*Figure 12.1* Particle motion in water waves. (a) Deep water, circular motion of water particles. (b) Shallow water, elliptical motion of water particles.

The formal analysis of water waves is difficult, but known; see Coulson and Jeffrey (1977) for standard theory. For deep water waves, frictional, surface tension and inertial forces are small compared with the two dominant forces of gravity and circular motion. As a result, the water surface always takes up a shape so that its tangent lies perpendicular to the resultant of these two forces, Figure 12.2.

It is of the greatest importance to realize that there is no net motion of water in deep water waves. Objects suspended in the water show the motions of Figure 12.1, which contrasts deep water waves with the kinds of motion occurring in shallower water.

A particle of water in the surface has a circular motion of radius $a$ equal to the amplitude of the wave (Figure 12.3). The wave height $H$ from the top of a crest to the bottom of a trough is twice the amplitude: $H = 2a$. The angular velocity of the water particles is $\omega$ (radian per second). The wave surface has a shape that progresses as a moving wave, although the water itself does not progress. Along the direction of the wave motion the moving shape results from the phase differences in the motion of successive particles of water. As one particle in the crest drops to a lower position, another particle in a forward position circles up to continue the crest shape and the forward motion of the wave.

The resultant forces $F$ on water surface particles of mass $m$ are indicated in Figure 12.4. The water surface takes up the position produced by this resultant, so that the tangent to the surface is perpendicular to $F$. A particle at the top of a crest, position P1, is thrown upwards by the centrifugal force $ma\omega^2$. A moment later the particle is dropping, and the position in

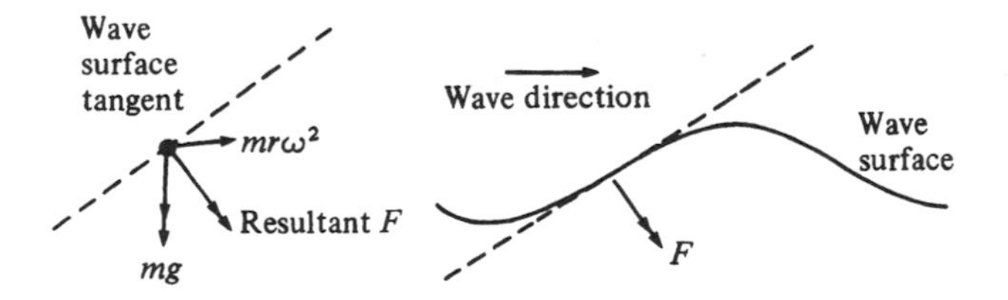


*Figure 12.2* Water surface perpendicular to resultant of gravitational and centrifugal force acting on an element of water, mass $m$.

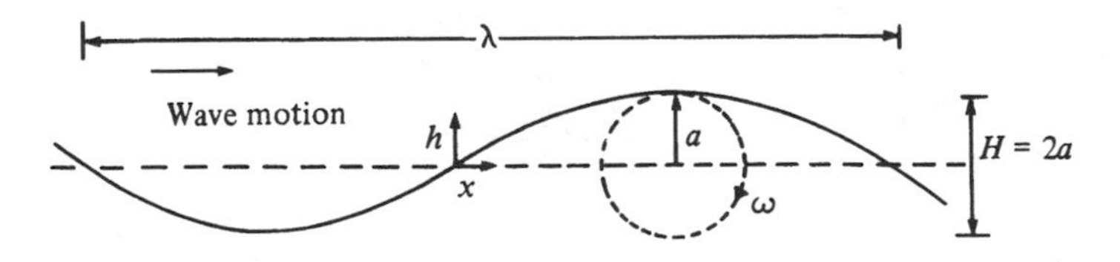


*Figure 12.3* Wave characteristics.

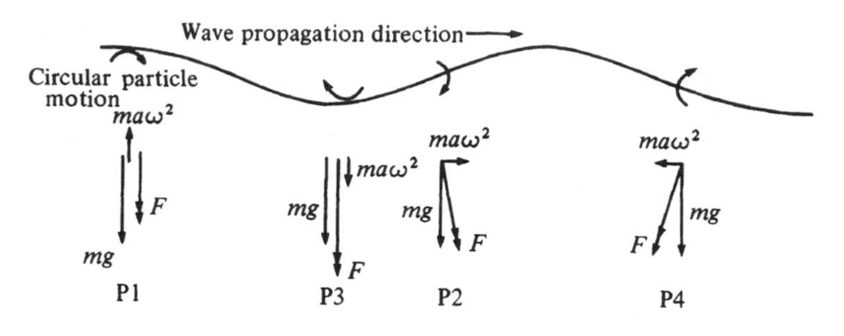


*Figure 12.4* Resultant forces on surface particles.

the crest is taken by a neighbouring particle rotating with a delayed phase. At P2 a particle is at the average water level, and the surface orientates perpendicular to the resultant force $F$. At the trough, P3, the downward force is maximum. At P4 the particle has almost completed a full cycle of its motion.

The accelerations of a surface particle are drawn in Figure 12.5(b). Initially $t = 0$, the particle is at the average water level, and subsequently:

$$\phi = \frac{\pi}{2} - \omega t \tag{12.1}$$

and

$$\tan s = \frac{a\omega^2 \sin \phi}{g + a\omega^2 \cos \phi} \approx \frac{a\omega^2 \sin \phi}{g} \tag{12.2}$$

since in practice $g \gg a\omega^2$ for non-breaking waves (e.g. $a = 2\,\text{m}$, $T$ (period) $= 8\,\text{s}$, $a\omega^2 = 1.2\,\text{m}\,\text{s}^{-2}$ and $g = 9.8\,\text{m}\,\text{s}^{-2}$). Let $h$ be the height of the surface above the mean level. The slope of the tangent to the surface is given by

$$\frac{\mathrm{d}h}{\mathrm{d}x} = \tan s \tag{12.3}$$

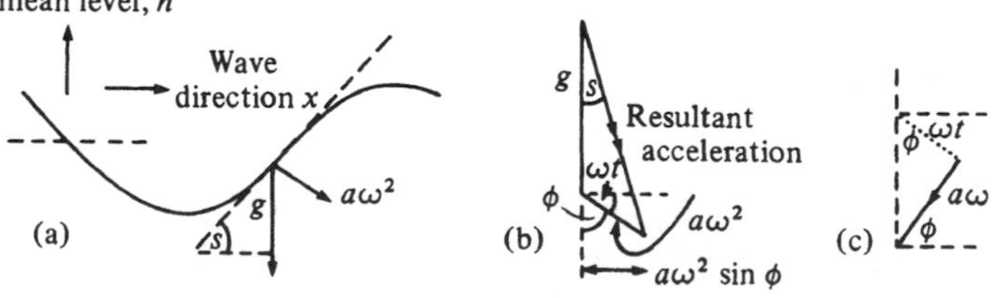


*Figure 12.5* Accelerations and velocities of a surface water particle. (a) Water surface. (b) Particle acceleration, general derivation. (c) Particle velocity.

From (12.1), (12.2) and (12.3),

$$\frac{\mathrm{d}h}{\mathrm{d}x} = \frac{a\omega^2}{g} \sin\phi = \frac{a\omega^2}{g} \cos\left(\frac{\pi}{2} - \phi\right) = \frac{a\omega^2}{g} \cos\omega t \tag{12.4}$$

From Figure 12.5(c), the vertical particle velocity is

$$\frac{\mathrm{d}h}{\mathrm{d}t} = a\omega \sin\phi = a\omega \cos\omega t \tag{12.5}$$

The solution of (12.4) and (12.5) is

$$h = a \sin\left(\frac{\omega^2 x}{g} - \omega t\right) \tag{12.6}$$

Comparing this with the general travelling wave equation of wavelength $\lambda$ and velocity $c$, we obtain

$$\begin{aligned} h &= a \sin\frac{2\pi}{\lambda}(x - ct) \\ &= a \sin\left(\frac{2\pi}{\lambda}x - \omega t\right) = a \sin(kx - \omega t) \end{aligned} \tag{12.7}$$

where $k = 2\pi/\lambda$ is called the wave number.

It is apparent that the surface motion is that of a travelling wave, where

$$\lambda = \frac{2\pi g}{\omega^2} \tag{12.8}$$

This equation is important; it gives the relationship between the frequency and the wavelength of deep water surface waves.

The period of the motion is $T = 2\pi/\omega = 2\pi/(2\pi g/\lambda)^{1/2}$. So

$$T = \left(\frac{2\pi\lambda}{g}\right)^{\frac{1}{2}} \tag{12.9}$$

The velocity of a particle at the crest of the wave is

$$v = a\omega = a\left(\frac{2\pi g}{\lambda}\right)^{\frac{1}{2}} \tag{12.10}$$

The wave surface velocity in the $x$ direction, from (12.7), is

$$c = \frac{\omega\lambda}{2\pi} = \frac{g}{\omega} = g\left(\frac{\lambda}{2\pi g}\right)^{\frac{1}{2}} \tag{12.11}$$

i.e.

$$c = \left(\frac{g\lambda}{2\pi}\right)^{\frac{1}{2}} = \frac{gT}{2\pi}$$

The velocity $c$ is called the phase velocity of the travelling wave made by the surface motion. Note that the phase velocity $c$ does not depend on the amplitude $a$, and is not obviously related to the particle velocity $v$.

*Example 12.1*
What is the period and phase velocity of a deep water wave of 100 m wavelength?

*Solution*
From (12.8),

$$\omega^2 = \frac{2\pi g}{\lambda} = \frac{(2\pi)(10\,\mathrm{m\,s^{-2}})}{100\,\mathrm{m}}, \qquad \omega = 0.8\,\mathrm{s^{-1}}$$

and so $T = 2\pi/\omega = 8.0\,\mathrm{s}$.
From (12.11)

$$c = \left[\frac{(10\,\mathrm{m\,s^{-2}})(100\,\mathrm{m})}{2\pi}\right]^{\frac{1}{2}} = 13\,\mathrm{m\,s^{-1}}$$

So

$$\lambda = 100\,\mathrm{m}, \quad T = 8\,\mathrm{s}, \quad c = 13\,\mathrm{m\,s^{-1}} \tag{12.12}$$

## 12.3 Wave energy and power

### 12.3.1 Basics

The elementary theory of deep water waves begins by considering a single regular wave. The particles of water near the surface will move in circular orbits, at varying phase, in the direction of propagation $x$. In a vertical column the amplitude equals half the crest to trough height at the surface, and decreases exponentially with depth.

The particle motion remains circular if the sea bed depth $D > 0.5\lambda$, when the amplitude becomes negligible at the sea bottom. For these conditions (Figure 12.6(a)) it is shown in standard texts that a water particle whose mean position below the surface is $z$ moves in a circle of radius given by

$$r = a\,e^{kz} \tag{12.13}$$

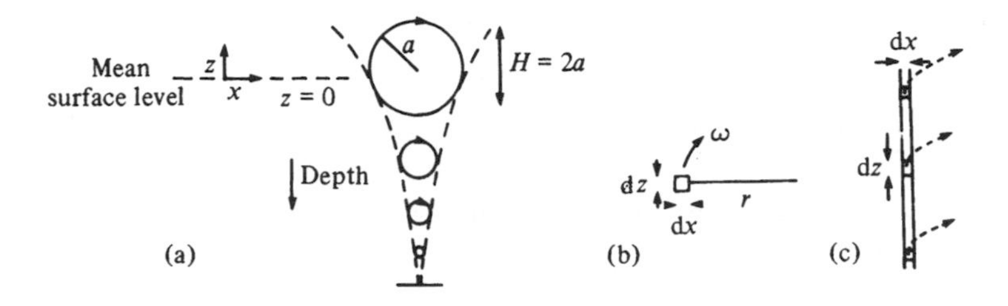


*Figure 12.6* Elemental motion of water, drawn to show the exponential decrease of amplitude with depth.

Here $k$ is the wave number, $2\pi/\lambda$, and $z$ is the mean depth below the surface (a negative quantity).

We consider elemental 'strips' of water across unit width of wavefront, of height $\mathrm{d}z$ and 'length' $\mathrm{d}x$ at position $(x, z)$ (Figure 12.6(b)). The volume per unit width of wavefront of this strip of density $\rho$ is

$$\mathrm{d}V = \mathrm{d}x\,\mathrm{d}z \tag{12.14}$$

and the mass is

$$\mathrm{d}m = \rho \mathrm{d}V = \rho \mathrm{d}x\,\mathrm{d}z \tag{12.15}$$

Let $E_K$ be the kinetic energy of the total wave motion to the sea bottom, per unit length along the $x$ direction, per unit width of the wave-front. The total kinetic energy of a length $\mathrm{d}x$ of wave is $E_K\,\mathrm{d}x$. Each element of water of height $\mathrm{d}z$, length $\mathrm{d}x$ and unit width is in circular motion at constant angular velocity $\omega$, radius of circular orbit $r$, and velocity $v = r\omega$ (Figure 12.6(b)). The contribution of this element to the kinetic energy in a vertical column from the sea bed to the surface is $\delta E_K \mathrm{d}x$, where

$$\delta E_K\,\mathrm{d}x = \frac{1}{2}mv^2 = \frac{1}{2}(\rho\,\mathrm{d}z\,\mathrm{d}x)r^2\omega^2 \tag{12.16}$$

Hence

$$\delta E_K = \frac{1}{2}\rho r^2\omega^2\,\mathrm{d}z \tag{12.17}$$

It is easiest to consider a moment in time when the element is at its mean position, and all other elements in the column are moving vertically at the same phase in the $z$ direction (Figure 12.6(c)).

From (12.13) the radius of the circular orbits is given by

$$r = ae^{kz} \tag{12.18}$$

where $z$ is negative below the surface.

Hence from (12.17),

$$\delta E_K = \frac{1}{2}\rho(a^2 e^{2kz})\omega^2\,dz \qquad (12.19)$$

and the total kinetic energy in the column is

$$E_K\,dx = \int_{z=-\infty}^{z=0} \frac{\rho\omega^2 a^2}{2} e^{2kz}\,dz\,dx = \frac{1}{4}\rho\frac{\omega^2 a^2}{k}\,dx \qquad (12.20)$$

Since $k = 2\pi/\lambda$, and from (12.8) $\omega^2 = 2\pi g/\lambda$, the kinetic energy per unit width of wave-front per unit length of wave is

$$E_K = \frac{1}{4}\rho a^2 \frac{2\pi g}{\lambda}\frac{\lambda}{2\pi} = \frac{1}{4}\rho a^2 g \qquad (12.21)$$

In Problem 12.1 it is shown that the potential energy per unit width of wave per unit length is

$$E_P = \frac{1}{4}\rho a^2 g \qquad (12.22)$$

Thus, as would be expected for harmonic motions, the average kinetic and potential contributions are equal.

The total energy per unit width per unit length of wavefront, i.e. total energy per unit area of surface, is

$$\text{total} = \text{kinetic} + \text{potential}$$

$$E = E_K + E_P = \frac{1}{2}\rho a^2 g \qquad (12.23)$$

Note that the root mean square amplitude is $\sqrt{(a^2/2)}$, so

$$E = \rho g(\text{root mean square amplitude})^2 \qquad (12.24)$$

The energy per unit wavelength in the direction of the wave, per unit width of wavefront, is

$$E_\lambda = E\lambda = \frac{1}{2}\rho a^2 g\lambda \qquad (12.25)$$

From (12.28) $\lambda = 2\pi g/\omega^2$, so

$$E_\lambda = \pi\rho a^2 g^2/\omega^2 \qquad (12.26)$$

Or, since $T = 2\pi/\omega$

$$E_\lambda = \frac{1}{4\pi}\rho a^2 g^2 T^2 \tag{12.27}$$

It is useful to show the kinetic, potential and total energies in these various forms, since all are variously used in the literature.

### *12.3.2 Power extraction from waves*

So far, we have calculated the total excess energy (kinetic plus potential) in a dynamic sea due to continuous wave motion in deep water. The energy is associated with water that remains at the same location when averaged over time. However, these calculations have told us nothing about the transport of energy (the power) across vertical sections of the water.

Standard texts, e.g. Coulson and Jeffrey (1977), calculate this power from first principles by considering the pressures in the water and the resulting displacements. The applied mathematics required is rigorous and comprehensive, and of fundamental importance in fluid wave theory. We can extract the essence of the full analysis, which is simplified for deep water waves.

Consider an element or particle of water below the mean surface level (Figure 12.7). For a surface wave of amplitude $a$ and wave number $k$, the radius of particle motion below the surface is

$$r = a\,e^{kz} \tag{12.28}$$

The vertical displacement by (Figure 12.7(b)) from the average position is

$$\Delta z = r \sin \omega t = a\,e^{kz} \sin \omega t \tag{12.29}$$

The horizontal component of velocity $u_x$ is given by

$$u_x = r\omega \sin \omega t = \omega a\,e^{kz} \sin \omega t \tag{12.30}$$

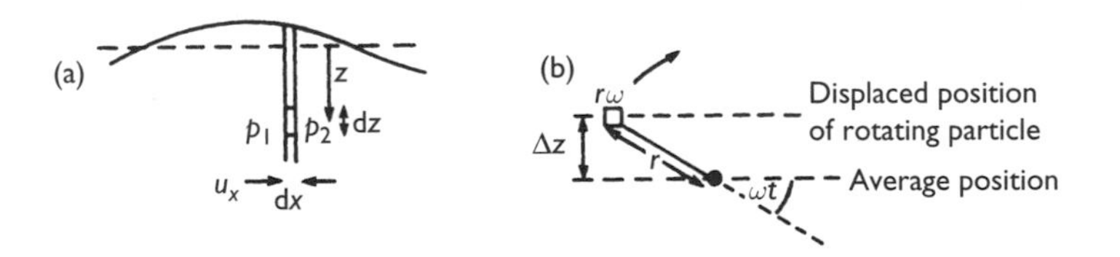


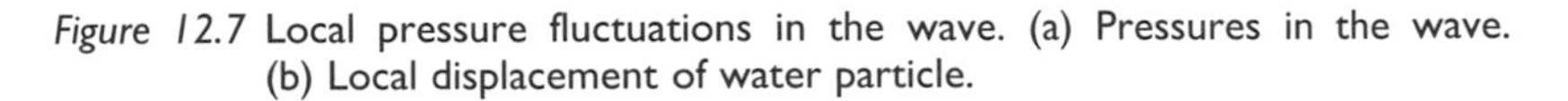
*Figure 12.7* Local pressure fluctuations in the wave. (a) Pressures in the wave. (b) Local displacement of water particle.

Therefore, from Figure 12.7(a), the power carried in the wave at $x$, per unit width of wave-front at any instant, is given by

$$P' = \int_{z=-\infty}^{z=0} (p_1 - p_2) u_x \, \mathrm{d}z \tag{12.31}$$

Where $p_1$ and $p_2$ are the local pressures experienced across the element of height d$z$ and unit width across the wavefront. Thus $(p_1 - p_2)$ is the pressure difference experienced by the element of width $\Delta y (=1\,\mathrm{m})$ in a horizontal direction. The only contribution to the energy flow that does not average to zero at a particular average depth in the water is associated with the change in potential energy of particles rotating in the circular paths; see Coulson and Jeffrey (1977). Therefore by conservation of energy

$$p_1 - p_2 = \rho g \Delta z \tag{12.32}$$

Substituting for $\Delta z$ from (12.29),

$$p_1 - p_2 = \rho g a \, e^{kz} \sin \omega t \tag{12.33}$$

In (12.31), and with (12.30) and (12.33),

$$\begin{aligned} P' &= \int_{z=-\infty}^{z=0} (\rho g a \, e^{kz} \sin \omega t)(\omega a \, e^{kz} \sin \omega t) \mathrm{d}z \\ &= \rho g a^2 \omega \int_{z=-\infty}^{z=0} e^{2kz} \sin^2 \omega t \, \mathrm{d}z \end{aligned} \tag{12.34}$$

The time average over many periods of $\sin^2 \omega t$ equals 1/2, so

$$P' = \frac{\rho g a^2 \omega}{2} \int_{z=-\infty}^{z=0} e^{2kz} \, \mathrm{d}z = \frac{\rho g a^2 \omega}{2} \frac{1}{2k} \tag{12.35}$$

The phase velocity of the wave is, from (12.7)

$$c = \frac{\omega}{k} = \frac{\lambda}{T} \tag{12.36}$$

So the power carried forward in the wave per unit width across the wave-front becomes

$$P' = \frac{\rho g a^2}{2} \frac{c}{2} = \frac{\rho g a^2 \lambda}{4T} \tag{12.37}$$

From (12.23) and (12.37) the power $P'$ equals the total energy (kinetic plus potential) $E$ in the wave per unit area of surface, times $c/2$. $c/2$ is called the group velocity of the deep water wave, i.e. the velocity at which the energy in the group of waves is carried forward. Thus, with the group velocity $u = c/2$,

$$P' = Eu = Ec/2 \tag{12.38}$$

where $E = \rho g a^2/2$.

From (12.8),

$$k = \omega^2/g \tag{12.39}$$

therefore, the phase velocity is

$$c = \frac{\omega}{k} = \frac{g}{\omega} = \frac{g}{(2\pi/T)} \tag{12.40}$$

This difference between the group velocity and the wave (phase) velocity is common to all waves where the velocity depends on the wavelength. Such waves are called dispersive waves and are well described in the literature, both descriptively, e.g. Barber (1969), and analytically, e.g. Lighthill (1978).

Substituting for $c$ from (12.11) into (12.37) gives

$$P' = \frac{\rho g a^2}{2}\frac{1}{2}\left(\frac{gT}{2\pi}\right)$$

So

$$P' = \frac{\rho g^2 a^2 T}{8\pi} \tag{12.41}$$

Therefore, the power in the wave increases directly as the square of the wave amplitude and directly as the period. The attraction of long period, large amplitude ocean swells to wave power engineers is apparent. This relationship is perhaps not obvious, and may be written in terms of wavelength using (12.9),

$$P' = \frac{\rho g^2 a^2}{8\pi}.\left(\frac{2\pi\lambda}{g}\right)^{\frac{1}{2}} \tag{12.42}$$

*Example 12.2*

What is the power in a deep water wave of wavelength 100 m and amplitude 1.5 m?

*Solution*

From (12.12) for Example 12.1, $c = 12.5\,\mathrm{m\,s^{-1}}$. With (12.38),

$$u = c/2 = 6.5\,\mathrm{m\,s^{-1}}$$

where $u$ is the group velocity of the energy and $c$ is the phase velocity.

The sea water waves have an amplitude $a = 1.5$ m ($H = 3$ m); realistic for Atlantic waves, so in (12.37)

$$P' = \frac{1}{2}(1025\,\mathrm{kg\,m^{-3}})(9.8\,\mathrm{m\,s^{-2}})(1.5\,\mathrm{m})^2(6.5\,\mathrm{m\,s^{-1}}) = 73\,\mathrm{kW\,m^{-1}}$$

Alternatively, $P'$ is obtained directly from 12.41b.

From Example 12.2, we can appreciate that there can be extremely large power densities available in the deep water waves of realistic ocean swells.

## 12.4 Wave patterns

Wave systems are not, in practice, the single sine wave patterns idealised in the previous sections. Very occasionally natural or contrived wave diffraction patterns, or channelled-waves, approach this condition, but normally a sea will be an irregular pattern of waves of varying period, direction and amplitude. Under the stimulus of a prevailing wind the wave trains may show a preferred direction, e.g. the south west to north east direction of Atlantic waves off the British Isles, and produce a significant long period sea 'swell'. Winds that are more erratic produce irregular water motion typical of shorter periods, called a 'sea'. At sea bottom depths ~30 m or less, significant focusing and directional effects can occur, however, possibly producing more regular or enhanced power waves at local sites. Wave power devices must therefore match a broad band of natural conditions, and be designed to extract the maximum power averaged over a considerable time for each particular deployment position. In designing these devices, it will be first necessary to understand the wave patterns of the particular site that may arise over a 50-year period.

The height of waves at one position was traditionally monitored on a wave-height analogue recorder. Separate measurements and analysis are needed to obtain the direction of the waves. Figure 12.8 gives a simulated trace of such a recorder. A crest occurs whenever the vertical motion changes

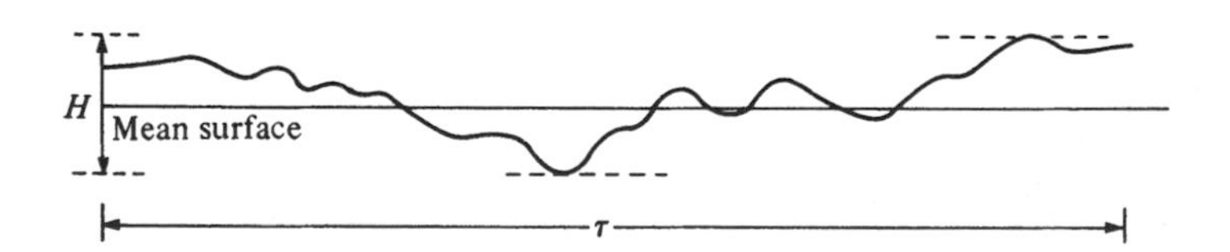


*Figure 12.8* Simulated wave height record at one position (with an exaggerated set of crests to explain terminology).

from upwards to downwards, and vice versa for a trough. Modern recorders use digital methods for computer-based analysis of large quantities of data. If $H$ is the height difference between a crest and its succeeding trough, there are various methods of deriving representative values, as defined in the following.

The basic variables measured over long intervals of time are:

1. $N_c$, the number of crests; in Figure 12.8 there are 10 crests.
2. $H_{1/3}$, the 'one-third' significant wave height. This is the average height of the highest one-third of waves as measured between a crest and subsequent trough. Thus $H_{1/3}$ is the average of the $N_c/3$ highest values of $H$.
3. $H_s$, the 'true' significant wave height. $H_s$ is defined as

$$H_s = 4a_{rms} = 4\left[\left(\sum_{i=1}^{n} h^2\right)/n\right]^{\frac{1}{2}} \tag{12.43}$$

   where $a_{rms}$ is the root mean square displacement of the water surface from the mean position, as calculated from $n$ measurements at equal time intervals. Care has to be taken to avoid sampling errors, by recording at a frequency at least twice that of the highest wave frequency present.
4. $H_{max}$ is the measured or most probable maximum height of a wave. Over 50 years $H_{max}$ may equal 50 times $H_s$ and so this necessitates considerable overdesign for structures in the sea.
5. $T_z$, the mean zero crossing period is the duration of the record divided by the number of upward crossings of the mean water level. In Figure 12.8, $T_z = \tau/3$.
6. $T_c$, the mean crest period, is the duration of the record divided by the number $N$ of crests. In Figure 12.8, $T_c = \tau/10$; in practice $N$ is very large, so reducing the error in $T_c$.
7. The spectral width parameter $\varepsilon$ gives a measure of the variation in wave pattern:

$$\varepsilon^2 = 1 - (T_c/T_z)^2 \tag{12.44}$$

   For a uniform single frequency motion, $T_c = T_z$, so $\varepsilon = 0$. In our example $\varepsilon = [1 - (0.3)^2]^{1/2} = 0.9$, implying a mix of many frequencies. The full information is displayed by Fourier transformation to a frequency spectrum, e.g. Figure 12.9.

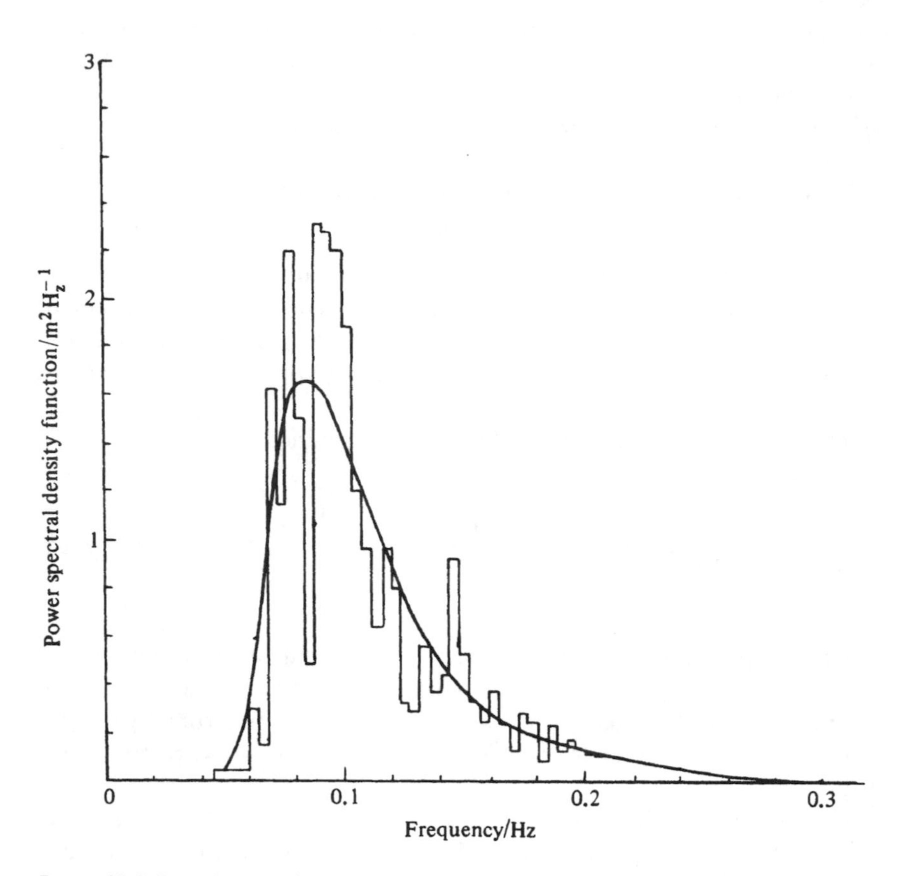


*Figure 12.9* Distribution of power per frequency interval in a typical Atlantic deep water wave pattern (Shaw 1982). The smoothed spectrum is used to find $T_e$, the energy period.

From (12.41) the power per unit width of wave-front in a pure sinusoidal deep water wave is

$$P' = \frac{\rho g^2 a^2 T}{8\pi} = \frac{\rho g^2 H^2 T}{32\pi} \tag{12.45}$$

where the trough to crest height is $H = 2a$. The root mean square (rms) wave displacement for a pure sinusoidal wave is $a_{max} = a/\sqrt{2}$, so in (12.45)

$$P' = \frac{\rho g^2 a_{rms}^2 T}{4\pi} \tag{12.46}$$

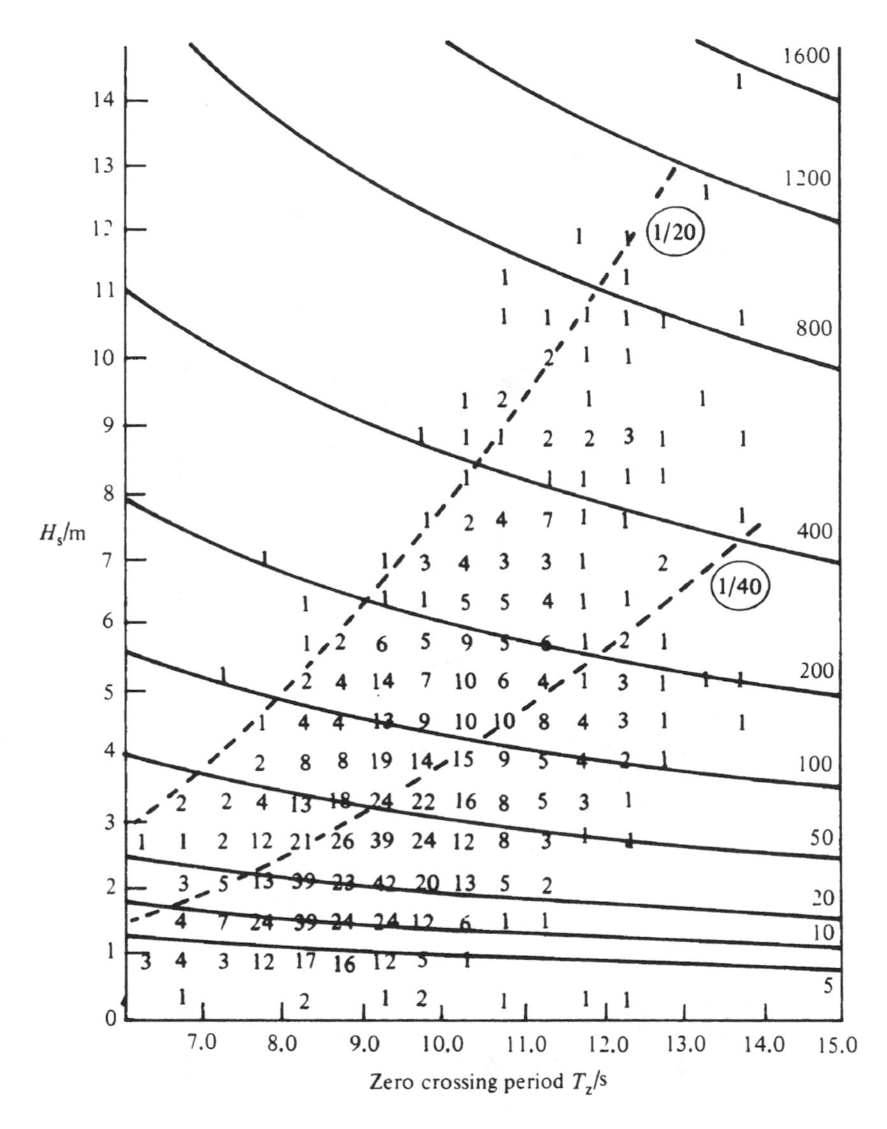


*Figure 12.10* Scatter diagram of significant wave height $H_s$ against zero crossing period $T_z$. The numbers on the graph denote the average number of occurrences of each $H_s$, $T_z$ in each 1000 measurements made over one year. The most frequent occurrences are at $H_s \sim 3$ m, $T_z \sim 9$ s, but note that maximum likely power occurs at longer periods.

- - - - - these waves have equal maximum gradient or slope

(1/n) the maximum gradient of such waves, e.g. 1 in 20

______ lines of constant wave power, kW m$^{-1}$

Data for 58°N19°W in the mid-Atlantic. After Glendenning (1977).

*Figure 12.11* Average annual wave energy (MWh $m^{-1}$) in certain sea areas of the world. After NEL (1976).

In practice, sea waves are certainly not continuous single frequency sine waves. The power per unit width of wavefront is therefore written in the form of (12.46) as

$$P' = \frac{\rho g^2 H_s^2 T_e}{64\pi} \quad (12.47)$$

Here $H_s$ is the significant wave height defined by (12.43), and $T_e$, called the 'energy period', is the period of the dominant power oscillations given by the peak in the power spectrum, see Figure 12.9. For many seas

$$T_e \approx 1.12 T_z \quad (12.48)$$

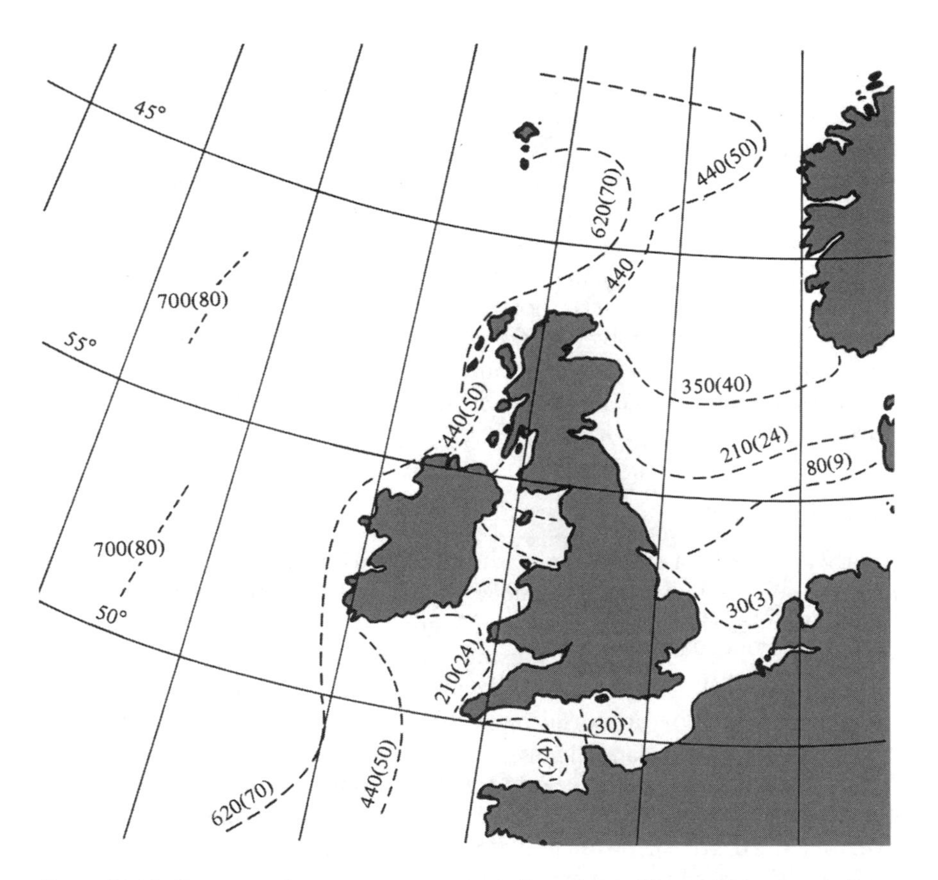


*Figure 12.12* Contours of average wave energy off north-west Europe. Numbers indicate annual energy in the unit of MWh, and power intensity (bracketed) in the unit of $kW\,m^{-1}$. Note that local effects are not indicated.

Until modern developments in wave power, only an approximate value of $P'$ could be obtained from analogue recording wave meters such that

$$P' \approx \frac{\rho g^2 H_{1/3}^2 T_z}{64\pi} \approx (490\,\mathrm{W\,m^{-1}\,m^{-2}\,s^{-1}}) H_{1/3}^2 T_z \tag{12.49}$$

However, with modern equipment and computer analysis, more sophisticated methods can be used to calculate (1) $a_{rms}$ and hence $H_s$ and (2), $T_z$ or $T_e$. Thus

$$\begin{aligned} P' &= (490\,\mathrm{W\,m^{-1}\,m^{-2}\,s^{-1}}) H_s^2 T_e \\ &= (550\,\mathrm{W\,m^{-3}\,s^{-1}}) H_s^2 T_z \end{aligned} \tag{12.50}$$

Since a wave pattern is not usually composed of waves all progressing in the same direction, the power received by a directional device will be significantly reduced.

Wave pattern data are recorded and tabulated in detail from standard meteorological sea stations. Perhaps the most important graph for any site is the wave scatter diagram over a year, e.g. Figure 12.10. This records the number of occurrences of wave measurements of particular ranges of significant wave height and zero-crossing period. Assuming the period is related to the wavelength by (12.9) it is possible to also plot on the diagram lines of constant wave height to wavelength. Contours of equal number of occurrences per year are also drawn.

From the wave data, it is possible to calculate the maximum, mean, minimum etc. of the power in the waves, which then can be plotted on maps for long-term annual or monthly averages. See Figures 12.11 and 12.12 for annual average power intensities across the world and north-west European.

## 12.5 Devices

As a wave passes a stationary position the surface changes height, water near the surface moves as it changes kinetic and potential energy, and the pressure under the surface changes. A great variety of devices have been suggested for extracting energy using one or more of these variations as input to the device. Included are devices that catch water at the crest of the waves, and allow it to run back into the mean level or troughs after extracting potential energy.

The Engineering Committee on Oceanic Resources (2003) described over 40 devices that have reached a stage of 'advanced development'. Of these over one-third are, or have been, 'operational', but only as one-off pilot projects. We describe here a representative sample, classified by their general principles. Wave-power has yet to reach the stage of widespread deployment of commercial devices that wind-power has attained.

### 12.5.1 Wave capture systems

These schemes are probably the simplest conceptually. They develop from a phenomenon often observed in natural lagoons. Waves break over a sea wall (equivalent to a natural reef) and water is impounded at a height above the mean sea level. This water may then return to the sea through a conventional low head hydroelectric generator. The system thus resembles a tidal range power system, Figure 13.7, but with a more continual and less regular inflow of water.

Figure 12.13 is a schematic diagram of the 350 kW *Tapchan* system demonstrated successfully in Norway in 1985. In this particular design, the waves were funnelled in through a tapered channel, whose concrete walls reached 2–3 m above mean sea level. This allows bigger waves to overtop the wall early, while smaller waves increase in height as they go up the channel, so that most of them also overtop the channel walls and supply water to the reservoir. Most of the engineering work was built into a natural gully in the rockface – a feature which enabled the system to withstand several storms over its 5 years of operation, one of which destroyed another less robust wave power device nearby.

A site for such a system needs to have the following features:

- Persistent waves with large average wave energy
- Deep water close to shore so the oncoming waves are not dissipated

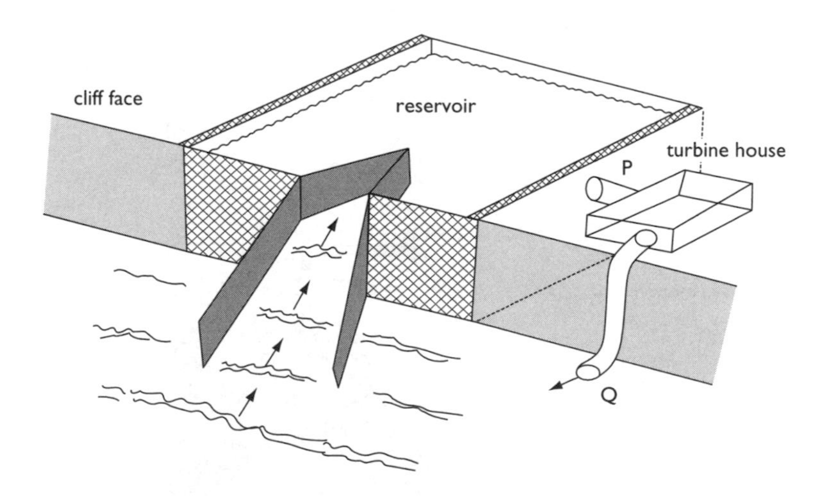


*Figure 12.13* Schematic diagram of the *Tapchan* wave capture system built in Norway (see text). Waves flow over the top of the tapered channel into the reservoir. Water flows from the pipe P near the top of the reservoir, though the conventional small-head hydro turbines, see Section 8.5, and then out to the sea at Q.

- A small tidal range (<1 m)
- A convenient and cheap means of constructing the reservoir, e.g. suitable local natural features.

### 12.5.2 Oscillating water column (OWC)

When a wave passes on to a partially submerged cavity open under the water, Figure 12.14, a column of water oscillates up and down in the cavity. This can induce an oscillatory motion in the air above the column, which may be connected to the atmosphere through an air turbine. Electrical power is usually derived from the oscillating airstream using a Wells turbine; such turbines, once started, turn in the same direction to extract power from air flowing in either axial direction, i.e. the turbine motion is independent of the fluid direction – see Problem 12.4 and Figure 12.18.

The first device of this kind has been developed by Professor Trevor Whittaker and his team of Queens University Belfast and operated without damage on the Scottish island of Islay for several years, but at less than expected power output. Based on that experience, a larger 500 kW device was installed on Islay in 2000, using robust construction techniques adaptable for other sites; named the '*Limpet*', after shellfish renowned for their firm attachment to rocks, it is contributing much of the island's power. The Limpet is one of the three initial wave power devices in Scotland for commercial operation; electricity is exported to the utility grid within the Scottish Administration's renewables obligation programme, see Chapter 17.

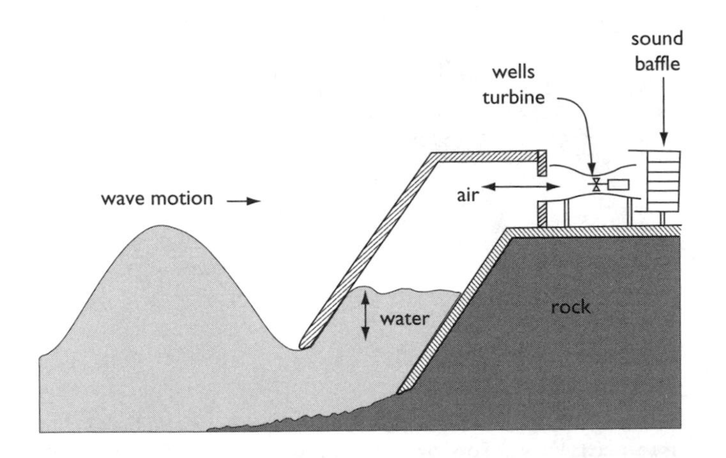


*Figure 12.14* Schematic diagram of an on-shore wave power system using an oscillating water column. Based on the LIMPET device operational on the island of Islay, west of Scotland, for grid connected electricity generation.

An advantage of using an oscillating water column for power extraction is that the air speed is increased by smooth reduction in the cross-sectional area of the channel approaching the turbine. This couples the slow motion of the waves to the fast rotation of the turbine without mechanical gearing. Another advantage is that the electrical generator is displaced significantly from the column of saline water. The air cavity's shape and size determine its frequency response, with each form and size of cavity responding best to waves of a particular frequency. In principle, the system efficiency is considerably improved if such devices have active tuning to the wide range of sea-wave frequencies encountered by using (a) multiple cavities, or (b) detectors of the incoming waves that feed-forward information for the cavity shape to be changed and the pitch of the turbine to be adjusted.

A majority of second generation wave-power devices have been shoreline OWC devices, broadly similar to the *Limpet*. However, the OWC mechanism is also used in offshore devices sitting on the sea bed, such as the *Osprey* (which operates in the near-shore zone off the islands of Orkney, northern Scotland) or in floating devices such as the Japanese *Whale*, Figure 12.15, and the Masuda wave-powered navigation buoys.

### 12.5.3 Wave profile devices

This class of devices float on or near the sea surface and move in response to the shape of the wave, rather than just the vertical displacement of water. Ingenious design is needed to extract useable power from the motion.

The *Pelamis*, Figure 12.16, is a semi-submerged articulated structure, which sits as a 'snake' aligned approximately head-on to the oncoming waves. The device consists of cylindrical sections linked by hinged joints, and wiggles like a snake in both the vertical and horizontal directions as

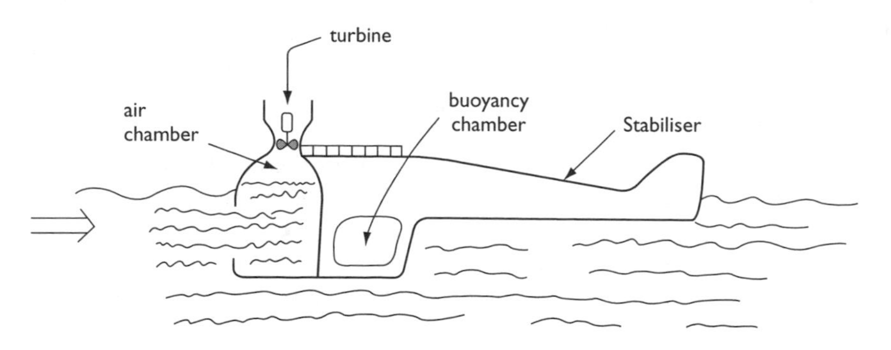


*Figure 12.15* The *Whale* device – a prototype floating wave-power system deployed off Gokasho Bay, Japan in 1998. The device is 50 m long and has 120 $kW_e$ generating capacity [after 'Japan's Marine Science and Technology Center (JAMSTEC)'].

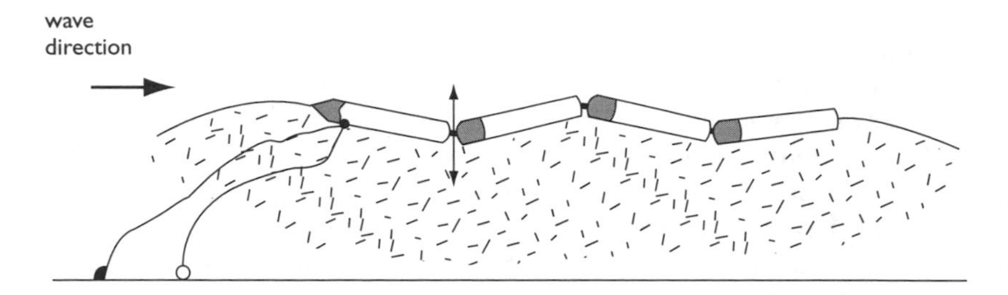


*Figure 12.16* Sketch of the *Pelamis* wave-power device, as seen from the side (not to scale in the vertical direction). Motion at the hinges produces hydraulic power fed to electrical generators. Note that the motion is not purely vertical; the device also 'wriggles' from side to side. The device is loosely moored to the sea floor by the mooring ropes as indicated. See <www.oceanpd.com/Pelamis/>.

a wave goes past. The wave-induced motion of its joints is resisted by hydraulic rams, which pump high-pressure oil through hydraulic motors via smoothing accumulators. The hydraulic motors drive electrical generators at each joint to produce electricity. Electrical power from all the joints in parallel is fed down a single umbilical cable to an electricity grid network connection on the sea bed. Several devices can be so connected in an array or 'farm', which itself is then linked to shore through a seabed cable.

Each *Pelamis* is held in position by a mooring system which maintains enough restraint to keep the device positioned but allows the machine to swing head-on to oncoming prevailing waves as the 'snake' spans successive wave crests. A 750 $kW_e$ prototype, 120 m long and 3.5 m in diameter, was installed in 2004 offshore of the main island of Orkney, northern Scotland. The length is such that it automatically 'detunes' from the longer-wavelength high-power waves, in order to enhance its survivability in storms.

## 12.6 Social and environmental aspects

Only about 10 to 15 devices were operational in sea conditions by 2005, so there is some uncertainty about their social economic and environmental aspects. Nevertheless, some generalisations can already be made on the basis of studies and pilot projects.

The grandiose schemes that were promoted by governments, especially in the UK, in response to the 'energy crisis' of the 1970s have faded from view as such, but did promote much useful research and marine data collection. The emphasis now is to obtain small to medium scale working systems that have proven operational experience and that, from the onset, export power into utility grid networks for commercial income. Development is from laboratory and simulation models to prototype installation, typically of 100–1000 $kW_e$ capacity and thence to 'farms' of multiple devices. Of importance is future interconnection and shared management with offshore wind farms.

From the experience of the initial plants, the projected cost of wave-power generated electricity power encourages optimism. For example the *Limpet* and *Pelamis* installations both accepted contracts to supply electricity for 15 years at less than 7p/kWh (≈US$0.15/kWh). It is reasonable to project that with greater deployment, which spreads development costs over multiple units, and with incremental engineering improvements from the pilot plants, these costs may halve within tens of years (an example of the 'learning curves' discussed in Section 17.6). The consequent predicted costs of ~3p/kWh would compare favourably with most alternatives for the isolated coastal and island communities, which offer the best initial opportunities for wave-power. Thereafter, electrical power connection will be to national power networks, possibly sharing sub-sea grids with offshore wind power.

Reliability and low operational costs are the most critical factors in achieving low average costs per kWh for systems which are capital intensive (see Chapter 17). This is particularly true for wave-power systems, which necessarily operate in vigorous sea conditions. If a system is destroyed by a storm in its first few years of operation, it will not pay its way, and power suppliers will not want to invest in further similar devices. Schemes therefore need to be designed for long lifetimes and with small number of moving parts to minimise failures. Fortunately, engineers can now draw on the experience of the offshore oil and wind industry to 'ruggedise' their designs and allow more confident installation and operation.

Onshore and near-shore wave-power devices, in general, are easier and cheaper to construct, maintain and connect to power networks than fully offshore devices. Yet offshore devices can tap into waves of greater power, with consequent larger and more sustained output, yet perhaps greater danger of malfunction.

Good efficiency requires the device to be matched to a wide frequency spectrum of waves, yet engineering for survivability may reduce this efficiency. Typically wave-power devices have average efficiencies ~30% and capacity factors also ~30%. (The capacity factor is the ratio of energy output over a period to that which would have been output if the device had operated throughout at its full rated power.)

As with other energy systems, the cost of energy from a wave-power system can be reduced if the cost of construction is shared with other benefits. For example, some systems can be integrated with conventional breakwaters. Others (particularly a string of floating devices like the *Whale*, aligned to face the waves) can in principle absorb energy across a wavefront to the extent that the water behind them is relatively calm; such wave-power systems act as breakwaters.

The main environmental benefit of wave-power systems is, as with all renewable energy, the mitigation of greenhouse gas emissions by substituting for fossil fuel use. Negative factors for offshore installation include hindrance to shipping and fishing. For near shore devices, acoustic noise may cause annoyance – the noise from some OWC devices has been likened to a rhinoceros giving birth.

## Problems

12.1 By considering elements of water lifted from depth $z$ below the mean sea level to a height $z$ above this level in a crest, show that the potential energy per unit length per unit width of wave front in the direction of the wave is

$$E_P = \frac{1}{4}\rho a^2 g$$

12.2 Figure 12.17(a) shows a conceptually simple device for extracting power from the horizontal movement of water in waves. A flat vane hinged about a horizontal axis at A (about $\lambda/8$ below the mean surface level) oscillates as indicated as waves impinge on it. Experiment indicates that such a device can extract about 40% of the energy in the incoming waves; about 25% of the energy is transmitted onwards (i.e. to water downstream of the vane) about 20% is reflected.

Salter (1974) designed the 'vane' shown in Figure 12.17(b) with a view to minimising these losses. It rotates about the central axis at O. Its stern is a half-cylinder (radius $a$) centred at O (lower dotted line continues the circular locus), but from the bottom point the shape changes into a surface which is another cylinder centred at O′, above O. This shape continues until it reaches an angle $\theta$ to the vertical, at which point it develops into a straight tangent which is continued to above the surface. For the case shown $OO' = 0.5a$ and $\theta = 15°$.

a By considering the movement of water particles that would occur in the wave in the absence of the device and relating this to the shape of the device, show that for wavelengths from $\sim 4a$ to $\sim 12a$ the device can absorb $\sim 70\%$ of the incoming energy.

b By 2004, this device, known as a *Duck* because of its bobbing motion, had undergone extensive laboratory and theoretical development. Figure 12.17(c) indicates how a full scale ($a \sim 8$ m) system might look in cross-section. The outer body moves (oscillates) relative to the inner cylinder. Suggest and justify (i) a way in which the inner cylinder could be made into a sufficiently stable reference point (ii) a way in which the irregular oscillatory motion could be harnessed into useable energy for distribution to the shore.

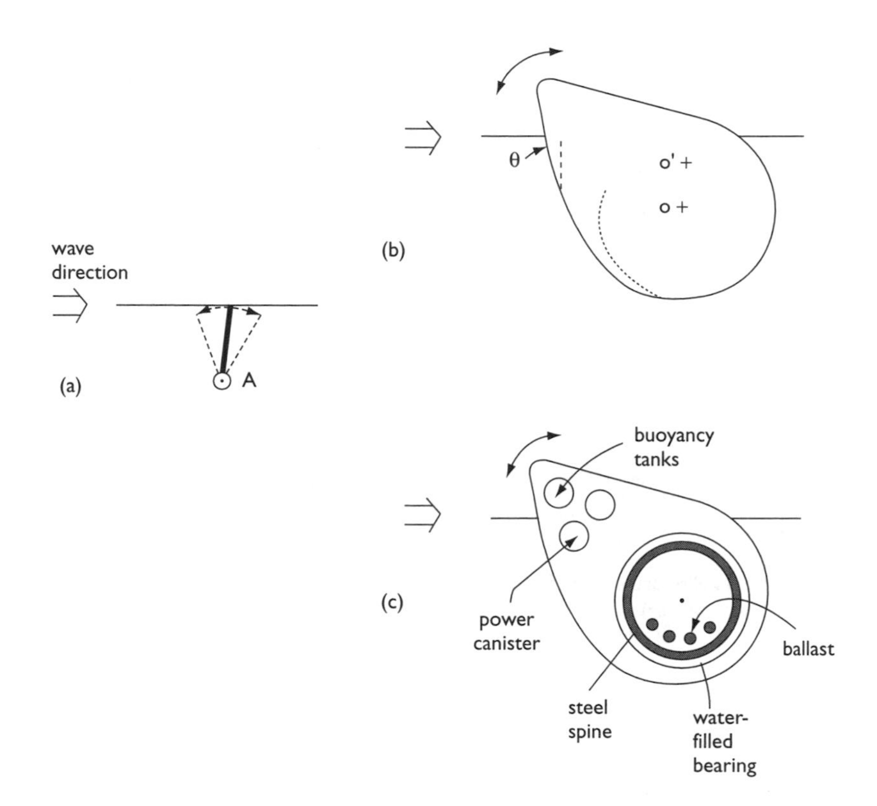


*Figure 12.17* (a) A vane oscillates as waves impinge on it from the left. (b) A more efficient 'vane' (Salter's *duck*) designed to extract more energy from the waves. See Problem 12.2 concerning its geometry. (c) A scheme for extracting energy from a full-scale *duck* (~10 m diameter).

12.3 Figure 12.18(a) is a perspective sketch of a *Wells turbine*; Figure 12.18(b) shows (schematically) a cross-section of its symmetrical blade and its movement as seen by a fixed observer. By drawing and analysing a blade diagram in the frame rotating with the turbine (see Figure 9.14) show that it is possible for the airflow to generate a net forward force on the blade if the lift and drag forces are of suitable magnitude.

12.4 Figure 12.19 shows a wave power device developed by Energetech (Australia). It features a parabolic wall of width ~50 m mounted in front of a breakwater. At the focus of the parabola is a 'tower' containing an oscillating water column, which is connected to an air turbine and generator mounted on the breakwater. The purpose of the parabola is to increase the amplitude of the vertical water movement in the OWC.

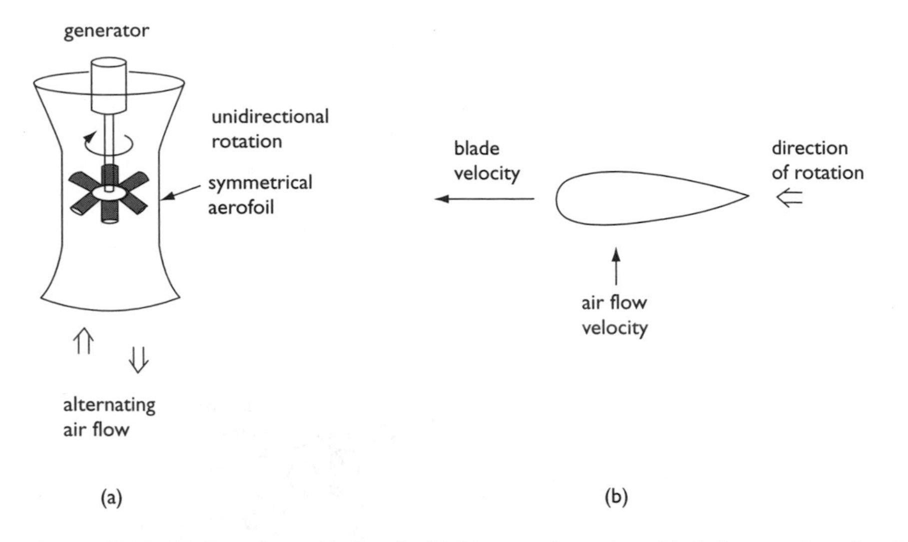


*Figure 12.18* Wells turbine. (a) Sketch. (b) Motion of a turbine blade (as seen by a fixed observer).

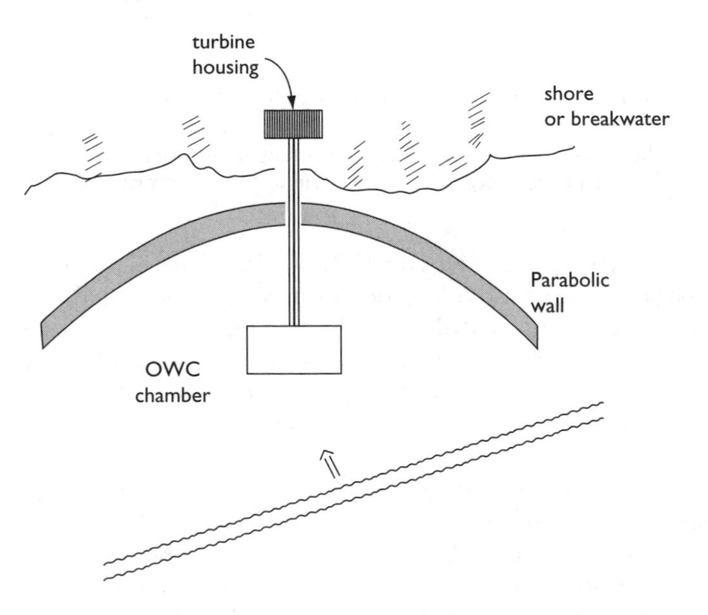


*Figure 12.19* Oscillating water column device with parabolic focussing, seen from above (after Energetech <www.energetech.com.au>.

What amplification might expected from a 50 m parabola for wavelengths of 25 m, 50 m, 250 m? Discuss the merits of using a bigger parabola (say 200 m span). Are there any natural geometries which might have a similar effect?

## Bibliography

### General

Barber, N.F. (1969) *Water Waves*, Wykecham, London.

Coulson, C.A. and Jeffrey, A. (1977) *Waves*, Longman, London. {An excellent theoretical text, partly considering water waves.}

Energy Technology Support Unit (1992) *Wave Energy Review*, ETSU, AEA Harwell, UK. Energy Technology Support Unit (1999) *A Brief Overview of Wave Energy*, ETSU, AEA Harwell, UK. {Continues the saga of UK wave power development.}

Engineering Committee on Oceanic Resources (2003) *Wave Energy Conversion*, Elsevier. {Excellent survey of state of the art with supporting theory and good descriptions of current installations and prototypes.}

Falnes, J. and Loveset, J. (1991) Ocean wave energy, *Energy Policy*, **19**, 768–775

Lighthill, M.J. (1978) *Waves in Fluids*, Cambridge University Press. {Advanced text with clear physical description.}

Ross, D. (1995) *Power from the Waves*, Oxford University Press. {Journalistic account of the history of wave energy, especially including the machinations of wave-energy politics in the UK. Incorporates and extends the author's earlier book 'Energy from the Waves' (1979), Pergamon.}

Salter, S.H. (1974) Wave power, *Nature*, **249**, 720–4. {Now seen as a classic paper for wave power. Later papers deal with the 'Salter Duck' developments.}

### Specific references

Glendenning, I. (1977) Energy from the sea, *Chemistry and Industry*, July, 592–599.

NEL (1976) *The Development of Wave Power – A Techno-economic Study*, by Leishman, J. M. and Scobie, G. of the National Engineering Laboratory, East Kilbride, Glasgow, Report EAU M25.

Shaw, R. (1982) *Wave Energy – A Design Challenge*, Ellis Horwood, Chichester, and Halsted Press, New York.

### Journals and websites

Wave-power development is published in a range of engineering and marine science journals. In addition most analysis is reported in conferences and specialist seminars. Particularly useful are the series of European Wave Energy Conferences, for which proceedings are usually published, e.g. Third European Wave Energy Conference, held at Patras, Greece, Sept. 1998, edited by W. Durstoff, published by the University of Hannover.

Commercial activity is being encouraged within the ambit of the British Wind Energy Association (BWEA)<www.bwea.com/marine.> since this is of interest to companies involved in offshore wind power.

The websites of device developers are often informative, e.g. Ocean Power Delivery (re the Pelamis) www.oceanpd.com Wavegen (re the Limpet) www.wavegen.co.uk

For more of an overview see the European Union energy site http://europa.eu.int/comm/energy_transport/atlas/htmlu/wave.html

Chapter 13

# Tidal power

## 13.1 Introduction

The level of water in the large oceans of the Earth rises and falls according to predictable patterns. The main periods $\tau$ of these tides are diurnal at about 24 h and semidiurnal at about 12 h 25 min. The change in height between successive high and low tides is the range, $R$. This varies between about 0.5 m in general and about 10 m at particular sites near continental land masses. The movement of the water produces tidal currents, which may reach speeds of $\sim 5\,\mathrm{m\,s^{-1}}$ in coastal and inter-island channels.

The seawater can be trapped at high tide in an estuarine basin of area $A$ behind a dam or barrier to produce tidal *range* power. If the water of density $\rho$ runs out through turbines at low tide, it is shown in Section 13.5 that the average power produced is

$$\bar{P} = \rho A R^2 g/(2\tau).$$

For example, if $A = 10\,\mathrm{km}^2$, $R = 4\,\mathrm{m}$, $\tau = 12\,\mathrm{h}$ 25 min, then $\overline{P} = 17\,\mathrm{MW}$. Obviously sites of large range give the greatest potential for tidal power, but other vital factors are the need for the power, and the costs and secondary benefits of the construction. The civil engineering costs charged to a tidal range power scheme could be reduced if other benefits are included. Examples are the construction of roads on dams, flood control, irrigation improvement, pumped water catchments for energy storage and navigation or shipping benefits. Thus the development of tidal power is very site-specific.

The power of tidal *currents* may be harnessed in a manner similar to wind power; this is also called '*tidal stream power*'. It is shown in Section 13.4 that the average power per unit area $\overline{q}$ in a current of maximum speed $u_0$ is $\overline{q} \sim 0.1\rho u_0^3$. For $u_0 = 3\,\mathrm{m\,s^{-1}}$, $\overline{q} \sim 14\,\mathrm{kW\,m^{-2}}$. In practice, tidal current is likely to be attractive for power generation only where it is enhanced in speed by water movement in straights between islands and mainland, or between relatively large islands. Therefore the opportunities for viable commercial sites are unusual. Where it is possible however, much of the discussion concerning the use of the power is in common with tidal range power. We may also note that the flow power in a river has similar characteristics, but without the temporal variation. Harnessing river-stream power in the same manner as tidal stream power is certainly possible, but seldom considered.

The harnessing of *tidal range power* (henceforward called 'tidal power' as distinct from 'tidal current/stream power') has been used for small mechanical power devices, e.g. in medieval England and in China. The best-known large-scale electricity generating system is the 240 $MW_e$ 'La Rance' system at an estuary into the Gulf of St Malo in Brittany, France, which has operated reliably since 1967, thereby proving the technical feasibility of this technology at large scale. However, economic and environmental constraints have meant that very few similar systems have been constructed since; see Section 13.6. Other sites with large tidal range, such as the Severn estuary in England and the Bay of Fundy on the eastern boundary between Canada and the United States, have been the subject of numerous feasibility studies over the past hundred years.

The range, flow and periodic behaviour of tides at most coastal regions are well documented and analysed because of the demands of navigation and oceanography. The behaviour may be predicted accurately, within an uncertainty of less than ±4%, and so tidal power presents a very reliable and assured form of renewable power. The major drawbacks are:

1 The mismatch of the principal lunar driven periods of 12 h 25 min and 24 h 50 min with the human (solar) period of 24 h, so that optimum tidal power generation is not in phase with demand.
2 The changing tidal range and flow over a two-week period, producing changing power production.
3 The requirement for large water volume flow at low head, necessitating many specially constructed turbines set in parallel.
4 The very large capital costs of most potential installations.
5 The location of sites with large range may be distant from the demand for power.
6 Potential ecological harm and disruption to extensive estuaries or marine regions.

For optimum electrical power generation from tides, the turbines should be operated in a regular and repeatable manner. The mode of operation will depend on the scale of the power plant, the demand and the availability of other sources. Very many variations are possible, but certain generalisations apply:

a If the tidal-generated electricity is for local use, then other assured power supplies must exist when the tidal power is unavailable. However, the tidal basin provides energy storage, so extending power generation times and being available for storage from other power sources.

b If the generated electricity can feed into a large grid and so form a proportionately minor source within a national system, then the predictable tidal power variations can be submerged into the national demand.

c If the immediate demand is not fixed to the human (solar) period of 24 h, then the tidal power can be used whenever available. For example, if the electrical power is for transport by charging batteries or by electrolysing water for hydrogen, then such a decoupling of supply and use can occur.

The following sections outline the physical background to tides and tidal power. Readers interested only in power-generating installations should turn directly to Sections 13.4 and 13.5. Section 13.6 briefly reviews the social and environmental aspects of the technology.

## 13.2 The cause of tides

The analysis of tidal behaviour has been developed by many notable mathematicians and applied physicists, including Newton, Airy, Laplace, George Darwin (son of Charles Darwin) and Kelvin. We shall use Newton's physical theory to explain the phenomena of tides. However, present day analysis and prediction depends on the mathematical method of harmonic analysis developed by Lord Kelvin in Glasgow. A complete physical understanding of tidal dynamics has not yet been attained owing to the topological complexity of the ocean basins.

The seas are liquids held on the solid surface of the rotating Earth by gravity. The gravitational attraction of the Earth with the Moon and the Sun perturbs these forces and motions so that tides are produced. Tidal power is derived from turbines set in this liquid, so harnessing the kinetic energy of the rotating Earth. Even if all the world's major tidal power sites were utilised, this would lead to an extra slowing of the Earth's rotation by no more than one day in 2000 years; this is not a significant extra effect.

### *13.2.1 The lunar induced tide*

The Moon and Earth revolve about each other in space (Figure 13.1), but since the mass of the Earth is nearly hundred times greater than the Moon's mass, the Moon's motion is more apparent. The centre of revolution is at O, such that

$$
\begin{aligned}
ML &= M'L' \\
L' &= MD/(M'+M) \qquad (13.1) \\
L' &= 4670\,\text{km}.
\end{aligned}
$$

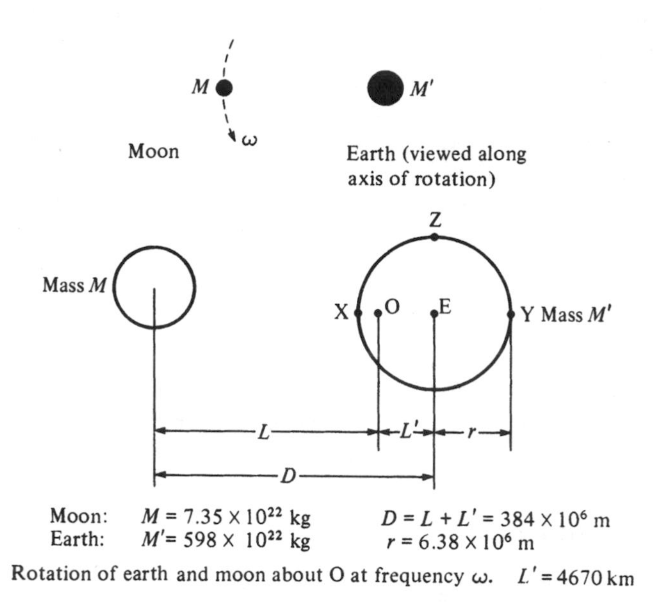


*Figure 13.1* Motion of the Moon and the Earth.

The Earth's mean radius is 6371 km, so the point of revolution O is *inside* the surface of the Earth.

A balance of gravitational attraction and centrifugal force maintains the Earth–Moon separation. If the gravitational constant is $G$,

$$\frac{GMM'}{D^2} = ML\omega^2 = M'L'\omega^2 \tag{13.2}$$

If all the mass of the Earth could be located at the centre of the Earth E, then each element of mass would be at the equilibrium position with respect to the Moon. However, the mass of the Earth is not all at one point, and so is not all in this equilibrium. Material furthest from the Moon at Y (see Figure 13.1), experiences an increased outward centrifugal force with distance of rotation $(r + L')$ and a decreased gravitational force from the Moon. Material nearest the Moon at X has an increased gravitational force towards the Moon, plus the centrifugal force, also towards the Moon but reduced, because of the reduced rotation distance $(r - L')$. The solid material of the Earth experiences these changing forces as the Moon revolves, but is held with only small deformation by the structural forces of the solid state. Liquid on the surface is, however, free to move, and it is this movement relative to the Earth's surface that causes the tides. If the Moon is in the equatorial plane of the Earth, the water of the open seas attempts to heap together to form peaks at points X and Y, closest to and furthest from the

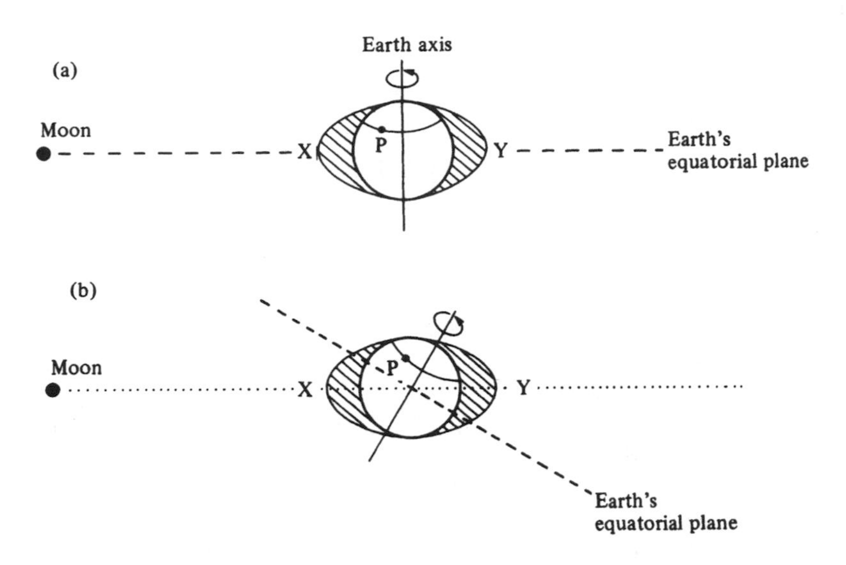


*Figure 13.2* Basic physical explanation of the semidiurnal and diurnal tide. (a) Simple theory of equilibrium tide with the Moon in the plane of the Earth's equator, P experiences two equal tides each day (semidiurnal tide). (b) Normally the Moon is not in the Earth's equatorial plane, and so P may experience only one tide each day (diurnal tide).

Moon. The solid Earth would rotate with a period of one day underneath these two peaks (Figure 13.2(a)). Thus with no other effect occurring, each sea-covered position of the Earth would experience two rises and two falls of the water level as the Earth turns through the two peaks. This is the semidiurnal (half daily) tide. Note that the daily rotation of the Earth on its own axis has no first order effect, as such, in producing tidal range.

We may estimate the resultant force causing the tides from (i) the centrifugal force about O at the lunar frequency $\omega$, and (ii) the force of lunar gravitational attraction (see Figure 13.1). For a mass $m$ of water, furthest from the Moon at Y,

$$F_Y = m(L' + r)\omega^2 - \frac{GMm}{(D+r)^2} \tag{13.3}$$

Nearest to the Moon at X

$$F_x = \frac{GMm}{(D-r)^2} + m(r - L')\omega^2 \tag{13.4}$$

At position E (see Figure 13.1), by definition of $L'$,

$$\frac{GMm}{D^2} = mL'\omega^2 \tag{13.5}$$

However, since $r \ll D$,

$$\frac{1}{(D \pm r)^2} = \frac{1}{D^2}\left(1 \pm \frac{2r}{D}\right)$$

So by substituting (13.5) in (13.3) and (13.4),

$$F_X = F_Y = mr\omega^2\left(1 + \frac{2L'}{D}\right) \tag{13.6}$$

We therefore expect two lunar tidal ranges each day of equal amplitude. This does indeed happen in the large oceans when the Moon is in the equatorial plane of the Earth.

At low tide on this equilibrium tide model the lunar-related force is $mr\omega^2$, and so the tide-raising force within (13.6) is $mr\omega^2 2L'/D$. It can be shown, see Problem 13.1, that this would produce a maximum equilibrium tidal range of 0.36 m.

There are three principal reasons why actual tidal behaviour is different from this simplistic 'equilibrium tide' explanation:

1 The explanation of the forcing function producing the tides is physically correct, but the theory has failed to determine if the peaks of water can move at about 1600 km h$^{-1}$ to keep up with the Earth's rotation. In practice, the tidal wave cannot move fast enough to remain in the meridian of the Moon (see Problem 13.3). Thus as the Moon passes over an ocean, tidal motion is induced, which propagates at a speed of about 500 km/h and lags behind the Moon's position. The time of this lag is the 'age of the tide'. Each ocean basin tends to have its own system of moon-induced tides that are not in phase from one ocean to another.
2 The Moon is not usually in the equatorial plane of the Earth (Figure 13.2(b)), and so a diurnal component of the tide occurs. Other minor frequency components of the tidal motion occur. For instance, the Moon–Earth distance oscillates slightly, from a maximum of $4.06 \times 10^8$ m at apogee to a minimum of $3.63 \times 10^8$ m at perigee in a period of 27.55 solar days (the anomalistic month). Also the Moon's plane of motion moves about 2° in and out of the Earth–Sun ecliptic plane. (Unrelated to tidal forces, it is interesting to realise that the Moon rotates on its own axis in a period of 27.396 Earth days, and is locked to show one 'face' to the Earth. No place on the Moon is permanently dark, so solar systems would function everywhere, but, for the same production, would require more storage capacity than on Earth!)
3 A great many other complications occur, mostly associated with particular ocean basins. For instance, resonances occur across oceans and

especially near continental shelves, which produce distinct enhancements of the tidal range. We will show in Section 13.3 that these resonant enhancements are of great importance for tidal power installations.

### 13.2.2 Period of the lunar tides

To calculate the period of the tides more precisely, we have to be more precise about what we mean by a 'day' (Figure 13.3). At a point A on the Earth, a solar day is the interval between when the Sun crosses the meridional plane at A on a specified day and when it does so the subsequent day. This period actually varies through the year because of the irregularities in the Earth's orbit, and so the common unit of time, the *mean solar day* $t_S$, is defined to be the interval averaged over a whole year. Its value is defined as exactly 24 h, i.e.

$$t_S = 86\,400\,\text{s} \tag{13.7}$$

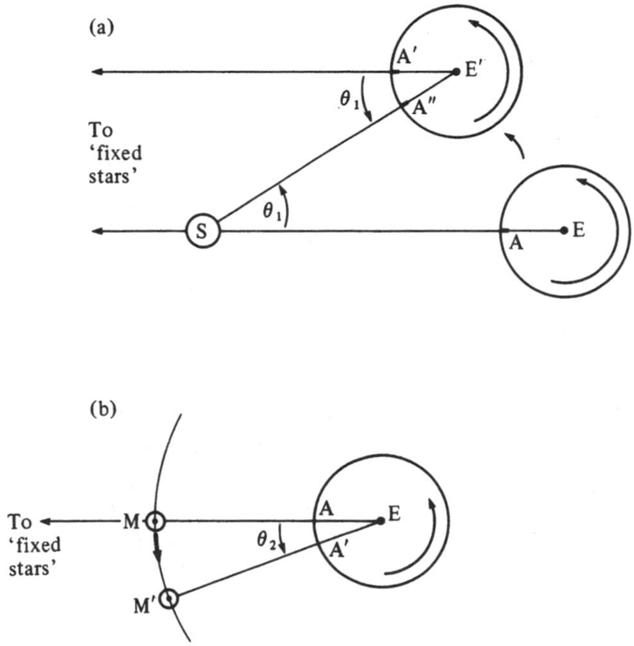

*Figure 13.3* Comparison of three different 'days' that can be observed from Earth. (a) Sidereal and solar day. (b) Sidereal and lunar. The solar day is 24 hours exactly by definition, the sidereal is slightly shorter and the lunar slightly longer. The diagrams are not to scale. See also Figure 4.4 for meridional plane.

The *sidereal day* $t^*$ is similarly defined to be the average interval between successive transits of a 'fixed star', i.e. one so distant that its apparent motion relative to the Earth is negligible. The sidereal day is therefore the 'true' period of rotation of the Earth, as seen by a distant observer.

Figure 13.3(a) shows how the difference between $t_S$ and $t^*$ is related to the revolution of the Earth around the Sun (period $T_S = 365.256t_S$). Suppose that at midday on a certain day the centre of the Earth E, point A on the Earth's surface, the Sun S and some fixed star are all aligned. One solar day later, the Sun, A and the Earth's centre are again aligned. In this time E has moved through an angle $\theta_1$ around the Sun to E′. Since $t_S$ is the mean solar day, for time-keeping purposes we can regard the Earth as moving uniformly around a circular orbit so that

$$\frac{\theta_1}{2\pi} = \frac{t_S}{T_S} \tag{13.8}$$

In this time A has rotated around E through an angle $(2\pi + \theta_1)$ to $A''$. Its time to rotate through an angle $2\pi$ (as seen by a distant observer) is just $t^*$, so that

$$\frac{\theta_1}{2\pi} = \frac{t_S - t^*}{t^*} \tag{13.9}$$

Equating (13.8) and (13.9) gives

$$\begin{aligned} t^* &= \frac{t_S}{1 + (t_S/T_S)} \\ &= 86\,164\,\text{s} \\ &= 23\,\text{h}\ 56\ \text{min}\ 4\,\text{s} \end{aligned} \tag{13.10}$$

Similarly the *mean lunar day* $t_M$ is defined to be the mean interval between successive alignments of E, A and the Moon's centre. Figure 13.3(b) shows the fictitious mean Moon M moving uniformly in a circular orbit around the Earth. In a time $t_M$, the Moon moves through an angle $\theta_2$ from M to M′, while A on the Earth rotates through $2\pi + \theta_2$. Thus as seen by a distant observer

$$\frac{\theta_2}{2\pi} = \frac{t_M}{T^*} = \frac{T_M - t^*}{t^*} \tag{13.11}$$

where $T^* = 27.32t_S$ (called the sidereal month, the 'true' lunar month) is the period of revolution of the Moon about the Earth's position as seen by a distant observer. This is shorter than the lunar month as recorded by an observer on Earth ($T_M = 29.53$ days) owing to the Earth moving around the Sun. Equation (13.11) implies that

$$t_M = \frac{t^*}{1-(t^*/T^*)} = 89\,428\,\text{s} = 24\,\text{h}\ 50\,\text{min}\ 28\ \text{s} \tag{13.12}$$

Such a period is called 'diurnal' because it is near to 24 h.

### 13.2.3 The solar induced tide and combined effects

The same Newtonian theory that explains the major aspects of the twice daily lunar tide can be applied to the Sun/Earth system. A further twice daily tide is induced with a period of exactly half the solar day of 24 h. Other aspects being equal, the range of the solar tide will be 2.2 times less than the range of the lunar tide, which therefore predominates. This follows from considering that the tidal range is proportional to the *difference* of the gravitational forces from the Moon and the Sun across the diameter $d$ of the Earth. If $M_M$ and $M_S$ are the masses of the Moon and the Sun at distances from the Earth of $D_M$ and $D_S$, then for either system:

$$\text{gravitational force} \propto \frac{M}{D^2}$$

$$\text{difference in force} \propto \frac{\partial F}{\partial D} d = \frac{-2Md}{D^3} \tag{13.13}$$

The range of the lunar tide $R_M$ and solar tide $R_S$ are proportional to the difference, so

$$\frac{R_M}{R_S} = \frac{(M_M/D_M^3)}{(M_S/D_S^3)} = \left(\frac{D_S}{D_M}\right)^3 \frac{M_M}{M_S} = \left(\frac{1.50\times10^{11}\,\text{m}}{3.84\times10^{8}\,\text{m}}\right)^3 \left(\frac{7.35\times10^{22}\,\text{kg}}{1.99\times10^{30}\,\text{kg}}\right) = 2.2 \tag{13.14}$$

The solar tide moves in and out of phase with the lunar tide. When the Sun, Earth and Moon are aligned in conjunction, the lunar and solar tides are in phase, so producing tides of maximum range. These are named '*spring tides*' of maximum range occurring twice per lunar (synodic) month at times of both full and new Moons (Figure 13.8(c)).

When the Sun/Earth and Moon/Earth directions are perpendicular (in quadrature) the ranges of the tides are least. These are named '*neap tides*' that again occur twice per synodic month. If the spring tide is considered to result from the sum of the lunar and solar tides, and the neap tide from their difference, then the ratio of spring to neap ranges might be expected to be

$$\frac{R_s(\text{spring})}{R_n(\text{neap})} = \frac{1+(1/2.2)}{1-(1/2.2)} = 3 \tag{13.15}$$

In practice, dynamical and local effects alter this rather naive model, and the ratio of spring to neap range is more frequently about 2. Spring tides at the Moon's perigee have greater range than spring tides at apogee, and a combination of effects including wind can occur to cause unusually high tides.

## 13.3 Enhancement of tides

The normal mid-ocean tidal range is less than one metre and of little use for power generation. However, near many estuaries and some other natural features, enhancement of the tidal range may occur by (i) funnelling of the tides (as with sound waves in an old-fashioned trumpet-shaped hearing aid), and (ii) by resonant coupling to natural frequencies of water movement in coastal contours and estuaries. This *local enhancement is essential* for tidal power potential; we stress this point most strongly.

The ordinary tidal movement of the sea has the form of a particular type of moving wave called a 'tidal wave'. The whole column of water from surface to sea bed moves at the same velocity in a tidal wave, and the wavelength is very long compared with the sea depth (Figure 13.4). (This is the same relative proportion as the so-called 'shallow' water waves, which may be a correct, but totally misleading, name given to tidal waves.) Motion of a continuously propagating natural tidal wave has a velocity $c$ related to the acceleration of gravity $g$ and the sea depth $h$ such that $c = \sqrt{(gh)}$, i.e. $\sim 750\,\text{km h}^{-1}$ across major oceans, which have depth $\sim 4000\,\text{m}$.

Underwater volcanic or earthquake activity can induce a freely propagating 'seismic sea wave' in deep oceans correctly called a *tsunami*, but

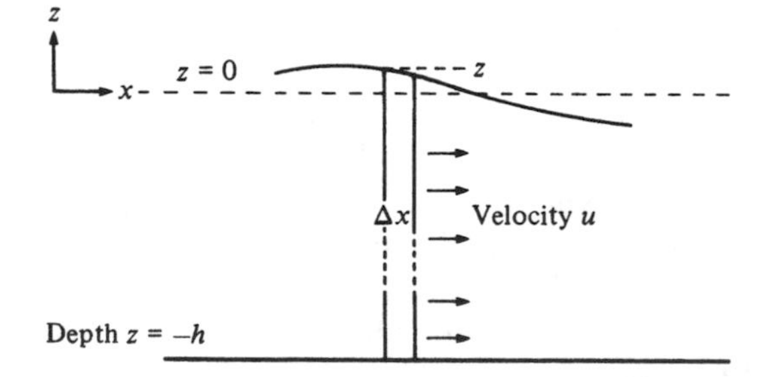


*Figure 13.4* Motion of water in a tidal wave; the elemental section of sea has thickness $\Delta x$, depth $h$ and width $b$ (along the $y$ axis).

sometimes incorrectly called a 'tidal wave' since there is no relationship to tides. A tsunami is initiated by a relatively localised, but extreme, sudden change in the height of the sea bottom, which injects an immense pulse of energy over a short relatively horizontal distance on the sea bed. The resulting 'shock' creates the physical equivalent of a 'shallow' depth wave (with $\lambda/4 >$ depth), where 'shallow' has to be interpreted as the ~4000 m sea depth, and wave movement encompasses the whole depth. The wave spreads rapidly at speed $c = \surd(gh)$ and wavelength $\lambda \approx 150\,\text{km}$. When the tsunami reaches the decreasing sea depth near shore, friction at the sea bed slows the wave and so shortens the wavelength, with the consequence of rapidly increased surface amplitude to perhaps 30 m. This amplitude will be apparent at the coast as perhaps an exceptional outflow of sea water followed quickly by huge and damaging breaking waves.

Considering the solar and the lunar forces, neither is in the form of a pulse, so no 'tsunami-like' behaviour occurs. The only possibility for enhanced motion is for the natural tidal motion to be in resonance with the solar and the lunar forces. But, as seen from Earth, the Sun moves overhead at ~2000 km h$^{-1}$ and the Moon at ~60 km h$^{-1}$, therefore, the tidal forcing motions for the lunar- and the solar-induced tides do not, in general, coincide with the requirements for a freely propagating tidal wave in the deep ocean, and so resonant enhancement of the forced motion does not occur in the open oceans.

In certain estuaries and bays, resonance can occur, however, and most noticeable changes in tidal motion therefore appear. We consider a slab of water of depth $h$, width $b$, thickness $\Delta x$ and surface level above the mean position $z$. The change in surface level over the thickness $\Delta x$ is $(\partial z/\partial x)\Delta x$, and this is small compared with $z$. The side area of the slab is $A = hb$.

The form of the wave is obtained by considering Newton's equation of motion for the slab, and the requirement for conservation of water mass. The pressure difference across each side of the slab arises from the small change in height of the surface:

$$\Delta p = -\rho g \left(\frac{\partial z}{\partial x}\right) \Delta x \tag{13.16}$$

So the equation of motion of the slab of velocity $u$ is

$$\text{force} = \text{mass} \times \text{acceleration}$$

$$\left[-\rho g \left(\frac{\partial z}{\partial x}\right) \Delta x\right] hb = \rho hb \Delta x \frac{\partial u}{\partial t} \tag{13.17}$$

$$\frac{\partial u}{\partial t} = -g \frac{\partial z}{\partial x} \tag{13.18}$$

The difference between the flow of water into and out of the slab must be accounted for by a change in volume $V$ of the slab with time, and this conservation of water mass leads to the equation of continuity

$$-\left[\frac{\partial}{\partial x}(A+bz)u\right]\Delta x=\frac{\partial V}{\partial t}=\frac{\partial[(A+bz)\Delta x]}{\partial t} \qquad (13.19)$$

Since $A=bh$ is constant and much larger than $bz$,

$$-A\frac{\partial u}{\partial x}=b\frac{\partial z}{\partial t}$$

$$\frac{\partial u}{\partial x}=-\frac{1}{h}\frac{\partial z}{\partial t} \qquad (13.20)$$

From (13.18) and (13.20),

$$\frac{\partial^2 u}{\partial t\partial x}=-g\frac{\partial^2 z}{\partial x^2}=\frac{\partial^2 u}{\partial x\partial t}=-\frac{1}{h}\frac{\partial^2 z}{\partial t^2} \qquad (13.21)$$

So

$$\frac{\partial^2 z}{\partial t^2}=gh\frac{\partial^2 z}{\partial x^2}=c^2\frac{\partial^2 z}{\partial x^2} \qquad (13.22)$$

This is the equation of a wave of speed $c$, with

$$c=\surd(gh) \qquad (13.23)$$

Resonant enhancement of the tides in estuaries and bays occurs in the same manner as the resonance of sound waves in open and closed pipes, for example as in Figure 13.5. Resonance with the open sea tide occurs when

$$L=\frac{j\lambda}{4},\ j \text{ an odd integer} \qquad (13.24)$$

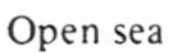


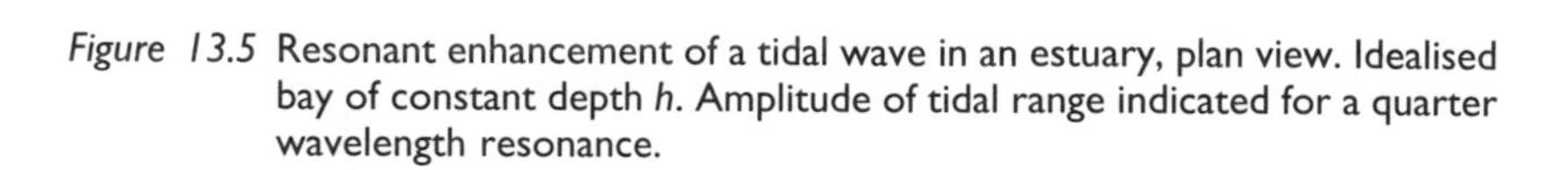

*Figure 13.5* Resonant enhancement of a tidal wave in an estuary, plan view. Idealised bay of constant depth $h$. Amplitude of tidal range indicated for a quarter wavelength resonance.

The natural frequency of the resonance $f_r$ and the period $T_r$ is given by

$$f_r = \frac{1}{T_r} = \frac{c}{\lambda} \tag{13.25}$$

So

$$T_r = \frac{\lambda}{c} = \frac{4L}{jc} = \frac{4L}{j\surd(gh)} \tag{13.26}$$

Resonance occurs when this natural period equals the forced period of the tides in the open sea $T_f$, in which case

$$T_f = \frac{4L}{j\surd(gh)}; \; \frac{L}{\surd h} = \frac{j}{4}\surd(g)T_f \tag{13.27}$$

The semidiurnal tidal period is about 12 h 25 min (45 000 s), so resonance for $j = 1$ occurs when

$$\frac{L}{\surd h} = \frac{45\,000\,\text{s}}{4}\surd(9.8\,\text{m}\,\text{s}^{-2}) = 36\,000\,\text{m}^{1/2} \tag{13.28}$$

Usually, if it occurs at all, such enhancement occurs in river estuaries and ocean bays, as in the Severn Estuary (Example 13.1). However, there is a small general enhancement for the whole Atlantic Ocean.

*Example 13.1 Resonance in the Severn Estuary*

The River Severn Estuary between Wales and England has a length of about ~200 km and depth of about 30 m, so

$$\frac{L}{\surd h} \approx \frac{200 \times 10^3\,\text{m}}{\surd(30\,\text{m})} \approx 36\,400\,\text{m}^{1/2} \tag{13.29}$$

As a result there is close matching of the estuary's resonance frequency with the normal tidal frequency, and large amplitude tidal motions of 10–14 m range occur.

In practice, estuaries and bays do not have the uniform dimensions implied in our calculations, and analysis is extremely complicated. It becomes necessary to model the conditions (i) in laboratory wave tanks using careful scaling techniques, and (ii) by theoretical analysis. One dominant consideration for tidal power installations is to discover how barriers and dams will affect the resonance enhancement. For the Severn estuary, some studies have concluded the barriers would reduce the tidal range and hence the power available: yet other studies have concluded the range will be increased! The construction of tidal power schemes is too expensive to allow for mistakes to occur in understanding these effects.

## 13.4 Tidal current/stream power

Near coastlines and between islands, tides may produce strong water currents that can be considered for generating power. This may be called tidal-current, tidal-stream or tidal-flow power. The total power produced may not be large, but generation at competitive prices for export to a utility grid or for local consumption may be possible.

The theory of tidal stream power is similar to wind power, see Chapter 9. The advantages are (a) predictable velocities of the fluid and hence predictable power generation, and (b) water density 1000 times greater than air and hence smaller scale turbines. The main disadvantages are (a) small fluid velocity and (b) the intrinsically difficult marine environment.
The power density in the water current is, from (9.2),

$$q = \frac{\rho u^3}{2} \tag{13.30}$$

For a tidal or river current of velocity, for example, $3\,\mathrm{m\,s^{-1}}$,

$$q = \frac{(1025\,\mathrm{kg\,m^{-3}})(27\,\mathrm{m^3\,s^{-3}})}{2} = 13.8\,\mathrm{kW\,m^{-2}}$$

Only a fraction $\eta$ of the power in the water current can be transferred to useful power and, as for wind, $\eta$ will not exceed about 60%. In practice, $\eta$ may approach a maximum of 40%.

Tidal current velocities vary with time approximately as

$$u = u_0 \sin(2\pi t/\tau) \tag{13.31}$$

where $\tau$ is the period of the natural tide, 12 h 25 min for a semidiurnal tide, and $u_0$ is the maximum speed of the current.

Generation of electrical power per unit cross section may therefore be on average (assuming 40% efficiency of tidal current power to electricity),

$$\bar{q} \approx \frac{0.4}{2}\rho u_0^3 \frac{\int_{t=0}^{t=\tau/4} \sin^3(2\pi t/\tau)\mathrm{d}t}{\int_{t=0}^{t/4} \mathrm{d}t} \tag{13.32}$$

$$= 0.2\rho u_0^3(\tau/3\pi)(4/\tau)$$

$$\approx 0.1\rho u_0^3 \tag{13.33}$$

for a device that could generate power in the *ebb* (out) and *flow* (in) tidal currents, and with a maximum current of $3\,\mathrm{m\,s^{-1}}$, $\bar{q} \sim 2.8\,\mathrm{kW\,m^{-2}}$. With a maximum current of $5\,\mathrm{m\,s^{-1}}$, which occurs in a very few inter-island

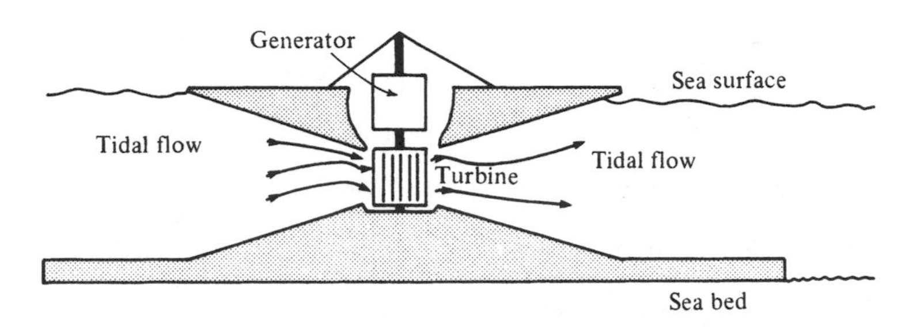


*Figure 13.6* Tidal current power device.

channels, $\bar{q} \sim 14\,\mathrm{kW\,m^{-2}}$; if the intercepted area is a circle of area $100\,\mathrm{m^2}$ (radius 5.6 m), then the total average power generation would be 1.4 MW. (We may note that in most cases a 4 MW capacity wind turbine would be expected to have a similar average power production, but the rotor radius would be ~60 m, see Table 9.2.)

The periodic nature of the power generation would lead to complications, but we note that tidal flow power lags about $\pi/2$ behind range power from a single basin, so the two systems could be complementary.

Few modern tidal flow power devices have been developed, but Figure 13.6 shows the design of one device. Some prototype devices have very similar form to wind turbines; visit <www.itpower.co.uk/OceanEnergy.htm> and <www.marineturbines.com/projects.htm>. The capital cost per unit capacity depends on eventual commercial series manufacture. If tidal flow devices are to be used at all, the best opportunities are obviously where unusually fast tidal stream flows occur, where alternative sources are expensive and where ships can be excluded. The predictable nature of the generated power is an advantage, despite the tidal rhythms.

## 13.5 Tidal range power

### *13.5.1 Basic theory*

The basic theory of tidal power, as distinct from the tides themselves, is quite simple. Consider water trapped at high tide in a basin, and allowed to run out through a turbine at low tide (Figure 13.7). The basin has a constant surface area $A$ that remains covered in water at low tide. The trapped water, having a mass $\rho AR$ at a centre of gravity $R/2$ above the low tide level, is all assumed to run out at low tide. The potential maximum energy available per tide if all the water falls through $R/2$ is therefore (neglecting small changes in density from the sea water value, usually $\rho = 1025\,\mathrm{kg\,m^{-3}}$)

$$\text{energy per tide} = (\rho AR)\, g \frac{R}{2} \tag{13.34}$$

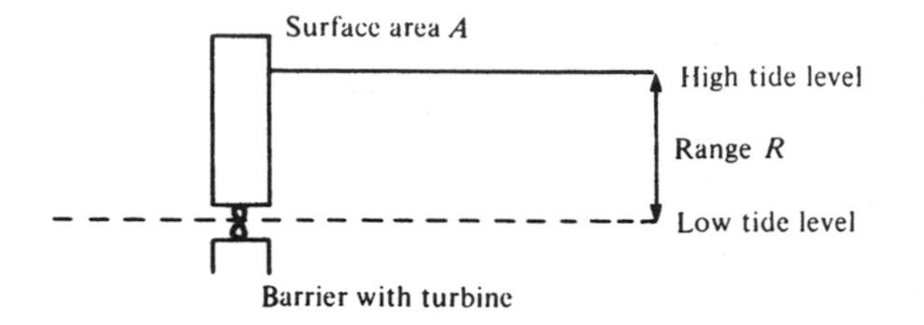


*Figure 13.7* Power generation from tides.

If this energy is averaged over the tidal period $\tau$, the average potential power for one tidal period becomes

$$\overline{P} = \frac{\rho A R^2 g}{2\tau} \tag{13.35}$$

The range varies through the month from a maximum $R_s$ for the *spring* tides, to a minimum $R_n$ for the *neap* tides. The envelope of this variation is sinusoidal, according to Figure 13.8, with a period of half the lunar month.

At any time $t$ after a mean high tide within the lunar month of period $T(=29.53$ days), the range is given by

$$\frac{R}{2} = \left(\frac{R_s + R_n}{4}\right) + \left(\frac{R_s - R_n}{4}\right) \sin(4\pi t/T) \tag{13.36}$$

If

$$R_n = \alpha R_s \tag{13.37}$$

then the range is given by

$$R = \frac{R_s}{2}[(1+\alpha) + (1-\alpha)\sin(4\pi t/T)] \tag{13.38}$$

The power is obtained from the mean square range:

$$\overline{R^2} = \frac{R_s^2}{4} \frac{\int_{t=0}^{T} [(1+\alpha) + (1-\alpha)\sin(4\pi t/T)]^2 \mathrm{d}t}{\int_{t=0}^{T} \mathrm{d}t} \tag{13.39}$$

Hence

$$\overline{R^2} = \frac{R_s^2}{8}(3 + 2\alpha + 3\alpha^2) \tag{13.40}$$

The mean power produced over the month is

$$\overline{P}_{\text{month}} = \frac{\rho A g}{2\tau} \frac{R_s^2}{8}(3 + 2\alpha + 3\alpha^2) \tag{13.41}$$

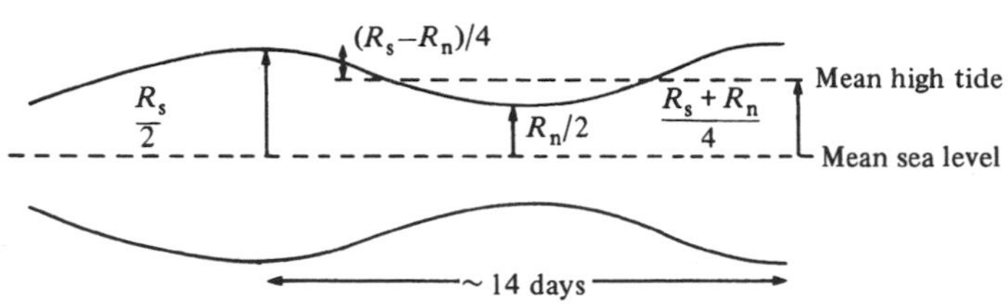


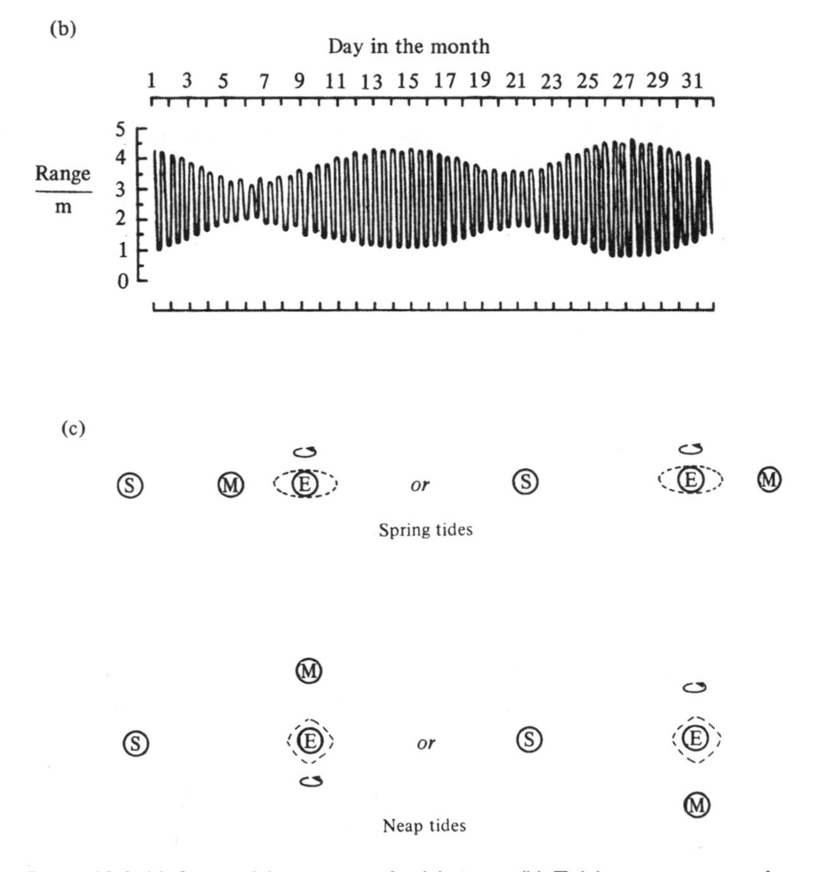


*Figure 13.8* (a) Sinusoidal variation of tidal range. (b) Tidal range variation for one month (from Bernshtein 1965) for a regular semidiurnal tide. Large range at spring tides, small range at neap tides. (c) Positions of the Sun (S), Moon (M) and Earth (E) that produce spring and neap tides twice per month.

where $R_n = \alpha R_s$ and $\tau$ is the intertidal period.

Since $\alpha \sim 0.5$, (13.41) differs little from the approximations often used in the literature, i.e.

$$\overline{P} \approx \frac{\rho A g}{2\tau}(\overline{R})^2 \tag{13.42}$$

where $\overline{R}$ is the mean range of all tides, and

$$\overline{P} \approx \frac{\rho A g}{2\tau}\frac{(R_{max}^2 + R_{min}^2)}{2} \tag{13.43}$$

where $R_{max}$ and $R_{min}$ are the maximum and minimum ranges.

*Example 13.2 Typical values of mean tidal power*

If $R_s = 5\,\text{m}, R_n = 2.5\,\text{m}, \alpha = 0.5, \overline{R} = 3.7\,\text{m}, R_{max} = 5\,\text{m}, R_{min} = 2.5\,\text{m}, A = 10\,\text{km}^2, \rho = 1.03 \times 10^3\,\text{kg m}^{-3}$ and $\tau = 12$ h 25 min $= 4.47 \times 10^4$ s,

then

$$\begin{aligned} &(13.41)\text{yields} \quad \overline{P} = 16.6\,\text{MW} \\ &(13.42)\text{yields} \quad \overline{P} = 15.4\,\text{MW} \end{aligned} \tag{13.44}$$

and

$$(13.43)\text{yields} \quad \overline{P} = 16.1\,\text{MW}$$

### 13.5.2 Application

The maximum potential power of a tidal range system cannot be obtained in practice, although high efficiencies are possible. The complications are:

1. Power generation cannot be maintained near to low tide conditions and so some potential energy is not harnessed.
2. The turbines must operate at low head with large flow rates – a condition that is uncommon in conventional hydropower practice, but similar to 'run-of-the-river' hydropower. The French have most experience of such turbines, having developed low head, large flow bulb turbines for generation from rivers and the Rance tidal scheme. The turbines are least efficient at lowest head.

3 The electrical power is usually needed at a near constant rate, and so there is a constraint to generate at times of less than maximum head.

Efficiency can be improved if the turbines are operated as pumps at high tide to increase the head. Consider a system where the range is 5 m. Water lifted 1 m at high tide can be let out for generation at low tide when the head becomes 6 m. Even if the pumps and generators are 50% efficient there will be a net energy gain of ~200%; see Problem 13.5.

In Figure 13.7, note that power can be produced as water flows with both the incoming ('flow') and the outgoing ('ebb') tide. Thus a carefully optimised tidal power system that uses reversible turbines to generate at both ebb and flow, and where the turbine can operate as pumps to increase the head, can produce energy of 90% of the potential given by (13.41).

## 13.6 World range power sites

The total dissipation of energy by water tides in the Earth is estimated to be 3000 GW, including no more than about 1000 GW dissipated in shallow sea areas accessible for large civil engineering works. The sites of greatest potential throughout the world are indicated in Figure 13.9 and detailed in Table 13.1. They have a combined total potential of about 120 GW. This is about 12% of near-shoreline potential and 10% of the total world

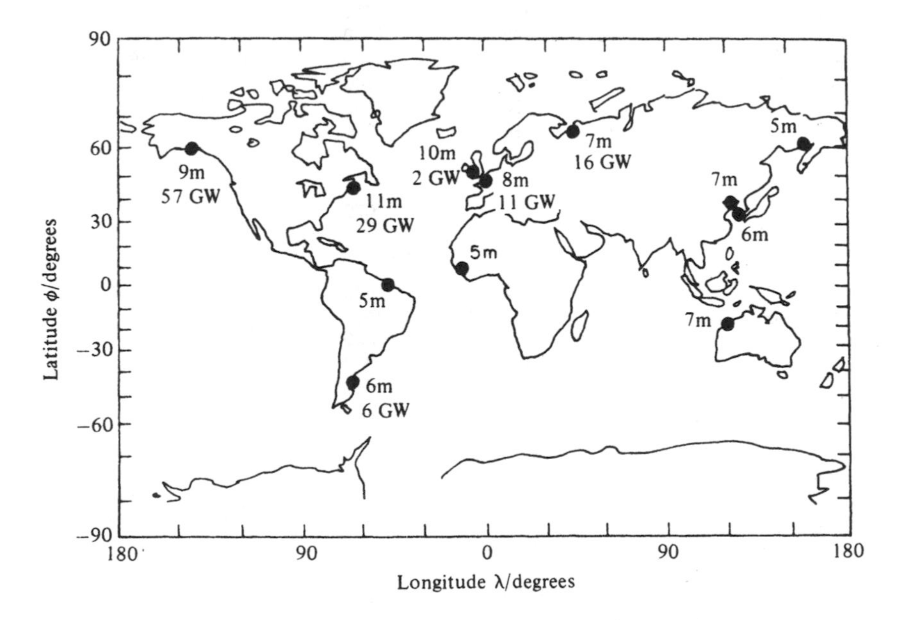


*Figure 13.9* Location of major world tidal power sites, showing the average tidal power range and power potential. From Sørensen (2000) with permission of Elsevier.

*Table 13.1* Major world tidal sites (authors' tabulation from various sources, including the classic tabulation of Hubbert, 1971)

| *Location* | *Mean range* | *Basin area* | *Potential Mean power* | *Potential Annual prodn* | *Actual Installed capacity* | *Date commissioned* |
|---|---|---|---|---|---|---|
| | *(m)* | *(km²)* | *(MW)* | *(GWh y⁻¹)* | *(MW)* | |
| *North America* | | | | | | |
| Passamaquoddy | 5.5 | 262 | 1800 | 15 800 | – | |
| Cobscook | 5.5 | 106 | 722 | 6330 | – | |
| Bay of Fundy | 6.4 | 83 | 765 | 6710 | 17.8 | 1985 |
| Minas-Cobequid | 10.7 | 777 | 19 900 | 175 000 | – | |
| Amherst Point | 10.7 | 10 | 256 | 2250 | – | |
| Shepody | 9.8 | 117 | 520 | 22 100 | – | |
| Cumberland | 10.1 | 73 | 1680 | 14 700 | – | |
| Petitcodiac | 10.7 | 31 | 794 | 6960 | – | |
| Memramcook | 10.7 | 23 | 590 | 5170 | – | |
| *South America* | | | | | | |
| San Jose Argentina | 5.9 | 750 | 5870 | 51 500 | – | |
| *UK* | | | | | | |
| Severn | 9.8 | 70 | 1680 | 15 000 | – | |
| Mersey | 6.5 | ~7 | 130 | 1300 | | |
| Solway Firth | 5.5 | ~60 | 1200 | 10 000 | | |
| Thames | 4.2 | ~40 | 230 | 1400 | | |
| *France* | | | | | | |
| Aber-Benoit | 5.2 | 2.9 | 18 | 158 | – | |
| Aber-Wrac'h | 5 | 1.1 | 6 | 53 | – | |
| Arguenon | 8.4 | 28 | 446 | 3910 | – | |
| Frenaye | 7.4 | 12 | 148 | 1300 | – | |
| La Rance | 8.4 | 22 | 349 | 3060 | 240 | 1966 |
| Rotheneuf | 8 | 1.1 | 16 | 140 | – | |
| Mont St Michel | 8.4 | 610 | 9700 | 85 100 | – | |
| Somme | 6.5 | 49 | 466 | 4090 | – | |
| *Ireland* | | | | | | |
| Strangford Lough | 3.6 | 125 | 350 | 3070 | – | |
| *Former Soviet Union* | | | | | | |
| Kislaya | 2.4 | 2 | 2 | 22 | 0.4 | 1966 |
| Lumbouskii Bay | 4.2 | 70 | 277 | 2430 | – | |
| WhiteSea | 5.65 | 2000 | 14 400 | 126 000 | – | |
| Mezen Estuary | 6.6 | 140 | 370 | 12 000 | – | |
| *Australia* | | | | | | |
| Kimberley | 6.4 | 600 | 630 | 5600 | – | |
| *China* | | | | | | |
| Baishakou | 2.4 | | | | 0.64 | 1978 |
| Jiangxia | 7.1 | 2 | | | 3.2 | 1980 |
| Xinfuyang | 4.5 | | | | 1.3 | 1989 |
| About five other small sites | ~5 | | | | 0.53 | 1961–76 |
| ***total*** | | | ~63 000 | ~570 000 | ~1000 | |

Note
* If no commissioning date indicated then only studies have been made at the site but no installation.

hydropower (river) potential. This is a significant power potential and of great potential importance for certain countries, e.g. the UK, where, in principle, about 25% of annual electricity could be generated by tidal power from known estuaries with enhanced tidal range. Unfortunately, as also indicated in Table 13.1, few tidal power stations have been constructed, mostly due to the large capital cost and small short-term financial gains.

## 13.7 Social and environmental aspects of tidal range power

Sites for tidal range power are chosen for their large tidal range; a characteristic that is associated with estuaries having large areas of mud flats exposed at lower tides. Tidal range power depends on the placing of a barrier for a height difference in water level across the turbines. In operation, (i) the level of water in the basin is always above the unperturbed low tide and always below the unperturbed high tide, (ii) the rates of flow of both the incoming and the outgoing tides are reduced in the basin, and (iii) sea waves are stopped at the barrier. These mechanical factors are the driving functions likely to cause the following effects:

1. The areas of exposed mud flats are reduced, so significantly reducing the food available for birds; usually including migratory birds habitually passing such special habitats. The change in flow, depth and sea waves can be expected to change many other ecological characteristics, many of which may be unique to particular sites.
2. Visual impact is changed, but with a barrier the only necessary construction.
3. River flow can be controlled to reduce flooding.
4. Access for boats to harbours in the basin is increased.
5. Controlled depth and flow of the basin allows leisure activities such as sailing.
6. The barrier can be used as a viaduct for transport and for placing other constructions, e.g. wind turbines.

Tidal barriers are large and expensive structures that may require years to construct. No power can be produced, and hence no income generated, until the last section of the barrier is complete. Difficulties in finance may lead to lack of environmental care. Although the installation at La Rance now features a flourishing natural ecosystem, it is noticeably different from that which was there before the dam, and took some years to establish itself. Therefore it has been observed that La Rance may not have been constructed if it had had to face today's environmental impact procedures.

A developer's main criterion for the success of a tidal power plant is the cost per unit kWh of the power produced. As with other capital-intensive energy technologies, the economic cost per kWh generated can be reduced (i) if other advantages can be costed as benefit to the project, including carbon abatement, (ii) if interest rates of money borrowed to finance the high capital cost are small, and (iii) if the output power can be used to decrease consumption of expensive fuels such as oil. (See Chapter 17 for a more general discussion of these issues.) With such economic and environmental complexity, large-scale (~1000 MW) tidal power plants may not be best. Smaller schemes may perhaps be more economic.

## Problems

13.1 a In Figure 13.1, consider the lunar-related force $F_Z$ on a mass $m$ of seawater along the Earth's radius EZ. Since $D \gg r$, show that $F_Z = mr\omega^2$.

b Hence show that the difference in the lunar-related force on this mass between high and low tide is the tide raising force

$$F_t = F_X - F_Z = 2MmGr/D^3$$

c The tide raising force must equal the difference in the Earth's gravitational attraction on m between low and high tide. Hence show that the tidal range $R$ is 0.36 m and is given by

$$R = \frac{Mr^4}{M'D^3}$$

13.2 The sidereal month $T^*$ is defined after (13.11). The synodic month $T_m$ is defined as the average period between two new Moons as seen by an observer on Earth. $T_m$ is greater than $T^*$ because of the motion of the Earth and the Moon together about the Sun that effectively 'delays' the appearance of the new Moon. What is the relation between $T^*$ and $T_m$?

13.3 a The Earth's oceans have an average depth of 4400 m. Show that the speed of a naturally propagating tidal wave is about 200 m s$^{-1}$(750 km h$^{-1}$).

b Compare this speed with the speed of the lunar tidal force passing round the Earth's equator.

c What is the effect of the difference between these two speeds?

13.4 A typical ocean on the Earth's surface has a depth of 4400 m.

a What is the speed of a freely travelling tidal wave for this depth?

b At this speed, how long would it take the tidal wave to circle the Earth?

c If the tidal wave is started by the influence of the Moon, can its motion be reinforced continually as the Earth rotates?

13.5 Water is pumped rapidly from the ocean at high tide to give an increased water level in a tidal power basin of 1.0 m. If the tidal range is 5.0 m and if the pump/generator system is only 50% efficient, show that the extra energy gained can be nearly twice the energy needed for pumping.

13.6 The *capacity factor* of a power system is the ratio of energy produced to the energy that would have been produced if the system ran continuously at its rated (maximum) power. By integrating over a day (i.e. two full tides), calculate the capacity factor of a tidal range station. (Assume that the range $R = \beta R_{max}$ with $0 < \beta < 1$ and that the rated power is the maximum output on a day when $\beta = 1$.)

## Bibliography


Baker, A.C. (1991a) *Tidal Power*, Peter Peregrinus, London. {Basic theory and account of UK tidal power studies; a comprehensive text.}

Baker, A.C. (1991b) 'Tidal Power', in *Energy Policy*, **19**, no. 8., pp. 792–797.

Bernshtein, L.B. (1965) *Tidal Energy for Electric Power Plants*, Israel Program for Scientific Translations (Translated from the Russian), Jerusalem. {A careful comprehensive text considering the important properties of tides and generating systems. Comparison of thirteen different basin arrangements in a model tidal bay. Discussion of low head generation and dam construction.}

Cavanagh, J.E., Clarke, J.H., and Price, R. (1993) Ocean energy systems, in Johansson, T.B., Kelly, H., Reddy, A.K.N., Williams, R.H and Burnham, L. (eds) *Renewable Energy-Sources for Fuels and Electricity*, Earthscan, London. {pp. 513–530 are a brief review of tidal power, including its economics.}

Charlier, R.H. (2001), Ocean alternative energy. The view from China—small is Beautiful, *Renewable and Sustainable Energy Reviews*, **3**, no. 3, pp. 7–15.

Charlier, R.C. (2003) Sustainable co-generation from the tides: A review, *Renewable and Sustainable Energy Reviews*, **7**, pp. 187–213. {Comprehensive, including work before 1980 and up to 2002.}

Clare, R. (ed.) (1992) *Tidal Power: Trends and Developments*, Thomas Telford, London. Conference papers, mostly studies of potential sites and installations in UK, and also including M. Rodier 'The Rance tidal power station: A quarter of a century in operation'. {Indicates that not much had changed since Severn *et al.* (1979).}

Cotillon, J. (1979) *La Rance tidal power station, review and comments*, in Severn, R.T. *et al.* (1979). {Useful review of the world's only operating installation.}

Energy Technology Support Unit (1996) *Review of Severn and Mersey Tidal Power*, ETSU, UK.

Hubbert, M.K. (1971) *Scientific American*, September, 60–87. {Classic estimates of global tidal estimates.}

Severn, R.T., Dineley, D. and Hawker, L.E. (eds) (1979) *Tidal Power and Estuary Management*, Colston Research Society 30th Meeting 1978, Colston Research Society, Bristol. {A set of useful research and development reviews, including economic and ecological aspects. Summaries of French (La Rance) and Russian experience, and suggested plans for the UK (Severn estuary).}

Shaw, T.L. (1980) *An Environmental Appraisal of Tidal Power Stations*, Pitman Publishing, London.

Sørensen, B. (2000) *Renewable Energy*, 2nd edition, Academic Press, London. ISBN 0–12–656152–4. {Useful but short summary of tidal power potential, p. 281 etc.}

Tricker, R.A.R. (1965) *Bores, Breakers, Waves and Wakes – An Introduction to the Study of Waves on Water*, Mills and Boon, London. {Includes explanation of seismic sea waves.}

Webb, D.J. (1982) Tides and tidal power. *Contemp. Phys.*, **23**, pp. 419–442. {Excellent review of tidal theory and resonant enhancement.}

Chapter 14

# Ocean thermal energy conversion (OTEC)

## 14.1 Introduction

The ocean is the world's largest solar collector. In tropical seas, temperature differences of about 20–25 °C may occur between the warm, solar-absorbing near-surface water and the cooler 500–1000 m depth 'deep' water at and below the thermocline. Subject to the laws and practicalities of thermodynamics, heat engines can operate from this temperature difference across this huge heat store. The term *ocean thermal energy conversion* (OTEC) refers to the conversion of some of this thermal energy into useful work for electricity generation. Given sufficient scale of efficient equipment, electricity power generation could be sustained day and night at 200 $kW_e$ from access to about 1 $km^2$ of tropical sea, equivalent to 0.07% of the solar input. Pumping rates are about 6 $m^3 s^{-1}$ of water per $MW_e$ electricity production. The technology for energy extraction is similar to that used for energy efficiency improvement in industry with large flows of heated discharge, but on a much larger scale.

The attractiveness of OTEC is the seemingly limitless energy of the hotter surface water in relation to the colder deep water and its potential for constant, base load, extraction. However, the temperature difference is very small and so the efficiency of any device for transforming this thermal energy to mechanical power will also be very small. Even for heating, warm seawater cannot be spilt on land due to its high salt content. Moreover, large volumes of seawater need to be pumped, so reducing the net energy generated and requiring large pipes and heat exchangers.

There have been hundreds of paper studies, and a few experimental demonstration plants, with the first as far back as 1930. These were mostly resourced from France (pre-1970s) and then the USA, Japan and Taiwan in the 1980s, but less activity since then; see Avery and Wu (1994) for a detailed history. This experience confirmed that the cost per unit of power output would be large, except perhaps on a very large scale, and led to other justifications for pumping up the cold, deeper waters, which contain nutrients and therefore increase surface photosynthesis of phytoplankton and hence fish population. It now appears that OTEC could be at best a secondary aspect of systems for deep-water nutrient enrichment for marine fisheries, for cooling buildings or for desalination (see Section 14.5). Such integrated technology is called Deep Ocean Water Application (DOWA).

## 14.2 Principles

Figure 14.1 outlines a system for OTEC. In essence it is a heat engine with a low boiling point 'working fluid', e.g. ammonia, operating between the 'cold' temperature $T_c$ of the water pumped up from substantial depth and the 'hot' temperature, $T_h = T_c + \Delta T$, of the surface water. The working fluid circulates in a *closed cycle*, accepting heat from the warm water and discharging it to the cold water through heat exchangers. As the fluid expands, it drives a turbine, which in turn drives an electricity generator. The working fluid is cooled by the cold water, and the cycle continues. Alternative 'open cycle' systems have seawater as the working fluid, but this is not recycled but condensed, perhaps for distilled 'fresh' water; the thermodynamic principles of the open cycle are similar to the closed cycle.

In an idealised system with perfect heat exchangers, volume flow $Q$ of warm water passes into the system at temperature $T_h$ and leaves at $T_c$ (the cold water temperature of lower depths). The power given up from the warm water in such an ideal system is

$$P_0 = \rho c Q \Delta T \tag{14.1}$$

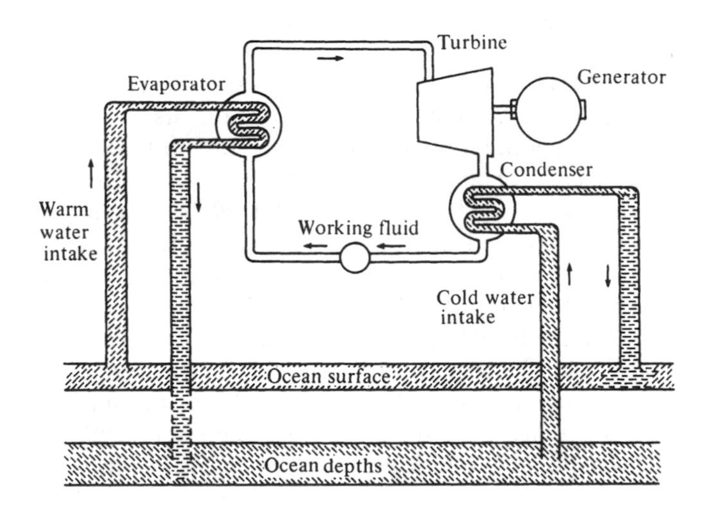


*Figure 14.1* Schematic diagram of an OTEC system. A heat engine operates between the warm water from the ocean surface and the cold water from the ocean depths.

where

$$\Delta T = T_h - T_c$$

The second law of thermodynamics dictates that the maximum output of work energy $E_1$ obtainable from the heat input $E_0$ is

$$E_1 = \eta_{Carnot} E_0 \tag{14.2}$$

Naively forgetting time dependence and the practicalities of heat exchangers, this is usually also given as

$$P_1 = \eta_{Carnot} P_0 \tag{14.3}$$

where

$$\eta_{Carnot} = \Delta T / T_h \tag{14.4}$$

is the efficiency of an ideal Carnot engine operating at an infinitely slow rate between $T_h$ and $T_c = T_h - \Delta T$. With $\Delta T$ only ~20 °C (=20 K), even this ideal Carnot efficiency is very small, ~7%. In practice, we cannot wait an infinite time for ideal thermal processes, so no practical system ever reaches Carnot efficiencies or a perfect heat exchange. So allowing for temperature drops of ~5 °C across each heat exchanger and for the internal power for pumping, the efficiency of a real system will be substantially less at about 2–3%. Nevertheless these equations suffice to illustrate the promise and limitations of OTEC. From (14.1)–(14.4) the ideal mechanical output power is

$$P_1 = (\rho c Q / T_h)(\Delta T)^2 \tag{14.5}$$

*Example 14.1 Required flow rate*

For $\Delta T = 20$°C the flow rate required to yield 1.0 MW from an ideal heat engine is (from (14.5))

$$Q_1 = \frac{(10^6\,\mathrm{J\,s^{-1}})(300\,\mathrm{K})}{(10^3\,\mathrm{kg\,m^{-3}})(4.2 \times 10^3\,\mathrm{J\,kg^{-1}\,K^{-1}})(20\,\mathrm{K})^2}$$
$$= 0.18\,\mathrm{m^3\,s^{-1}}$$
$$= 650\,\mathrm{t\,h^{-1}}$$

Example 14.1 shows that a substantial flow is required to give a reasonable output, even at the largest $\Delta T$ available in any of the world's oceans. Such a system requires large, and therefore expensive, machinery.

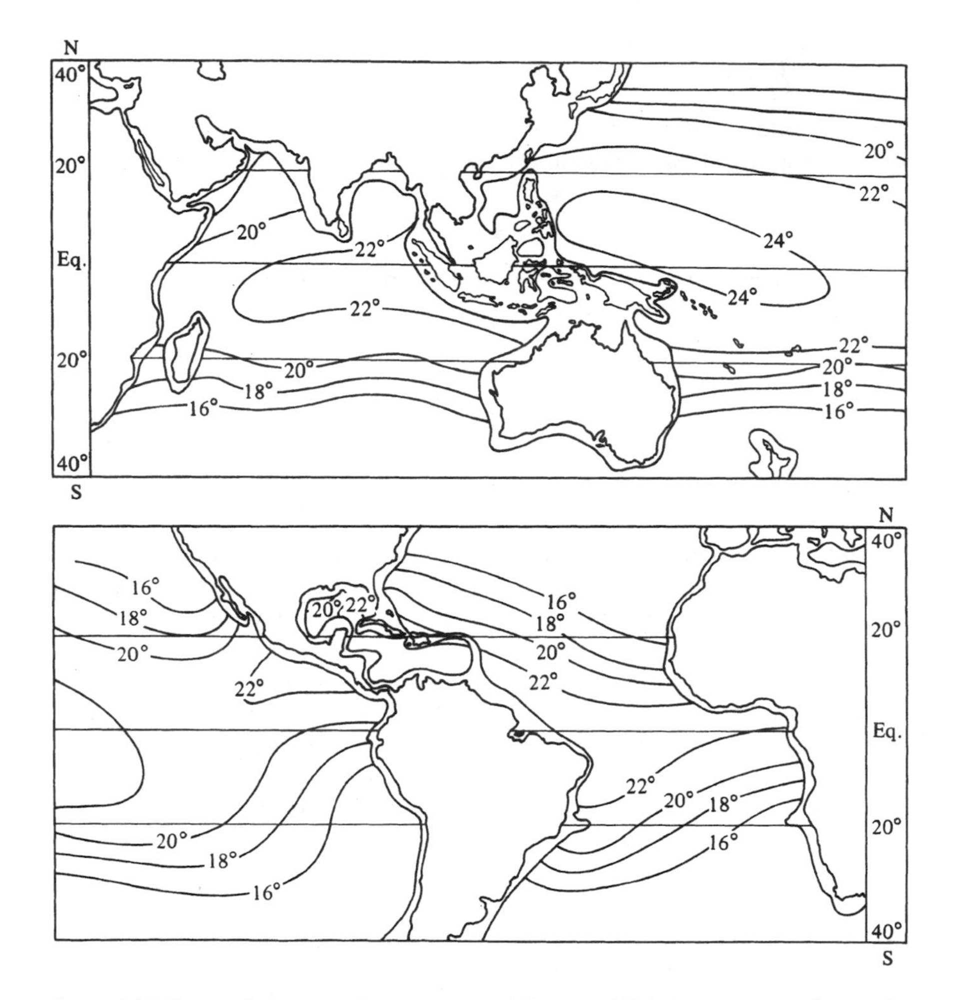



*Figure 14.2* Seasonal average of temperature difference $\Delta T$ between sea surface and a depth of 1000 m. Zones with $\Delta T \geq 20°C$ are most suitable for OTEC. These zones all lie in the tropics. Source: US Department of Energy.

Since $P_1$ depends *quadratically* on $\Delta T$, experience shows that only sites with $\Delta T \geq 20°C$ may possibly be economic. Figure 14.2 indicates that such sites are confined to the tropics, and Figure 14.3 suggests that the cold water has to come from a depth $>\sim 400$ m.

Sites investigated include Hawaii (20°N, 160°W), Nauru (0°S, 166°E) and the Gulf Stream off Florida (30°N, 80°E). Tropical sites have the added advantage that both $T_h$ and $T_c$ have little seasonal variation, so the potential output of the system is constant through the year.

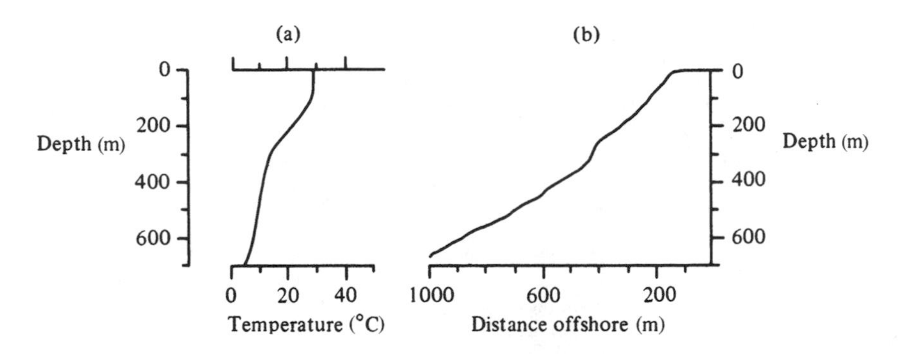


*Figure 14.3* Ocean conditions offshore from the island of Nauru, in the Central Pacific Ocean (0°S, 166°E). (a) Water temperature. (b) Cross-section of sea bottom. The water temperatures are typical of those at good OTEC sites, and the steeply sloping sea floor allows a land-based system. Data from Tokyo Electric Power Services Co. Ltd.

Indeed steadiness and independence of the vagaries of weather are major advantages of OTEC as a renewable source of energy. Its other major advantages as a possible technology are:

1. At a suitable site, the resource is essentially limited only by the size of the machinery.
2. The machinery to exploit it economically requires only marginal improvements in such well-tried engineering devices as heat exchangers and turbines. No dramatically new or physically impossible devices are required.

The major disadvantages are cost and scale. Even if the ideal power $P_1$ of (14.5) was obtainable, the costs per unit output would be large, but resistances to the flow of heat and to fluid motion reduce the useful output considerably and therefore increase unit costs. Sections 14.3 and 14.4 estimate the energy losses due to imperfect heat exchangers and pipe friction. The installed costs of the best experimental OTEC plants (1980s to 1990s) were as large as \$40 000 per $kW_e$ of electricity capacity, in comparison with about \$1000 per $kW_e$ for conventional generating capacity in remote areas. However, the theory of Sections 14.2 to 14.4 suggests that even larger systems would be more economical, which maintains interest in OTEC. However, a large scale-up in a single step from small demonstration plants is imprudent engineering and therefore difficult to finance.

Factors increasing the cost of offshore OTEC are maintenance at sea and submarine cabling, as discussed further in Section 14.5. However, there are a few especially favourable coastal sites where the sea bed slopes down so steeply that all the machinery can be placed on dry land. The island of

*Figure 14.4* Experimental land–based OTEC plant on Nauru, built by Tokyo Electric Power Services Company in 1981 for research. It was a 'closed cycle' system, rated at 100 $kW_e$ output. On the photograph the vertical framework to the rear contains the condenser, the nearer large horizontal cylinder is the evaporator, the turbine house is at the left, the cold water pipe runs out to sea (in the background), and cylinders in the foreground contain spare working fluid.

Nauru in the South Pacific has such topography. Figure 14.3 shows a section of the sea bottom there, and Figure 14.4 is a photograph of an experimental OTEC installation on the shore. Experience showed (i) the beach and submarine pipes must be buried or fixed extremely well to survive wave current forces, (ii) biofouling could be mitigated by 24-hourly pulses of chlorination, and (iii) in the pipes, both thermal losses and friction decreased efficiency significantly.

## 14.3 Heat exchangers

These need to be relatively large to provide sufficient area for heat transfer at low temperature difference, and are therefore expensive (perhaps 50% of total costs). In calculating the ideal output power $P_1$ as calculated in (14.5), we have assumed perfect heat transfer between the ocean waters and the working fluid. In practice, there is significant thermal resistance, even with the best available heat exchangers and with chemical 'cleaning' to lessen internal biofouling.

### 14.3.1 General analysis

A heat exchanger transfers heat from one fluid to another, while keeping the fluids apart. Many different designs are described in engineering handbooks, but a typical and common type is the shell-and-tube design (Figure 14.5). Water flows one way through the tubes while the working fluid flows through the shell around the tubes.

Figure 14.6 shows some of the resistances to heat transfer. The most fundamental of these arises from the relatively small thermal conductivity of water. As in Section 3.4, one can think of heat being carried by blobs of water to within a fraction of a millimetre of the metal surface, but, even with clean surfaces, the last transfer from liquid to solid has to be by pure conduction through effectively still water. Similarly the heat flow through both the metal and the adhering scum and biological growth is by pure conduction. A temperature difference $\delta T$ is required to drive the heat flow across these conductive resistances.

Let $P_{wf}$ be the heat flow from water (w) to working fluid (f). Then

$$P_{wf} = \delta T / R_{wf} \tag{14.6}$$

where $R_{wf}$ is the thermal resistance between water and fluid. If it is assumed that there will be a similar temperature drop $\delta T$ in the other heat exchanger,

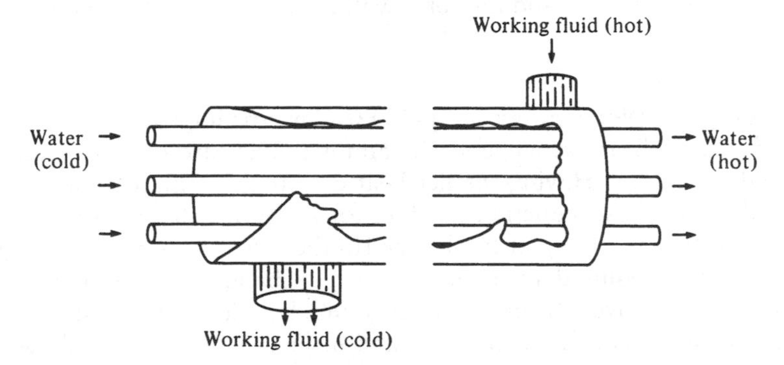


*Figure 14.5* Shell-and-tube heat exchanger (cut-away view).

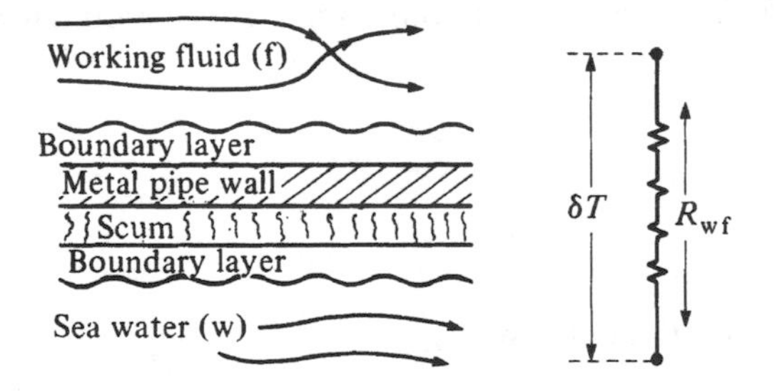


*Figure 14.6* Resistances to heat flow across a heat exchanger wall.

the temperature difference actually available to drive the heat engine is not $\Delta T$ but

$$\Delta_2 T = \Delta T - 2\delta T \tag{14.7}$$

With an idealised Carnot engine the mechanical power output would be

$$P_2 = \left( \frac{\Delta T - 2\delta T}{T_h} \right) \frac{\delta T}{R_{wf}} \tag{14.8}$$

Equation (14.8) implies that $\delta T / R_{wf}$ should be large to increase output power. Yet $\delta T$ must be small to obtain maximum engine efficiency, so it is crucial to minimise the transfer resistance $R_{wf}$ by *making the heat exchanger as efficient as possible*. Therefore the tubes must be made of metal (good conductor) and there must be many of them, perhaps hundreds, to provide a large total surface area. Other refinements may include fins or porous surfaces on the tubes, and baffles within the flow. With such an elaborate construction, it is not surprising that the heat exchangers constitute one of the major expenses of an OTEC system. This is the more so since the tube material has to be resistant to corrosion by seawater and the working fluid, all joints must be leakproof and all pipes capable of internal cleaning.

The overall thermal resistance can be analysed in terms of the thermal resistivity of unit area $r_{wf}$ and the total wall area $A_{wf}$, as in Section 3.6:

$$R_{wf} = r_{wf} / A_{wf} \tag{14.9}$$

Much of the development work in OTEC concerns improvements in the design of existing heat exchangers. The aim is to decrease $r_{wf}$, and thereby decrease the area $A_{wf}$. Having smaller heat exchangers with less metal can lead to substantial cost reductions. Values for $r_{wf}$ of $3 \times 10^{-4}\,\mathrm{m^2\,K\,W^{-1}}$ (i.e. $h = 1/r = 3000\,\mathrm{W\,m^{-2}\,K^{-1}}$) can be obtained by the best of existing technology.

The flow rate required through the heat exchanger is determined by the power $P_{wf}$ removed from the water, and by the heat transfers and temperatures involved. These are indicated in Figure 14.7, which shows a counterflow heat exchanger on each side of the working fluid circuit. At each point along the heat exchanger, the temperature difference between the working fluid and the water is $\delta T$. Thus the hottest point in the working fluid is at

$$T_{hf} = T_{hw}^{(in)} - \delta T$$

and the coldest is at

$$T_{cf} = T_{cw}^{(in)} + \delta T$$

Therefore the power given up by the hot water is

$$P_{wf} = \rho c Q \left( T_{hw}^{(in)} - T_{hw}^{(out)} \right) \tag{14.10}$$

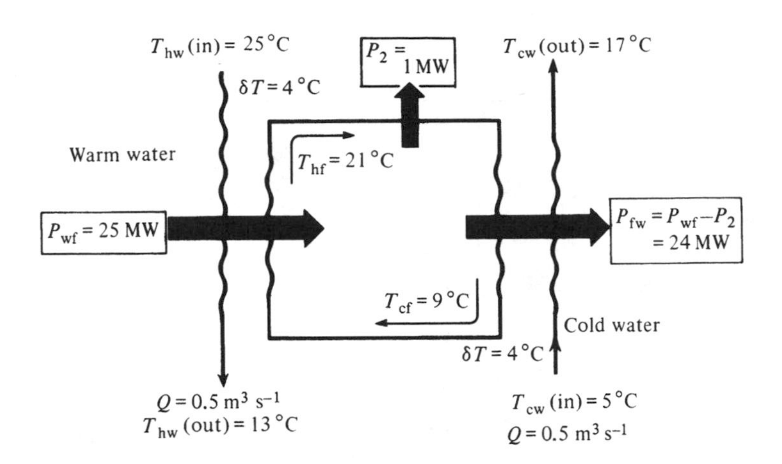


*Figure 14.7* Temperatures and heat flows in the OTEC system of Example 14.2. The other quantities are calculated from $T_{cw}^{(in)}$, $T_{hw}^{(in)}$, $\delta T$, $P_2$.

with the temperature drop

$$T_{hw}^{(in)} - T_{hw}^{(out)} = \Delta T - 2\delta T \qquad (14.11)$$

### *14.3.2 Size*

*Example 14.2 Heat exchanger dimensions*

Find a set of working dimensions for a shell-and-tube heat exchanger suitable for an OTEC system set to produce 1 MW. Assume a Carnot cycle for the working fluid, but allow for temperature reductions in non-perfect heat exchangers.

Assume $r_{wf} = 3 \times 10^{-4}\,m^2\,K\,W^{-1}$, $\Delta T = 20\,°C$, $\delta T = 4\,°C$, etc. as in Figure 14.7.

*Solution*

1 *Surface area*
From (14.9),

$$A_{wf} = r_{wf}/R_{wf}$$

From (14.8),

$$1/R_{wf} = \frac{P_2 T_h}{(\Delta T - 2\delta T)\delta T}$$

so

$$A_{wf} = \frac{(1 \times 10^6\,\mathrm{W})(300\,\mathrm{K})(3 \times 10^{-4}\,\mathrm{m^2\,K\,W^{-1}})}{(20-8)\,\mathrm{K}(4\,\mathrm{K})}$$
$$= 1.9 \times 10^3\,\mathrm{m}^2$$

This is a very large area of transfer surface.

2 *Flow rate*
For the parameters of Figure 14.7,

$$\eta_{carnot} = \frac{(21-9)^\circ\mathrm{C}}{(273+21)\,\mathrm{K}} = \frac{12}{294}$$
$$P_{wf} = P_2/\eta_{carnot} = (1\,\mathrm{MW})(294/12) = 25\,\mathrm{MW}$$

Therefore, from (14.10), and (14.11), the flow rate is

$$Q = \frac{(25 \times 10^6\,\mathrm{W})}{(10^3\,\mathrm{kg\,m^{-3}})(4.2 \times 10^3\,\mathrm{J\,K^{-1}\,kg^{-1}})(12\,\mathrm{K})}$$
$$= 0.50\,\mathrm{m^3\,s^{-1}}$$

3 *Thermal resistance of the boundary layers*
We suppose that each fluid boundary layer of Figure 14.6 contributes about half of $r_{wf}$. In particular, assume that the thermal resistivity of the boundary layer (of water) on the inside of the pipe is given by

$$r_v = 1.5 \times 10^{-4}\,\mathrm{m^2\,K\,W^{-1}}$$

Let $d$ be the diameter of each tube in the heat exchanger. The convective heat transfer to the inside wall of a smooth tube is given by (C.14), see Section 3.4:

$$\mathcal{N} = 0.027\mathcal{R}^{0.8}\mathcal{P}^{0.33}$$

By definition of the Nusselt number, $\mathcal{N} = d/(r_v k)$. Thus the Reynolds number in each tube is

$$\mathcal{R} = [d/(0.027 r_v k \mathcal{P}^{0.33})]^{1.25}$$
$$= [(0.027)(0.6\,\mathrm{W\,m^{-1}\,K^{-1}})(7.0)^{0.33}]^{-1.25}(d/r_v)^{1.25}$$
$$= ad^{1.25}$$

where $a = 4.67 \times 10^6\,\mathrm{m}^{-1.25}$ and the properties of water are from Appendix B.

4 *Diameter of tube*
As an initial estimate, suppose $d = 0.02\,\text{m}$. Then $\mathcal{R} = 3.5 \times 10^4$. Hence, flow speed in each tube is

$$u = \mathcal{R}\frac{\nu}{d}$$

$$= \frac{(3.5 \times 10^4)(1.0 \times 10^{-6}\,\text{m}^2\,\text{s}^{-1})}{(0.02\,\text{m})} = 1.7\,\text{m}\,\text{s}^{-1}$$

Since the total flow through $n$ tubes is $Q = nu\pi d^2/4$ the number of tubes required is

$$n = \frac{(0.50\,\text{m}^3\,\text{s}^{-1})(4)}{(1.7\,\text{m}\,\text{s}^{-1})(3.14)(0.01\,\text{m})^2}$$

$$= 3600$$

5 *Length of tubes*
To make up the required transfer area $A = n\pi dl$, each tube must have length

$$l = \frac{(1.9 \times 10^3\,\text{m}^2)}{(3600)\pi(0.02\,\text{m})} = 32\,\text{m}$$

This example makes it clear that large heat exchangers, with substantial construction costs, are required for OTEC systems. Indeed the example underestimates the size involved because it does not allow for imperfections in the heat engines etc., which increase the required $Q$ to achieve the same power output. Also the example assumes that the pipe is clean and smooth.

### *14.3.3 Biofouling*

The inside of the pipe is vulnerable to encrustation by marine organisms, which will increase the resistance to heat flow (Figure 14.6), and thereby reduce the performance. Such *biofouling* is one of the major problems in OTEC design, since increasing the surface area available for heat transfer also increases the opportunity for organisms to attach themselves. Among the methods tried to keep this fouling under control are mechanical cleaning by continual circulation of close fitting balls and chemical cleaning by additives to the water.

The effect of all these complications is that the need for cost saving encourages the use of components working at less than optimal performance, e.g. undersized heat exchangers.

## 14.4 Pumping requirements

Work is required to move large quantities of hot water, cold water and working fluid around the system against friction. This will have to be supplied from the gross power output of the OTEC system, i.e. it constitutes yet another loss of energy from the ambient flow $P_0$. Example 14.3 shows how the work may be estimated numerically using the methods of Section 2.6, although analytic calculations are difficult. The effect of cooling the water in the hydrostatic 'circuit' is small, but does encourage circulation.

*Example 14.3 Friction in the cold water pipe*
The OTEC system of Example 14.2 (Figure 14.7) with $P_2 = 1\,\text{MW}$, $\Delta T = 20\,°\text{C}$ has a cold water pipe with $L = 1000\,\text{m}$, diameter $D = 1\,\text{m}$. Calculate the power required to pump water up the pipe.

*Solution*
The mean speed is

$$u = Q/A$$
$$= \frac{(0.50\,\text{m}^3\,\text{s}^{-1})}{\pi(0.5\,\text{m})^2} = 6.3\,\text{m}\,\text{s}^{-1}$$

Therefore the Reynolds number is

$$R = \frac{uD}{v}$$
$$= \frac{(0.63\,\text{m}\,\text{s}^{-1})(1\,\text{m})}{(1.0 \times 10^{-6}\,\text{m}^2\,\text{s}^{-1})} = 6.3 \times 10^5$$

In practice, many varieties of marine organisms brought up from the depths will adhere to the pipe, giving an equivalent roughness height $\xi \sim 20\,\text{mm}$, i.e. $\xi/D = 0.02$. Thus, from Figure 2.6, the pipe friction coefficient is $f = 0.012$
From (2.14), the head loss is

$$H_f = 2fLu^2/Dg = 1.0\,\text{m}$$

To overcome this requires the same power as to lift a mass $\rho Q$ per second through a height $H_f$, i.e.

$$P_f = \rho Q g H_f = 4.7\,\text{kW}$$

From Example 14.3, we see that the cold water pipe can be built large enough to avoid major friction problems. However, because the head loss varies as (diameter)$^{-5}$ (See problem 2.6), friction loss can become appreciable in the smaller piping between the cold water pipe and the heat exchanger, and in the heat exchanger itself. Indeed, because the same turbulence carries both heat and momentum from the heat exchanger surfaces, all attempts to increase heat transfer by increasing the surface area necessarily increase fluid friction in the heat exchangers.

In addition, the flow rate required in practice to yield a given output power is greater than that calculated in Example 14.2, because a real heat engine is less efficient than a Carnot engine in converting the input heat into work. This increases the power lost to fluid friction. Fouling of the heat exchanger tubes makes the situation worse, both by further raising the $Q$ required to yield a certain power output, and by decreasing the tube diameter. As a result, in some systems over 50% of the input power may be lost to fluid friction. Power used by the pumps themselves is another 'loss' from the output power.

## 14.5 Other practical considerations

The calculations of the previous sections confirm that there are no *fundamental* thermodynamic difficulties that prevent an OTEC system from working successfully. Although there remain a number of practical, engineering and environmental difficulties, we shall see that none of these appears insuperable from a technical point of view.

### 14.5.1 The platform

American designers drew up conceptual plans for large systems, generating electricity at about 400 $MW_e$, based on a large floating offshore platform, similar to those used in oil drilling. Since such a platform would be heavy and unwieldy, there would be a major problem in connecting it to the cold water pipe (CWP), because of the stresses from surface waves and currents.

One response to this problem is to make the platform neutrally buoyant and moor it underwater (Figure 14.8) thereby avoiding the major stresses at the surface.

### 14.5.2 Construction of the cold water pipe

The pipe is subject to many forces in addition to the stresses at the connection. These include drag by currents, oscillating forces due to vortex shedding, forces due to harmonic motion of the platform, forces due to drift of the platform and the dead weight of the pipe itself. It is debatable whether a rigid, e.g. steel, or flexible material, e.g. polythene, would

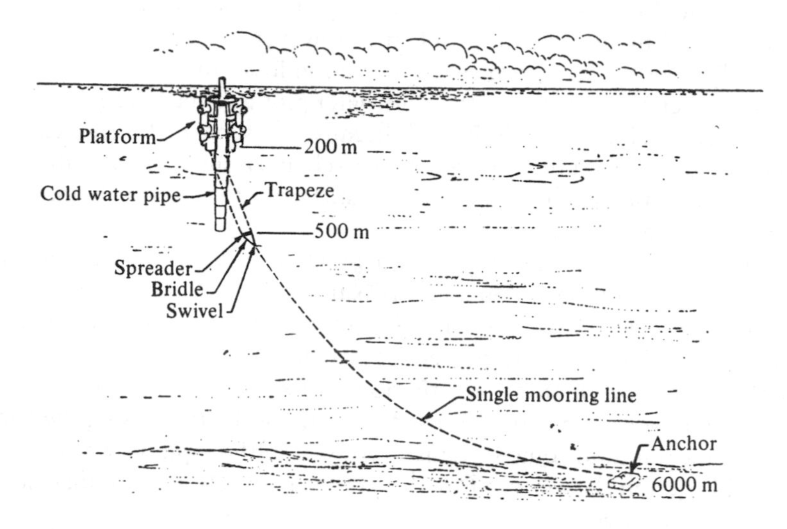


*Figure 14.8* Underwater platform for 400 $MW_e$ systems; proposed by Lockheed for the US Department of Energy. The platform can be moored in position in any depth of water.

withstand these forces better. In addition, there are substantial difficulties involved in assembling and positioning the pipe. Some engineers favour bringing out a prefabricated pipe and slowly sinking it into place; however, transporting an object several meters in diameter and perhaps a kilometre long is difficult. Premature failure of the CWP, e.g. from storm damage, caused the failure of several demonstration projects.

### 14.5.3 Link to the shore

High voltage, large power, submarine cables are standard components of electrical power transmission systems. They are expensive, as with all marine engineering, but a cable about 50 km long is quite practicable, with power loss about 0.05% per km for AC and 0.01% per km for DC. There is now considerable experience with such cables for offshore wind power and for underwater connections in power-grid networks.

Alternatively it has been suggested that large OTEC plants, which might be hundreds of kilometres away from energy demand, could use the electricity on board to produce a chemical store of energy, e.g. $H_2$, Section 16.3.

Land-based systems, like that of Figure 14.4, are possible at certain favourable locations, where the sea bed slopes sharply downward. Their main advantage is reduced cost, since the link to shore, assembly and maintenance are much simplified. The CWP is also not so subject to stress, since it rests on the sea bottom; however, it is still vulnerable to storm damage from wave motion to a depth of about 20 m.

### 14.5.4 The turbine

Even though the turbine has to be large, standard designs can be used. For example, engines for working across relatively small temperature differences have been developed and used in Israel in connection with solar ponds, Section 6.7. As with all practical heat engines, the efficiency will not be greater than 50% of an ideal Carnot engine with the same heat input to the working fluid.

### 14.5.5 Choice of working fluid

There are many common fluids having an appropriate boiling point, e.g. ammonia, freon or water, but many of these are environmentally unacceptable, since leaks increase greenhouse or ozone-depleting gases. By applying a partial vacuum, i.e. reducing the pressure, the boiling point of water can be reduced to the temperature of the warm water intake. This is the basis of the *open cycle* system, in which the warm seawater itself is used as the working fluid. Such a system provides not only power but also substantial quantities of distilled water.

### 14.5.6 Related technologies

OTEC is one of several possible deep ocean water applications (DOWA) associated with pumping seawater from depths of at least 100 m. Others are listed below. All have dimensional scaling factors encouraging large equipment, unlike the modular operation and smaller scale of most renewable energy options.

a *Marine farming*. Seawater from the depths below about 500 m is rich in nutrients, and these may be pumped to the surface, as from an OTEC plant. This encourages the growth of algae (phytoplankton), which feed other marine creatures higher up the food chain and so provides a basis for commercial fish farming.

b *Cooling*. Deep, cool water pumped to the surface may be used to cool buildings, tropical horticultural 'greenhouses' or engineering plants as in chemical refineries.

c *Fresh water*. Flash evaporation of upper surface sea water onto condensers cooled by deep water produces 'distilled' 'fresh' water for drinking, horticulture, etc. This process may be integrated with solar distillation.

d *$CO_2$ injection*. The aim is to absorb $CO_2$ emitted from large-scale fossil fuel combustion by absorption into surface sea water and pumping to depth. This is almost the reverse of the technology for the OTEC CWP, and would be on a very large scale. Environmental impact on the biota at depth is an issue, as are cost and sustainability.

e *Floating industrial complexes*. Concepts exist to match the large scale of OTEC and DOWA with industry on very large, km scale, floating rafts, e.g. for ammonia and hydrogen production for shipping to land-based markets. Talk is cheap!

If OTEC, or similar technologies, are ever to become accepted commercially, it seems inevitable that an integrated set of operations will be used for a combination of several benefits.

## 14.6 Environmental impact

The main environmental impacts of OTEC-like technologies relate to:

- small thermodynamic efficiencies of engineering plant which in turn relate to the relatively small temperature differences of about 25°C between surface and deep water;
- leakage, and likely pollution, from engineering plant, especially of the working fluids and antifouling chemicals;
- consequent large volumes of pumped marine water;
- forced mixing of deep nutrient-rich (nitrate, phosphate and silicate) water with upper, solar irradiated, water;
- location of engineering plant.

Local pollution must always be avoided. Otherwise none of these impacts appear to be particularly grave on a *global* scale unless very large numbers of OTEC systems are deployed. The hypothetical location of very many OTEC plants, say 1000 stations of 200 $MW_e$ each in the Gulf of Mexico, has been calculated to reduce surface sea temperature by 0.3°C. Such a reduction, even at such an unlikely scale, is not considered physically significant. Of more significance locally would be impacts from onshore OTEC or DOWA engineering plant with local waters, including local circulation and currents.

The total biological effects of releasing large quantities of cool, nutrient-rich water into the warmer surface environment are not fully known. The effects may or may not be desirable, and have to be estimated from small-scale trials and computer modelling. Large deployment of OTEC plant, say 100 stations at 10 km separation, would cause the upwelling of nitrate to a concentration found naturally off Peru, where fish populations are much increased. Consequently immediate impacts need not all be negative to mankind, and certainly the prospect of enriching fisheries with deep water nutrients is considered as potentially positive. As cold, deep water reaches the ocean surface, a proportion of dissolved $CO_2$ passes into the atmosphere. If 50% of the excess $CO_2$ is emitted, the rate would be about 0.1 kg $kW_e^{-1}$, as compared with about $0.8\,\mathrm{kg\,kW_e^{-1}}$ from electricity generation by fossil fuel. Only if the OTEC energy produced is used to abate the use of fossil fuels are global emissions of $CO_2$ reduced.

The social impacts of OTEC would be similar to those of running an offshore oil rig or an onshore power station, i.e. minimal.

## Problems

14.1 Calculate the dimensions of a shell-and-tube heat exchanger to produce an output power $P_2 = 10\,\mathrm{MW}$. Assume $r_v = 3 \times 10^{-4}\,\mathrm{m^2\,K\,W^{-1}}$, $\delta T = 4\,°\mathrm{C}$. and tube diameter $D = 5\,\mathrm{cm}$.
*Hint*: Follow Example 14.2.

14.2 Calculate the power lost to fluid friction in the heat exchanger of Example 14.2.

14.3 *Heat engine for maximum power*. As shown in textbooks of thermodynamics, no heat engine could be more efficient than the ideal concept of the Carnot engine. Working between temperatures $T_h$ and $T_c$, its efficiency is

$$\eta_{\mathrm{Carnot}}(\Delta T) = (T_h - T_c)/T_h$$

However, the power output from a Carnot engine is zero. Why?

Use (14.8) to show that the engine which produces the greatest power, for constant thermal resistance of pipe, has $\delta T = 1/4\Delta T$, i.e. it 'throws away' half the input temperature difference. What is the efficiency of this engine as an energy converter compared with an ideal Carnot engine?

14.4 If $P \propto \Delta T^2/T_h$ (14.4), calculate the rate of change of efficiency with respect to temperature difference $\Delta T$. What is the percentage improvement in power production if $\Delta T$ increases from 20 to 21°C?

## Bibliography

### Monographs

Avery, W.H. and Wu, C. (1994) *Renewable Energy from the Ocean – A Guide to OTEC*, Oxford University Press (John Hopkins University series). {A substantial and authoritative study of the science, engineering and history of OTEC.}

Ramesh, R., Udayakumar, K. and Anandakrishnan, M. (1997) *Renewable Energy Technologies: Ocean Thermal Energy and Other Sustainable Energy Options*, Narosa Publishing, London and Delhi. {Collection of optimistic papers from a conference on OTEC in Tamil Naidu, India.}

### *Articles*

d'Arsonval, Jacques (1881) *Revue Scientifique*, **17**, pp. 370–372. {Perhaps the earliest published reference to the potential of OTEC.}

Gauthier, M., Golman, L. and Lennard, D. (2000) Ocean Thermal Energy Conversion (OTEC) and Deep Water Applications (DOWA) – market opportunities for European Industry, in Proc. Euro. Conf. *New and Renewable Technologies for Sustainable Development*, Madeira, June 2000. {Excellent review of working plant since the 1930's to 2000, with future industrial market potential.}

Johnson, F.A. (1992) Closed cycle thermal energy conversion, in Seymour, R.J. (ed.), *Ocean Energy Recovery: The State of the Art*, American Society of Civil Engineers (1992). {Useful summary of thermodynamics, economics and history.}

Masutani, S.M. and Takahashi, P.K. (1999) Ocean Thermal Energy Conversion, in J.G. Webster (ed.) *Encyclopaedia of Electrical and Electronics Engineering*, **18**, pp. 93–103, Wiley. {Authoritative summary.}

McGowan, J.G. (1976) Ocean thermal energy conversion – a significant solar resources, *Solar Energy*, **18**, pp. 81–92. {Reviewed US design philosophy at a historically important time.}

Ravidran, M. (1999) *Indian 1 MW Floating Plant: An overview*, in *Proc. IOA '99 Conf.*, IMARI, Japan.

UN (1984) *A guide to Ocean Thermal Energy Conversion for Developing Countries*, United Nations Publications, New York.

Wick, G.I. and Schmidt, W.R. (1981) (eds) *Harvesting Ocean Energy*, United Nations, Paris.

Zener, C. (1974) Solar sea power, in *Physics and the Energy Problem – 1974*, American Institute of Physics Conference Proceedings no. 19, pp. 412–419. {Useful for heat exchanger thermodynamics. Whole volume makes interesting reading.}

### *Thermodynamics of real engines*

Curzon, F.L. and Ahlborn, B. (1975) Efficiency of a Carnot engine at maximum power output, *Amer. J. Phys.*, **43**, pp. 22–24.

Chapter 15

# Geothermal energy

## 15.1 Introduction

The inner core of the earth reaches a maximum temperature of about 4000 °C. Heat passes out through the solid submarine and land surface mostly by conduction – geothermal heat – and occasionally by active convective currents of molten magma or heated water. The average geothermal heat flow at the Earth's surface is only $0.06\,\mathrm{W\,m^{-2}}$, with a temperature gradient of $<30\,°\mathrm{C\,km^{-1}}$. This continuous heat current is trivial compared with other renewable supplies in the above surface environment that in total average about $500\,\mathrm{W\,m^{-2}}$; see Figure 1.2. However, at certain specific locations, increased temperature gradients occur, indicating significant geothermal resources. These may be harnessed over areas of the order of square kilometres and depths of ~5 km at fluxes of $10–20\,\mathrm{W\,m^{-2}}$ to produce ~100 MW (thermal) $\mathrm{km^{-2}}$ in commercial supplies for at least 20 years of operation.

Geothermal heat is generally of low quality, and is best used directly for building or process heat at about 50–70 °C, or for preheating of conventional high temperature energy supplies. Such supplies are established in several parts of the world and many more projects are planned. Occasionally geothermal heat is available at temperatures above about 150 °C, so electrical power production from turbines can be contemplated. Several important geothermal electric power complexes are fully established, especially in Italy, New Zealand and the USA.

It is common to use heat from the near-surface ground or from lakes, etc. as input to a *heat pump*. Although this may be interpreted as a 'geothermal' source, we do not include such systems as geothermal supplies for the purposes of this chapter. It is probably more meaningful to consider such sources as stored heat from sunshine, since replenishment will be more from the environment above than below.

In Chapter 1 renewable energy was defined as currents of energy occurring naturally in the environment. By this definition some sources of geothermal energy can be classed as renewable, because the energy would otherwise be dissipated in the environment, e.g. from hot springs or geysers. In other geothermal sites, however, the current of heat is increased artificially, e.g. by fracturing and actively cooling hot rocks, or by drilling into hot aquifers, and so the supply is not renewable at the extraction rate on a long time scale. Such finite supplies are included in this text only because they are usually included with other 'alternative' supplies.

## 15.2 Geophysics

A section through the earth is shown in Figure 15.1. Heat transfer from the semi- fluid mantle maintains a temperature difference across the relatively thin crust of 1000 °C, and a mean temperature gradient of $\sim 30\,°\mathrm{C\,km^{-1}}$. The crust solid material has a mean density $\sim 2700\,\mathrm{kg\,m^{-3}}$, specific heat capacity $\sim 1000\,\mathrm{J\,kg^{-1}\,K^{-1}}$ and thermal conductivity $\sim 2\,\mathrm{W\,m^{-1}\,K^{-1}}$. Therefore the average geothermal flux is $\sim 0.06\,\mathrm{W\,m^{-2}}$, with the heat stored in the crust at temperatures greater than surface temperature being $\sim 10^{20}\,\mathrm{J\,km^{-2}}$. If just 0.1% of this heat was removed in 30 years, the heat power available would be $100\,\mathrm{MW\,km^{-2}}$. Such heat extraction would be replenished in time from the mantle below. These calculations give the order of magnitude of the quantities involved and show that geothermal sources are a large potential energy supply.

Heat passes from the crust by (1) natural cooling and friction from the core, (2) radioactive decay of elements such as uranium and thorium, and (3) chemical reactions. The time constants of such processes over the whole Earth are so long that it is not possible to know whether the Earth's temperature is presently increasing or decreasing. The radioactive elements are concentrated in the crust by fractional recrystallisation from molten

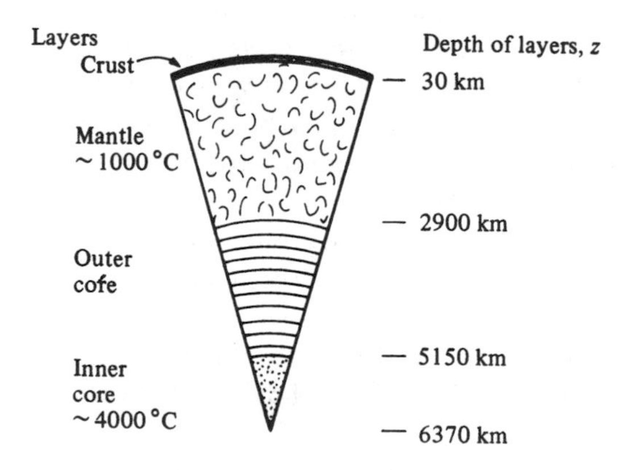


*Figure 15.1* Section through the Earth, showing average lower depths of named layers. The crust has significant variation in composition and thickness over a local scale of several kilometres.

material, and are particularly pronounced in granite. However, the production of heat by radioactivity or chemical action is only significant over many millions of years, see Problem 15.2; consequently geothermal heat extraction relies on removing stored heat in the thermal capacity of solid material and water in the crust, rather than on replenishment. If conduction through uniform material was the only geothermal heat transfer mechanism, the temperature gradient in the crust would be constant. However, if convection occurs 'locally', as from water movement, or if local radioactive or exothermic chemical heat sources occur, there are anomalous temperature gradients.

On a global perspective, the Earth's crust consists of large plates, Figure 15.2. At the plate boundaries there is active convective thermal contact with the mantle, evidenced by seismic activity, volcanoes, geysers, fumaroles and hot springs. The geothermal energy potential of these regions is very great, owing to increased anomalous temperature gradients (to $\sim 100\,°\mathrm{C\,km^{-1}}$) and to active release of water as steam or superheated liquid, often at considerable pressure when tapped by drilling.

Moderate increases in temperature gradient to $\sim 50\,°\mathrm{C\,km^{-1}}$ occur in localized regions away from plate boundaries, owing to anomalies in crust composition and structure. Heat may be released from such regions naturally by deep penetration of water in aquifers and subsequent convective water flow. The resulting hot springs, with increased concentrations of dissolved chemicals, are often famous as health spas. Deep aquifers can be tapped by drilling, to become sources of heat at temperatures from ~50 to ~200 °C. If the anomaly is associated with material of small thermal conductivity,

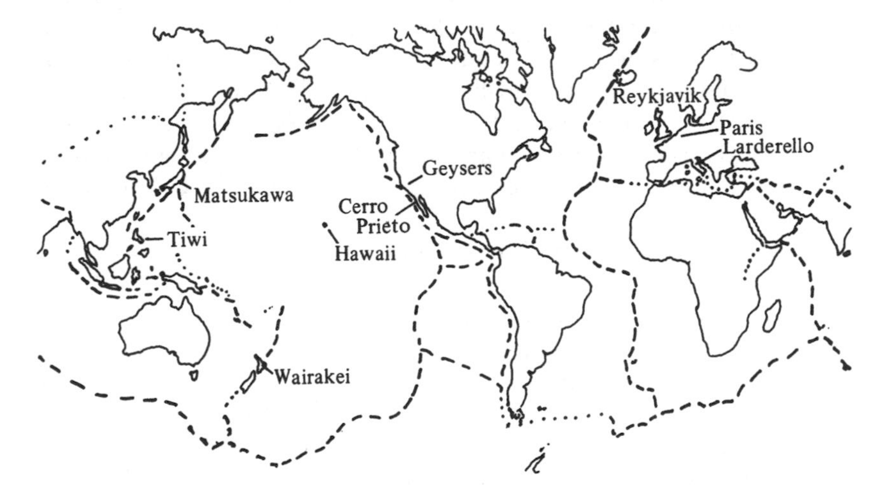


*Figure 15.2* World geothermal activity. Some well-known geothermal fields for heat production and/or electricity generation are indicated. _ _ _ _ _ main plate boundaries, . . . . . . . . . . regions of strain.

*Table 15.1* Installed electricity generating capacity ($MW_e$) using only geothermal sources. The countries with major experience are listed (after Goodman and Love, 1980 and the International Geothermal Association, 2004)

| *Country* | *Key regions* | *1980* | *1990* | *2000* |
|---|---|---|---|---|
| China | | | 19 | 29 |
| El Salvador | | 100 | 95 | 161 |
| Iceland | Namafjall | 40 | 45 | 170 |
| Indonesia | | | 145 | 590 |
| Italy | Larderello | 420 | 545 | 785 |
| Japan | Matsukawa | 250 | 215 | 546 |
| Kenya | Rift Valley | | 45 | 45 |
| Mexico | Cerro Prieto | 150 | 700 | 755 |
| New Zealand | Wairakei | 250 | 283 | 437 |
| Nicaragua | | | 70 | 70 |
| Philippines | | 250 | 891 | 1909 |
| Russia | | | 11 | 23 |
| Turkey | | | 20 | 20 |
| USA | Geysers, California | 700 | 2775 | 2228 |
| Total (2 significant figures) | | 2200 | 5900 | 8000 |

i.e. dry rock, then a 'larger than usual' temperature gradient occurs with a related increase in stored heat.

Geothermal information has been obtained through mining, oil exploration and geological surveys; therefore, geothermal information is available for most countries (Table 15.1). The most important parameter is temperature gradient; accurate measurements depend on leaving the drill hole undisturbed for many weeks so that temperature equilibrium is re-established after drilling. Deep drilled survey wells commonly reach depths of 6 km, and the technology is available to drill to 15 km or more. The principal components of a geothermal energy plant are the boreholes and so heat extraction from depths to 15 km can be contemplated.

There are three classes of geothermal region:

1 *Hyperthermal.* Temperature gradient $\geq 80\,°C\,km^{-1}$. These regions are usually on tectonic plate boundaries. The first such region to be tapped for electricity generation was at Larderello in Tuscany, Italy in 1904. Nearly all geothermal power stations are in such areas.
2 *Semithermal.* Temperature gradient $\sim 40–80\,°C\,km$. Such regions are associated generally with anomalies away from plate boundaries. Heat extraction is from harnessing natural aquifers or fracturing dry rock. A well-known example is the geothermal district heating system for houses in Paris.

3 *Normal.* Temperature gradient <40 °C km. These remaining regions are associated with average geothermal conductive heat flow at $\sim 0.06\,\mathrm{W\,m^{-2}}$. It is unlikely that these areas can ever supply geothermal heat at prices competitive to present (finite) or future (renewable) energy supplies.

In each class it is, in principle, possible for heat to be obtained by:

1 *Natural hydrothermal circulation.* In this, water percolates to deep aquifers to be heated to dry steam, vapour/liquid mixtures or hot water. Emissions of each type can be observed in nature. If pressure increases by steam formation at deep levels, spectacular geysers may occur, as at the Geysers near Sacramento in California and in the Wairakei area near Rotorua in New Zealand. Note, however, that liquid water is ejected, and not steam.
2 *Hot igneous systems.* These are associated with heat from semi-molten magma that solidifies to lava. The first power plant using this source was the $3\,\mathrm{MW_e}$ station in Hawaii, completed in 1982.
3 *Dry rock fracturing.* Poorly conducting dry rock, e.g. granite, stores heat over millions of years with a subsequent increase in temperature. Artificial fracturing from boreholes enables water to be pumped through the rock to extract the heat.

In practice, geothermal energy plants in hyperthermal regions are associated with natural hydrothermal systems; in semithermal regions both hydrothermal and hot rock extraction is developed; and normal areas have too small a temperature gradient for commercial interest.

## 15.3 Dry rock and hot aquifer analysis

### 15.3.1 Dry rock

We consider a large mass of dry material extending from near the earth's surface to deep inside the crust, (Figure 15.3). The rock has density $\rho_r$, specific heat capacity $c_r$ and cross-section $A$. With uniform material and no convection, there will be a linear increase of temperature with depth. If $z$ increases *downward* from the surface at $z = 0$,

$$T = T_0 + \frac{\mathrm{d}T}{\mathrm{d}z} z = T_0 + Gz \tag{15.1}$$

Let the minimum useful temperature be $T_1$ at depth $z_1$, so

$$T_1 = T_0 + Gz_1; \qquad z_1 = (T_1 + T_2)/G \tag{15.2}$$

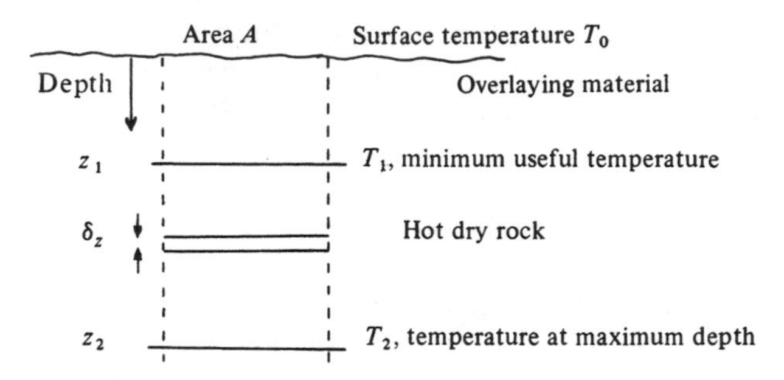


*Figure 15.3* Profile of hot dry rock system for calculating the heat content. Density $\rho$, specific heat capacity $c$, temperature gradient $dT/dz = G$.

The useful heat content $\delta E$, at temperature $T(>T_1)$, in an element of thickness $\delta z$ at depth $z$ is

$$\delta E = (\rho_r A \delta z) c_r (T - T_1) = (\rho_r A \delta z) c_r G(z - z_1) \tag{15.3}$$

The total useful heat content of the rock to depth $z_2$ becomes

$$\begin{aligned} E_0 &= \int_{z=z_1}^{z_2} \rho_r A c_r G(z - z_1) \mathrm{d}z \\ &= \rho_r A c_r G(z_2 - z_1)^2/2 \end{aligned} \tag{15.4}$$

Alternatively, let the average available temperature greater than $T_1$ be $\theta$:

$$\theta = (T_2 - T_1)/2 = G(z_2 - z_1)/2 \tag{15.5}$$

then $E_0 = C_r\theta$, where $C_r$ is the thermal capacity of the rock between $z_1$ and $z_2$,

$$C_r = \rho_r A c_r (z_2 - z_1) \tag{15.6}$$

so as with (15.4),

$$E_0 = \rho_r A c_r G(z_2 - z_1)^2/2 \tag{15.7}$$

Assume heat is extracted from the rock uniformly in proportion to the temperature excess over $T_1$ by a flow of water with volume flow rate $\dot{V}$, density $\rho_w$ and specific heat capacity $c_w$. The water will be heated through a temperature difference of $\theta$ in the near perfect heat exchange process.

Thus

$$\dot{V}\rho_w c_w \theta = -C_r \frac{d\theta}{dt} \tag{15.8}$$

$$\frac{d\theta}{\theta} = -\frac{\dot{V}\rho_w c_w}{C_r} dt = -\frac{dt}{\tau} \tag{15.9}$$

where the time constant $\tau$ is given by

$$\begin{aligned}\tau &= \frac{C_r}{\dot{V}\rho_w c_w} \\ &= \frac{\rho_r A c_r (z_2 - z_1)}{\dot{V}\rho_w c_w} \quad \text{using (15.6)}\end{aligned} \tag{15.10}$$

Hence

$$\theta = \theta_0 e^{-t/\tau} \tag{15.11}$$

The useful heat content is $E = C_r\theta$, so

$$E = E_0 e^{-t/\tau} \equiv E_0 \exp(-t/\tau) \tag{15.12}$$

and

$$\frac{dE}{dt} = \frac{E_0}{\tau} e^{-t/\tau} \tag{15.13}$$

*Example 15.1 after Garnish 1976*

1. Calculate the useful heat content per square kilometre of dry rock granite to a depth of 7 km. Take the geothermal temperature gradient at 40 °C km$^{-1}$, the minimum useful temperature as 140 K above the surface temperature $T_0$, $\rho_r = 2700\,\text{kg}\,\text{m}^{-3}$, $c_r = 820\,\text{J}\,\text{kg}^{-1}\,\text{K}^{-1}$.
2. What is the time constant for useful heat extraction using a water flow rate of $1\,\text{m}^3\,\text{s}^{-1}\,\text{km}^{-2}$?
3. What is the useful heat extraction rate initially and after 10 years?

*Solution*

1. At 7 km the temperature $T_2 = Tot280\,\text{K}$. The minimum useful temperature of 140 K more than $T_0$ occurs at 3.5 km.

So by (15.7),

$$
\begin{aligned}
E_0/A &= \rho_r c_r (z_2 - z_1)(T_2 - T_1)/2 \\
&= (2.7 \times 10^3\,\mathrm{kg\,m^{-3}})(0.82 \times 10^3\,\mathrm{J\,kg^{-1}\,K^{-1}})(3.5\,\mathrm{km})(70\,\mathrm{K}) \\
&= 5.42 \times 10^{17}\,\mathrm{J\,km^{-2}} \qquad (15.14)
\end{aligned}
$$

2 Substituting the appropriate values in (15.10)

$$
\begin{aligned}
\tau &= \frac{\rho_r c_r A (z_2 - z_1)}{\dot{V} \rho_w c_w} \\
&= \left(\frac{1}{1\,\mathrm{m^3\,s^{-1}\,km^{-2}}}\right)\left(\frac{2700}{1000}\right)\left(\frac{820}{4200}\right) \qquad (15.15) \\
&\quad \times (1\,\mathrm{km^{-2}})(3.5\,\mathrm{km}) \\
&= 1.84 \times 10^9\,\mathrm{s} = 58\,\mathrm{y}
\end{aligned}
$$

3 By (15.13),

$$
\left(\frac{\mathrm{d}E}{\mathrm{d}t}\right)_{t=0} = \frac{5.42 \times 10^{17}\,\mathrm{J\,km^{-2}}}{1.84 \times 10^9\,\mathrm{s}} = 294\,\mathrm{MW\,km^{-2}} \qquad (15.16)
$$

$$
\left(\frac{\mathrm{d}E}{\mathrm{d}t}\right)_{t=20\,\mathrm{y}} = 294 \exp(-10/58) = 247\,\mathrm{MW\,km^{-2}} \qquad (15.17)
$$

### *15.3.2 Hot aquifers*

In a hot aquifer, the heat resource lies within a layer of water deep beneath the ground surface, (Figure 15.4). We assume that the thickness of the

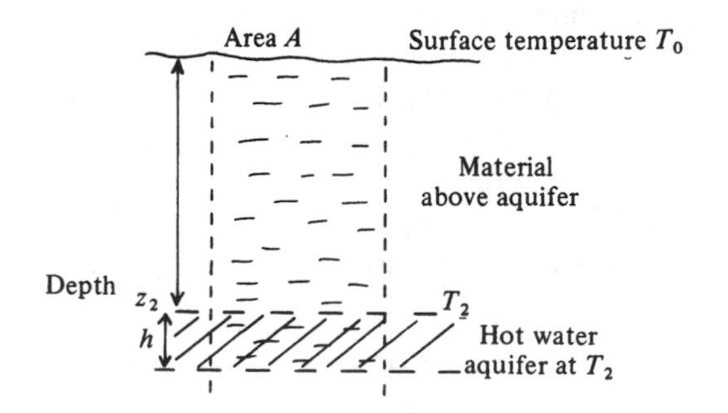

*Figure 15.4* Profile of hot aquifer system for calculating the heat content.

aquifer ($h$) is much less than the depth ($z_2$) below ground level, and that consequently the water is all at temperature $T_2$. The porosity, $p'$, is the fraction of the aquifer containing water, assuming the remaining space to be rock of density $\rho_r$. The minimum useful temperature is $T_1$. The characteristics of the resource are calculated similarly to those for dry rock in Section 15.3.1.
Then

$$T_2 = T_0 + \frac{dT}{dz}z = T_0 + Gz \tag{15.18}$$

$$\frac{E_0}{A} = C_a(T_2 - T_1) \tag{15.19}$$

where

$$C_a = [p'\rho_w c_w + (1-p')\rho_r c_r]h \tag{15.20}$$

As with (15.8) onwards, we calculate the removal of heat by a water volume flow rate $\dot{V}$ at $\theta$ above $T_1$:

$$\dot{V}\rho_w c_w \theta = -C_a \frac{d\theta}{dt} \tag{15.21}$$

So

$$E = E_0 \exp(-t/\tau_a) \tag{15.22}$$

$$\frac{dE}{dt} = -(E_0/\tau_a)\exp(-t/\tau_a) \tag{15.23}$$

and

$$\tau_a = \frac{C_a}{\dot{V}\rho_w c_w} = \frac{[p'\rho_w c_w + (1-p')\rho_r c_r]h}{\dot{V}\rho_w c_w} \tag{15.24}$$

*Example 15.2 (after Garnish, 1976)*

1. Calculate the initial temperature and heat content per square kilometre above 40 °C of an aquifer of thickness 0.5 km, depth 3 km, porosity 5%, undersediments of density $2700\,\mathrm{kg\,m^{-3}}$, specific heat capacity $840\,\mathrm{J\,kg^{-1}\,K^{-1}}$ and temperature gradient $30\,°\mathrm{C\,km^{-1}}$. Suggest a use for the heat if the average surface temperature is 10 °C.
2. What is the time constant for useful heat extraction with a pumped water extraction of $0.1\,\mathrm{m^3\,s^{-1}\,km^{-2}}$?
3. What is the thermal power extracted initially and after 10 years?

*Solution*

1 Initial temperature

$$T_2 = 10\,°\mathrm{C} + (30 \times 3)\mathrm{K} = 100\,°\mathrm{C} \tag{15.25}$$

From (15.20),

$$\begin{aligned} C_a &= [(0.05)(1000)(4200) + (0.95)(2700)(840)] \\ &\quad \times (\mathrm{kg\,m^{-3}\,J\,kg^{-1}\,K^{-1}})(0.5\,\mathrm{km}) \\ &= 1.18 \times 10^{15}\,\mathrm{J\,K^{-1}\,km^{-2}} \end{aligned} \tag{15.26}$$

With (15.19),

$$\begin{aligned} E_0 &= (1.18 \times 10^{15}\,\mathrm{J\,K^{-1}\,km^{-2}})(100 - 40)\,°\mathrm{C} \\ &= 0.71 \times 10^{17}\,\mathrm{J\,km^{-2}} \end{aligned} \tag{15.27}$$

The quality of the energy is suitable for factory processes or household district heating.

2 In (15.24),

$$\begin{aligned} \tau_a &= \frac{(1.2 \times 10^{15}\,\mathrm{J\,K^{-1}\,km^{-2}})}{(0.1\,\mathrm{m^3\,s^{-1}\,km^{-2}})(1000\,\mathrm{kg\,m^{-3}})(4200\,\mathrm{J\,kg^{-1}\,K^{-1}})} \\ &= 2.8 \times 10^9\,\mathrm{s} = 90\,\mathrm{y} \end{aligned} \tag{15.28}$$

3 From (15.23),

$$\begin{aligned} \left(\frac{\mathrm{d}E}{\mathrm{d}t}\right)_{t=0} &= \frac{(0.71 \times 10^{17}\,\mathrm{J\,km^{-2}})}{(2.8 \times 10^9\,\mathrm{s})} \\ &= 25\,\mathrm{MW\,km^{-2}} \end{aligned} \tag{15.29}$$

Check:

$$\begin{aligned} \left(\frac{\mathrm{d}E}{\mathrm{d}t}\right)_{t=0} &= \dot{V}\rho_w c_w (T_2 - T_1) \\ &= (0.1\,\mathrm{m^3\,s^{-1}\,km^{-2}})(1000\,\mathrm{kg\,m^{-3}})(4200\,\mathrm{J\,kg^{-1}\,K^{-1}})(60\,\mathrm{K}) \\ &= 25\,\mathrm{MW\,km^{-2}} \end{aligned}$$

From (15.23),

$$\begin{aligned} \left(\frac{\mathrm{d}E}{\mathrm{d}t}\right)_{t=10\,\mathrm{y}} &= 25\,\mathrm{MW\,km^{-2}}\exp(-10/90) \\ &= 22\,\mathrm{MW\,km^{-2}} \end{aligned} \tag{15.30}$$

## 15.4 Harnessing Geothermal Resources

Geothermal power arises from heat sources having a great range of temperatures and local peculiarities. In general, available temperatures are much less than from furnaces; therefore although much energy is accessible, the thermodynamic quality is poor. The sources share many similarities with industrial waste heat processes and OTEC, see Chapter 14. In this section we shall review the strategy for using geothermal energy.

### *15.4.1 Matching supply and demand*

With a geothermal source it is always sensible to attempt electricity generation since this is a valued product, and the rejected heat can be used in a combined heat and power mode. Electricity can be distributed on a widely dispersed grid and integrates with other national power supplies. Nevertheless, the energy demand for heat at $<100\,°C$ is usually greater than that for electricity, and so the use of geothermal energy as heat is important. Electricity generation will probably be attractive if the source temperature is $>300\,°C$, and unattractive if $<150\,°C$.

Heat cannot be distributed easily over distances greater than $\sim 30\,km$, and so concentrated uses near to the point of supply are needed. In cold climates, household and business district heating schemes make sensible loads if the population density is $\geq 350$ people per $km^2$ ($>100$ premises per $km^2$). Thus a $100\,MW_{th}$ geothermal plant might serve an urban area $\sim 20\,km \times 20\,km$ at $\sim 2\,kW$ per premises. Such geothermal schemes have been long established in Iceland and, on a smaller scale, in New Zealand. Other heating loads are for glasshouse heating (at $60\,MW_{th}\,km^{-2}$ in one installation in northern Europe), fish farming, food drying, factory processes, etc.

Several factors fix the scale of geothermal energy use. The dominant costs are capital costs, especially for the boreholes whose costs increase exponentially with depth. Since temperature increases with depth, and the value of the energy increases with temperature, most schemes settle on optimum borehole depths of $\sim 5\,km$. Consequently, the scale of the energy supply output is usually $\geq 100\,MW$ (electricity and heat for high temperatures, heat only for low temperatures), as shown in Examples 15.1 and 15.2.

The total amount of heat extracted from a geothermal source can be increased by reinjecting the partially cooled water from the above ground heat exchanger. This has the extra advantage of disposing of this water, which may have about $25\,kg\,m^{-3}$ of solute and be a substantial pollutant, see Section 15.5. Nevertheless a substantial extra cost is incurred.

### 15.4.2 Extraction techniques: hydrothermal

The most successful geothermal projects have boreholes sunk into natural water channels in hyperthermal regions, Figure 15.5. This is the method used at Wairakei, New Zealand (Figure 15.9) and at the Geysers, California. Similar methods are used for extraction from hot aquifers in semithermal regions where natural convection can be established from the borehole without extra pumping.

### 15.4.3 Extraction techniques: hot dry rock

Sources of 'hot dry rock' (HDR) are much more abundant than are hydrothermal regions: temperatures of 200 °C are accessible under a significant proportion of the world's landmass. This has motivated considerable research in the USA and Europe on techniques to harness this heat for electricity power generation. One result has been the recognition that few basement rocks are completely dry, but there are many regions where utilisation of their geothermal heat requires 'enhanced geothermal systems', in which reinjection is necessary to maintain commercial production.

In the 1980s, the research group at Los Alamos Scientific Laboratory, USA pioneered methods of fracturing the rock with pressurised cold water around the end of the injection borehole (Figure 15.6). After the initial fracturing, water was pumped down the injection bore to percolate through the hot rock at depths of ~5 km and temperatures ~250 °C before returning through shallower return pipes. Using such methods, complex arrays of injection and return boreholes might, in principle, enable gigawatt supplies of heat to be obtained. However, the technical difficulties and large costs have meant that only a few pilot plants, including a European joint venture at Soultz in the upper Rhine Valley, have exploited HDR. Despite these efforts, by 2004 there was still no commercial power station based on HDR.

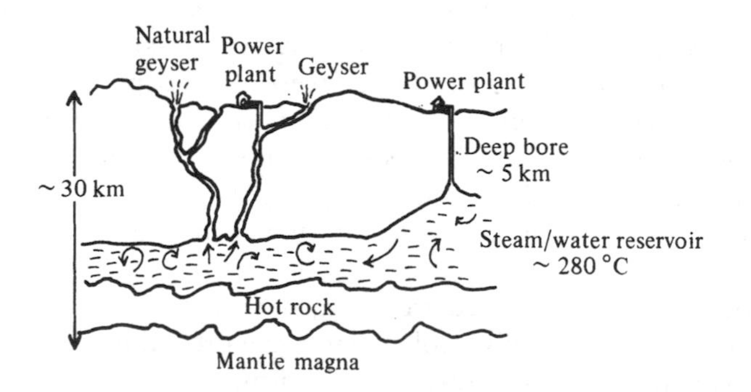


*Figure 15.5* Schematic diagram, not to scale, of hydrothermal power stations in a hyperthermal region, e.g. the Geysers geothermal field, California.

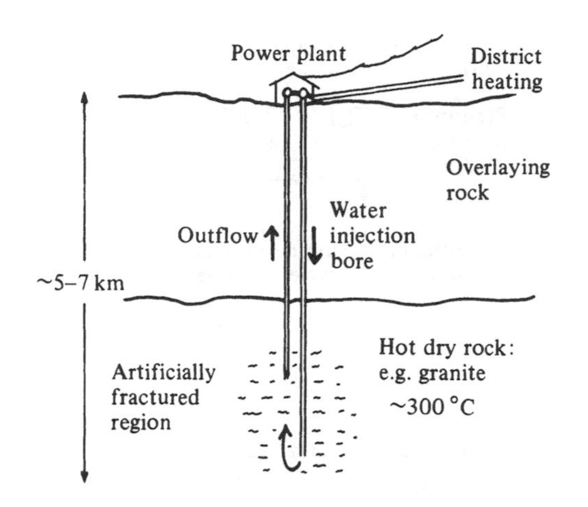


*Figure 15.6* Schematic diagram of heat extraction from a hot dry rock system.

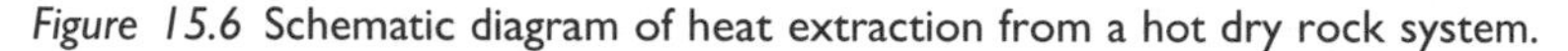

### *15.4.4 Electricity generating systems*

The choice of the heat exchange and turbine system for a particular geothermal source is complex, requiring specialist experience. In 1977, Milora and Tester provided one of the first extensive reviews of the subject, from which Figure 15.7 outlines some of the possible arrangements for the generating plant.

The small temperature increment of the source requires that compounds other than water may be considered to drive the turbines (e.g. toluene, or substitutes for the environmentally unacceptable freons), and novel techniques are needed to improve efficiency. Particular difficulties may occur with heat exchangers, owing to the large concentration of chemicals in the borehole water. In Chapter 14, similar problems with heat exchangers are discussed for OTEC.

## 15.5 Social and environmental aspects

Geothermal power from hydrothermal regions has a proven record of providing safe and very reliable power at costs that are fully competitive with conventional (brown) sources, even without allowing for the external costs of the latter. Capital costs of new systems are about $US 2500 per installed kilowatt (electric) capacity, which are competitive with those of nuclear and hydro power stations. Perhaps the greatest virtue of geothermal energy for a power company is that the power can be provided almost continuously at full rating and does not depend on an intermittent or purchased source of energy. Maintenance requirements are moderate and not expensive.

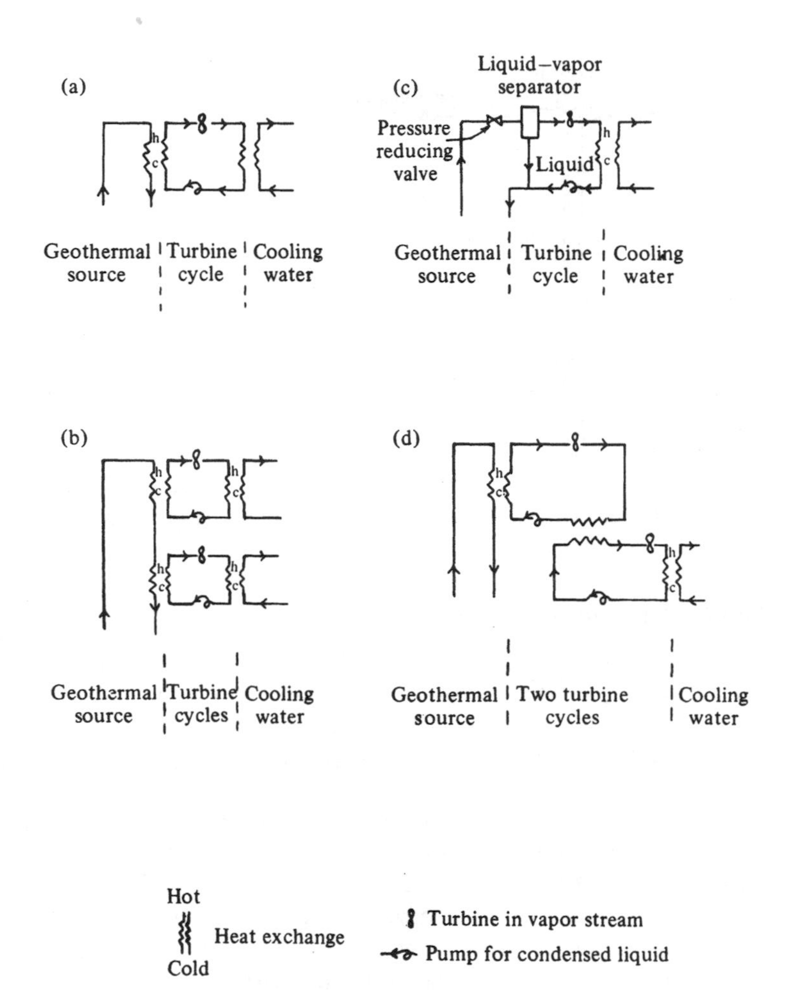


*Figure 15.7* Example of geothermal turbine cycles for electricity generation. (a) Single-fluid cycle, e.g. water, freon substitute. (b) Two-fluid series cycles, e.g. first water, second freon substitute. (c) Direct steam flashing cycle. (d) Topping/bottoming cycles.

In New Zealand, for example, geothermal stations operate at a capacity factor of well over 90% (i.e. their annual output is over 90% of what it would be if run at full rating for 8760 h); consequently they provide about 7% of the national electricity supply from only 4% of the national installed power capacity.

These advantages have encouraged the increase of geothermal capacity in the world's established hydrothermal regions listed in Table 15.1. Figure 15.8 outlines the growth in worldwide geothermal power.

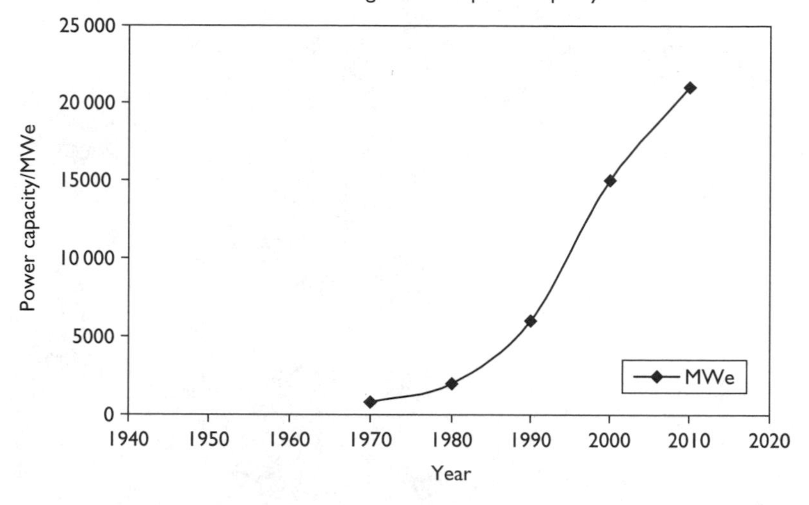


*Figure 15.8* Growth of world geothermal electricity generating capacity/$MW_e$. Data accurate to year previous to the publication of this book, and extrapolated beyond.

McLarty *et al.* (2000) estimated that up to 6 $GW_e$ in the USA and 72 $GW_e$ worldwide could be produced with current technology at known hydrothermal sites, despite hydrothermal regions being relatively rare worldwide, as indicated in Figure 15.2.

Resources of 'hot dry rock' (HDR) are much more abundant than hydrothermal resources: temperatures of 200 °C are accessible under a large proportion of the world's landmass. The study in 2000 by McLarty *et al.* indicated that if these resources could be exploited, e.g. with enhanced technology as Figure 15.6, the potential electricity generating capacity from geothermal sources could be doubled to 19 $GW_e$ in the USA and 138 $GW_e$ worldwide. Unfortunately, even after several decades of technical development, the technology of 'enhanced geothermal systems' to exploit HDR economically is still only at the 'pilot plant' stage.

We illustrate the environmental impacts of geothermal power through the example of the 140 $MW_e$ Wairakei power station in New Zealand (Figure 15.9). The station was built in the 1950s in one of the most geologically active areas in the world. Natural geysers and hot springs abound in the region to the delight of tourists; a major volcanic eruption in 1886 re-formed the landscape about 50 km away, and earthquakes are not uncommon. The wells, top left of the photo, tap into a mixture of water and steam; the hot water is separated with the high-pressure steam

*Figure 15.9* The Wairakei geothermal power station in New Zealand. Well-heads are at top of photo; condensed steam is discharged into the Waikato River at bottom. [Photo by courtesy of Contact Energy, New Zealand.]

being directed through the pipes to the power station at bottom right. At Wairakei there is a considerable overpressure in the boreholes. The clouds of steam at top left come from the hot water boiling as the pressure on it is released.

Removal of the hot water from the ground through the power station resulted in subsidence affecting some local buildings. Consequently, some of the output water flow was re-injected into the area, alleviating the difficulty. There has also been a diminution in the intensity of some of the natural geysers of the area, though most remain substantially unaffected. Note that such negative impact on natural geothermal phenomena is an issue inhibiting the wider use of geothermal power in Japan.

At the bottom of the photograph of Figure 15.9 is the Waikato River, which both provides cooling water and receives the condensed steam and other emissions at discharge. The common emission of $H_2S$ is treated before discharge. The Waikato is one of the largest rivers in the country, so the discharged heat and remaining chemicals are rapidly diluted. An environmental study in 2001 confirmed that downstream concentrations of the chemical elements As, B and Hg, and of dissolved ammonia, were all much less than the limits for native fish.

Geothermal systems also emit the greenhouse gas $CO_2$. Wairakei's emission of 0.03 kg $CO_2$ per $kW_eh$ is less than the average concentration for geothermal power stations of around 0.1 kg $CO_2$ per $kW_eh$ produced, which is much less than the typical value of 1.0 kg $CO_2$ per $kW_eh$ from a coal-fired power station. The benefit/cost ratio of geothermal systems is improved by making use of the low-grade heat leaving the power station. At Wairakei, a prawn farm benefits from this; shown at the left of the photograph.

## Problems

15.1 a A cube of 'hot rock' of side $h$ has its top surface at a depth $d$ below the earth's surface. The rock has a density $\rho$ and specific heat capacity $c$. The material above the cube has thermal conductivity $k$. If the rock is treated as an isothermal mass at temperature $T$ above the earth's surface with no internal heat source, show that the time constant for cooling is given by $\tau = \rho hcd/k$

b Calculate $\tau$ for a cubic mass of granite (of side 10 km, density $2.7 \times 10^3\,\mathrm{kg\,m^{-3}}$, specific heat capacity $0.82 \times 10^3\,\mathrm{J\,kg^{-1}\,K^{-1}}$), that is 10 km below ground under a uniform layer material of thermal conductivity $0.40\,\mathrm{J\,m^{-1}\,s^{-1}\,K^{-1}}$.

c Compare the natural conductive loss of heat from the granite with commercial extraction at 100 MW from the whole mass.

15.2 a Calculate the thermal power produced from the radioactive decay of $^{238}U$ in 5 $\mathrm{km^3}$ of granite. ($^{238}U$ is 99% of the uranium in granite, and is present in average at a concentration of $4 \times 10^{-3}$%. The heat produced by pure $^{238}U$ is $3000\,\mathrm{J\,kg^{-1}\,y^{-1}}$.)

b $^{238}U$ radioactivity represents about 40% of the total radioactive heat source in granite. Is the total radioactive heat a significant continuous source of energy for geothermal supplies?

## Bibliography

### *General*

Armstead, H.C.H. (1983, 2nd edn) *Geothermal Energy – Its Past, Present and Future Contributions to the Needs of Man*, E&FN Spon, London. {An authoritative and fundamental account.}

Armstead, H.C.H. and Tester, J.W. (1987) *Heat Mining: A New Source of Energy*, E&FN Spon. {Treatise on extraction of power from hot dry rock.}

Bowen, R. (1989, 2nd edn) *Geothermal Resources*, Elsevier Applied Science.

Schlumberger Oilfield Review (1989) *The Earth's Heat*, Elsevier Science.

Wilbur, L.C. (1985) *Handbook of Energy Systems Engineering*, Wiley Interscience. ISBN 0–471–86633–4.

## *Specific references*

Environment Waikato (2004) Environmental consent hearing on Contact Energy-Wairakei geothermal operations. On internet at: http://www.ew.govt.nz/resourceconsents/hearingsdecisions/contact.htm#Bookmark_techappendices>{See especially the technical appendices which give detail about its history and present operation}.

Garnish, J.D. (1976) *Geothermal Energy: The Case for Research in the UK*, Department of Energy paper no. 9, HMSO, London. {Succinct evaluation with basic analysis.}

Goodman, L.J. and Love, R.N. (1980) *Geothermal Energy Projects: Planning and Management*, Pergamon Press, New York. {Useful for case studies of completed projects by 1980, especially of Wairakei, New Zealand.}

McLarty, L., Grabowski, P., Etingh, D., and Robertson-Tait, A. (2000) Enhanced geothermal systems R&D in the United States, in *Proceedings of the World Geothermal Congress*, Kyushu, Japan.

Milora, S.L. and Tester, J.W. (1977) *Geothermal Energy as a Source of Electric Power*. MIT Press, Cambridge, Mass. {Evaluates and analyzes the varied and specialised turbine cycles and systems suitable for geothermal heat sources.}

## *Websites and periodicals*

International Geothermal Association (in 2000 <www.demon.co.uk/geosci or http://iga.igg.cnr.it> c/o ENEL DP-PDG, Via A. Pisano 120, 56122 Pisa, Italy.

World Geothermal Congress: conference held every ~5years (e.g. 2000 congress at Kyushu, Japan); proceedings available from the International Geothermal Association.

Chapter 16

# Energy systems, storage and transmission

## 16.1 The importance of energy storage and distribution

Energy is useful only if available when and where it is wanted. Carrying energy to *where* it is wanted is called *distribution* or *transmission*; keeping it available until *when* it is wanted is called *storage*. Within natural ecology, biomass is an energy store for animals and parasites, with seeds becoming a form of distribution. Within society and technology, energy storage, local distribution and long-distance transmission are not new concepts. Fossil and nuclear fuels are effectively energy stores, whose energy density is large, and high-voltage cables allow transmission of electricity. However, as renewable supplies increase, there is a need to develop other storage methods, including secondary fuels, and to sustain and improve distribution and transmission, especially for electricity. As discussed in Chapter 1, renewable energy supplies have different requirements for storage and distribution than do fossil and nuclear energy supplies. Usually the low intensity and widespread location of most renewable sources favour decentralised end-use, and the variable time dependence favours integration of several supplies with storage in a common system. Nevertheless, some renewable sources are of relatively large scale, e.g. large hydroelectric, geothermal and offshore windfarms, and so suitable for relatively intensive use (e.g. aluminium smelting and high-voltage power transmission).

Since the use of renewable energy supplies constitutes a diversion of a continuing natural *flow* of energy, there are problems in matching supply and demand in the time domain, i.e. in matching the *rate* at which energy is used. This varies with time on scales of months (e.g. house heating in temperate climates), days (e.g. artificial lighting) and even seconds (e.g. starting motors). In contrast to fossil fuels and nuclear power, the primary input power of renewable energy sources is outside our control. As discussed more fully in Chapter 1, we have the choice of either *matching the load* to the availability of renewable energy supply or *storing the energy* for future use. Energy can be stored in many forms, i.e. chemical, heat, electric, potential or kinetic energy. Moreover by linking supplies and consumption in a grid (e.g. hot water, gas pipe, vehicle transportation and networked electricity), the controlled system has access to forms of *virtual storage*, e.g. as pressurised gas in pipes, export and import of electricity. The extent of this virtual storage varies as the grid-intensive parameters are adjusted (e.g. voltage, temperature, pressure or speed). Incorporation of small and intermediate scale renewables sources in a widespread grid system is called *embedded generation*, especially for electricity.

Table 16.1 and Figure 16.1 summarise the performance of various storage mechanisms. 'Performance' can be measured in units such as $MJ\,\$^{-1}$, $MJ\,m^{-3}$ and $MJ\,kg^{-1}$. Of these, the first unit (cost effectiveness) is usually the deciding factor for commerce, but is the hardest to estimate, see Chapter 17; note that 'cost' here is wholesale cost before taxes and that taxation, especially of transport fuels, varies greatly between countries. The second unit is important when space is at a premium, e.g. in buildings of fixed size. The third unit is considered when weight is vital, e.g. in aircraft. In this chapter we show how these performance figures are estimated.

## 16.2 Biological storage

The growth of plants by photosynthesis, and the consequent storage of the solar energy, is fully discussed in Chapter 10. This energy is released in the combustion of biological and, from prehistory, of fossil fuel material. The heats of combustion of many of these materials are listed in Appendix B, Table B.6. The many chemical paths for producing biofuels are described in Chapter 11, see Figure 11.2.

Some of the biofuels are liquids and gases that may be used in internal combustion engines, and are therefore important to replace and augment conventional petroleum fuels, especially for vehicles. The generation of electricity by such fuels in diesel and spark-ignition engines and in gas turbines is also important. These and other aspects of biological energy storage and use are fully discussed in Chapter 11.

## 16.3 Chemical storage

### 16.3.1 Introduction

Energy can be held in the bonds of many chemical compounds and released by exothermic reactions, notably combustion. Sometimes it is necessary to apply heat or other catalysts, e.g. enzymes, to promote the desired reaction. Biochemical compounds are a special case. Here we discuss the most important inorganic compounds which have been suggested as practical energy stores.

*Table 16.1* Storage devices and their performance[a]

| *Store* | *Energy density* | | *Operating temp* | *Likely commercial Development time* | *Operating value* | *Conversion Type* | *Efficiency* |
|---|---|---|---|---|---|---|---|
| | (MJ kg$^{-1}$) | (MJ L$^{-1}$) | (°C) | *per year* | (MJ \$US$^{-1}$) | | (%) |
| *conventional fuels* | | | | | | | |
| Diesel oil | 45 | 39 | ambient | in use | 200[b] | chem->work | 30 |
| Coal | 29 | 45 | ambient | in use | 500[b] | chem->work | 30 |
| Wood | 154 | 7 | ambient | in use | $200 - \infty$ | chem ->heat | 60 |
| *other chemicals* | | | | | | | |
| Hydrogen gas | 140 | 1.7[d] | (−253) to (−30) | 10 | 0.1 – 10[b] | elec->chem | 60 |
| Ammonia (to $N_2 + H_2$) | 2.9 | 0.3[d] | 0–700 | 10 | ~1[b] | heat ->chem | 70 |
| $FeTiH_{1.7}$ | 1.8 | 20 | 100 | 10 | 1[b] | chem ->chem | 90 |
| *Sensible heat* | | | | | | | |
| Water | 0.2 | 0.2 | 20–100 | in use | 3–100[c] | heat ->heat | 50 – 100[e] |
| Cast iron | 0.05 | 0.4 | 20–400 | in use | 0.1[c] | heat ->heat | 50 – 90[e] |
| *Heat (phase change)* | | | | | | | |
| Steam | 2.2 | 0.02[f] | 100–300 | in use | 10 | heat ->heat | 70[d] |
| $Na_2SO_4.10H_2O$ | 0.25 | 0.29 | 32 | in use | 2[c] | heat ->heat | 80 |

*Table 16.1* (Continued)

| *Store* | *Energy density* | | *Operating temp* | *Likely commercial Development time* | *Operating value* | *Conversion Type* | *Efficiency* |
|---|---|---|---|---|---|---|---|
| | (MJ kg$^{-1}$) | (MJ L$^{-1}$) | (°C) | *per year* | (MJ \$US$^{-1}$) | | (%) |
| *Electrical* | | | | | | | |
| Capacitors | – | $10^{-6}$ | | unlikely[g] | | | |
| Superconducting magnets | – | $10^{-3}$ | | unlikely[g] | | | |
| Batteries (in practice) | | | | | | | |
| lead-acid | 0.15 | 0.29 | ambient | in use | 0.02 | elec ->elec | 80 |
| sodium-sulphur | 0.7 | | 350 | 5 | 0.01 | elec ->elec | 85 |
| NiCd | 0.2 | | ambient | in use | 0.01 | elec ->elec | 75 |
| ZnBr | 0.3 | | ambient | 5 | 0.02 | elec ->elec | 70 |
| *Mechanical* | | | | | | | |
| Pumped hydro | 0.001 | 0.001 | ambient | in use | 1–50 | elec ->elec | 80 |
| Flywheel (steel disc) | 0.05 | 0.4 | ambient | in use | 0.05[c] | elec ->elec | 80 |
| Flywheel (composite) | 0.05 | 0.15 | ambient | in use | 0.05[c] | elec ->elec | 80 |
| Compressed air | 0.2–2 | 2[f] | 20–1000 | in use | 0.3[c] | elec ->elec | 50 |

Notes

a These figures are for 'typical' operation and are only approximations for a particular application. This is especially true for those relating to commercial applications and to costs. Here 'costs' are approximate wholesale prices before taxation. Table adapted and updated for this edition from Jensen and Sørensen (1984).

b Energy throughput per unit cost.

c Energy capacity per unit cost – e.g. If a battery cycles 10 times to 50% discharge, its throughput per unit cost would be (0.1 MJ US\$$^{-1}$)(10)(50%) because it may be used repeatedly.

d At 150 atmospheres pressure.

e Depends on time and heat leaks.

f At 20 atmospheres pressure.

g Unlikely for *large-scale* energy storage, although in use in 2004 for the avoidance of rapid voltage-change transients on electricity supply, i.e. 'spiking'.

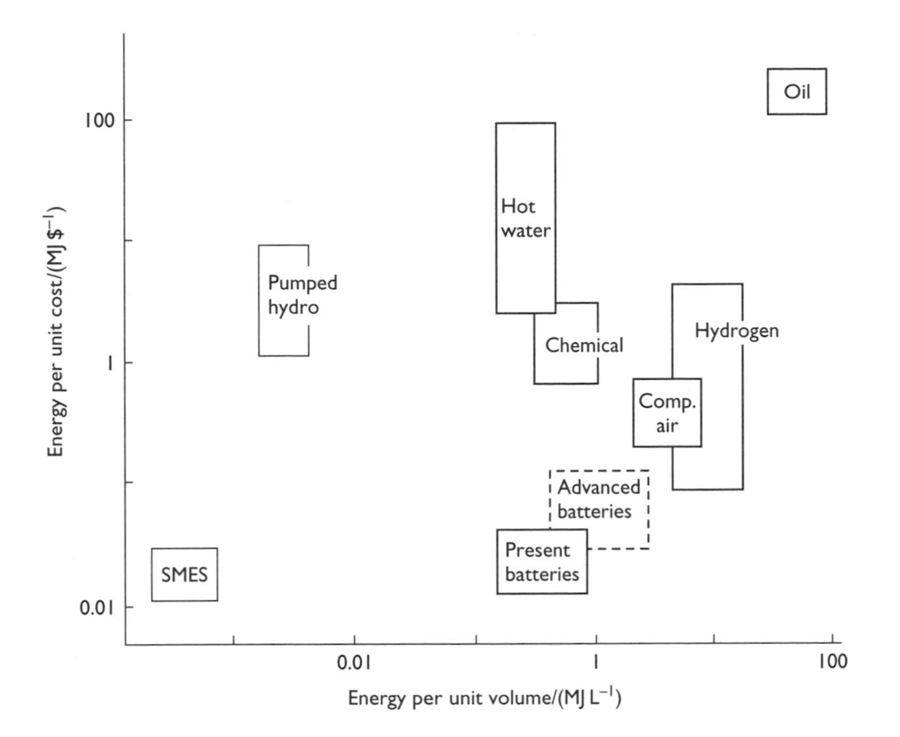


*Figure 16.1* Energy per unit cost versus energy per unit volume of storage methods (indicative 2004 prices in $US). NB: logarithmic scales are used. For more details see Table 16.1. SMES – superconducting electromagnetic energy storage. Note the superiority of 'oil'; the range shown for oil includes petroleum and most liquid biofuels.

### 16.3.2 Hydrogen

Hydrogen can be made from water by electrolysis, using any source of dc electricity. The gas can be stored, distributed and burnt to release heat. The only product of combustion is water, so at end-use no pollution results. The enthalpy change is $\Delta H = -242\,\mathrm{kJ\,mol^{-1}}$; i.e. 242 kJ is released for every mole (18 g) of $H_2O$ formed. Hydrogen (with CO in the form of 'town gas' made from coal) was used for many years as an energy store and supply, and there is no overriding technical reason why hydrogen-based systems could not come into wide use again. Note, however, that most hydrogen is made now from fossil fuels.

Electrolysis is a well-established commercial process yielding pure hydrogen, but generally efficiencies have been only ~60%. Some of this loss is due to electrical resistance in the circuit, especially around the electrodes where the evolving bubbles of gas block the current carrying ions in the water. Electrodes with 'bubble-removing mechanisms' should be advantageous. The best electrodes have large porosity, so giving a greater effective area and thus allowing a larger current density, which implies having fewer cells and reduced cost for a given gas output. Efficiencies of ~80% have been so obtained, and can be increased further by using, usually expensive, catalysts. High temperatures also promote the decomposition of water. The change in Gibbs free energy associated with a reversible electrochemical reaction at absolute temperature $T$ is

$$\begin{aligned}\Delta G &= nF\xi \\ &= \Delta H - T\Delta S\end{aligned} \tag{16.1}$$

where $\xi$ is the electrical potential, $\Delta H$ is the enthalpy change and $\Delta S$ is the entropy change, $F = 96\,500\,\mathrm{C\,mol^{-1}}$ is Faraday's constant, and $n$ is the number of moles of reactant.
The decomposition reaction

$$H_2O \rightarrow H_2 + \frac{1}{2}O_2 \tag{16.2}$$

has $\Delta G$, $\Delta H$ and $\Delta S$ all positive. Therefore from (16.1), increasing $T$ decreases the electric potential $\xi$ required for decomposition. Problem 16.1 shows that $\xi = 0$ for $T \sim 2700\,\mathrm{K}$, so it is impracticable to decompose water by straightforward heating. A more promising strategy is to replace some of the input electrical energy by heat from a cheaper source. Heat at $T \sim 1000\,\mathrm{K}$ from solar concentrators may be cheaper than electricity, and this may be the cheapest route to produce hydrogen.

A technical difficulty in the electrolysis of seawater is that chlorine may also be evolved at the 'oxygen' electrode; approximate chemical calculations suggest that $O_2$ can be kept pure if the applied voltage per cell is less than 1.8 V, but unfortunately this limits the current density, so electrodes of large surface area would be needed. Several other methods of producing hydrogen without using fossil fuels have been tried in the laboratory, including special algae which 'photosynthesise' $H_2$, see Section 10.7; but none has yet shown worthwhile efficiencies.

To store hydrogen in large quantities is not trivial. Most promising is the use of underground caverns, such as those from which natural gas is now extracted, but storage of gas, even if compressed, is bulky. Hydrogen can be liquefied, but since its boiling point is 20 K, i.e. −253 °C, such stores are expensive to build and operate, requiring continued refrigeration. Chemical storage as metal hydrides, from which the hydrogen can be released by heating, is more manageable and allows large volumes of $H_2$ to be stored, see Table 16.1. For example

$$\mathrm{FeTiH_{1.7}} \overset{T\sim 50\,^\circ\mathrm{C}}{\rightarrow} \mathrm{FeTiH_{0.1}} + 0.8\mathrm{H_2} \tag{16.3}$$

This reaction is reversible, so a portable hydride-store can be replenished with hydrogen at a central 'filling station'. The heat released in this process can be used for district heating, and the portable hydride-store can be used as the 'fuel tank' of a vehicle. The main difficulty is the weight and cost of the metals used (Table 16.1). Hydrogen could also be distributed through the extensive pipeline networks already used for delivering natural gas in many countries, although hydrogen carries less energy per unit volume than methane. Note that fuel cells generate electricity from hydrogen and oxygen from air (Section 16.6).

### 16.3.3 Ammonia

Unlike water, ammonia can be dissociated at realisable temperatures:

$$N_2 + 3H_2 \rightleftharpoons 2NH_3 \tag{16.4}$$

In conjunction with a heat engine, these reactions form the basis of systems that may be the most efficient way to generate continuous electrical power from solar heat. The system proposed by Carden is described in Section 6.9.1.

## 16.4 Heat storage

A substantial fraction of world energy use is as low temperature heat. For example, Figure 16.2 shows the demand in Britain for total energy and for space heating. Although details change from year to year, the conclusion remains that in winter over half of the national energy consumption is for space heating in buildings at temperatures of about $18 \pm 3\,°\text{C}$. It is usually not sensible to meet this demand for *heat* from the best thermodynamic quality energy supplies (Section 1.4.3), since these should be saved for electricity generation, engines and motor drives. Thus, for example, it is better to capture solar heat gains (Section 6.3), and then to keep buildings within comfortable temperatures using the averaging and heat-storage characteristics of the building mass. Heat storage also provides a way of fruitfully using 'waste' energy utilised or recovered from other processes, e.g. by load control devices (Section 1.5.4).

In the higher latitudes, solar heat *supply* is significantly greater in summer than in winter, see Figures 4.7 and 4.10, yet the *demand* for heat is greatest in winter. Therefore the maximum benefit from solar heat requires heat storage for at least 3 months, say in hot water in an underground enclosure. To consider this possibility, we estimate the time, $t_{\text{loss}}$, for such a heat store to have 50% of its content withdrawn while maintaining a uniform temperature $T_{\text{s}}$. Assume that the immediate environment, e.g. the soil temperature, has constant temperature $T_{\text{a}}$. The heat balance equation is

$$mc\frac{dT_{\text{s}}}{dt} = -\frac{T_{\text{s}} - T_{\text{a}}}{R} \tag{16.5}$$

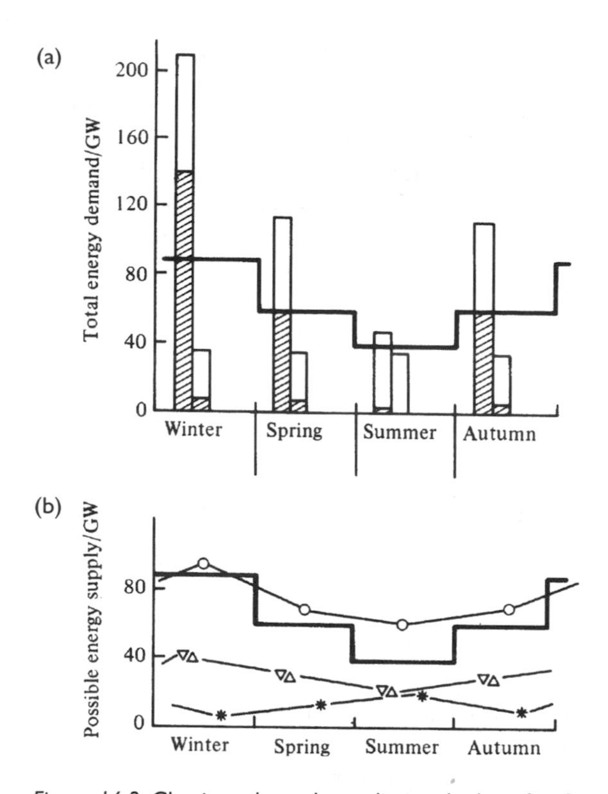


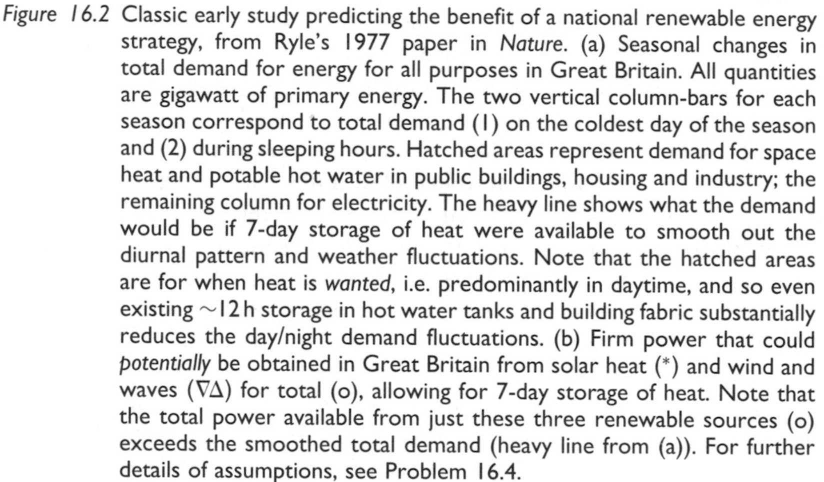


*Figure 16.2* Classic early study predicting the benefit of a national renewable energy strategy, from Ryle's 1977 paper in *Nature*. (a) Seasonal changes in total demand for energy for all purposes in Great Britain. All quantities are gigawatt of primary energy. The two vertical column-bars for each season correspond to total demand (1) on the coldest day of the season and (2) during sleeping hours. Hatched areas represent demand for space heat and potable hot water in public buildings, housing and industry; the remaining column for electricity. The heavy line shows what the demand would be if 7-day storage of heat were available to smooth out the diurnal pattern and weather fluctuations. Note that the hatched areas are for when heat is *wanted*, i.e. predominantly in daytime, and so even existing ~12 h storage in hot water tanks and building fabric substantially reduces the day/night demand fluctuations. (b) Firm power that could *potentially* be obtained in Great Britain from solar heat (*) and wind and waves (∇Δ) for total (o), allowing for 7-day storage of heat. Note that the total power available from just these three renewable sources (o) exceeds the smoothed total demand (heavy line from (a)). For further details of assumptions, see Problem 16.4.

where $mc$ is its heat capacity, and $R$ is the thermal resistance between the store and the surroundings. The solution of (16.5) is

$$\frac{T_s - T_a}{T_s(0) - T_a} = \exp\left(-\frac{t}{mcR}\right) \tag{16.6}$$

from which it follows that

$$t_{loss} = 1.3mcR \tag{16.7}$$

If the store is a sphere of radius $a$, the thermal resistance is $R = r/4\pi a^2$, where $r$ is the thermal resistivity of unit area, and $m = 4\pi a^3 \rho/3$, so for a sphere,

$$t_{loss} = 0.43\rho cra \tag{16.8}$$

*Example 16.1 Size and insulation of a domestic heat store*
A small well-insulated passive solar house requires an average internal heat supply of 1.0 kW. Together with the free gains of lighting, etc., this will maintain an internal temperature of 20°C. It is decided to build a hot water store in a rectangular tank whose top forms the floor of the house, and of area 200 m$^2$. The heating must be adequate for 100 days as all the heat loss from the tank passes by conduction through the floor, and as the water cools from an initial 60°C to a final 40°C.

1 Calculate the volume of the tank.
2 Calculate the thermal resistivity of the heat path from the tank to the floor.
3 Suggest how the tank should be enclosed thermally.
4 What is the energy density of storage?

*Solution*
1

$$\text{Heat required} = (1\,\text{kW})(100\,\text{day})(24\,\text{h}\,\text{day}^{-1})(3.6\,\text{MJ}\,\text{kWh}^{-1})$$
$$= 8640\,\text{MJ}$$
$$\text{Volume of water} = \frac{(8640\,\text{MJ})}{(1000\,\text{kg}\,\text{m}^{-3})(4200\,\text{J}\,\text{kg}^{-1}\,\text{K}^{-1})(20\,\text{K})}$$
$$= 103\,\text{m}^3$$
$$\text{Depth of tank} = (103\,\text{m}^3)/(200\,\text{m}^2) = 0.5\,\text{m}$$

2 Assume the heat only leaves through the top of the tank. From (16.7),

$$R = \frac{(100\,\text{day})(86\,400\,\text{s day}^{-1})}{(1.3)(103\,\text{m}^3)(1000\,\text{kg m}^{-3})(4200\,\text{J kg}^{-1}\,\text{K}^{-1})} = 0.0154\,\text{K W}^{-1}$$

From (3.6) the thermal resistivity $r$

$$\begin{aligned} &= R \cdot (\text{area}) \\ &= (0.0154\,\text{K W}^{-1})(200\,\text{m}^2) \\ &= 3.1\,\text{m}^2\,\text{K W}^{-1} \end{aligned}$$

3 Insulating material (e.g. dry expanded polystyrene) has a thermal conductivity $k \sim 0.04\,\text{W m}^{-1}\,\text{K}^{-1}$. A satisfactory layer on top of the tank, protected against excess pressure, would have a depth

$$d = (3.1\,\text{m}^2\,\text{K W}^{-1})(0.04\,\text{W m}^{-1}\,\text{K}^{-1}) = 12\,\text{cm}$$

To avoid unwanted heat loss, the base and sides should be insulated by the equivalent of 50 cm of dry expanded polystyrene.

4 Energy density of the used storage above 40°C = $(8640\,\text{MJ})/(103\,\text{m}^3) = 84\,\text{MJ m}^{-3}$

Energy density above ambient house temperature at 20°C = $168\,\text{MJ m}^{-3}$

*Note*: An active method of extracting the heat by forced convection through a heat exchanger would enable better control, a smaller initial temperature and/or a smaller tank.

Example 16.1 shows that 3-month heat storage is realistic if this forms part of the initial design criteria, and if other aspects of the construction are considered. These include the best standards of thermal insulation with damp-proof barriers, controlled ventilation (best with recycling of heat), and the inclusion of free gains from lighting, cooking and metabolism. Examples exist of such high technology houses, and the best also have imaginative architectural features so that they are pleasant to live in, see Section 6.3. Many such buildings utilise rock bed storage, rather than the water system of the example. It follows from Example 16.1 that short-term heat storage of about 4 days is easily possible, with the fabric of the building used as the store. Similarly thermal capacity and cold storage can have important implications for building design in hot weather conditions.

Materials that change phase offer a much larger heat capacity, over a limited temperature range, than systems using sensible heat. For example, Glauber's salt ($Na_2SO_4 \cdot 10H_2O$) has been used as a store for room heating. It decomposes at 32 °C to a saturated solution of $Na_2SO_4$ and an anhydrous residue of $Na_2SO_4$. This reaction is reversible and evolves $250\,kJ\,kg^{-1}$ or $\sim 650\,MJ\,m^{-3}$. Since much of the cost of a store for house heating is associated with the construction, such stores may be cheaper overall than simple water tanks of less energy density per unit volume. Nevertheless, this seemingly simple method requires practical difficulties to be overcome. In particular, the solid and liquid phases often separate spatially, so recombination is prevented; consequently, without mixing, the system becomes inefficient after many cycles.

## 16.5 Electrical storage: batteries and accumulators

Electricity is a high quality form of energy, and therefore great effort is made to find cheap and efficient means for storing it. A device that has electricity both as input and output is called an (electrical) accumulator or (electrical) storage battery. Batteries are an essential component of most autonomous power systems (especially with photovoltaic and small wind turbine generation), of standby and emergency power systems, and of electric vehicles. There is also a research expectation that solar energy could be stored directly in commercial photochemical cells (Section 10.7.3).

### *16.5.1 The lead acid battery*

Although many electrochemical reactions are reversible in theory, few are suitable for a practical storage battery, which will be required to cycle hundreds of times between charging and discharging currents of 1–100 A or more. The most widely used storage battery is the lead acid battery, invented by Planté in 1860 and continuously developed since.

Such a battery is built up from cells, one of which is shown schematically in Figure 16.3. As in all electrochemical cells, there are two electrode 'plates' immersed in a conducting solution (electrolyte). In this case the electrodes are in the form of grids holding pastes of lead (Pb) and lead dioxide respectively; the pastes are made from powders to increase surface area in 'spongy' form. Electrodes shaped as tubes give added mechanical strength and resist 'shedding' (see later), and so are suitable for deep discharge. The electrolyte is sulphuric acid, which ionises as follows:

$$H_2SO_4 \rightarrow H^+ + HSO_4^- \tag{16.9}$$

During *discharge*, the reaction at the negative electrode is

$$Pb + HSO_4^- \rightarrow PbSO_4 + H^+ + 2e^- \tag{16.10}$$

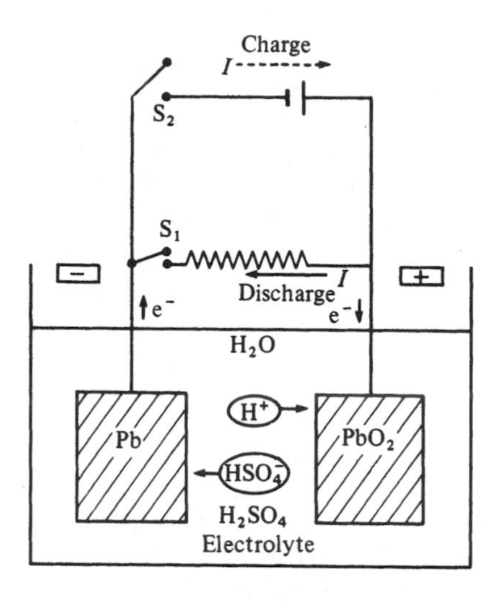


*Figure 16.3* Schematic diagram of lead acid cell. The charge carriers move in the direction shown during the discharge reactions (16.10) and (16.11). The reactions and carrier movements are reversed during charging (switch S1 open and S2 closed).

Spongy lead is oxidised to $Pb^{2+}$, which is deposited as $PbSO_4$ crystals. The smaller density sulphate takes the place of the Pb paste in the plate and, having larger molecular form, causes mechanical expansion. The electrons so liberated travel through the external circuit to the positive electrode, where they contribute to the reaction:

$$PbO_2 + HSO_4^- + 3H^+ + 2e^- \rightarrow PbSO_4 + 2H_2O \qquad (16.11)$$

Thus $PbSO_4$ replaces the $PbO_2$ in that plate, with similar, but less disruptive, mechanical effects than in the negative plate. The electrical current through the solution is carried by $H^+$ and $HSO_4^-$ ions from the sulphuric acid electrolyte, which themselves take part in the plate reactions. Transportable 'gelled cells' have this electrolyte immobilised in pyrogenic silica, with the fibrous glass mat separator giving open gas paths for the release of hydrogen and oxygen in overcharge. Although this makes them relatively expensive, they are safer to use and transport, since there is no danger of spilling highly corrosive sulphuric acid, and are 'maintenance free'.

Knowing the reactions involved and the corresponding standard electrode potentials (given in chemical tables), the theoretical energy density of any proposed battery can be calculated.

*Example 16.2 Theoretical energy density of lead acid battery*
The reactions (16.9) and (16.10) show that to transfer 2 mol of electrons requires

1 *mol* Pb $= 207\,g$
1 *mol* $PbO_2$ $= 239\,g$
2 *mol* $H_2SO_4 = 196\,g$
Total active material 642 g

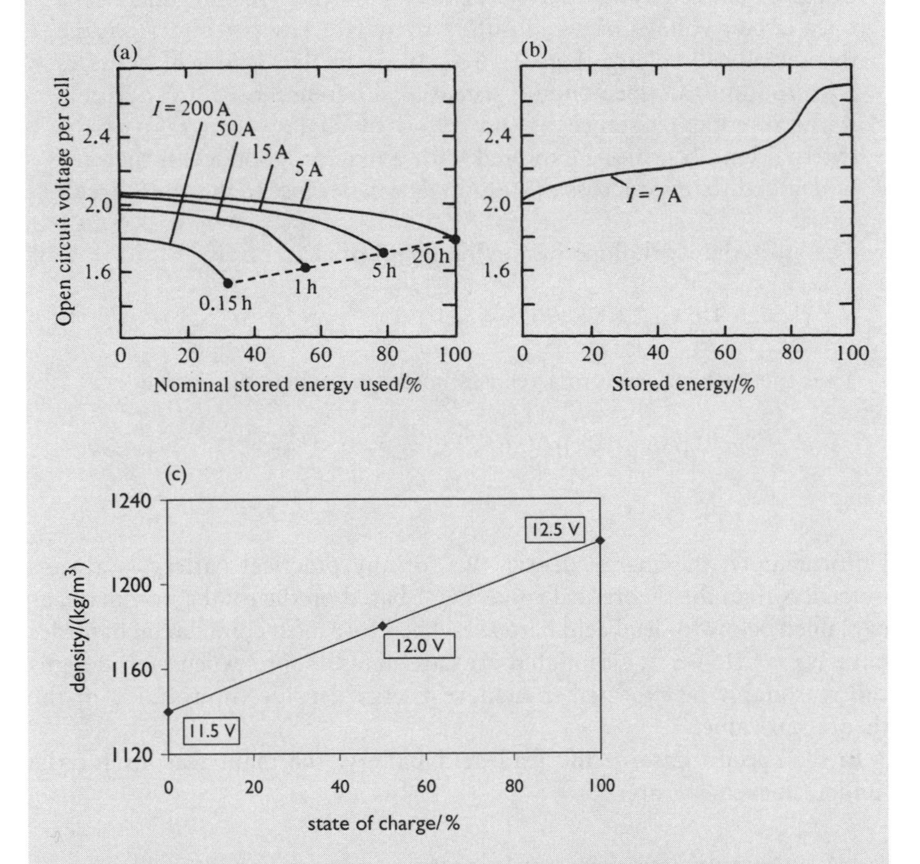


*Figure 16.4* Operating characteristics of a typical lead acid battery (SLI type of about 100 Ah nominal capacity). (a) *Discharge*. The curves are labelled by the discharge current (assumed steady) and by the time taken to 'fully' discharge at that current. (b) *Charge*. The curve is for charging at a constant low current. (c) *Density of electrolyte* as function of state of charge. After Crompton (2000).

But 2 mol of electrons represent a charge:
$(2\ mol)(-1.60 \times 10^{-19}\,C)(6.02 \times 10^{23}\ mol^{-1})$
$= -(2)(9.6 \times 10^4)\,C = -1.93 \times 10^5\,C$

The electrode potential, under standard conditions of concentration, for ($Pb/PbSO_4$) is 0.30 V and for ($PbSO_4/Pb_4^+$) is −1.62 V. So the theoretical cell-EMF at standard conditions for ($Pb/PbSO_4/H_2SO_4/PbSO_4/PbO_2$) is $\xi_{cell} = +1.92\,V$, with the $PbO_2$ plate positive, according to the IUPAC sign convention.

The actual cell EMF depends on the concentration of reagents, and can be calculated by standard electrochemical methods. In general, the open-circuit voltage of a cell differs by only a few per cent from the theoretical cell voltage (Figure 16.4). In particular, lead acid batteries are produced at open-circuit potential difference of 2.0 V per cell. If the internal resistance of the cell is much less than that of the external load (as can be expected with a new or 'good' cell), then the potential difference across the terminals will be close to the open-circuit value.

Therefore the work done in moving 2 mol of electrons is

$$(1.93 \times 10^5\,C)(2.0\,V) = 0.386 \times 10^6\,J$$

Thus the energy stored in 1.0 kg of active ingredients is, in theory,

$$W_m^{(0)} = (0.386 \times 10^6\,J)/(0.642\,kg) = 0.60\,MJ\,kg^{-1}$$

Unfortunately, the energy density $W_m$ of any practical battery is always much less than the theoretical value $W_m^{(0)}$ based on the total active mass, as explained below for lead acid batteries. Therefore most commercial batteries have $W_m \sim 0.15 W_m^{(0)}$, although more careful (and more expensive!) designs can reasonably be expected to achieve energy densities up to 25% of the theoretical values.

In the specific case of the lead acid battery, the main reasons for the 'underachievement' are:

1 A working battery necessarily contains non-active materials, e.g. the case, the separators (which prevent the electrodes short-circuiting) and the water in which the acid is dissolved. Moreover, the acid concentration must not be too large, since the battery would then discharge itself. Since the mass of actual battery contents exceeds the mass of the active ingredients, the energy density based on the mass of the whole

battery is less than the theoretical value based on the active mass alone. However, this factor is not of great importance for stationary batteries.

2 The reactions cannot be allowed to go to completion. If all the lead were consumed by reaction (16.10) there would be no electrode left for the reverse reaction to operate, i.e. the battery could not be cycled. Similarly, if the concentration of $H_2SO_4$ is allowed to reduce too much, the electrolyte ceases to be an adequate conductor. In practice, many battery types should not be allowed to discharge more than about 50% of total potential stored energy, or they may be ruined. However, specially designed batteries do allow 'deep discharge' beyond 50%.

A further limitation of real batteries is familiar to all motor vehicle owners: they do not last forever. Solid Pb is almost twice as dense as the $PbSO_4$ found in the discharge reaction (16.10). Therefore it is difficult to fit the $PbSO_4$ crystals into the space originally occupied by the Pb paste in the negative electrode. After many charge/discharge cycles, the repeated expansion and contraction causes plate material and some $PbSO_4$ to fall to the bottom of the cell. This constitutes an irreversible loss of active material. This loss is worse if the battery is allowed to fully discharge; indeed, it may rapidly become impossible to recharge the battery. In addition, the 'shed' material may provide an electrically conducting link between plates, so increasing 'self-discharge'. Storage batteries should have a generous space below the plates so debris can accumulate without short-circuiting the electrodes.

The other main factor limiting the life of even a well-maintained battery is self-discharge of the positive electrode. This is particularly acute in vehicle SLI (Starting, Lighting and Ignition) batteries in which the grid is not pure Pb but usually a lead–antimony–calcium alloy. Electrode plates with antimony are physically stronger and better able to stand the mechanical stresses during motion. Unfortunately antimony promotes the reaction

$$5PbO_2 + 2Sb + 6H_2SO_4 \rightarrow (SbO_2)_2SO_4 + 5PbSO_4 + 6H_2O \qquad (16.12)$$

which also slowly, but irreversibly, removes active material from the battery. Thus batteries designed for use with motor vehicles do not usually perform well in photovoltaic and wind-power systems.

Batteries for stationary applications, e.g. photovoltaic lighting systems, can use Sb-free plates, and have longer life (usually at least 8 y and perhaps as long as 20 y), but only if charged in a controlled manner and if not excessively and frequently discharged. The performance of a battery depends on the current at which it is charged and discharged, and the depth to which it is regularly discharged. Figure 16.4(a) shows the *discharge characteristics* of a typical lead acid car battery. Its nominal capacity is $Q_{20} = 100\,\text{Ah}$, which is the charge which can be extracted if it is discharged at a constant current over 20 h (usually labelled $I_{20}$). The voltage per cell of a new battery during this discharge should drop only slightly from 2.07 to 1.97 V, as the first 60% of $Q_{20}$ is discharged. This discharge removes dense $HSO_4^-$ ions from the electrolyte solution, and stores them as solid $PbSO_4$ in the electrodes, by reactions (16.10) and (16.11), thereby reducing the density of the electrolyte solution as in Figure 16.4(c). Thus the density of the 'battery acid', measured with a hydrometer, can be used as a measure of the state of charge of the battery. If the same battery is discharged between the same voltages over about 1 h, its voltage drops much more sharply, and the total charge which can be removed from it may be only about $0.5Q_{20}$. This is because the rate of reaction of the electrodes is limited by the rate at which the reactants can diffuse into contact with each other. A rapid build-up of reaction products ($PbSO_4$ in particular) can block this contact. Moreover the internal resistance across this $PbSO_4$ layer reduces the voltage available from the cell.

A set of *charging characteristics* for the same battery is shown in Figure 16.4 (b). To commence charging, an EMF of at least 2.1 V per cell is required. The voltage required initially increases slowly but increases rapidly to about 2.6 V per cell as the battery nears full charge (if constant charging current is maintained). This is because the water in the cell begins to electrolyse.

When the cell is overcharged, $H_2$ gas will be released. Such 'bubbling' can benefit the battery by mixing the electrolyte and so lessening battery stratification; indeed sophisticated charge controllers arrange for this to happen periodically. However, excessive gas release from the electrolysis requires the electrolyte to be 'topped up' with distilled water, and the emitted $H_2$ may produce an explosive mixture with air and so has to be ducted away. *Sealed batteries* – sometimes sold as 'maintenance free' – have a catalyst in the top of the battery over which electrolysed hydrogen can combine with oxygen to reform water within the battery casing, so that 'topping up' the electrolyte with distilled water is not necessary. Extreme overcharging may cause mechanical damage within the cell and may raise the concentration of acid to the point where the ions are not mobile enough to allow the battery to work. Many cycles of mild charging and discharging, e.g. as in small photovoltaic power systems, cause large $PbSO_4$ crystals to develop within the plates and effectively remove active material, as well as causing mechanical damage. In such conditions, occasional deep-discharging may reactivate the battery.

The overall lesson is that charge/discharge control is essential for long battery life; at least charging at constant voltage and at best having a sophisticated controller allowing occasional de-stratification bubbling, controlled charging and discharging currents, voltage cut-offs and, perhaps, occasional deep-discharging. A good battery has extremely small internal impedance (<0.1 Ω) and is capable of delivering large currents at high frequency. The 'farad capacity' is very small, despite the 'charge capacity' being large, so do not be misled by the two distinct meanings of the word 'capacity'.

Development of improved lead acid batteries still continues, producing a variety of models with performance optimised for different applications, in terms of reliability, long life, cost, power/weight ratio, etc. Key developments over the last few decades include polypropylene for inert leak-proof enclosures, 'absorbent glass mat' technology for plate separators, valve regulated lead acid batteries (sealed to prevent air ingress but allowing excess gas to escape and having internal reformation of overcharge electrolysed hydrogen and oxygen), and a wide range of 'recipes' with small concentration additives for specialist plates and separators and electronically controlled charging.

### 16.5.2 Other storage batteries

Development of other electrochemical systems for storage batteries is also active, especially for applications where weight is more of a constraint than cost, e.g. for aerospace or solar-powered cars. Among the more promising systems are nickel–cadmium, sodium–sulphur, and zinc–bromide batteries; see Table 16.1.

### 16.5.3 Superconducting electromagnetic energy storage (SMES)

A superconducting electromagnetic energy storage (SMES) system is a device for storing and very quickly discharging large quantities of electric power, e.g. 10 MW for <1 s. It stores energy in the magnetic field created by the flow of DC in a coil of superconducting material that has been cryogenically cooled to ~4 K). At these very low temperatures, certain materials have essentially zero resistance to electric current and can maintain a DC current for years without appreciable loss. SMES systems have been in use for some years to improve industrial power quality and to provide a premium quality service for those electricity users who are particularly vulnerable to voltage fluctuations. An SMES recharges within minutes and can repeat the charge/discharge cycle thousands of times without any degradation of the magnet.

Systems with large high-performance capacitors are being developed for similar power-conditioning uses. Although there have been proposals to use SMES more generally for storing large amounts of electrical energy, the cost appears to be prohibitive, see Table 16.1.

## 16.6 Fuel cells

A fuel cell converts chemical energy of a fuel into electricity directly, with no intermediate combustion cycle. Since there is no intermediate 'heat to work' conversion, the efficiency of fuel cells is not limited by the second law of thermodynamics, unlike conventional 'fuel → heat → work → electricity' systems. The efficiency of conversion from chemical energy to electricity by a fuel cell may theoretically be 100%. Although not strictly 'storage' devices, fuel cells are treated in this chapter because of their many similarities to batteries, Section 16.5, and their possible use with $H_2$ stores, Section 16.3.2. In a 'hydrogen economy', fuel cells would be expected to be used not only for stationary electricity generation but also for powering electric vehicles. Therefore we shall discuss only fuel cells using $H_2$, although other types exist.

Like a battery, a fuel cell consists of two electrodes separated by an electrolyte, which transmits ions but not electrons. In the fuel cell, hydrogen (or another reducing agent) is supplied to the negative electrode and oxygen (or air) to the positive electrode (Figure 16.5). A catalyst on the porous anode causes hydrogen molecules to dissociate into hydrogen ions and electrons. The $H^+$ ions migrate through the electrolyte, usually an acid, to the cathode, where they react with electrons, supplied through the external circuit, and oxygen to form water.

The efficiencies of practical fuel cells, whether hydrogen/oxygen or some other gaseous 'couple', are much less than the theoretical 100%, for much the same reasons as for batteries. In practice the efficiency is perhaps 40% for the conversion of chemical energy to electricity, but this is not dependent

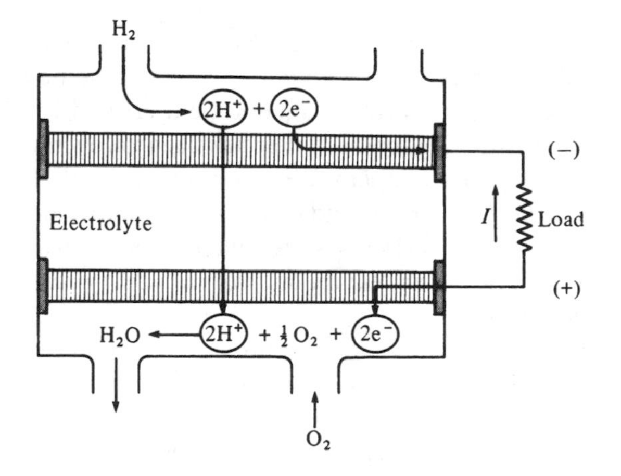


*Figure 16.5* Schematic diagram of a fuel cell. Hydrogen and oxygen are combined to give water and electricity. The porous electrodes allow hydrogen ions to pass.

on whether or not the cell is working at its full rated power. This contrasts with most diesel engines, gas turbines and other engines.

Since the efficiency of an assembly of fuel cells is nearly equal to that of a single cell, there are few economies of large scale. Therefore small localised plants of 1–100 kW capacity are a promising proposition. Using the fuel cell as a CHP source, a single building could be supplied with both electricity and heat (from the waste heat of the cells), for the same amount of fuel ordinarily required for the thermal demand alone. The main reason why fuel cells are not in wide use for such applications is their capital cost ($>$\$2000 kW$^{-1}$).

## 16.7 Mechanical storage

### *16.7.1 Water*

Hydropower systems draw on a natural flow of energy $P_0 = \rho g Q_0 H$, where $Q_0$ is the natural flow rate of water at the site and $H$ is the vertical distance through which it falls (Section 8.2). Since the natural flow $Q_0$ depends on rainfall, which has a different rhythm to the demand for power, all large hydroelectric systems incorporate energy storage by means of a dam; see Figures 1.5(b) and 8.6. Water is stored at the high elevation $H$, and released to the turbine below at a controlled flow rate $Q$. The potential energy stored in a dam at 100 m head has an energy density $W_v = 1.0\,\text{MJ}\,\text{m}^{-3}$. Although this is a relatively small energy density, the total energy stored in a hydro dam can still be very large. A *pumped hydro* system uses two reservoirs, an upper and a lower. When power is available and not otherwise required, water is pumped uphill. When demand occurs, the water is allowed to fall again, driving a hydroelectric turbine at the bottom and thereby generating power. In practice, the same machine, usually a Francis turbine, Figure 8.6(b), is used successively for both pumping and generating. Some very large systems of this type have been built to smooth the fluctuating demand on conventional power stations, thereby allowing them to run at constant load and greater overall efficiency. Nuclear power plants especially need such support.

Since about 15% of the input power is used to keep the turbines/pumps spinning to allow quick response and since a further 15% is lost in friction and distribution, it may be argued that the large capital cost of such schemes would have been better spent on control of demand (Section 1.4.5).

### *16.7.2 Flywheels*

The kinetic energy of a rotating object is

$$E = \frac{1}{2} I \omega^2 \tag{16.13}$$

where $I$ is the moment of inertia of the object about its axis and $\omega$ is its angular velocity ($\mathrm{rad\,s^{-1}}$). In the simplest case, the mass $m$ is concentrated in a rim of radius $r = a$, so $I = ma^2$. However, for a uniform disc of the same mass, $I$ is less ($ma^2/2$) because the mass nearer the shaft contributes less to the inertia than at the rim.

From (16.13), the energy density of a uniform disc is

$$W_{\mathrm{m}} = E/m = \frac{1}{2}a^2\omega^2 \tag{16.14}$$

For a flywheel to be a useful store of energy and not just a smoothing device, it follows from (16.14) that it must rotate as fast as possible. However, its angular velocity is limited by the strength of the material resisting the centrifugal forces tending to fling it apart. For a uniform wheel of density $\rho$, the maximum tensile stress is

$$\sigma^{\mathrm{max}} = \rho\omega^2 a^2 \tag{16.15}$$

In general $I = Kma^2/2$ for a particular shape, where $K$ is a constant $\sim 1$. So

$$W_{\mathrm{m}} = \frac{Ka^2\omega^2}{2} \tag{16.16}$$

and

$$W_{\mathrm{m}}^{\mathrm{max}} = \frac{K\sigma^{\mathrm{max}}}{2\rho} \tag{16.17}$$

Conventional materials, such as steel, have relatively small energy densities.

*Example 16.3 Maximum energy density of a rotating steel disc*

For a fairly strong steel, (16.17) gives, with $K = 1$,

$$\begin{aligned} W_{\mathrm{m}}^{\mathrm{max}} &= \frac{(1000 \times 10^6\,\mathrm{N\,m^{-2}})}{(2)(7800\,\mathrm{kg\,m^{-3}})} \\ &= 0.06\,\mathrm{MJ\,m^{-3}} \end{aligned}$$

Much larger energy densities can be obtained by using lightweight fibre composite materials, such as fibreglass in epoxy resin, which have larger tensile strength $\sigma^{\mathrm{max}}$ and smaller density $\rho$. To make the best use of these materials, flywheels should be made in unconventional shapes with the strong fibres aligned in the direction of maximum stress. Such systems can have energy densities of $0.5\,\mathrm{MJ\,kg^{-1}}$ (better than lead acid batteries) or even greater (Problem 16.3).

For smoothing demand in large electricity networks, flywheels have the advantage over pumped hydro-systems that they can be installed anywhere and take up little land area. Units with a 100 tonne flywheel would have a storage capacity of about 10 MWh. Larger storage demands would probably best be met by cascading many such modular 'small' units. Flywheels also offer a theoretical, but not commercially utilised, alternative to storage batteries for use in electrically powered vehicles, especially since the energy in a flywheel can be replenished more quickly than in a battery; see Problem 16.2.

### 16.7.3 Compressed air

Air can be rapidly compressed and slowly expanded, and this provides a good way of smoothing large pressure fluctuations in hydraulic systems. The energy densities available are moderately large. A small-scale example occurs within the hydraulic ram pump of Section 8.7.

Consider, for example, the slow compression of $V_1 = 1.0\,\mathrm{m^3}$ of air, at pressure $p_1 = 2.0$ atmosphere $= 2.0 \times 10^5\,\mathrm{N\,m^{-2}}$, to $V_2 = 0.4\,\mathrm{m^3}$, at constant temperature. For $n$ moles of the air, considered as a perfect gas.

$$pV = nR_0T \tag{16.18}$$

from which it follows that the work done (energy stored) is

$$\begin{aligned} E &= -\int_{V_1}^{V_2} p\mathrm{d}V = -nR_0T\int_{V_1}^{V_2} \frac{\mathrm{d}V}{V} \\ &= p_1V_1\log(V_1/V_2) \\ &= 0.19\,\mathrm{MJ} \end{aligned} \tag{16.19}$$

for the figures quoted.

In the compressed state, $W_v = E/V_2 = 0.48\,\mathrm{MJ\,m^{-3}}$. For a system operating under less idealised conditions, $W_v$ will be less but of similar magnitude. A major difficulty is to decrease the losses of heat production during the compression.

## 16.8 Distribution of energy

### 16.8.1 Introduction

Table 16.2 outlines the major methods for commercial energy to be distributed to its points of use. The methods are categorised by whether they

*Table 16.2* Summary of major means and flows for distributing energy

| | *Long distance* (>1000 km) | *Flow* (MW per unit) | *Flow* (MJ user$^{-1}$ day$^{-1}$) | *Medium distance* (1–1000 km) | *Flow* (MW per unit) | *Flow* (MJ user$^{-1}$ day$^{-1}$) | *Short distance* (10 m–1 km) | *Flow* (MW per unit) | *Flow* (MJ user$^{-1}$ day$^{-1}$) |
|---|---|---|---|---|---|---|---|---|---|
| continuous | oil pipeline | 15 000 | 60 | oil pipeline | 10 000 | 60 | | | |
| | gas pipeline (high pressure) | 500 | 20 | gas pipeline (high pressure) | 500 | 20 | gas pipeline (low pressure) | | 7 |
| | | | | electricity (high voltage) | 100 | 20 | electricity (low voltage) | | 10 |
| | | | | | | | heat in gas, vapour or liquid | | |
| batch | oil tanker | 1200 | | oil (or substitute, e.g. ethanol) | | | | | |
| | | | | in vehicle as cargo | 200 | | | | |
| | | | | in vehicle as fuel | | 28 | | | |
| | coal in ships | | | coal in trains | | | | | |
| | | | | biomass on lorry | 15 | | biomass on lorry | 15 | |
| | | | | | | | wood by hand | 0.03 | 15 |

involve continuous flow, e.g. pipelines, or batch movement, e.g. lorries, and by their suitability for use over long, medium or short distances. Also shown are numerical values representing the energy flows in a typical unit of the type in question, e.g. an individual gas pipeline. Although these flows vary greatly in scale, the energy flow through each system *per end-user per day* is, in practice, remarkably similar, i.e. $\sim 10\,\mathrm{MJ\,user^{-1}\,day^{-1}} = \sim 100\,\mathrm{W\,user^{-1}}$. The following subsections, and Problems 16.6–16.8, give the basis of the numerical values in the table, except for electrical transmission, which is discussed in Section 16.9.

For renewable energy supplies, short distance distribution is more important than long distance, because the sources themselves are usually widespread and of relatively small capacity. In particular, short haul carriage of biomass, and movement of heat, e.g. through a building, is most significant. The renewable energy supplies that are mechanical in origin, e.g. hydro, wave and wind, are usually best distributed by electricity. In this way electricity is a carrier or vector of energy, and not necessarily the main end-use requirement. Movement of gas, perhaps on the large scale of natural gas pipelines today, would be required if hydrogen becomes a common store of energy.

### *16.8.2 Pipelines*

In the pipelines usually used for carrying fuel gases, the flow is turbulent but not supersonic. Therefore the theory of Section 2.6 applies, although its interpretation is affected by the compressibility of the gas, as follows.

According to (2.12) the pressure gradient along a small length of pipe of diameter $D$ is

$$\frac{\mathrm{d}p}{\mathrm{d}x} = -2f\frac{\rho u^2}{D} \tag{16.20}$$

where $f$ is the friction coefficient, and $\rho$ and $u$ are respectively the density and mean speed of the fluid.

In a steady flow of *gas*, both $\rho$ and $u$ vary along the length of a long pipe, but the mass flow rate

$$\dot{m} = \rho u A \tag{16.21}$$

is constant; $A = \pi D^2/4$ is the cross-sectional area. Also the density $\rho$ varies with the pressure $p$:

$$p = \left(\frac{\mathrm{R}_0 T}{M}\right)\rho \equiv K\rho \tag{16.22}$$

with $K$ more or less constant for a given gas. $\mathrm{R}_0$ is the universal gas constant, $T$ is absolute temperature, and $M$ is the gram molecular mass/1000, so as to have units of $\text{kg mol}^{-1}$. If the Reynolds number ($\mathcal{R} = uD/\nu$) is large, $f$ will not vary appreciably along the pipe, and we can integrate (16.20) between stations $x_1$ and $x_2$ to get

$$p_1^2 - p_2^2 = \frac{64 f R_0 T \dot{m}^2 (x_2 - x_1)}{\pi^2 M D^5} \tag{16.23}$$

Thus the pressure falls off rapidly along the length of the pipe, and frequent pumping (recompression) stations are needed to maintain the flow. As a numerical example, a pipe of diameter 30 cm, carrying methane at a mean pressure about 40 times atmospheric, holds an energy flow of about 500 MW, which is very substantial, see Problem 16.7. According to (16.23), larger pipes (bigger $D$) will require much less pumping. The most economical balance between pipe size (capital cost) and pump separation (running cost) depends largely on the accessibility of the pipe. Construction costs are very variable, but are unlikely to be less than US\$0.2 $\text{GJ}^{-1}$ $(1000\,\text{km})^{-1}$.

The compressibility of the gas offers another benefit. The pipe itself can be used as a store of adjustable capacity by simply pumping gas *in* faster than it is taken *out*, so the compressed gas accumulates in the pipe. For the pipe considered above, the energy 'stored' in the 100 km length might be:

$$(32\,\text{kg m}^{-3})(50\,\text{MJ kg}^{-1})(10^5\,\text{m})\pi(0.15\,\text{m})^2 = 11 \times 10^6\,\text{MJ per 100 km}$$

Such storage is very substantial.

The rate of flow of energy (i.e. the power) can likewise be very large in liquid fuel pipelines; see Table 16.2 and Problem 16.6.

### 16.8.3 Batch transport

Biomass can be transported in suitable vehicles by road, rail, river or sea. However, the small density and bulky nature of most biomass as harvested means that it is rarely economic to distribute it over long distances ($\gtrsim$500 km). Even over medium distances (100–500 km), it is unlikely to be economic to distribute such biomass for its energy value alone. The guiding principle for the economic and ecological use of biomass is to interact with a 'flow' of harvested biomass which is already occurring for some other purpose. An excellent example is the extraction of sugar from sugarcane, so leaving the cane residue (bagasse) to fuel the factory, as described in Section 11.3.1. In this case the transport of the fuel may be regarded as 'free', or nearly so; Section 11.1(4) amplifies this argument. Biofuels can however be transported over medium to long distances after conversion from raw biomass, e.g. by pyrolysis (Section 11.5), or as biodiesel (Section 11.2.3). In all countries, firewood is usually used close to its source ($<$100 km).

### *16.8.4 Heat*

The movement of heat within a building, either through hot air 'ducts' or through open doorways, and through steam pipes is a major means of distributing energy over short distances. This is especially true in cold climates, where space heating dominates energy use, e.g. Figure 16.2. Heat transport by steam is also used in many industrial processes. Heat transport is limited to short distances by heat losses from the sides of the duct.

*Example 16.4 Heat loss from a steam pipe*
A pipe 6 cm in diameter is to deliver heat over a distance of 100 m. It is insulated with glass wool of thickness $\Delta x = 1.0\,\text{cm}$. Estimate the heat loss along the path. (Take $T_a = 10\,°\text{C}$.)

*Solution* As a first approximation, assume the steam is at $T_v = 100\,°\text{C}$ along the whole pipe. (Steam at higher pressure will actually be at higher temperature: see most books on engineering thermodynamics.) The conductivity of mineral wool is $k = 0.04\,\text{W}\,\text{m}^{-1}\,\text{K}^{-1}$ (similar to that of other insulators, using trapped air). The major resistance to heat loss is by conduction through the insulation, so from (3.12)

$$\begin{aligned} P_{loss} &= -kA\Delta T/\Delta x \\ &= (-0.04\,\text{W}\,\text{m}^{-1}\,\text{K}^{-1})(100\,\text{m})\pi(0.06\,\text{m})(100 - 10)\,°\text{C}/(0.01\,\text{m}) \\ &= 6.8\,\text{kW} \end{aligned}$$

The loss calculated in Example 16.4 is independent of the flow rate in the pipe. Obviously, very large heat flows (~10 MW) are needed if the losses are to be proportionately small. *District heating* of this kind operates successfully in many cities. The *heat pipe* offers another way to move large quantities of heat over very short distances. It is a tube containing vapour with the condensate recycled by a wick, and offers an effective conductivity much greater than that of copper, see Section 3.7.2.

## 16.9 Electrical power

### *16.9.1 Electricity transmission*

As considered earlier, renewable energy supplies that are mechanical in origin, e.g. hydro, wave and wind, are usually best distributed by electricity. In this way electricity is a carrier or vector of energy, and not necessarily the main end-use requirement.

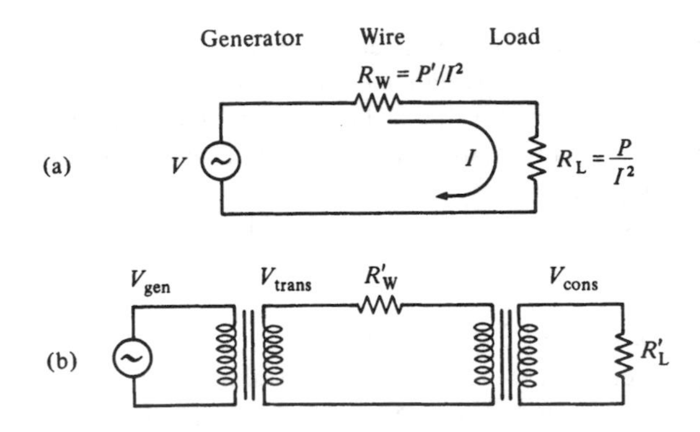


*Figure 16.6* Electrical transmission. (a) Power transmission to a load of resistance $R_L$, through a wire of finite resistance $R_w$. (b) More likely realisation, with generated voltage transformed up for transmission and then down for consumption.

Consider two alternative systems transmitting the same useful power $P$ to a load $R_L$ at different voltages $V_1$, $V_2$ in a wire of the same resistance per unit length (Figure 16.6). The corresponding currents are $I_1 = P/V_1$ and $I_2 = P/V_2$, and therefore the ratio of power lost in the two systems is

$$\frac{P_2'}{P_2'} = \frac{I_1{}^2 R_w}{I_2{}^2 R_w} = \left(\frac{P}{V_1}\right)^2 \left(\frac{V_2}{P}\right)^2 = \frac{V_2{}^2}{V_1{}^2} \tag{16.24}$$

Thus significantly less power is dissipated in the system working at high voltage. The low voltage system can have the same loss as the high voltage system only with thick, and therefore expensive, cable. For instance if electricity is to be transmitted at domestic mains voltage (~110 or ~220 V), the cost of cabling becomes prohibitive for distances greater than about 200 m. The difficulty becomes even greater at very low voltage, ~12 V.

These factors govern the design of all electrical power networks. Normal rotating generators work best at voltages $\leq$10 kV. The ease with which alternating current (AC) can be transformed to larger or smaller voltage explains why AC transmission systems have been standard for all but the smallest networks. As indicated in Figure 16.6(b), power is generated at low voltage, stepped up for transmission, and then down again to a safer voltage for consumption. It is important to note, however, that solid-state power electronic components increasingly allow DC/AC/DC conversion at large power and reasonable cost, consequently transforming DC voltage is not so difficult as in the past.

The transmission voltage is limited by dielectric breakdown of the air around the overhead cables and by the insulation of the cables from the metal towers that are at earth potential. Improvements in insulation have allowed transmission voltages for long lines to increase from 6000 V in the year 1900 to over 200 000 V today. Grids using even larger voltages are now being constructed, but will probably make only a marginal improvement in costs. The same is true for very high voltage direct current (DC) systems, which have certain advantages in transmission, for instance having no inductive loss and larger power density for the same peak voltage, but require more expensive interconnection equipment. Superconducting lines of zero resistance are, in principle, attractive but can operate only at very low 'cryogenic' temperatures. Maintaining such low temperatures is difficult over large distances and such lines are not yet a commercial proposition.

### 16.9.2 Electricity grids (networks)

Electrical power generation usually links to the load demand by a common regional or national network, often called 'the grid'. The generators may be centralised power stations or smaller capacity embedded generation, such as gas turbines or windfarms. The grid allows the sharing of generation and consumption, and so provides a reliable and most cost-effective general means of supply. Thus when one generating node reduces its supply (e.g. for lack of wind, for maintenance or for fuel saving) its demand can be met by the other generators and electricity storage plant in the network. Since all major grid systems can cater for rapid demand fluctuations of ~20%, it is also possible for the same systems to cater for rapid supply fluctuations of ~20%. Thus the inclusion of variable supplies, such as wind power, is possible. If hydropower, which can be switched on or off in seconds, is available, then rapid control of the grid is easy to adjust for changes in other generation or in demand.

Despite their name and original intention, since about 1930, electricity 'grids' have been characterised by centralised despatch of electricity at high voltage (~100 kV) from a small number of very large power stations (~1000 MW capacity), usually by overhead lines insulated only by the air around them. Near the point of end-use, the electricity is transformed down to ~10 000 V, e.g. for a suburb, and then to ~230 V or ~110 V, e.g. for a street of houses. Older rural grids from central dispatch may lose 10–20% of power in transmission, distribution and transformation, with the best urban grids having about 5% losses. An alternative model for a grid, in which many relatively small generators are embedded, can make better use of renewable energy supplies, as discussed in Section 16.9.4.

### 16.9.3 Electrical generators

#### (a) Basics

The basic operation of all generators is simple, but many complexities and variations are used to give particular properties and improvements in efficiency. Essentially a magnetic field is arranged to cut a wire with a relative velocity, so inducing an electric current by the Faraday Effect. Every generator has a stator (coils of wire that stay static) and a rotor (coils of wire that rotate within the stator); one of these has a coil (winding) to produce the generated current, and the other has other windings or permanent magnets to produce the magnetic fields. We give a brief account here; for further detail see any textbook on electrical machines.

A magnet moved across an electrical conductor will induce an electrical potential difference in the conductor (an EMF; or a 'voltage'). If the conductor forms a closed loop, then an electrical current will have been induced. The EMF induced is:

$$V = -N\frac{d\phi}{dt} \tag{16.25}$$

where there are $N$ conductors in series, each cut by a magnetic flux $\phi$ of rate of change $d\phi/dt$. A coil of wire (solenoid) carrying an electric current produces a magnet field, as does a magnet. If the coil has a ferromagnetic core, e.g. iron, then the magnetic field is very considerably enhanced (by a factor of $\sim$1000). Therefore coils with ferromagnetic cores are predominantly used in electric machines, i.e. generators and motors. The equivalent north and south poles of such coils are called 'salient poles'.

In the *basic AC generator*, the stator produces a stationary magnetic field, $B_{stator}$. A single coil turning in this has an induced EMF, which produces a current in an external circuit connected by brushes on fully circular slip rings at the commutator. Because the wires of the coil alternately cut up and down, the current is AC. The induced current itself produces a magnetic field, $B_{rotor}$, which rotates. The shaft is driven by the external mechanical torque $\Gamma_m$, e.g. a wind turbine shaft. However, the induced rotor magnetic field $B_r$ sets up an opposing torque $\Gamma_{em}$ from the electromechanical effects. Equilibrium is reached when the shaft mechanically driven torque, $\Gamma_m$, equals the induced electromagnetic torque $\Gamma_{em}$. Moreover, the mechanical power input equals the electrical power generated, less frictional and resistive losses. The electricity generated is AC, with voltage and current varying sinusoidally with time, commonly at $\sim$50 Hz but otherwise usually 60 Hz. Practical AC generators are variations of this model, either 'synchronous' or 'induction' described below.

Another important variation on the basic AC generator is the *basic DC generator*. Again, the stator produces a stationary magnetic field, $B_{stator}$. But this time, the rotor coils are connected by semi-circular slip rings at the commutator. This type of commutator reverses the current flow for each rotor coil as it passes perpendicular to the stator field, so there is a stationary rotor field, $B_r$, perpendicular to the stationary stator field (despite the rotating rotor). Thus the output current and power are unidirectional but varying as $|\sin \omega t|$. With appropriate electronic circuitry this can be converted to a more or less steady DC output.

*(b) Some features of alternating current*

If there is only resistance in the external circuit, then the voltage and current are in phase, in which case the power dissipated as heat in the external circuit is $P = V_p I_p \sin^2 \omega t$ and the average power is $P_{av} = V_p I_p / 2$. This is the *active power*, which, in this case, peaks at a frequency twice that of the voltage and current (half the period). It is common practice to call $V_p/\sqrt{2} = V$ the voltage (strictly, the root-mean-square voltage) and $I_p/\sqrt{2} = I$ the current, so that the power $P = VI$ as for direct current.

If the external circuit is purely inductive with zero resistance, then the voltage (push) and current (flow) are $\varphi = 90°$ out of phase (push maximum with zero flow; flow maximum with zero push). The phase difference is $\varphi$, and the product of $V_p \sin \omega t$ and $I_p \sin(\omega t + \varphi)$ has frequency $2\omega t$ and amplitude $Q$, where

$$Q = VI \sin \varphi \tag{16.26}$$

This function $Q$ has the units of power, but is not dissipated as heat; it is power oscillating backwards and forwards in the magnetic field of the inductance. The timed averaged power into the external circuit is zero, and is called the *reactive power*; it does not appear as heat or work.

If the external circuit is inductive and also resistive, then $\varphi$ is not 90°. The product of $V_p \sin \omega t$ and $I_p \sin(\omega t + \varphi)$ can now be separated into a reactive part (oscillating with equal amplitude between positive and negative) and an active part (always positive). The average value of the active power $P_{av} = VI \cos \varphi$, and the average value of the reactive power is zero as before.

Similar effects occur with capacitance in the external circuit, but with the current leading the voltage. It is common practice to adjust a compensating capacitance to negate the reactive power effects of inductance, and vice versa. Electrical engineers speak of 'active' and reactive' power as separate parameters; each separately instrumented and quantified. In general, reactive power is not wanted, and so grid connected users may be charged for reactive power consumption that they cause, despite it not being usable power.

As noted in Section 16.9.1, the voltage of alternating current can easily be altered by a *transformer*, see Figure 16.6(b). Essentially a transformer consists of two coils of wire (with different number of turns $N$) on the same ferromagnetic core. Since the magnetic flux $\phi$ is effectively confined to the core, for each loop $\phi$ is the same, and so from (16.25) the voltage $V$ in each winding is proportional to $N$. Because of this ease of transformation, and also ease of generation and its suitability for electric motors, AC (rather than DC) is used in all grid systems; DC is, however, used for many large power underground and under-water grid interconnections, such as between France and the UK.

### *(c) Synchronous generator (alternator)*

If the magnetic fields are created by permanent magnets or DC currents, then the current will be induced at an AC frequency ($f_1$) in synchronisation with the rotating shaft frequency ($f_s$) of the generator. With $n$ pole pairs, each acting as a single magnetic 'north/south pole pair', $nf_s = f_1$ exactly. Such a generator is a *synchronous AC generator* with the output frequency locked to the shaft frequency. This is fine if the shaft frequency is closely controlled, as in a diesel-powered system, but raises complications with wind turbines.

Usually power will be taken from stationary coils on the stator, and the coils on the rotor producing the magnetic field are connected, via slip rings, to a DC generator. Synchronous generators have large efficiency (commonly 97%), with losses mainly only resistive in the windings. Note that power is extracted from the stator, which is connected to the grid. A benefit of synchronous generators is that the reactive power can be controlled and minimised, so maximising real power. In the most common arrangement, the stator coils are directly connected to the grid, in which case power is only exported when the rotor turns, so $nf_s = f_1$ and $f_1$ exactly equals the grid network frequency, e.g. 50 Hz, or 60 Hz in North America. This control requirement makes it uncommon for wind turbines to use synchronous generators, since induction generators are cheaper and easier to use on electrically strong grids.

### *(d) Induction generator (asynchronous generator)*

The *induction AC generator* is strictly an *induction machine*, since the same device can be a motor or a generator. This generality of design allows induction generators to be cheaper than synchronous generators. The usual arrangement is that the stator windings (coils) are connected to the AC grid, so producing a rotating magnetic field around the shaft of the machine. The rotor is a 'squirrel cage' with copper bars set parallel to the axis, and connected together by rings at each end. Currents are thereby induced within the short-circuited coils on the shaft. These induced currents themselves produce magnetic fields, which in turn generate power into the stator coils, but only if the rates of rotation of the shaft magnetic field and the stator coils differ. The phase relationships are such that power can be transferred between the mechanical rotor shaft and electrical power in the stator circuit.

For the induction machine (i) there are $n$ pairs of windings on the rotor simulating $n$ pole-pairs, (ii) the induced rotor currents have a frequency $f_2$ ($f_2 = 0$ if stator rotating field and shaft coils move together in phase; $f_2$ is negative if the shaft coils rotate faster than the stator field, (iii) if the grid supply frequency is $f_1$, and the shaft frequency is $f_s$, then $f_2 = f_1 - nf_s$, (iv) the slip $s = f_2/f_1 = (f_1 - nf_s)/f_1$ is defined as negative for a generator, and positive when the same device acts as a motor. Generator slip is usually less than 10% and between 0.5% and 5% for a motor.

An induction generator can only generate when the induced closed loop currents have been initiated and maintained. There are generally two methods for this: (1) reactive power is drawn from the live grid to which the generator output is connected, or (2) autonomous, self-excited, generation is made possible by capacitors connected between the output and earth. The benefits of method (1) (grid linking) are: (a) the simplicity and cheapness of the system, (b) safety, since the generator should not generate if the grid power is off, and (c) the grid can be used to export power when the shaft is turning, e.g. in windy conditions, and import power at other times. In method (2) there has to be some residual magnetism in the framework or surroundings of the generator to provide the initial current, with the capacitors maintaining the correct phase relationships. For small systems, it is also possible to maintain and control generation from an induction generator by running an idling synchronous generator in parallel as a 'synchronous compensator'. This system may be attractive for small autonomous systems since the same synchronous generator can be used for, say, diesel generation at times of inadequate renewable input, e.g. of wind.

When used with wind turbines, it is possible to increase the slip of the induction generator and thus allow an increased variation of rotor speed to maintain more constant tip–speed ratio. The earliest method involves impedance change of the otherwise unconnected rotor windings, but at the expense of increased generator heating. The more modern method (doubly fed induction generator) has the rotor windings connected through slip rings to an external power-electronic control of the voltage and phase. In this way (a) the rotational speed of the rotor may vary considerably from synchronism with the mains AC frequency and (b) power then be taken from the rotor circuits as well as the stator. Such doubly fed induction generators, with the associated power electronics, allow wind turbines to have variable rotor speed and hence match the wind speed for the most efficient power extraction.

### *16.9.4 Autonomous and 'in-between' systems*

A distinguishing feature of renewable energy sources is their distributed and localised nature; see Section 1.4.4 and Table 1.1. Some renewables can be used as a primary source in a centralised electricity grid, e.g. large-scale hydro. Some are suitable as the sole source of electricity for (usually small scale) applications remote from a grid ('autonomous' or 'stand-alone' systems), especially with suitable energy storage in the system, e.g. photovoltaic powered lighthouses using battery storage (see Chapter 7). But almost all renewable energy sources can contribute significant electricity as distributed inputs to a wider grid. Section 9.8.2 examines some of the practical considerations applying when the renewable input is (a) dominant, (b) about the same as the rest of a grid, or (c) only a small contributor to the grid. Although that discussion is phrased in terms of wind systems, similar considerations apply more generally.

## 16.10 Social and environmental aspects

Present civilisation ('the age of fossil fuels') depends on distribution of energy on a very large scale. International trade in fossil fuels (coal, oil and gas) from the relatively few countries that export in large quantities represented about 4% of world trade in 2003. The concentrated lines of supply are vulnerable to disruption. Several wars can be attributed to the desire of oil-consuming counties to secure their supplies, see Yergin's notable 1992 study. The fact that oil and coal are cheap stores of large quantities of easily accessible energy (see Figure 16.1) allowed the rapid growth of cities with little initial attention to the environmental consequences (McNeill 2000). Occasional failures of the distribution systems themselves had severe environmental consequences, notably large-scale oil spills.

Potentially the most harmful environmental impact of utilising the stored energy from fossil fuel combustion may be climate change from the release of extra $CO_2$. This threat is recognised worldwide and the majority of culpable countries have tentative policies to abate the harm; see Chapters 4 and 17. However, the international drive for alternative forms of energy supply and storage, including biofuels, is not yet as intense as it could be.

As discussed in Chapter 1, the less concentrated and dispersed nature of renewable energy sources allows a major shift away from the centralised energy systems and their vulnerable distribution systems, including the possibility of less centralised social systems. Because the sources are more distributed and may be closer to the end-uses, the energy distribution system becomes a more meaningful 'gridiron' network for the exchange back and forth of power and so much less vulnerable to disruption of any particular component , be it by war, overload or natural disaster (e.g. lightning). We see this as socially, economically and environmentally positive. Although high-voltage power lines have a good safety record, some see them as a blot on the visual environment. In a less centralised electricity system, they could be less obtrusive since the transmission voltage, and hence the height of the poles and pylons, would generally be less.

Because of the intermittent nature of most renewable energy sources, stand-alone (autonomous) systems based on renewables would be impossible without energy storage. The possibility of using, for example, photovoltaic lighting systems (with battery storage) to increase amenity in rural areas of developing countries without the hazards and cost of stringing in long power lines or transporting diesel fuel offers a prospect of decreasing urban drift with its attendant social and environmental consequences (see Mandela 2000).

Nevertheless, there are relatively minor environmental downsides to some of the storage mechanisms described in this chapter. In particular, batteries of all kinds are filled with noxious chemicals, so that their disposal is an issue. Lead acid batteries, however, are so widespread for vehicles that there is a thriving recycling business for them in most countries. Although lead metal is poisonous, it is also expensive and has a low melting point so that it is relatively easy and economically worthwhile to salvage lead from 'dead' batteries and reform it for new ones.

Contrary to popular impression, hydrogen gas, although capable of forming an explosive mixture with air, is no more hazardous than the more familiar natural gas (methane), for which the same is true. Thus the safety and social issues about moving to a 'hydrogen economy' are not a serious issue given professional safety provision; a transition to a hydrogen economy is much more restricted by the economics, the infrastructure and the need to adapt most end-use devices.

A final comment is to reinforce the need for thermal mass in buildings as a form of energy storage. 'Heavy construction', with appropriate external insulation, allows passive solar and other variable (perhaps variably priced also) heat gains to be stored intrinsically from at least day to night and from day to day. Alternatively, the 'coolness' of a building from losing heat in the night can be 'stored' through the day. Such simple energy storage has major implications for comfort and more efficient energy supplies in buildings, which generally utilise at least 30% of national energy supplies, as considered in Section 16.4. The widespread re-introduction of such 'heavy buildings' has considerable implications for energy efficiency, planning regulations and constructional resources.

## Problems

16.1 The changes in enthalpy, free energy, and entropy in the formation of water

$$H_2 + \frac{1}{2}O_2 \rightarrow H_2O_{(gas)}$$

are respectively

$$\Delta H = -242\,\text{kJ}\,\text{mol}^{-1}$$
$$\Delta G = -228\,\text{kJ}\,\text{mol}^{-1}$$
$$\Delta S = -64\,\text{J}\,\text{K}^{-1}\,\text{mol}^{-1}$$

Estimate the temperature above which $H_2O$ is thermodynamically unstable. *Hint*: consider (16.1).

16.2 A passenger bus used in Switzerland derived its motive power from the energy stored in a large flywheel. The flywheel was brought up to speed, when the bus stopped at a station, by an electric motor that could be attached to the electric power lines. The flywheel was a solid steel cylinder of mass 1000 kg, diameter 180 cm, and would turn at up to 3000 rev $min^{-1}$.

a At its top speed, what was the kinetic energy of the flywheel?
b If the average power required operating the bus was 20 kW, what was the average time between stops?

16.3 A flywheel of three uniform bars, rotating about their central points as spokes of a wheel, is made from fibres of 'E' glass with density $\rho =$ 2200 kg $m^{-3}$, and tensile strength 3500 MN $m^{-2}$. The fibres are aligned along the bars and are held together by a minimal quantity (10%) of resin of negligible tensile strength and similar density. Calculate the maximum energy density obtainable. If $a = 1.0$ m, what is the corresponding angular velocity?

16.4 Verify that the data plotted in Figure 16.2 (taken from the classic paper of Ryle 1977) are in fact 'reasonable'.

a The population of Great Britain was then about 50 million. How does the total energy demand per person compare with the world average of 2.0 t oil equivalent (TOE) per person per year (UN statistic, 1980)?
b How does the non-heat demand vary with season? What types of industrial usage does this correspond to?
c The peak demand shown in winter is for a daytime temperature of −3°C. How much heat is being used per household? Does this seem likely? (The figure quoted for 'heating' in Figure 16.2 includes non-household use.)
d Use the data in Chapter 4 (especially Figure 4.16) to estimate the solar heat input on 1 $m^2$ of horizontal surface, and on 1 $m^2$ of (south-facing) vertical surface in each season. (The latitude of Britain is about 50° N.) What is a typical efficiency of a solar heater? What collector area would be required to supply the power required for heating indicated in Figure 16.2? How many square metre per house does this represent? Is this reasonable? Would passive solar energy techniques, combined with thermal insulation, ventilation control, and the use of free gains, be of significance?
e Approximately what is the electrical power obtainable from 1 $m^2$ of swept area in a mean wind of 8 m $s^{-1}$, see (9.74). The land and shallow seawaters of Britain can be treated very approximately as two rectangles 1000 km × 200 km, with the longer sides facing

the prevailing wind. Consider large 100 m diameter wind turbines with mean wind speed 8 m s$^{-1}$ at hub height. How many wind turbines would be needed to produce an average power of 30 GW for the whole country? What would be the average spacing between them if half were on land and half at sea?

f Use the wave power map in Figure 12.12 to estimate the length of a barrage to generate a mean power of 30 GW off the north-west coast of Britain. How does this length compare with the length of the coast?

16.5 The largest magnetic field that can be routinely maintained by a conventional electromagnet is $B_0 \sim 1\,\mathrm{Wb\,m^{-2}}$. The energy density in a magnetic field is $W_v = B^2/(2\mu_0)$. Calculate $W_v$ for $B = B_0$.

16.6 Calculate the energy flows in the following cases:

a About 30 million barrels of oil per day being shipped out of the Persian Gulf area (1 barrel = 160 litre).

b The TAP crude oil pipeline from Iraq to the Mediterranean carries about 10 million tons of oil per year.

c A family of four in a household uses (for cooking) one cylinder of LPG (gas) (13 kg) per month.

d The same family runs a car that covers 8000 km y$^{-1}$, with a petrol consumption of 7 km per litre.

e A villager in Papua New Guinea takes 2 h to bring one load of 20 kg wood from the bush, carrying it on her back.

f A 3 t lorry carries fuelwood into town at a speed of 30 km h$^{-1}$.

g A 40-litre car fuel tank being filled from empty in 2 min.

16.7 A steel pipeline of diameter 30 cm carries methane gas ($CH_4$). Recompression stations are sited at 100 km intervals along the pipeline. The gas pressure is boosted from 3 to 6 MN m$^{-2}$ at each station. (These are typical commercial conditions.) Calculate (a) the mass flow and (b) the energy flow. (c) What volume per day of gas at STP would this correspond to? *Hint*: Refer to (16.3) and Fig 2.6, then make a first estimate of $f$, assuming $\mathcal{R}$ is 'high enough'. Then find $\dot{m}$ and check for consistency. Iterate if necessary. Viscosity of methane at these pressures is

$$\mu = 10 \times 10^{-6}\,\mathrm{N\,s\,m^{-2}}$$

16.8 An electrical transmission line links a 200 MW hydroelectric installation A to a city B 200 km away, at 220 kV. The cables are designed to dissipate 1% of the power carried. Calculate the dimensions of wire required, and explain why losses of 1% may be economically preferable to losses of 10 or 0.1%.

16.9 Considering a 6 pole-pair induction generator (Section 16.9.3), if $s = 0.1$ at generation into a 50 Hz grid, determine the induced rotor current frequency $f_2$ and $f_s$

## Bibliography

### General

Institution of Mechanical Engineers (2000) *Renewable Energy Storage: Its Role in Renewables and Future Electricity Markets*, Professional Engineering Publications, Bury St Edmunds. {Set of conference papers, including short articles on 'regenerative fuel cells', flywheels and superconducting magnetic energy storage.}

Jensen, J. and Sørensen, B. (1984) *Energy Storage*, Wiley. {One of the few books specifically on this topic. Good coverage at about same level as this book.}

Ter-Gazarian, A. (1994) *Energy Storage for Power Systems*, P. Peregrinus, London (on behalf of Institution of electrical Engineers), 232pp. {Emphasis on storage suitable for electric power.}

Most books on particular renewable energy sources (referred to in the appropriate chapters) include some discussion of storage media applicable to that source, e.g. heat stores for solar, batteries for wind.

### Chemical storage

Carden, P.O. (1977) Energy corradiation using the reversible ammonia reaction, *Solar Energy*, **19**, 365–378. {Sets out the main features of a solar/ammonia system using distributed collectors. Many later papers elaborate on details and similar systems, e.g. Luzzi, A. and Lovegrove, K. (1997) A solar thermochemical power plant using ammonia as an attractive option for greenhouse gas abatement, *Energy*, **22**, 317–325.}

Goel, N., Miraball, S., Ingley, H.A. and Goswami, D.Y. (2003) *Hydrogen Production*, Advances in Solar Energy, **15**, 405–451. {Emphasis on production by renewable energy; includes cost estimates.}

Yergin, D. (1992) *The Prize: the Epic Quest for Our Money and Power*, Simon and Schuster. {This book won the Pulitzer Prize for non-fiction in 1992 for its authoritative reporting and comment}

Wald, M.L. (2004) Questions about a hydrogen economy, *Scientific American*, **290**, 42–48. {Looks at 'wells to wheels' energy analysis.}

Any of the many textbooks on physical chemistry will give a thermodynamic analysis of the heat release in chemical reactions, e.g. Atkins, P.W. and de Paul, J. (2002), *Atkins' Physical Chemistry*, Oxford UP.

### Heat storage

Duffie, J.A. and Beckman, W.A. (1991 2nd edn,) *Solar Engineering of Thermal Processes*, Wiley, New York. {Chapter 9 is specifically concerned with heat storage.}

Ryle, M. (1977) Economics of alternative energy sources, *Nature*, **267**, 111–116. {Classic paper, cogently argueing that storage for about seven days enables wind/wave/solar to match most fluctuations in UK demand.}

### Electrical storage

Any of the many textbooks on physical chemistry will give an introduction to the elementary electrochemistry used in this chapter, e.g. Atkins, P.W. and de Paul, J. (2002), *Atkins' Physical Chemistry*, Oxford UP.

Crompton, T.R. (3rd edn, 2000) *Battery Reference Book*, Newnes, Oxford.

ITDG (2004) *Batteries*, ITDG technical brief on line at www.itdg.org. {Very down-to-earth guide on how to use and look after batteries.}

Lindsay, T.J. (1999) *Secrets of Lead Acid Batteries*, Lindsay Publications Inc, Il 60915, USA. {48 pages of practical explanations and guidance that is hard to find elsewhere.}

Rand, D.A.J., Woods, R. and Dell, R.M. (1998) *Batteries for Electric Vehicles*, Society of Automotive Engineers, Pennsylvania. {Covers all types of storage batteries.}

Vincent, C.A. (1984) *Modern Batteries: An Introduction to Electrochemical Power Sources*. Edward Arnold, London. {General and practical.}

### Fuel cells

Alleau, T. (2003) A state of the art of hydrogen and fuel cell technologies: diffusion perspectives and barriers, in A. Avadikyan, P. Cohendet and J.-A. Heraud (eds), *The Economic Dynamics of Fuel Cell Technologies*, Springer, Berlin.

Hoogers, G. (ed.) (2003) *Fuel Cell Technology Handbook*, CRC Press {Review of technology and its applications in power systems and vehicles.}

### Flywheels

Genta, G. (1985) *Kinetic Energy Storage: Theory and Practice of Advanced Flywheel Systems*, Butterworths, London. {Book-length detail.}

### Distribution and transmission and social aspects

*BP Statistical Review of World Energy* (annual).
Data and maps showing production, consumption and trade, especially in fossil fuels.

El-Hawary, M.A. (2000) *Electrical Energy Systems*, CRC Press. {Engineering text-book covering both electric machines and power distribution systems.}

Mandela, F. (2000) *Rural Electricity for Southern Africa, Refocus*, August 2000.

Nasar, S. (1996) *Electric Power Systems*, Prentice Hall. {Engineering textbook covering both electric machines and power distribution systems.}

Chapter 17

# Institutional and economic factors

## 17.1 Introduction

The impression may have been given in this book so far that everything can be understood by science and performed by engineering. Such an opinion is naïve in the extreme. The reality is that practical developments in energy concern about 75% 'institutional factors' and only about 25% of science and engineering. Scientists and engineers as such are minor influences, with the key parts played by others, including politicians, planners, financiers, lawyers, the media, the public and, because of ethical and cultural values, philosophers. Nevertheless when scientists and engineers themselves enter the other areas of influence, they may become more influential.

This chapter first briefly reviews some of the socio-political and economic factors that affect the choice of energy systems. It emerges that environmental costs are not well incorporated into the current prices of conventional energy systems, so choices are biased against more sustainable energy systems, including renewables. Some policy tools that could redress this are outlined. Section 17.5 outlines some methods used by economists and accountants to quantify choice between alternative projects, including the use of discounted cash flows.

The chapter, and the book, concludes with an examination of how the technological, socio-political and economic environment for renewable energy has evolved over the past 30 years and how it might evolve in future. We conclude that renewables are growth areas of development, with potential to supply much of the world's energy from millions of local and appropriate sites, but that to make this happen will require knowledge, vision, experience, finance, markets, and individual and collective choice.

## 17.2 Socio-political factors

Action within society depends on many factors, including culture, traditions, political frameworks and financing. Such influences vary greatly between localities, and also change with time; they also depend on the availability and awareness of technologies.

### 17.2.1 National energy policy

The key socio-political factors influencing policy on energy supply, especially from renewables, include, in approximate order of importance:

*Security of supply*. It is the duty of politicians to secure their nation's energy supply. Having the ability to utilise indigenous supplies is therefore strategically important to guard against international disruption. Having indigenous fossil and nuclear energy supplies (which many countries have only to a limited extent) does support national security; even so, the large-scale centralisation of such resources leaves them vulnerable. Using renewables provides the necessary dispersed security and does not deplete finite resources. Every nation has its own set of renewable resources and there is generally common consensus, at least in principle, that these should be assessed and harnessed as a major component of energy supply. Large-scale resources concentrated at a few sites, such as large hydro power, tend to be recognised, but not, unfortunately, the sum total of widely dispersed low intensity supplies, such as rooftop photovoltaics.

*Diversity of supply*. Having 'all my eggs in one basket' is not a robust strategy. Both individuals and nations can increase security of supply by having several operational options in parallel. There is therefore perceived value in diversity, but no common method of quantifying such diversity. The accounting method of discounting (see below, Section 17.5) allows financial optimisation, but includes no evaluation of risk and gives no credit for diversity. For energy supplies, there should be diversity in all aspects of energy supply and use, including transport, fuels, electricity generation and heating. Clearly renewables are able to give great diversity of supply in these aspects, and with geographical variation.

*Economic supply* is usually taken to mean 'low price to the consumer within a competitive market'. This price is heavily influenced by taxes, subsidies, monopoly influences and supplier profits, as well as the more obvious material supply costs; see below regarding economic conditions and energy markets. Evaluating what is 'economic' is attempted by various forms of analysis, usually based on 'discounting' (Section 17.5), but the actual price paid per unit tends to dominate once a supply is available. Renewables, by definition, utilise energy from the environment, which usually arrives without payment as with sunshine, wind and rain. The major cost of renewables is therefore the initial capital cost of the equipment, and so the method of integrating capital and operational costs is vital for economic comparisons with fossil and nuclear fuel systems.

*Sustainability and climate change*. As discussed in Section 1.2, many environmental issues have risen in public and political consciousness in the last 50 years. In particular, global concern for sustainable development and climate change led to international concern, expressed most notably by the United Nations Framework Convention on Climate Change (FCCC 1990) and its associated Kyoto Protocol (1997). Almost all countries have agreed to accept obligations under the FCCC to reduce, or at least reduce the increase of, their greenhouse gas emissions and to report on their progress for this. Since the principal source of greenhouse gas emissions is $CO_2$ from burning fossil fuels, this constitutes an incentive both to use energy more efficiently, and to substitute renewable energy for fossil energy.

*Health and Safety.* We all have a duty to prevent accidents to others and a desire to prevent damage to ourselves and our families. Governments and responsible organisations have many regulations to safeguard and improve the safety and health of citizens. Like other energy installations, such as nuclear power stations, oil refineries, and high-voltage transmission lines, renewable energy installations can be dangerous, with recognisable difficulty in maintaining safety at the many and dispersed locations. Working near rotating machinery and electrical power systems, climbing structures and handling combustible materials present dangers. In practice, many renewable installations have relatively small-scale operation, so personnel are involved in many varied tasks; although providing interesting and responsible work, such variation presents dangers.

Pollution may be defined as negative impacts, usually chemical emissions, not present in the natural environment. Therefore, since its sources are energy flows in the natural environment, renewable energy is intrinsically pollution-free. This contrasts with fossil and nuclear energy, whose sources are intrinsically pollutants. It is the mechanisms for harnessing renewable energy that may introduce negative impacts, such as smoke and noise, not the sources themselves. Fossil and nuclear energy processes concentrate and then emit chemicals and ionising radiation whose precursors are already present in the primary materials. Therefore, in general, renewables avoid the widespread pollution hazards to health associated with brown energy supplies. The obvious exception is incomplete combustion of biomass, which is common from burning firewood or in poorly regulated machines using biofuels.

*Legislation.* Governments tend to have much legislation concerning energy supply to regulate security, diversity, costs and safety. Specific legislation is needed for renewables, e.g. rights to hydropower, sunshine and wind. To increase security and diversity of supply, governments may enforce (obligate) energy suppliers to include a certain proportion from renewables. Benefits may be enforced, e.g. by net metering for householders so imported electricity from the grid is offset by exporting from embedded photovoltaic, wind and hydro generation.

*Planning.* Governments establish planning legislation and procedures, which may vary greatly between nations and states, but are usually considered and decided upon by local authorities. However, central government will legislate and pronounce upon on general strategy (such as the need for sustainable development) and will reserve the power to decide upon appeals. So although it may involve itself closely in large and influential developments, e.g. large-scale hydropower and offshore wind power, the decisions about medium and small developments tend to reside with local government. Therefore overall strategy and legislation can be expected from national governments, with implementation mostly regulated by local government officials and elected representatives. Democratic rights may give individual citizens considerable influence within planning procedures, but usually only to present arguments to the decision-makers.

*Structure of energy markets.* Until the 1990s, most governments granted the electricity supplier in each region a regulated monopoly, in order to allow economies of scale and discourage wasteful duplication of distribution systems. By 2000, legislation in most industrialised countries encouraged a trend to allow independent generators and suppliers to compete in liberalised markets, but under the control of a Regulator, who acts as a judge to manage prices and services within the bounds set by government. In general, this led to reductions in the price of conventional electricity and has also encouraged many new companies to set up as renewable generators and suppliers. The same governmental mechanisms have been used to obligate the supply of increasing amounts of renewable energy, usually in tandem with financial inducements to renewable energy generators paid from levies on all consumers. A particular mechanism that has been very successful in producing rapid expansion of renewable energy generation is the 'feed law' legislation whereby private generators, often individuals or cooperatives, are guaranteed attractive tariff payments for energy exported to the grid or energy network; such feed laws have been influential in Denmark, Germany and Spain especially.

*Economic conditions.* The relatively large capital costs and initial loans for renewables, together with low fossil fuel price competition, require relatively long payback periods (often 10–15 years and more). Settled economies with small present and predicted inflation rates (and the associated small interest rates) favour such investment. Unsettled economies with large interest rates (reflecting larger risk) discourage capital investment.

### 17.2.2 Developing countries

While the factors outlined above are important for policy everywhere, some extra social and institutional factors bear on energy development in developing countries. Most obviously, developing countries do not have already in place a wide-scale established energy infrastructure, such as a widespread national electricity grid, nor the economic and technical support to establish and maintain such an infrastructure. Energy supplies tend to be installed primarily in the major cities as contrasted with the rural areas, yet even in the cities, supply may be irregular due to excessive demand. So although national consumption of commercial energy may be relatively small on a per capita basis, potential and projected growth in energy consumption raises issues of economic and ecological sustainability; as considered in Chapter 1. A path to the future in which renewable energy and energy efficiency feature strongly should alleviate many of these concerns.

In the rural communities, where probably most people live, many basic energy needs are either unmet or inefficiently met, e.g. cooking with fuelwood and lighting with kerosene. The potential demand for electricity from households is often too small to justify the cost of grid electrification, particularly in remote or island communities. However, the most essential services, such as lights (which enable students to study at night), television, radio and appliances such as hand tools, sewing machines and water pumps, can almost certainly be met from locally appropriate renewable sources. Small household photovoltaic sets, often comprising a single panel, a few lights, a battery, a controller and a 12 V outlet, are technically very well suited to meet this distributed stand-alone demand. Throughout the world there are hundreds of thousands of such 'solar homes', but continued satisfaction and expansion depends upon careful planning and institutional arrangements for their financing and maintenance. As with all renewables, the initial capital cost will be relatively large compared with regular income and so beyond the means of most potential customers. In such situations, microfinancing arrangements can spread the cost over time at an amount comparable to the cost of the alternative (e.g. kerosene) and are widely used in Africa, Sri Lanka, the Pacific Islands and elsewhere. In many industrialised countries, grants and other mechanisms are used to help consumers with the initial capital cost.

## 17.3 Economics

### *17.3.1 Essential questions*

Economics as a discipline aims to develop tools that will help people and governments make rational decisions about the allocation of scarce resources. It does this by reducing all options to monetary terms. From an economist's perspective, all actions have both costs and benefits; the question to be answered is which of several alternative choices, including the status quo, has the most favourable balance. We acknowledge that economics includes more than finance (the root meaning of the word 'economics' is 'the good health of the household'), but we are aware that financial criteria tend to dominate economic choice. Several questions immediately arise in the context of renewable energy systems:

a *Whose* financial costs and benefits are to be assessed: the owners, the end-users or those of the nation or the world as a whole? For example, the actual costs of damage from pollution emitted by a centralised coal-burning electricity power station (corrosion from acid rain, climate change from greenhouse gases, cleaning contaminated effluents, etc.) are mostly not included in the internal financial accounts of the electricity generating company or its customers, but are paid by others. These are called 'external' costs, as described in Box 17.1 on the external costs of energy. Moreover there are other unwanted effects of emissions, e.g. loss of biodiversity, which may not be identifiable in any financial accounting. In contrast, photovoltaic power produces no emissions and has low external costs. If the PV power abates (substitutes for) the use of coal, then real savings are made in society; yet these savings may not accrue to the PV generator whose financial challenge is to pay for the capital costs of the PV system. This comparison favours the fossil-fuel power station because its full costs are not included.

There is much controversy about how to put a monetary value on many of the factors relevant to renewable energy sources, such as having a cleaner environment than otherwise. Because such factors have been hard to quantify, they have often been left out of account in the past, to the detriment of those promoting renewable energy systems.

b *Which* parameters or systems should be assessed; the primary energy sources or the end-use service? For example, householders lighting their house at night are interested in the cost and amount of illumination rather than energy as such. The cost of a clock battery is never considered in terms of Wh delivered, but always in terms of the *service* provided to know the time.

c *Where* does the assessment apply? The costs of renewable energy systems (RES) are very site-specific. Since they are designed to tap into existing natural flows of energy (Chapter 1), it is obvious that a particular RES will be cheapest where the appropriate flow exists. Thus hydroelectric systems are only practical and economically viable where there is a flow of water. The cost of a biomass-based system depends on the availability and cost of the biomass; if this is already on site as waste, as in a cane-sugar mill (Section 11.3.1), the operation is much cheaper than when biomass has to purchased and transported to site.

d *When* are the costs and benefits to be assessed? Renewable energy systems generally have small operational costs and large initial, capital cost. Fossil fuel plant has the reverse, especially if there is no emissions prevention. Economists have developed tools for combining future and continuing known costs with initial costs. These tools are discussed in Section 17.4 below.

From a longer-term perspective, a critical question is: what extra cost should be attributed to the use *now* of a resource which may become severely limited for future generations? There are as yet no agreed answers.

Varying the points of reference (a)–(d) gives very different answers to the question of whether a particular renewable energy system is 'economic'. As one economics professor is reported to have said (when challenged about repeating identical examination questions from year to year): 'in economics, it's the answers that change each year, not the questions!'

### *Box 17.1* External costs of energy

*External costs* are actual and real costs resulting from a process, but which are not included in the price of the product and therefore have to be paid by the public. Electricity generation from coal or nuclear can have significant external costs. For example, burning high-sulphur coal produces $SO_2$ emissions, which give 'acid rain', which causes damage to forests, metal structures and heritage stone buildings. Particulate emissions can cause lung diseases. The costs of disposal of nuclear waste are usually a significant external cost (since, in practice, costs are paid by government from general taxation and not from the sale of electricity), as are the R&D costs of reactors (which in many countries were counted as defence expenditures). The widespread use of motor cars (automobiles) similarly has major external costs, arising from climate change, smog, the productive land 'lost' to roads, and the health and productivity losses caused by injuries and deaths in road accidents.

In the early 1990s there were several large studies which attempted to evaluate numerically such externalities, especially in electricity production. Some indicative results are shown in Table 17.1. The results cover a wide range, reflecting not only methodological difficulties but also the fact that in some countries power stations are in populous areas (so a given amount of pollution will cause more damage to humans and buildings), and in others coal has less sulphur content or is burnt more efficiently (so causing less emissions). Table 17.1 also includes some later estimates of the potential cost of climate change, based on IPCC estimates of the carbon taxes needed to reduce emissions to meet the targets of the Kyoto Protocol, which also cover a wide range. It is also possible in principle to estimate the costs from the damages due to climate change (but enormously sensitive to the discount rate over 100 years or more!) or from the costs of adaptation (constructing sea walls and dykes, etc.).

Table 17.1 Some estimates of the external costs of electricity generation from coal or nuclear (in USc $kWh^{-1}$). Compare these to typical energy retail prices of 3–7 USc $kWh^{-1}$. Based on ORNL (1994), European Commission (1995) and Hohmeyer (1988). See Box 17.1 for discussion

| *Effect* | *USc* $kWh^{-1}$ | *Notes* |
|---|---|---|
| *Coal-fired Electricity* | | |
| Acid rain (from $SO_2$) | 0.02–20 | Larger estimate is for high sulphur coal in urban areas. |
| Climate change (from $CO_2$) | 0.4–12 | Larger estimate assumes no emissions trading. |
| *Nuclear Power* | | |
| Subsidies for R&D | 1.2 | |
| Health impact of accidents | 0.1–10 | |
| Cost of safeguarding waste | unknown | Over thousands of years |

### *17.3.2 Life cycle analysis*

The total of all measurable factors involved in, or resulting from, a process is 'the system of interest'. If these factors are analysed comprehensively and to the full lifetime of the process and its consequences, the study is called 'life cycle analysis'. Comprehensive analysis includes both internal and external factors. Thus for a wind turbine, life cycle analysis would include the manufacture of components from recycled metal and from the mining of metallic ores, the environmental impact of such mining, likewise for other materials, manufacture, per unit factory construction and maintenance, energy supplies in construction, decommissioning, etc. Reducing the analysis to quantifiable amounts for mathematical analysis requires common units, which are often money, but may also be mass or embodied energy; see Section 11.3.4. The full extent of such factors is enormous, but the 'per unit' impact becomes less the further removed the factor. In practice, only significant influences are included, but usually there is debate about what these are and how they are valued and quantified, e.g. visual impact, assessed perhaps by change in local house prices, or employment, assessed perhaps by cancellation of unemployment entitlement. A particular difficulty for finite resources is to assess the value of *not* using them, since, on one hand, pollution is abated by leaving the resource underground, and yet, on the other hand, the resource could be used by future generations. For renewables, a difficulty is how to assess the natural variability of the resource, e.g. for wind power the cost of generating the electricity supply on windless days.

## 17.4 Some policy tools

Table 17.1 clearly suggests that the external costs of electricity generation from non-renewable sources may be comparable to the prices currently charged for electricity. Not taking them into account means that the price signals (the 'market') do not push society as strongly as they should to move to the efficient use of energy and to cleaner patterns of electricity production and consumption, including wider use of renewable energy.

Several policy tools are available to governments to redress this balance, taking into account the socio-political factors outlined in Section 17.2. Many of them use economic means to encourage technological change. We illustrate these, again using the case of electricity generation, though the concepts are more widely applicable to other external costs.

1 *Technological removal of the pollutant.* The pollutant emissions may be abated technologically by removing the pollution at source and adding the cost to the price of electricity, thus *internalising the cost.* In the case of $SO_2$ emissions, this has been done in many countries by legislation making 'flue gas desulphurisation' compulsory. Although this measure does remove the pollutant effectively, it can be unnecessarily costly; especially if the technological tool is mandated and imposed immediately (there may be cheaper equally effective technologies on the way).
2 *Environmental taxes.* The external costs may be *internalised* by, for example, imposing a tax on electricity generated equal to the external cost and using the revenue to pay for the damage. A tax of this kind on $CO_2$ emissions is called a *carbon tax*. In this way, consumers are made more conscious of the true costs of their consumption patterns, and may change them themselves, e.g. by using less energy or by lobbying through political channels for a less polluting source of production. Among the difficulties with this method are (a) determining an appropriate numerical charge, given the ranges in Table 17.1, although even a small positive charge is a step in the right direction, (b) reluctance of consumers (voters) to pay more, (c) competitiveness for businesses if an industry in Country A has to pay such a charge and their competitors in Country B do not, and (d) if the charge is too small, the pollution continues.
3 *Tradable emission permits (certificates).* With taxes, the cost is fixed, but the effect is determined by the market. With tradable permits (also called certificates), the government sets the total amount of pollutant that is to be accepted, and leaves it to the market to determine the least cost to prevent such pollution. For example, in the USA, power companies were obligated collectively to halve their $SO_2$ emissions between 1980 and 2000. So permits were issued for this amount of

$SO_2$, which companies could buy and sell among themselves. Thus if Company A reduced its pollution, say by changing from an obsolete $SO_2$-emitting coal plant to hydropower, then this company could sell its permits to Company B which wished for business reasons to continue using its polluting plant. In practice, the cost of such permit or certificate trading to producers, and thus to consumers, is considerably less than a punitive taxation mechanism. The Kyoto Protocol envisages the use of such permits internationally for greenhouse gas emissions.

4 *Removal of subsidies to polluting sources.* In some countries, particular non-renewable energy supplies are *subsidised* for social reasons, i.e. consumers pay less than the cost of production. For example, coal may be sold to power stations at less than the cost of mining it, with the government paying the difference, in order to keep coal miners in employment (as for many years in Germany). Such subsidies distort the energy market, biasing it against cleaner sources such as renewables. Removing the subsidies removes a barrier to cleaner sources. The social objectives can be accomplished at the same price through welfare payments, but without encouraging pollution.

5 *Subsidies to renewables.* If significant costs remain external to conventional generation, then renewables can still be encouraged by subsidising them, i.e. reducing their costs to consumers or giving a capital grant to generators in order to offset some or all of the external costs of the conventional sources. This can be done by tax concessions for renewable energy generators, as often used in the USA, or by legislating (obligating) that a certain proportion of electricity generation has to be supplied from renewable generation (as in the UK and Australia). In the later case, there is often (i) a competitive tendering process so that least cost renewables are preferentially adopted, and/or (ii) a mechanism to provide a maximum 'ceiling' price. The upshot is larger-scale production and use of renewable energy systems that encourage improvements in the efficiency and cost-effectiveness of renewable plant manufacture and use. This is an example of a mechanism to promote a 'learning curve'; see Section 17.6.

6 *Public research and development.* A first step in most developments, taken particularly by many governments around the 1980s when renewables were at a much earlier stage of development, is to subsidise companies for innovative R&D, often including pilot or demonstration projects. Although helpful for renewables in the 1970s to 1980s, this measure on its own for renewables proved inadequate to overcome the institutional and financial biases against renewables; see Section 17.6. Of course governments and other institutions continue to pay for academic research in many aspects of science and technology, including renewable energy.

## 17.5 Quantifying choice

Installing renewable energy, as with any development, requires commitment of money, time and effort. Choices have to be made, some financial and others ethical. There will be benefits, disadvantages and many other impacts. Some decisions will be taken personally, others on business and political criteria. This section considers the various methods used to analyse and quantify such decisions. However, it is vital to realise that there are no absolute or 'perfect' methods, in the sense used in science and engineering. The choices are made by humans who may decide as they wish, within a huge set of varying values. Analysis proceeds first by discussing values and then by using mathematics to quantify decision-making.

**Box 17.2 Definitions**

*Developer* – Person or organisation planning and co-ordinating a project (in this case, the supply and use of renewable energy).

*Embodied energy (of a product or service, i.e. of a 'good')* – The total of commercial energy expended in all processes and supplies for a good, calculated per unit of that good. Note that by this definition sunshine onto crops is not part of embodied energy, but the heat of combustion of commercial biofuels is included.

*Equity* – Funds in the ownership of the developer; usually obtained from shares sold to shareholders by a limited company. Loans are not equity, so project finance is the sum of equity and loans.

*External costs* – Actual and real costs resulting from a process, which are not included in the price of the product and so have to be paid by the public otherwise (e.g. acid rain from coal-burning power stations damaging metal structures). See Box 17.1.

*Inflation* – A general decrease in the value of money (increase in prices), usually measured by a national average annual rate of inflation $i$.

*Internal costs* – Costs included in the price of a product or service.

*Levelised cost* – (E.g. for the production of electricity.) The average cost of production per unit over the life of the system, allowing for discounting over time.

*Loan* – Money made available to a developer and requiring, usually, the payment of loan interest to the lender in addition to repayment of the sum borrowed . The contracts usually stipulate that, in the event of bankruptcy, loaned money has to be repaid as a priority over the interests of others, e.g. banks providing loans are repaid in preference to shareholders and suppliers of goods. *Operational and maintenance costs(O&M)*: These may be *fixed* (e.g. ground rent, regular staff) or *variable* (e.g. replacement parts, contract staff).

*Price = cost + profit + taxes.*

*Rate* – In accounting, it means a proportion of money exchanged per time period, usually per year. Accountants and economists usually leave out the 'per year' e.g. 'interest rate of 5%' means '5%/y'.

*Retail price index* – a measure of inflation (or deflation) made by the periodic costing of a fixed set of common expenditures.

Of the methods of choice outlined below, methods (1)–(4) are basically 'back of the envelope' sums especially suitable for individual consumers, and for preliminary evaluation. Only if a project looks promising on those bases, is it worth turning to the mathematically more sophisticated methods involving discounted cash flows as used by accountants and bankers. These latter techniques are always used for commercial scale projects that require borrowing from a bank. Box 17.2 is a collection of some definitions of quantities used in such analyses.

1. *Gut-feel.* Most personal and family decisions, and a surprising number of business and political choices, are taken because an individual or a group reach a conclusion instinctively or after discussion. Usually, but not always, the consequences of failure are small, so other methods are needless. Having a vegetable garden or installing a wood-burning stove in a sitting room may be an example; but another was the decision in Paris to construct the Eiffel Tower. Satisfaction and pleasure are obtained, in addition to perceived benefit. Often, a 'statement' is made by the development, e.g. having photovoltaic modules on the entrance roof of a prestigious office as a mark of autonomy and sustainability.
2. *Non-dimensional matrix analysis.* Decide on $n$ criteria or 'values' (e.g. price, noise, aesthetics, lifetime, etc.), each with weight $w_j$ (say $n = 1$ to 10). Then assess each possibility by awarding a mark $m_j$ within each criteria. The total score for each possible choice is:

$$S = \sum_{j=1}^{n} w_j m_j \tag{17.1}$$

Then accept the choice with the largest score, or reassess the weightings and criteria for a further score. Such non-dimensional methods are useful if the criteria cannot all have the same unit of account, e.g. money.

3 *Simple capital payback time* $T_p$. The first step is to decide the actual (internalised) money per year, $I_i$, gained or saved (abated) by a project of capital cost $C$, e.g. an advertisement increases income, or a solar water heating abates electricity purchase. Then $T_p = C/I_i$, which provides an initial criteria for further discussion or analysis. Business may expect $T_p$ of 2 years, whereas a private individual may accept 10 years.

4 *Simple return on investment (simple rate of return)* $R_s$. Expressed as a percentage per year, this is the inverse of payback time; $R_s = 1/T_p$; e.g. $T_p$ of 10 y gives $R_s$ of 10%/y. However, there are more meaningful and 'professional' methods for calculating financial return, so use of $R_s$ may be misleading.

*Example 17.1 Payback of a solar water heater*

In Perth (Australia), a sunny city at latitude 32°S, a typical household uses about 160 litre/day of hot (potable) water. An integrated roof-top solar water heater of collector area 4.0m$^2$ and storage 320 litre supplies this amount at 60 °C throughout the year with about 30% supplementary electric heating. Hence the 'solar fraction' is 70% (i.e. 70% of the energy for hot water is solar. (Australian regulations require all potable hot water to be heated >55 °C to safeguard against Legionnaires disease). Such a solar heater (including the electrical 'boost' heater) costs about Australian A$2750 installed, less a government carbon-abatement grant of A$800, i.e. A$1950 net. Whereas an entirely electric heater and tank for the same hot water supply costs A$670 installed. Assuming no change in prices with time, electricity cost 10c(A)/kWh and that water is heated from 10 °C, what is the payback time for the solar water heater? (Conversion rates at the time of the example were 1.0A$ ~ 0.75US$ ~ 0.8 Euro)

*Solution*:

Although solar water heater characteristics can be calculated with adequate radiation and engineering data, the calculation is beyond the scope of chapter 5 of this book; see Duffie & Beckman for further detail. Therefore we take the characteristics as empirical data.

To heat 160 litre of water though 50 °C (i.e. 10 °C to 60 °C) requires an energy input of:

$$160\,\text{kg} \times 4.2\,\text{kJ/kgK} \times 50\,\text{K} = 33\,600\,\text{kJ}$$

$$= 33.6\,\text{MJ} \times (1\,\text{kWh}/3.6\,\text{MJ}) = 9.3\,\text{kWh}.$$

Supplying this energy by electricity only at 80% efficiency and at A\$0.10/ kWh, costs annually:

$$9.3\,\text{kWh/day} \times 365\ \text{days/y} \times (1/0.8) \times \text{A\$0.10/kWh} = \text{A\$424}.$$

With a solar fraction of 0.7, the cost of electricity for the solar 'boost' system is $(1.0 - 0.7) = 30\%$ of this, i.e. A\$127.

Hence,

$$\text{payback time} = \frac{(\text{capital cost difference, solar} - \text{conventional})}{(\text{annual savings, solar} - \text{conventional})}$$

$$= \frac{(2750 - 800 - 670)\text{A\$}}{(424 - 127)\text{A\$/y}} = 4.3\,\text{y}$$

5 *External payback and benefit criteria.* There may be other benefits or disbenefits than actual internalised money gain or loss, e.g. reduction in pollution. It is entirely reasonable for the developer to include such factors in choice, perhaps for ethical values alone or for proclaiming a good example. If the factors can be quantified in monetary units, even approximately, then the external benefits can be internalised as $I_e$ and added to actual money gain and savings. The payback time then becomes $T_p = C/(I_i + I_e)$. For instance, for a system treating piggery waste to generate biogas for energy, $I_e$ would be the benefits of avoiding pollution from untreated waste, perhaps measured by the fine that would have been applied to the polluter. (In a sense, such a fine internalises the cost of pollution.)

6 *Discounted cash flow (DCF) techniques: net present value.* The word 'discount' in accountancy was originally used in the 17th century to mean 'to give or receive the present worth of a transaction before it is due'. Thus by paying early, less money was paid because a 'discount' was allowed. The amount of the discount was negotiated between the parties, each with different motivations. The corollary is that keeping ownership of money allows it to be increased, e.g. by interest of money from a savings bank. However, there need be no strict relationship between the rate of discount and the rate of bank interest. For instance, paying early may be attractive because of high inflation rate or because the payee wishes to avoid debt; neither factor necessarily relates to the interest rate of a particular bank.

This concept concerning the present value of future transactions provides a powerful accountancy tool for project analysis. If different transactions at different times can be brought to their present monetary values, these can be added as one sum for the '*present value* of the project'. So money of present value, now $V_0$, is treated of future value $V_1 = V_0(1+d)$ after 1 year, where $d$ is the discount rate. After $n$ years, the future value is $V_n = V_0(1+d)^n$ if the discount rate is considered constant. So for future transactions and allowing for the discounting:

$$\text{Present Value} = V_P = \text{sum of all present values} = \sum V_0 \tag{17.2}$$

where for each transaction at each specified future year $n$, present value $V_0 = V_n/(1+d)^n$. The factor $(1+d)^{-n}$ is called the *discount factor.*

Present and future transactions may be positive or negative, i.e. either income or expenditure.

If a fixed sum has to be transacted each year, say as an annuity $A$ for a loan repayment over $N$ years, then

$$V_{P,\text{ annuity}} = A\sum_{n=1}^{N}(1+d)^{-n} = A[1-(1+d)^{-N}]/d \tag{17.3}$$

Of course, agreement has to be reached for the value of discount rate $d$, e.g. governments may specify 8%. In which case a transaction valued at \$1000 in 3 years time (say a maintenance task) has $V_P = \$1000/(1.08)^3 = \$794$.

Note that the longer the time ahead of the transaction, the less is the present value if $d \neq 0$. So the analysis reflects our practical concern for the present and near future, rather than the distant future. Likewise, the less the discount rate, the more important is the future. Such implications of accountancy methods have significant meaning for sustainability and engineering quality.

It is possible to include national inflation rates into the calculation of present values, and so include the actual (real) sums transacted. However, since future inflation is not known, an alternative is to enumerate all transactions at the equivalent for a particular year, e.g. in US$ (year 2000).

With inflation, a future transaction in year $n$ of monetary amount $S_n$ will purchase less of a quantity than now; its current purchasing power, $S'_n$, is therefore reduced. If the inflation rate, $i$, has been constant, then:

$$S'_n = S_n/(1+i)^n \tag{17.4}$$

Discounting this sum, the present value of the inflated transaction becomes:

$$V_0 = \frac{S'_n}{(1+d)^n} = \frac{S_n}{[(1+d)^n \cdot (1+i)^n]} \tag{17.5}$$

If both the discount rate and inflation are $< 10\%/\text{y}$, then

$$V_0 \approx \frac{S_n}{(1+d+i)^n} = \frac{S_n}{(1+p)^n} \tag{17.6}$$

where the sum of discount and inflation rate, $p = d + i$, is the 'market rate of interest'. Note that investors in savings will expect their savings to earn interest rates of at least $p$, and that these banks will in turn loan money above such a rate. Such mixed expectations explain how discount and interest rates may differ.

Both income (say positive) and expenditure (say negative) can have present value, so if a whole complexity of present and future expenditures and incomes are entered into a spreadsheet, the total of all present values (the net present value, NPV) may be calculated. Specialist computer software is available for rapid calculation and variation of input. If the total is positive, then this is taken as a sign of success.

Nevertheless the whole calculation is sensitive to the somewhat arbitrary value given to discount rate $d$. Therefore these techniques are of most value in making comparisons between alternative projects (one of which may well be the *status quo*), where they offer the advantage of making the assumptions used in the alternatives explicit and comparable.

*Example 17.2 Domestic solar water heater*

For the water heater of example 17.1, compare the net present value (NPV) of the solar system to that of the conventional system (CEWH) from the year of installation to 15 years, at a discount rate of 5% and no inflation. What is the payback time? The equipment lifetime should be at least 20 years, so is the solar water heater a good investment?

*Solution*

Table 17.2 sets out the calculations, using the data from Example 17.1.

*Table 17.2* Present value of solar water heater and alternative from year 1 onwards. For discount rate $d = 5\%$, hence discount factor in year $n$ of $(1 + 0.05)^{-n}$

| | *Year (n)* | *SWH* | *CEWH* | *Difference (D)* | *Discount factor (F)* | PV = (D) × (F) | *NPV of (D)* $= \sum_n (PV)$ |
|---|---|---|---|---|---|---|---|
| Installed cost | 0 | 1950 | 670 | 1280 | 1.000 | 1280 | 1280 |
| with grant | 1 | 127 | 424 | −297 | 0.952 | −283 | 997 |
| annual | 2 | 127 | 424 | −297 | 0.907 | −269 | 728 |
| cost of | 3 | 127 | 424 | −297 | 0.864 | −257 | 471 |
| electricity | 4 | 127 | 424 | −297 | 0.823 | −244 | 227 |
| | 5 | 127 | 424 | −297 | 0.784 | −233 | −6 |
| | 6 | 127 | 424 | −297 | 0.746 | −222 | −228 |
| | 7 | 127 | 424 | −297 | 0.711 | −211 | −439 |
| | 8 | 127 | 424 | −297 | 0.677 | −201 | −640 |
| | 9 | 127 | 424 | −297 | 0.645 | −192 | −832 |
| | 10 | 127 | 424 | −297 | 0.614 | −182 | −1014 |
| | 11 | 127 | 424 | −297 | 0.585 | −174 | −1188 |
| | 12 | 127 | 424 | −297 | 0.557 | −165 | −1353 |
| | 13 | 127 | 424 | −297 | 0.530 | −157 | −1510 |
| | 14 | 127 | 424 | −297 | 0.505 | −150 | −1660 |

*Notes:*

(i) In calculating simple payback (as in Example 17.1), effectively the assumed discount rate $d = 0$, so the discount factor $(1 + d)^{-n}$ is 1.00 for all $n$.

(ii) In this case the discount factor is the same for both the alternatives, so it has been applied to the difference (D) to calculate the NPV. That is, the NPV of the difference between the alternatives equals the difference of the NPVs.

(iii) For $n < 5$, the NPV of the solar heater is greater than that of the electrical; for $n > 5$, the NPV of the solar heater is less than that of the alternative. That is, the solar system is costing less than the alternative after 5 years, so its payback according to this analysis at a discount rate of 5% is 5 years.

(iv) In practice, the boosting for the solar system would probably use cheaper 'off peak' electricity than the conventional system, which would improve the payback time against a full-price electrical system, though a fairer comparison might be with an off-peak non-solar system.

(v) For a larger discount rate, the payback time of the capital-intensive alternative is longer; see problem 17.1.

(vi) You may wish to re-work the calculations without the government grant, to see how the grant significantly reduces the payback times.

*Example 17.3 Wind farm*

A wind farm is located on an open plain in New South Wales (Australia). It comprises 10 turbines, each rated at 600 kW and with a cut-in speed of $4\,\mathrm{m\,s^{-1}}$. The wind is fairly steady, so that the turbines are observed to be nearly always turning, but is of only moderate speed ($u_m \sim 6\,\mathrm{m\,s^{-1}}$); the actual capacity factor of the turbines is 0.25. For each turbine, the cost ex-factory is US$440 000 and the cost of installation (including civil and electrical engineering) is $300 000. Operation and maintenance (O&M) costs are about $18 000y per turbine. (These costs exclude the cost of land.)

a Calculate the average ('levelised') cost of production of electricity at the site for a discount rate of 5% and an assumed life of the system of 15 years.

b Under an emissions-trading regime, the generator receives credit for carbon dioxide abated at US$30/t$CO_2$. What is now the effective net cost of production?

c The farmer on whose grazing land the wind farm is constructed, continues to graze his cows there. Three maintenance workers are employed on the farm, and 100 extra visitors come per year to the district to view the installation. What is the cost (actually benefit) to the local region?

d A similar system is installed at another site where the wind is stronger (e.g. Orkney, Scotland, whose wind regime is described in Section 9.6). The capacity factor there is 0.49 (according to Problem 9.11). What is the cost of electricity generated under similar financial assumptions?

*Solution*

a A capacity factor of 0.25 means that each turbine produces a fraction 0.25 of what it would produce if run at full rating for a full year. Thus the electricity produced per year per turbine

$$= 0.60\,\text{MW} \times 8760\,\text{h}\,\text{y}^{-1} \times 0.25 = 1314\,\text{MWh}\,\text{y}^{-1}$$

If this electricity is sold at $q\$\,\text{kWh}^{-1}$, then the stream of costs and benefits from the system will be as shown in Table 17.3.

Therefore the levelised price at which the total benefits will match the total costs in present value terms is

$$q = 918.2/13\,007 = \$0.071\,\text{kWh}^{-1}$$

*Table 17.3* Cost and benefit streams from a wind farm (US$'000 per 600 kW turbine). Benefits are given in terms of the unit price of electricity to the consumer, qUS$/kWh (which defines q in the Benefits Column).

| | *Costs (US$1000)* | | | | *BENEFITS (US$1000)* | | |
|---|---|---|---|---|---|---|---|
| | *Year* | *Capital* | *Annual* | *PV* | *Discount factor* | *Cash* | *PV* |
| Machinery ex-factory | | 440 | | | | | |
| Site engineering | | 300 | | | | | |
| | 0 | 740 | | 740.0 | 1.000 | | |
| | 1 | | 18 | 17.1 | 0.952 | 1314q | 1251q |
| | 2 | | 18 | 16.3 | 0.907 | 1314q | 1192q |
| | 3 | | 18 | 15.5 | 0.864 | 1314q | 1135q |
| | 4 | | 18 | 14.8 | 0.823 | 1314q | 1081q |
| | 5 | | 18 | 14.1 | 0.784 | 1314q | 1030q |
| | 6 | | 18 | 13.4 | 0.746 | 1314q | 981q |
| | 7 | | 18 | 12.8 | 0.711 | 1314q | 934q |
| | 8 | | 18 | 12.2 | 0.677 | 1314q | 889q |
| | 9 | | 18 | 11.6 | 0.645 | 1314q | 847q |
| | 10 | | 18 | 11.1 | 0.614 | 1314q | 807q |
| | 11 | | 18 | 10.5 | 0.585 | 1314q | 768q |
| | 12 | | 18 | 10.0 | 0.557 | 1314q | 732q |
| | 13 | | 18 | 9.5 | 0.530 | 1314q | 697q |
| | 14 | | 18 | 9.1 | 0.505 | 1314q | 664q |
| present value (PV) | | | | 918.2 | 10.899 | | 13 007q |

b For electricity produced from coal, Problem 17.4 shows that each kWh produced entails the emission of 1.0 kg $CO_2$. Therefore the annual carbon credit from each turbine in this system is

$$1314\,\text{MWh} \times (1.0\,\text{t}\,CO_2/\text{MWh}) \times (\$30/\text{t}\,CO_2) = \$39\,400.$$

This can be taken into account in Table 17.3 by subtracting this amount from the annual running cost. (i.e. by replacing 18 by $18 - 39.4 = -21.4$). Doing this the PV of costs changes from \$915 000 to \$528 000, and the cost per unit becomes

$$q' = 528/13\,007 = \$0.041\,\text{kWh}^{-1}$$

c Land costs vary hugely from place to place, but are likely to be relatively small in wind-swept rural areas, especially as the wind-farm operator does not require exclusive use of the land. Since normal agricultural activity can continue underneath the turbines, any rental charge for the facility is entirely gain to the landholder. Consequently, at most wind farm sites, land costs are substantially less than O&M costs, which are the principal annual running cost. Usually more than half the O&M costs is wages paid, which represent a benefit to the local community from extra cash flow.

d With all other financial factors unchanged, the unit cost is inversely proportional to the total kWh generated, i.e. to the capacity factor. Hence the unit cost at the windier site is 3.6 c$\,\text{kWh}^{-1}$. (Note that the capacity factor is not directly proportional to the mean wind speed; see Sections 9.7.)

Example 17.3 illustrates the points in Section 17.3 about the costs assessed depend on *who* is assessing, *what* costs are included, and *where* and *when* the assessment is made. In particular:

i For a capital intensive project, such as this, the levelised cost depends strongly on the assumed life of the system, since with a shorter life there are fewer units of energy produced over which to 'average' the initial cost. (See Problem 17.2).

ii Internalising the external benefits can make a very significant difference to 'the cost' of production.

iii A table like Table 17.3 is easily adapted to the situation where the annual costs vary significantly from year to year, e.g. if major components are replaced every 5 years.

## 17.6 The way ahead

The modern history of renewable energy systems, summarised in Table 17.4, shows an evolution of their status. From being (apart from hydro-power) small-scale 'curiosities' promoted by idealists, renewables have become mainstream technologies, produced and operated by companies competing in an increasingly open market where consumers and politicians are very conscious of sustainability issues. Out to about 2010, the trends are clear. The projections in the table beyond about 2010 are more speculative and should be taken as an expression of the authors' hopes, although they are consistent with several of the range of technically possible 'low emission' scenarios discussed by the IPCC and other reputable bodies.

Implicit in Table 17.4 is the way in which the costs of energy from the more promising renewable energy technologies have steadily decreased. Figure 17.1 illustrates this for the case of wind power. Such decreases in cost per unit of output are common as new technologies are developed from the stage of research prototypes to wide commercial use in a competitive environment. Curves like those in Figure 17.1 are called *learning curves* because they reflect how producers learn by experience how to make the technology more reliable, more efficient, and users learn how to integrate the new technology into their practices; in this case into electricity grids.

In the case of RES, much of this technological learning stems from R&D funded in the 1970s and 1980s, which has led to the technical developments described in earlier chapters, many of them involving new materials and microelectronic control; see Table 17.4. These have contributed to the 'push' for such modern technology. The realisation that substantial application of RESs produces a cleaner environment by the abatement of fossil and nuclear fuels has provided a matching 'pull'. Political and economic measures to encourage wide take-up of a technology can have positive feedback: as more are used, the price reduces. This is because of 'economies of scale'; for instance the cost of designing and tooling for a new model is much the same whether 1 or 100 turbines are being produced, but if more are sold, the cost can be spread more thinly so that the price per unit is less, so more users find it 'economic'. Consequently yet more systems are brought into production, in turn driving further technical and economic improvement in a virtuous cycle.

Figure 17.1 shows a band of costs for a technology. This reflects primarily the site dependence of RESs; for example, wind power is obviously cheaper in a place with stronger prevailing wind speeds. The range also arises from variations in the particular technology assumed (i.e. the precise turbine type), and in the discount rate assumed. Such cost curves can therefore be used only for general guidance. As emphasised throughout this book, appraisal of a particular project at a particular site requires appropriate

*Table 17.4* Evolution of the technological, economic and political environment for 'new' renewable energy systems (RES), from the 1960s to the 2030s

| *Period* | *Technological environment* | *Economic environment* | *Social/political environment* |
|---|---|---|---|
| 1960–1973 | | | |
| | • Traditional and elementary technologies<br>• RES promoted as 'Intermediate technology', especially for developing countries | • RES almost never cost-effective. | • Proponents seen as 'hippies', often living in small idealistic communities. |
| 1973–c1987 | | | |
| | • Public funding for research<br>• Many 'outlandish' ideas<br>• RES begin to incorporate composite materials and microelectronics | • Development of RES seen as an 'insurance' against unavailability and/or increased costs of conventional energy.<br>• High interest rates discourage capital-intensive projects. | • Fright prompted by 'oil crisis' (price increase) of 1973 (OPEC).<br>• Great concern in poorer countries about cost of energy imports<br>• First Ministries of Energy established |
| c1987–c1999 | | | |
| | • Development consolidates around the most (economically) promising technologies | • Commercial-scale projects begin with assistance of grants and other incentives<br>• Externalities considered | • Much talk of 'Sustainable development' following Bruntland (1987) and UNCED, The Earth Summit (Rio) (1992)<br>• Nuclear power falls out of favour<br>• Many of the former 'hippies' now managers |

| | | | |
|---|---|---|---|
| 2000–c2030 | | | |
| | • RES part of 'mainstream' technology<br>• Energy efficiencies improve<br>• Most R&D on RES now by industry itself<br>• Many RES embedded in grids | • Open markets for energy<br>• Cheaper capital<br>• 'Polluter pays' (environmental costs of fossil systems becoming internalised)<br>• Carbon abatement trading | • Sustainability a guiding principle in practice, not just in theory<br>• Diversity of energy supplies seen as important<br>• Climate change policies |
| 2030– | | | |
| | • Efficient and distributed RES embedded as major part of national energy systems<br>• Integrated technologies accepted (but mainly wind, biomass, solar thermal and solar PV) | • Externalities fully internalised<br>• GDP growth no longer seen as centre of well-being | • Climate change and related treaties having significant effect |

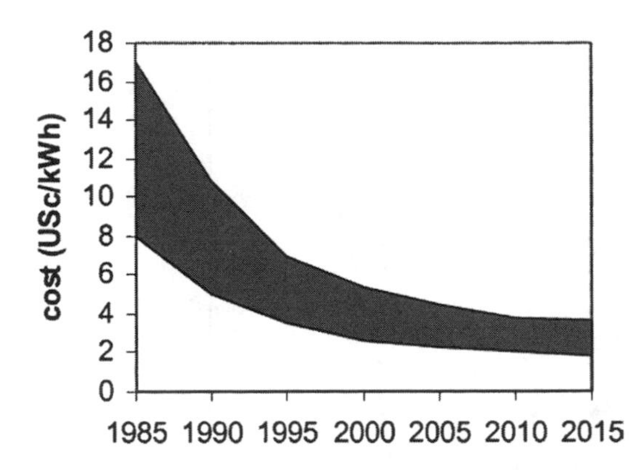


*Figure 17.1* An example of a 'learning curve' – the falling cost of renewable energy. This example is for the cost of electricity from wind-power in the USA. The shaded band reflects the dependence on site (wind-speed in this case), discount rate and other factors.

(*sources*: based on EPRI 1999 and US-DOE 1997).

assessment of the energy resource at that site and the specific characteristics of the system proposed.

Figure 17.2 compares a generic cost curve for renewable energy, i.e. some composite of the curves of Figure 17.1, to two generic cost curves for energy from conventional ('brown') sources. While the costs of renewable energy reduce over time, those from brown energy can be expected to increase over time. For fossil fuels, this reflects the producer's preference to bring to market first those resources which are more readily, and thus more cheaply, extracted. Also, innovation in fuel extraction technologies and in fuel-use technologies continues even though these are relatively mature technologies, driven by commercial pressure to keep these technologies as cost-competitive as possible. Moreover, as the ample literature on oil attests, the actual price increase is not steady, due to competition, political factors, producers trying to undercut others, etc. For nuclear power, the costs have increased over time, as the long-term costs associated with the complete nuclear cycle become increasingly apparent, including security and waste treatment and disposal.

The point at which the decreasing 'green' energy cost curve intersects the increasing 'brown' energy curve in Figure 17.2 represents the crossover point at which that form of renewable energy becomes economically favoured. Though no numerical values are indicated in this schematic diagram, the actual values in Figure 17.1 show that such crossovers occur soon after there is sustained initial trading. Indeed for hydropower in suitable locations, the crossover has long been passed, as is also the case for photovoltaics

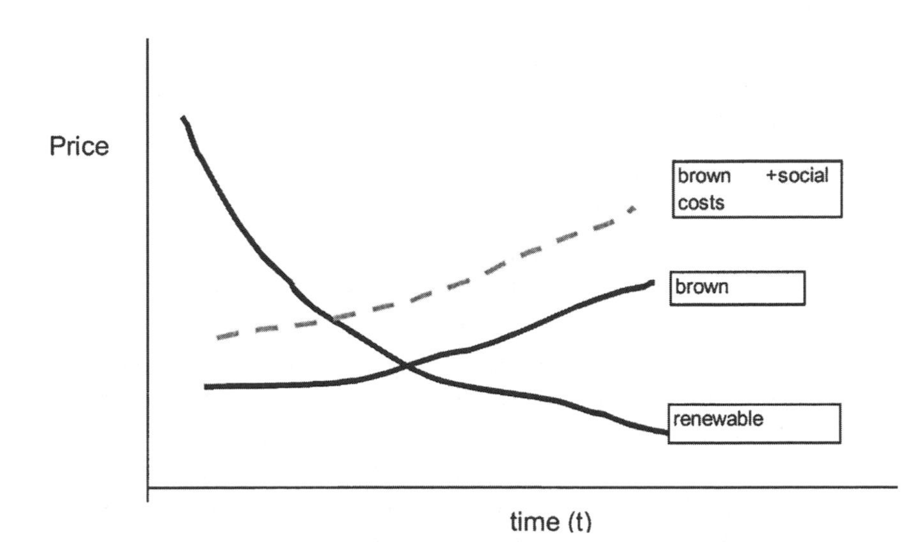


*Figure 17.2* Schematic cost curves for renewable energy, conventional (brown) energy (costed conventionally), and brown energy including external (social) costs. (After Hohmeyer 1988).

and biomass in many areas where conventional supply is difficult or out of scale.

There are two curves for brown energy in Figure 17.2: the more expensive includes the social and environmental costs, which are currently not included in the prices charged, i.e. it includes the 'externalities' of Box 17.1. As Table 17.4 indicates, society is already making some allowance for these externalities, and can be expected to make more allowance in the future. In that case, as Figure 17.2 suggests, renewables become not only the environmentally favoured option but, even sooner, also the economically favoured option.

We conclude that renewables are growth areas of development, with potential to supply much of the world's energy from millions of local and appropriate sites, but success requires knowledge, vision, experience, finance, markets and individual and collective choice. However, we caution that for a national energy system to be truly sustainable, not only must its energy sources be sustainable but also its pattern of energy consumption. That is, close attention needs to be paid to the efficiency and purposes of energy end-use. Unless this is done, even RESs will not be sufficient to meet the growing demand for energy services for heat, transport, cooking, electronics, etc. This is a key point in the low emission scenarios developed by the IPCC.

## Problems

17.1 For the solar water heating system of Example 17.2, calculate the payback time against the conventional electric system, using a discount rate of 10%.

17.2 For the wind farm of Example 17.3, calculate the levelised cost of electricity under the following assumptions:

a a discount rate of 10% and a system life of 15 years,
b a discount rate of 5% and a system life of 6 years,
c a discount rate of 10% and a system life of 6 years.

17.3 Discuss how your country stands in relation to the socio-political factors outlined in Section 17.2. Identify any forces that are acting to change this position.

17.4 Estimate the $CO_2$ emissions per unit of electricity produced by a conventional coal-fired power station. (*Hint*: Coal is about 80% carbon. Make reasonable assumptions about the efficiency of conversion from heat to power).

## Bibliography

### *Policy and institutional issues*

Edinger, R. and S. Kaul (2000) *Renewable Resources for Electric Power*, Quorum Books, Connecticut, USA. {Non-technical account of technologies, emphasising importance of institutional factors and end-use efficiency.}

Flavin, C. and Lenssen, N. (1995) *Power Surge: A Guide to the Coming Energy Revolution*, Earthscan/James and James, London. {Argues that, contrary to most 'incremental analysis', an energy revolution to a sustainable system based on greater efficiency and renewable sources is both possible and desirable.}

Goldemberg, J. (1996) *Energy, Environment and Development*, Earthscan/James and James, London. {Wide-ranging and readable exposition of the links between energy and social and economic development and sustainability, with consideration of equity within and between countries by a Brazilian expert.}

Hunt, S. and Shuttleworth, G. (1996) *Competition and Choice in Electricity*, Wiley, Chichester. {Explains how national electricity supply industries are changing by liberalisation and privatisation.}

International Energy Agency (1997) *Key Issues in Developing Renewables*, Paris. {Institutional and economic issues as seen by governments in industrialised countries.}

International Solar Energy Society (2004) *Transitioning to a Renewable Energy Future – A White Paper* [available at www.ises.org] {Focuses on technology commercialisation and policy shifts required.}

Scheer, H. (2001) *The Solar Economy: Renewable Energy for a Sustainable Global Future*, Earthscan/James and James, London. {A well phrased polemic arguing that an energy system based almost totally on renewables is feasible but will require 'creative destruction' of the old fossil-based system.}

UNCED (1992) *The Earth Summit*, several resulting United Nations publications, including: Agenda 21, the Rio Declaration on Environment and Development, the Statement of Forest Principles, the United Nations Framework Convention on Climate Change and the United Nations Convention on Biological Diversity.

Wilkins, G. (2002) *Technology Transfer for Renewable Energy*, Earthscan/James and James. {Examines the practicalities of bringing renewable energy into wider use in developing countries, with reference to the Kyoto Protocol mechanisms and case studies of biomass co-generation and household photovoltaic systems.}

Yergin, D. (1992) *The Prize: The Epic Quest for Oil Money and Power*, Simon and Schuster, New York. {Entertaining account of the politics and personalities of the oil world.}

### *Environmental economics textbooks*

Common, M. (1996) *An Introduction to Environmental Economics*, Longmans, London. {Clear exposition at introductory level of relevant economic principles and tools.}

Gilpin, A. (2000) *Environmental Economics: A Critical Overview*, Wiley. {Another introduction for non-economists, entertaining style coupled with technological optimism.}

### *Externalities*

Oak Ridge National Laboratory & Resources for the Future (1994) *Fuel Cycle Externalities*, USA.

Hohmeyer, O. (1988) *Social costs of energy consumption: External effects of electricity generation in the Federal Republic of Germany*, Springer Verlag (1988).

European Commission (1995) *ExternE: Externalities of Energy*, 7 vols (1995).

### *Tools for present value analysis*

Awerbuch, S. (1996) 'The problem of valuing new energy technologies', *Energy Policy*, **24**, pp. 127–128. {Introduction to new valuation techniques going beyond 'traditional' utility present value techniques.}

Boyle, G. (ed.) (2nd edn 2004) *Renewable Energy*. Oxford University Press. {The appendix on 'investing in renewable energy' is strongly recommended.}

International Energy Agency (1991) *Guidelines for the Economic Analysis of Renewable Energy Technology Applications*, Paris. {Very detailed account with worked examples, though ignoring externalities.}

### *Scenarios for the future*

Aitken, D.W., Billman, L.L. and Bull, S.R. (2004) 'The Climate stabilisation challenge: can renewable energy sources meet the target?', *Renewable Energy World*,

December, pp. 56–69. {A review of various published scenarios, which concludes that RE could make 50% of all energy supply by 2050.}
Bruntland, G.H. (Chairperson) (1987) World Commission on Environment and Development, *Our Common Future*, Oxford University Press (The 'Bruntland report'). {A seminal work, warning about the key issues in plain language for politicians.}
Intergovernmental Panel on Climate Change (2001) *Third Assessment Report*, 3 vols, Cambridge University Press. {Volume 3 considers mitigation and has a special report on emission scenarios; however this simulation analysis has little relevance to practical engineering opportunities. The IPCC is taken as an authoritative source on the science, impacts and mitigation of climate change. Summaries available on the web at www.ipcc.ch.}
Sims, R. (2004) Renewable energy: A response to climate change, *Solar Energy*, **76**, pp. 9–17. {Summary of renewable energy scenarios from the IPCC Third Assessment report (2001).}
US Department of Energy and Electric Power Research Institute (1997) *Technical Characterisation of Renewable Energy*. {Detailed projections of expected technical developments and projected costs.}

### Some case studies

Ling, S., Twidell, J. and Boardman, B. (2002) 'Household photovoltaic market in Xining, Qinghai Province, China: The role of local PV business', *Solar Energy*, **73**, pp. 227–240.
Lipp, J. (2001) 'Micro-financing solar power: the Sri Lankan SEEDS model', *Refocus*, October issue, pp. 18–21.
Mandela, M. (2000) 'Support for renewables: A perspective of the Development Bank of Southern Africa', *Refocus*, August issue, pp. 15–17.

### Journals and websites

*Refocus* magazine (published by Elsevier on behalf of the International Solar Energy Society) has numerous articles on rural electrification, e.g. Mandela (August 2000) on southern Africa, Lipp (October 2001) on Sri Lanka, and on policy developments.
*Energy Policy* (published by Elsevier) is an academic journal focussed on economic, policy and institutional aspects, e.g. the impact of climate change policies, of all forms of energy including renewable energy.

Appendix A

# Units and conversions

## A.1 Names and symbols for the SI units

### *Base units*

| *Physical quantity* | *Name of SI unit* | *Symbol for SI unit* |
|---|---|---|
| Length | metre | m |
| Mass | kilogram | kg |
| Time | second | s |
| Electric current | ampere | A |
| Thermodynamic temperature | kelvin | K |
| Amount of substance | mole | mol |
| Luminous intensity | candela | cd |

### *Supplementary units*

| *Physical quantity* | *Name of SI unit* | *Symbol for SI unit* |
|---|---|---|
| Plane angle | radian | rad |
| Solid angle | steradian | sr |

## A.2 Special names and symbols for SI derived units

| *Physical quantity* | *Name of SI unit* | *Symbol for SI unit* | *Definition of SI unit* | *Equivalent form(s) of SI unit* |
|---|---|---|---|---|
| Energy | joule | J | $m^2\,kg\,s^{-2}$ | N m |
| Force | newton | N | $m\,kg\,s^{-2}$ | $J\,m^{-1}$ |
| Pressure | pascal | Pa | $m^{-1}\,kg\,s^{-2}$ | $N\,m^{-2}$, $J\,m^{-3}$ |
| Power | watt | W | $m^2\,kg\,s^{-3}$ | $J\,s^{-1}$ |
| Electric charge | coulomb | C | s A | A s |
| Electric potential difference | volt | V | $m^2\,kg\,s^{-3}\,A^{-1}$ | $J\,A^{-1}\,s^{-1}$, $J\,C^{-1}$ |

(Continued)

| | | | | |
|---|---|---|---|---|
| Electric resistance | ohm | $\Omega$ | $m^2\,kg\,s^{-3}\,A^{-2}$ | $V\,A^{-1}$ |
| Electric capacitance | farad | F | $m^{-2}\,kg^{-1}\,s^4\,A^2$ | $A\,s\,V^{-1}$, $C\,V^{-1}$ |
| Magnetic flux | weber | Wb | $m^2\,kg\,s^{-2}\,A^{-1}$ | $V\,s$ |
| Inductance | henry | H | $m^2\,kg\,s^{-2}\,A^{-2}$ | $V\,A^{-1}\,s$ |
| Magnetic flux density | tesla | T | $kg\,s^{-2}\,A^{-1}$ | $V\,s\,m^{-2}$, $Wb\,m^{-2}$ |
| Frequency | hertz | Hz | $s^{-1}$ | |

## A.3 Examples of SI derived units and unit symbols for other quantities

| *Physical quantity* | *SI unit* | *Symbol for SI unit* |
|---|---|---|
| Area | square metre | $m^2$ |
| Volume | cubic metre | $m^3$ |
| Wave number | per metre | $m^{-1}$ |
| Density | kilogram per cubic metre | $kg\,m^{-3}$ |
| Speed; velocity | metre per second | $m\,s^{-1}$ |
| Angular velocity | radian per second | $rad\,s^{-1}$ |
| Acceleration | metre per second squared | $m\,s^{-2}$ |
| Kinematic viscosity | square metre per second | $m^2\,s^{-1}$ |
| Amount of substance concentration | mole per cubic metre | $mol\,m^{-3}$ |

## A.4 Other units

| *Physical Quantity* | *Unit* | *Unit symbol* | *Alternative representation* |
|---|---|---|---|
| Energy | electron volt | eV | $1\,eV = 1.602 \times 10^{-19}\,J$ |
| Time | year | y | $365.26\,d = 3.16 \times 10^7\,s$ |
| Time | minute | min | 60 s |
| Time | hour | h | 60 min = 3600 s |
| Time | day | d | 24 h = 86 400 s |
| Angle | degree | ° | $(\pi/180)$ rad |
| Angle | minute | ′ | $(\pi/10\,800)$ rad |
| Angle | second | ″ | $(\pi/648\,000)$ rad |
| Volume | litre | L | $10^{-3}\,m^3 = dm^3$ |
| Volume | gallon (US) | | $3.785 \times 10^{-3}\,m^3$ |
| | gallon (Brit.) | | $4.546 \times 10^{-3}\,m^3$ |
| Mass | tonne | t | $10^3\,kg = Mg$ |
| Celsius temperature* | degree Celsius | °C | K, kelvin |
| Area | acre (Brit.) | | $4.047 \times 10^3\,m^2$ |
| | hectare | | $10^2 \times (10^2\,m^2) = 10^4\,m^2$ |

* The Celsius temperature is the excess of the thermodynamic temperature more than 273.15 K.

A useful mnemonic for power conversions is: 'the number of seconds per year equals $\pi$ times ten to the number of days in the week' (i.e. $3.14 \times 10^7$). Allowing for leap years, this is accurate to 2 significant figures!

## A.5 SI prefixes

| *Multiple* | *Prefix* | *Symbol* | *Multiple* | *Prefix* | *Symbol* |
|---|---|---|---|---|---|
| $10^{-1}$ | deci | d | 10 | deca | da |
| $10^{-2}$ | centi | c | $10^2$ | hecto | h |
| $10^{-3}$ | milli | m | $10^3$ | kilo | k |
| $10^{-6}$ | micro | $\mu$ | $10^6$ | mega | M |
| $10^{-9}$ | nano | n | $10^9$ | giga | G |
| $10^{-12}$ | pico | p | $10^{12}$ | tera | T |
| $10^{-15}$ | femto | f | $10^{15}$ | peta | P |
| $10^{-18}$ | atto | a | $10^{18}$ | exa | E |

## A.6 Energy equivalents

1 kWh = 3.6 MJ
1 Btu = 1055.79 J
1 therm = $10^5$ Btu = 105.6 MJ = 29.3 kWh
1 calorie = 4.18 J
1 tonne coal equivalent = 29.3 GJ (UN standard)
1 tonne oil equivalent = 42.6 GJ (UN standard) = $11.833 \times 10^3$ kWh

## A.7 Power equivalents

1 Btu $s^{-1}$ = 1.06 kW
1 Btu $h^{-1}$ = 0.293 W
1 horsepower = 746 W
1 (tonne oil equivalent) $y^{-1}$ = 1.350 kW
1 Mtoe $y^{-1}$ = 1350 MW

## A.8 Converting units and labelling graphical axes etc.

This book uses the convention of 'algebraic' use of units, i.e.:

a quantity is a number of units

so... quantity = number × unit

and... number = quantity ÷ unit = quantity/unit

Therefore in tables, headings for columns and rows with numbers are labelled by:

(symbol of quantity)/unit

Likewise for the labels on axes of graphs.

In calculations, there is often confusion about whether to multiply or divide by a numerical conversion factor, but the technique shown here (and used in the examples throughout this book) is nearly fool-proof. In brief:

- Express all physical quantities as (number) × (unit); i.e. retain units explicitly throughout the working.
- Cancel out the same unit in the denominator and numerator to simplify
- Multiply quantities in such a way that only the desired units remain, with the undesired ones 'cancelled out'.

The trick is to plug in an appropriate expression for 1!

For example, the equality $1\,\text{kW} = 10^3\,\text{W}$ can be expressed as

$$1 = \left(\frac{1\,\text{kW}}{10^3\,\text{W}}\right) \quad \text{or} \quad \text{as } 1 = \left(\frac{10^3\,\text{W}}{1\,\text{kW}}\right)$$

*Example A.1*
Express 1 kWh in MJ

*Solution*

$$1.0\,\text{kWh} = 1.0\,\text{kWh} \times \left(\frac{10^3\,\text{W}}{1\,\text{kW}}\right) \qquad \text{using A.5}$$

$$= 1.0\,\text{kWh} \times \left(\frac{10^3\,\text{W}}{1\,\text{kW}}\right) \times \left(\frac{1\,\text{J}\,\text{s}^{-1}}{1\,\text{W}}\right) \qquad \text{using A.2}$$

$$= 1.0\,\text{kWh} \times \left(\frac{10^3\,\text{W}}{1\,\text{kW}}\right) \times \left(\frac{1\,\text{J}}{1\,\text{W}\,\text{s}}\right) \times \left(\frac{3600\,\text{s}}{1\,\text{h}}\right) \text{ using A.4}$$

$$= 1.0\,\text{kWh} \times \left(\frac{10^3\,\text{W}}{1\,\text{kW}}\right) \times \left(\frac{1\,\text{J}}{1\,\text{W}\,\text{s}}\right) \times \left(\frac{3600\,\text{s}}{1\,\text{h}}\right) \times \left(\frac{1\,\text{MJ}}{10^6\,\text{J}}\right)$$

$$= 3.6\,\text{MJ}$$

In Example A.1, each of the expressions in parentheses equals 1.0 since the numerator and denominator are identical, by definition of the units. With practice, one can go directly to the long expression at the end. Notice how the expressions are arranged so that the 'undesired' units (in this case, J, s etc.) 'cancel out' (i.e. appear in the numerator of one bracket and the denominator of another).

The technique can fruitfully be extended to cases where the equalities can be interpreted more loosely as 'yields' or 'produces' or 'is equivalent to', as in Example A.2.

*Example A.2*
Calculate the electricity output (in kWh) from burning 100 litre of diesel fuel in a diesel generator with 20% efficiency.

*Solution*
Let $E$ be the electrical output.
From Table B.6 (in Appendix B), diesel fuel has a heat content of $38\,\text{MJ L}^{-1}$, and from Example A.1,

$$1 = \left(\frac{1.0\,\text{kWh}}{3.6\,\text{MJ}}\right)$$

Hence

$$\begin{aligned} E &= 20\% \times (\text{heat output from } 100\,\text{L diesel}) \\ &= \left(\frac{20}{100}\right) \times 100\,\text{L diesel} \times \left(\frac{38\,\text{MJ heat}}{1.0\,\text{L diesel}}\right) \\ &\quad \times \left(\frac{1.0\,\text{kWh(electricity)}}{3.6\,\text{MJ(electricity)}}\right) \\ &= 210\,\text{kW}_e\text{h (to 2 significant figures)} \end{aligned}$$

Appendix B

# Data

The following tables give sufficient physical data to follow the examples and problems in this book. They are not intended to take the place of the standard handbooks listed in the following bibliography, from which the data have been extracted. These handbooks themselves use databases, such as those maintained by the US National Institute of Standards and Technology.

Only two or three significant figures are given except in the few cases where the data and their use in this book justify more accuracy.

## Bibliography


Mills, A.F. (1999, 2nd edn) *Basic Heat and Mass Transfer*, Prentice-Hall. {Includes a full appendix of accessible data.}

Monteith, J. and Unsworth, M. (1990, 2nd edn) *Principles of Environmental Physics*, Edward Arnold, London. {Extremely useful set of tables for data on air and water vapor, and on heat transfer with elementary geometrical shapes.}

Osborne, P.D. (1985) *Handbook of Energy Data and Calculations*, Butterworths, London. {Pragmatic information for heating and cooling engineers; uses S.I. units.}

Rohsenow, W.M., Hartnett, J.P. and Cho, J. (eds) (1997, 3rd edn) *Handbook of Heat Transfer*, McGraw-Hill, New York. {Chapter 2 by T.F. Irvine is an extensive compilation of thermophysical data.}

*Handbook of Physics and Chemistry*, CRC Press (annual). {Chemical emphasis, but useful for all scientists.}

Wong, H.Y. (1977) *Handbook of Essential Formulae and Data on Heat Transfer for Engineers*, Longman, London. {Student- oriented short compilation, but out of print. Has a useful 20 pages of thermophysical data (the rest is like Appendix C). Highly recommended if you can find a copy.}

*Table B.1* Physical Properties of dry air at atmospheric pressure. ($\mathscr{A}$ is the Raleigh number, $X$ the characteristic dimension)

| *Temperature* $T$ | *Density* $\rho$ | *Specific heat* $c_{(p)}$ | *Kinematic viscosity* $\nu = \mu/\rho$ | *Thermal diffusivity* $\kappa$ | *Thermal conductivity* $k$ | *Prandtl number* $\mathscr{P}$ | $\mathscr{A}/X^3\Delta T$ |
|---|---|---|---|---|---|---|---|
| °C | $kg\,m^{-3}$ | $10^3\,J\,kg^{-1}\,K^{-1}$ | $10^{-6}\,m^2\,s^{-1}$ | $10^{-6}\,m^2\,s^{-1}$ | $10^{-2}\,W\,m^{-1}$ | | $10^8\,m^{-3}\,K^{-1}$ |
| 0 | 1.30 | 1.01 | 13.3 | 18.4 | 2.41 | 0.72 | 1.46 |
| 20 | 1.20 | 1.01 | 15.1 | 20.8 | 2.57 | | 1.04 |
| 40 | 1.13 | 1.01 | 16.9 | 23.8 | 2.72 | | 0.78 |
| 60 | 1.06 | 1.01 | 18.8 | 26.9 | 2.88 | 0.70 | 0.58 |
| 80 | 1.00 | 1.01 | 20.8 | 29.9 | 3.02 | | 0.45 |
| 100 | 0.94 | 1.01 | 23.0 | 32.8 | 3.18 | 0.69 | 0.34 |
| 200 | 0.75 | 1.02 | 34.6 | 50 | 3.85 | 0.68 | 0.12 |
| 300 | 0.62 | 1.05 | 48.1 | 69 | 4.50 | | 0.052 |
| 500 | 0.45 | 1.09 | 78 | 115 | 5.64 | | 0.014 |
| 1000 | 0.28 | 1.18 | 174 | 271 | 7.6 | 0.64 | 0.0016 |

Other properties of air:

Velocity of sound in air (at 15°C) $= 340\,m\,s^{-1}$
Coefficient of diffusion of water vapor in air (at 15°C) $= 25 \times 10^{-6}\,m^2\,s^{-1}$
Coefficient of self-diffusion of $N_2$ or $O_2$ in air (at 15°C) $= 18 \times 10^{-6}\,m^2\,s^{-1}$
Coefficient of thermal expansion (at 27°C) $\beta = (1/T) = 0.0033\,K^{-1}$

*Table* B.2 Physical properties of water (at moderate pressures)

(a) *Liquid*

| *Temperature* $T$ | *Density* $\rho$ | *Kinematic viscosity* $\nu = \mu/\rho$ | *Thermal diffusivity* $\kappa$ | *Thermal conductivity* $k$ | *Prandtl number* $\mathscr{P}$ | $\mathscr{A}/X^3\Delta T$ | *Expansion coefficient* $\beta$ | *Specific heat capacity* $c_{(p)}$ |
|---|---|---|---|---|---|---|---|---|
| °C | $10^3 kg\,m^{-3}$ | $10^{-6} m^2 s^{-1}$ | $10^{-6} m^2 s^{-1}$ | $Wm^{-1}K^{-1}$ | | $10^{10} m^{-3}K^{-1}$ | $10^{-4} K^{-1}$ | $JKg^{-1}K^{-1}$ |
| 0 | 0.9998* | 1.79 | 0.131 | 0.55 | 13.7 | −0.24* | Changes | 4217 |
| 20 | 0.9982 | 1.01 | 0.143 | 0.60 | 7.0 | +1.44 | sign* | 4182 |
| 40 | 0.9922 | 0.66 | 0.151 | 0.63 | 4.34 | 3.81 | 3.0† | 4178 |
| 60 | 0.9832 | 0.48 | 0.155 | 0.65 | 3.07 | 6.9 | 4.5† | 4184 |
| 80 | 0.9718 | 0.37 | 0.164 | 0.67 | 2.23 | 10.4 | 5.7† | 4196 |
| 100 | 0.9584 | 0.30 | 0.168 | 0.68 | 1.76 | 14.9 | 6.7† | 4215 |

* The maximum density of water occurs at 3.98 °C and is 1000.0 $kg\,m^{-3}$. Therefore $\beta$ is negative in the range 0 °C < $T$ < 4 °C.
† These values of $\beta$ apply to the range from the line above, e.g. $3.0 \times 10^{-4} K^{-1}$ is the mean value between 20 and 40 °C.

(b) *Water vapour in air*

| *Temperature* $T$ | *Saturated vapour pressure* $p_v$ | *Mass of* $H_2O$ *in* 1 $m^3$ *of saturated air* $\chi$ |
|---|---|---|
| °C | $kN\,m^{-2}$ | $g\,m^{-3}$ |
| 0 | 0.61 | 4.8 |
| 10 | 1.23 | 9.4 |
| 20 | 2.34 | 17.3 |
| 30 | 4.24 | 30.3 |
| 40 | 7.38 | 51.2 |
| 50 | 12.34 | 82.9 |
| 60 | 19.9 | 130 |
| 70 | 31.2 | 197 |
| 80 | 47.4 | 291 |
| 90 | 70.1 | |
| 100 | 101.3 | |

Note $\chi = (2.17 \times 10^{-3}\,\text{kg K m}^2\,\text{N}^{-1})p_v/T$.

(c) *Other properties of water*

Latent heat of freezing $\Lambda_1 = 334\,\text{kJ kg}^{-1}$

Latent heat of vaporization $\Lambda_2 = 2.45\,\text{MJ kg}^{-1}(20°\text{C})$

$= 2.26\,\text{MJ kg}^{-1}(100°\text{C})$

Surface tension (against air) $= 0.073\,\text{N m}^{-1}(20°\text{C})$

*Table B.3* Density and conductivity of solids (at room temperature)

| *Material* | *Density* $\rho$ ($kg\,m^{-3}$) | *Thermal conductivity* $k$ ($W\,m^{-1}\,K^{-1}$) |
|---|---|---|
| Copper | 8795 | 385 |
| Steel | 7850 | 47.6 |
| Aluminum | 2675 | 211 |
| Glass (standard) | 2515 | 1.1 |
| Brick (building) | 2300 | 0.6–0.8 |
| Brick (refractory fireclay) | 2400 | 1.1 |
| Concrete (1:2:4) | 2400 | 1.5–1.7 |
| Granite | 2700 | 2 |
| Ice (−1°C) | 918 | 2.26 |
| Gypsum plaster (dry, 20°C) | 881 | 0.17 |
| Oak wood (14% m.c.) | 770 | 0.16 |
| Pine wood (15% m.c.) | 570 | 0.138 |
| Pine fiberboard (24°C) | 256 | 0.052 |
| Asbestos cement, sheet (30°C) | 150 | 0.319 |
| Cork board (dry, 18°C) | 144 | 0.042 |
| Mineral wool, batts | 32 | 0.035 |
| Polyurethane (rigid foam) | 24 | 0.025 |
| Polystyrene, expanded | 16 | 0.035 |
| Still air (27°C, 1 atmos.) | 1.18 | 0.026 |

Note: Approximate only for manufactured materials, whose properties vary.

*Table B.4* Emittance of common surfaces

| *Material* | *Temperature* $T$ | *Emittance* $\varepsilon$ |
|---|---|---|
| | °C | % |
| Aluminum | | |
| polished | 100 | 9.5 |
| roughly polished | 100 | 18 |
| Iron, roughly polished | 100 | 17 |
| Tungsten filament | 1500 | 33 |
| Brick (rough, red) | 0–90 | 93 |
| Concrete (rough) | 35 | 94 |
| Glass (smooth) | 25 | 94 |
| Wood (oak, planed) | 90 | 90 |

*Table B.5* Miscellaneous physical fundamental constants, to 3 significant figures

| | |
|---|---|
| Avogadro constant | $N_0 = 6.02 \times 10^{23}\ \mathrm{mol^{-1}}$ |
| Elementary charge | $e = 1.60 \times 10^{-19}\ \mathrm{C}$ |
| Gas constant | $R_0 = 8.31\ \mathrm{J\ K^{-1}\ mol^{-1}}$ |
| Gravitational constant | $G = 6.67 \times 10^{-11}\ \mathrm{N\ m^2\ kg^{-2}}$ |
| Permeability of free space | $\mu_0 = 4\pi \times 10^{-7}\ \mathrm{H\ m^{-1}}$ |
| Permittivity of free space | $\epsilon_0 = 8.85 \times 10^{-12}\ \mathrm{F\ m^{-1}}$ |
| Planck constant | $h = 6.63 \times 10^{-34}\ \mathrm{J\ s}$ |
| Speed of light in vacuum | $c = 3.00 \times 10^{8}\ \mathrm{m\ s^{-1}}$ |
| Stefan-Boltzmann constant | $\sigma = 5.67 \times 10^{-8}\ \mathrm{W\ m^{-2}\ K^{-4}}$ |

*Table B.6* Calorific values of various fuels

| *Fuel* | *Gross calorific value* [a] | | *Remarks* |
|---|---|---|---|
| | $MJ\,kg^{-1}$ | $MJ\,L^{-1}$ [b] | |
| *Crops* | | | |
| Wood | | | |
| Green | ~8 | ~6 | Varies more with moisture |
| Seasonal | ~13 | ~10 | content than species of wood |
| Oven dry | ~16 | ~12 | |
| Vegetation: dry | ~15 | | Examples: grasses, hay |
| *Crop residues* | | | |
| Rice husk | | | For dry material. |
| Bagasse (sugarcane solids) | 12–15 | | In practice residues may be |
| Cow dung | | | very wet |
| Peat | | | |
| *Secondary fuels* | | | |
| Ethanol | 30 | 25 | $C_2H_5OH$: $789\,kg\,m^{-3}$ |
| Hydrogen | 142 | $12 \times 10^{-3}$ | |
| Methanol | 23 | 18 | $CH_3OH$ |
| Biogas | 28 | $20 \times 10^{-3}$ | 50% methane +50% $CO_2$ |
| Producer gas | 5–10 | $(4–8) \times 10^{-3}$ | Depends on proportion of CO and $H_2$ |
| Charcoal | | | |
| Solid pieces | 32 | 11 | |
| Powder | 32 | 20 | |
| Coconut oil | 39 | 36 | |
| Biodiesel (1) | 39 | 33 | Ethyl esters of coconut oil |
| Biodiesel (2) | 40 | 35 | Methyl esters of soya oil |
| *Fossil fuels* | | | |
| Methane | 56 | $38 \times 10^{-3}$ | Also called 'natural gas' |
| Petrol | 47 | 34 | Motor spirit, gasoline |
| Kerosene | 46 | 37 | |
| Diesoline | 46 | 38 | Automotive distillate, derv, diesel |
| Crude oil | 44 | 35 | |
| Coal | 27 | | Black, coking grade |

Notes

a *Gross calorific value* (also called *heat of combustion*) is the heat evolved in a reaction of the type

$$CH_2O + O_2 \rightarrow CO_2(\text{gas}) + H_2O\ (\text{liquid})$$

Some authors quote instead the net (or lower) calorific value, which is the heat evolved when the final $H_2O$ is gaseous, without latent heat recovery.

b At 15 °C.

# Appendix C

# Some heat transfer formulas

For notation, definitions and sources see Chapter 3. $X$ is the characteristic dimension for calculation of Nusselt Number $\mathcal{N}$; Reynolds number $\mathcal{R}$ and Rayleigh number $\mathcal{A}$. These formulas represent averages over the range of conditions likely to be met in solar engineering. In particular $0.01 < P < 100$, where $\mathcal{P}$ is the Prandtl number.

*Table C.1* General

| | | *Text references* |
|---|---|---|
| Heat flow | $P = \Delta T/R$ | (3.2) |
| Heat flux density | $q = P/A = \Delta T/r$ | (3.4), (3.5) |
| Thermal resistance of unit area, thermal resistivity | $r = 1/h = RA$ | (3.6), (3.9) |
| Conduction | $r_n = \Delta x/k$ | (3.14), |
| | $R_n = \Delta x/kA$ | (3.13) |
| Convection | $r_v = X/\mathcal{N}k$ | (3.19) |
| | $R_v = X/A\mathcal{N}k$ | (3.18) |
| Radiation: in general | $r_r = (T_1 - T_2)/q$ | (3.4) |
| | $R_r = (T_1 - T_2)/P_{12}$ where $P_{12}$ is given in Table C.5 | (3.3) |
| Radiation: for small $(T_1 - T_2)$ | $R_r = 1/[4\sigma A_1 F'_{12}(\bar{T})^3]$ | (3.49) |
| Heat by mass transfer | $P_m = \dot{m}c\Delta T$ | (3.54) |
| | $R = 1/(\dot{m}c)$ | (3.55) |
| Nusselt number | $\mathcal{N} = \frac{XP}{kA\Delta T}$ | (3.17) |
| Reynolds number | $\mathcal{R} = uX/\nu$ | (2.11), (3.21) |
| Rayleigh number | $\mathcal{A} = \frac{g\beta X^3 \Delta T}{\kappa\nu}$ | (3.25) |
| Prandtl number | $\mathcal{P} = \nu/\kappa$ | (3.22) |
| Grashof number | $\mathcal{G} = \mathcal{A}/\mathcal{P}$ | (3.26) |
| Thermal diffusivity | $\kappa = k/\rho c$ | (3.15) |

*Table C.2* Free convection. Comparative tables in other texts may refer to Grashof number $\mathscr{G} = \mathscr{A}/\mathscr{P}$

| Shape | Case | Overall Nusselt number | Equation no. |
|---|---|---|---|
| 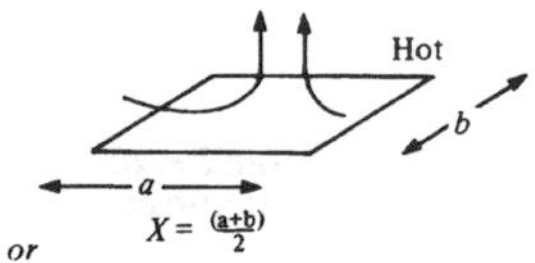 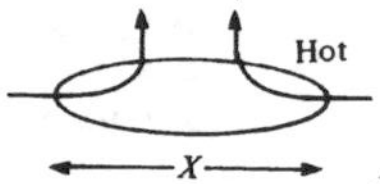 Horizontal flat plate ($X = \frac{(a+b)}{2}$) | Laminar ($10^2 < \mathscr{A} < 10^5$) | $\mathscr{N} = 0.54\mathscr{A}^{0.25}$ | (C.1) |
| or  Hot / or Cold | Turbulent ($\mathscr{A} > 10^5$) | $\mathscr{N} = 0.14\mathscr{A}^{0.33}$ | (C.2) |
| Horizontal cylinder | laminar ($10^4 < \mathscr{A} < 10^9$) | $\mathscr{N} = 0.47\mathscr{A}^{0.25}$ | (C.3) |
| | turbulent ($\mathscr{A} > 10^9$) | $N = 0.10\mathscr{A}^{0.33}$ | (C.4) |
| Vertical flat plate 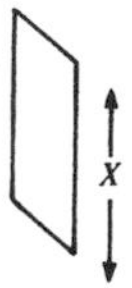 or Vertical cylinder 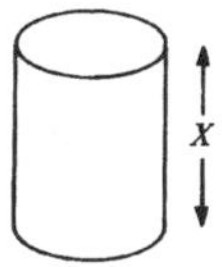 | If laminar, ($10^4 < \mathscr{A} < 10^9$) | $\mathscr{N} = 0.56\mathscr{A}^{0.25}$ | (C.5) |
| | If turbulent, ($10^9 < \mathscr{A} < 10^{12}$) | $\mathscr{N} = 0.20\mathscr{A}^{0.40}$ | (C.6) |
| Parallel plates (slope <50°) ($T_1$, $T_1 + \Delta T$) | Turbulent ($\mathscr{A} > 10^5$) | $\mathscr{N} = 0.062\mathscr{A}^{0.33}$ | (C.7) |

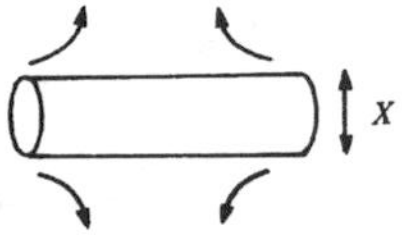

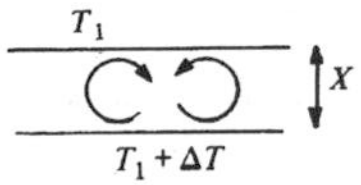

*Table C.3* Forced convection

| Shape | Case | Overall Nusselt number | Equation no. |
|---|---|---|---|
| Flow over flat plate | Laminar ($\mathscr{R} < 5 \times 10^5$) | $\mathscr{N} = 0.664\,\mathscr{R}^{0.5}\mathscr{P}^{0.33}$ | (C.8) |
| | *Turbulent* ($\mathscr{R} > 5 \times 10^5$) | $\mathscr{N} = 0.37\,\mathscr{R}^{0.8}\mathscr{P}^{0.33}$ | (C.9) |
| Flow over circular cylinder | Laminar ($0.1 < \mathscr{R} < 1000$) | $\mathscr{N} = (0.35 + 0.56\,\mathscr{R}^{0.52})\mathscr{P}^{0.3}$ | (C.10) |
| | Turbulent ($10^3 < \mathscr{R} < 5 \times 10^4$) | $\mathscr{N} = 0.26\,\mathscr{R}^{0.6}\mathscr{P}^{0.3}$ | (C.11) |
| Flow *inside* a circular pipe: from Wong, 1977 (see Appendix B, Bibliography) | Laminar flow, short pipe ($\mathscr{R} < 2300$, $\mathscr{G}_1 > 10$) | Graetz number $\mathscr{G}_1 = \mathscr{R}\mathscr{P}(D/L) = 4Q/\kappa\pi L$ | (C.12) |
| | | $\mathscr{N} = 1.86\,\mathscr{G}_1^{0.33}$ | (C.13) |
| | Turbulent flow ($\mathscr{R} > 2300$) | $\mathscr{N} = 0.027\,\mathscr{R}^{0.8}P^{0.33}$ | (C.14) |
| | General | $P = \rho c Q(T_2 - T_1)$ | (2.6) |



⁓⁓⁓ Indicates section of an extended shape

*Table C.4* Mixed convection (forced and free together)

| Shape | Case | Formula | |
|---|---|---|---|
| Air over flat plate | 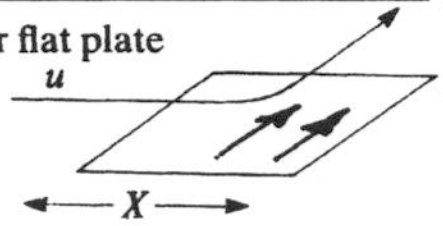 $X > 0.1\,\text{m}$, $u < 20\,\text{m s}^{-1}$ | $h = a + bu$ <br> $[a = 5.7\,\text{W m}^{-2}\text{K}^{-1}$ <br> $b = 3.8\,(\text{W m}^{-2}\text{K}^{-1})/(\text{m s}^{-1})]$ | (C.15) |
| General | | $\mathcal{N}_1 = \max(\mathcal{N}_{forced}, \mathcal{N}_{free})$ <br> $\mathcal{N}_1 < \mathcal{N}_{mixed} < \mathcal{N}_{forced} + \mathcal{N}_{free}$ | (C.16) |

*Table C.5* Net radiative heat flow between two diffuse grey surfaces. For definitions and notation, see Chapter 3, especially Sections 3.5.6 and 3.5.7. In general $P_{12} = \sigma A_1 F'_{12}(T_1^4 - T_2^4)$, where $F'_{12}$ is the exchange factor of (3.46). NB In these formulas $T$ is the *absolute temperature* (i.e. in kelvin, unit K)

| System | Schematic presentation | Net radiative heat flow | Equation no. |
|---|---|---|---|
| Gray surface to surroundings ($A_1 << A_2$) | 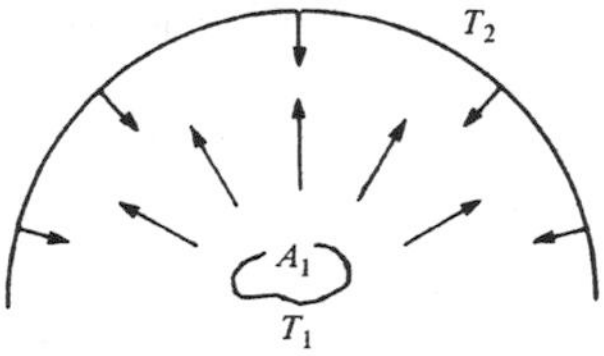 | $P_{12} = \epsilon_1 \sigma A_1 (T_1^4 - T_2^4)$ | (C.17) |
| Two closely spaced parallel planes ($L/D \to \infty$) | 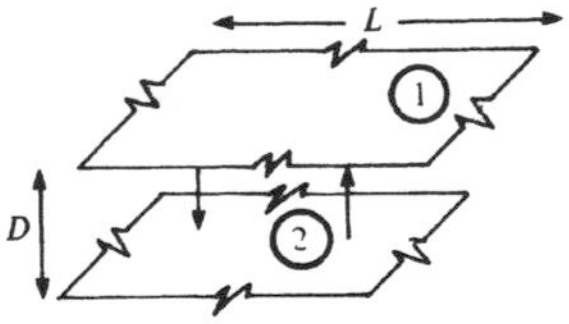 | $P_{12} = \dfrac{\sigma A_1 (T_1^4 - T_2^4)}{(1/\epsilon_1) + (1/\epsilon_2) - 1}$ | (C.18) |
| Closure formed by two surfaces (surface 1 convex or flat) | 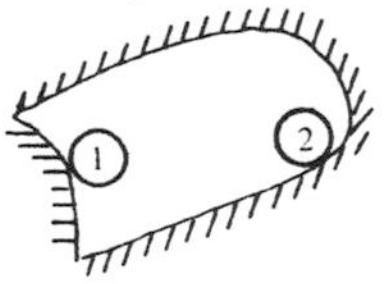 | $P_{12} = \dfrac{\sigma A_1 (T_1^4 - T_2^4)}{\dfrac{1}{\epsilon_1} + \left(\dfrac{A_1}{A_2}\right)\left(\dfrac{1}{\epsilon_2} - 1\right)}$ | (C.19) |
| General two-body system (neither surface receives radiation from a third surface) | 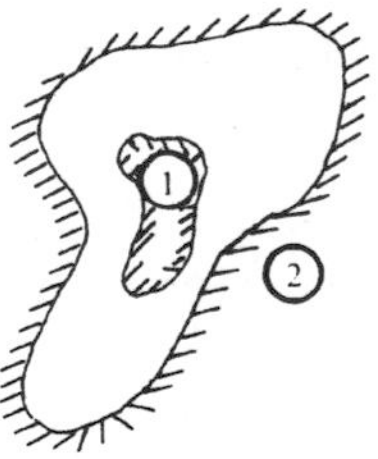 | $F_{12}$ = shape factor | (3.36) |
| | | $P_{12} = \dfrac{\sigma (T_1^4 - T_2^4)}{\dfrac{1-\epsilon_1}{\epsilon_1 A_1} + \dfrac{1}{A_1 F_{12}} + \dfrac{1-\epsilon_2}{\epsilon_2 A_2}}$ | (C.20) |

# Solution guide to problems at the end of chapters

## Chapter 1

1.1 a Average irradiance = absorbed flux/total area of Earth = $1.2 \times 10^{17}\,\mathrm{W}/5.1 \times 10^{14}\,\mathrm{m}^2$

1.2 Consider costs over 10 000 h., with an electricity price of €0.10 $\mathrm{kWh}^{-1}$.

a Direct cost = (10 × €0.5) + (10 000 h × 0.100 kW × €0.10 $\mathrm{kWh}^{-1}$) = €(5 + 100) = €105; dominant cost is the electricity.

b Direct cost = (1 × €4.0) + (10 000 × (5/22) 0.100 kW × €0.10 $\mathrm{kWh}^{-1}$) = €(4 + 23) = €27 = 18% of (a); so direct savings 70% of (a) with higher proportion of capital cost.

c Savings with (b) average 70% of cost of (a), i.e. savings average €74 per CFL. Savings equal cost of (b) when: (€74/(10 000 h)$T$ = €4 + (€23/(10 000 h)$T$, so $T$ = 780 h; i.e. at less than 10% of the life of the CFL, and less than the lifetime of one incandescent lamp.

## Chapter 2

2.1 a Mass: $u_2 = u_1 A_1/A_2$. Bernoulli equation (2.2) as written. Simple algebra.

b $Q$

c Geometric difference in levels.

d Calculated flow too high because some kinetic energy at 1 is degraded to turbulence and heat at 2.

2.2 a All $p$ equal in (2.4) and $g = 0$.

b Zero, because without viscosity fluid cannot apply shear force. Mass reduces to $A_j = A_1 + A_2$. Momentum (*vector*):

$$\rho A_j u_j^2 \cos\alpha = \rho A_1 u_1^2 - \rho A_2 u_2^2$$

Simple algebra.

c $F = \rho Q_j u_j \sin\alpha = 86\,\mathrm{N}$, pushing plane down and to the right.

2.3 a Difference in pressure forces at $x, x+\Delta x$ balances difference between forward (for $y < (1/2)D$) shear force on top and backward shear force on bottom. Use (2.9).

b $u = 0$ at $y = 0, y = D$. Simple algebra.

c $Q = \int u$ (Wdy).

2.4 a $[(\partial p/\partial x)\Delta x][\pi(r+\Delta r)^2 - \pi r^2] = \tau(r)2\pi r\Delta x - \tau(r+\Delta r)2\pi(r+\Delta r)\Delta x$

b Symmetry: $\partial u/\partial r = 0$ at $r = 0$, since $u = 0$ at $r = R$.

c $Q = \int u 2\pi r \mathrm{d}r$

d Simple algebra.

2.5 a Following Example 2.1 gives $H_\mathrm{f} = 12\,\mathrm{km} \gg L$.

b $f = 0.006$ (more or less independent of $\mathcal{R}$ in this range) gives $u = 4.4\,\mathrm{m\,s^{-1}}$, $Q = 8\,\mathrm{L\,s^{-1}}$.

2.6 Note $u = \frac{Q}{\left(\frac{1}{4}\pi D^2\right)}$.

## Chapter 3

| 3.3 $\mathcal{A}/\mathcal{A}_c$: | 0 | 1.48 | 3.8 | 16 |
|---|---|---|---|---|
| $\mathcal{N}$(expt): | 1 | 1.5 | 2.1 | 3.3 |

3.4 $26\,\mathrm{kJ\,h^{-1}}$, using $\varepsilon = 0.10$, $T_1 = 373\,\mathrm{K}$ in (C.17).

3.5 (a) 5.7 (b) 25 (c) $44\,\mathrm{W\,m^{-2}\,K^{-1}}$, using (C.15). Error ~50%.

3.6 $h_\mathrm{total} = h_\mathrm{forced} + h_\mathrm{free} = 4.8, 13.7, 43.2\,\mathrm{W\,m^{-2}\,K^{-1}}$, using (C.2), (C.8).

3.7 a 10 kW

b $r_\mathrm{v} = 0.032\,\mathrm{m^2\,K\,W^{-1}}$ on both sides;

$R_{14} = 0.029\,\mathrm{K\,W^{-1}}$; $P_{14} = 0.7\,\mathrm{kW}$.

$\mathrm{U}_{14} = 34\,\mathrm{W\,K^{-1}}$.

## Chapter 4

4.1 a $G_0 = \sigma T_s^4(4\pi R_s^2)/4\pi l^2 = 1366\,\mathrm{W\,m^{-2}}$

4.3 a Use (4.8) with $\phi = -18°$, $\delta = 20°$, $\beta = 0$, $\gamma$ arbitrary but irrelevant ($\gamma$ terms all multiplied by $\sin\beta$), $\omega = 45°$. Hence $\theta_1 = \cos^{-1}(0.526) = 58°$, with $G_\mathrm{d} = 0$, $G_\mathrm{b} = G_\mathrm{b}^* = G_\mathrm{h}/\cos\theta_1 = 0.9\,\mathrm{MJ\,h^{-1}\,m^{-2}}$.

b As for (a), with $\beta = +30°$, $\gamma = -180°$; $\theta_z = 40°$.

c Assume $G_\mathrm{bh} = G_\mathrm{dh} = \frac{1}{2}G_\mathrm{h} = 0.5\,\mathrm{MJ\,m^{-2}\,h^{-1}}$. Then $G_\mathrm{b}^* = G_\mathrm{bh}/\cos\theta_1 = 0.9$, and so

$$G^* = G_\mathrm{b}^* + G_\mathrm{d}^* = 0.9 + 0.5 = 1.4$$

(assuming $G_d^* = G_{dh} = G_{dc}$) and

$$G_c = G_b^* \cos\theta_c + G_{dc} = 1.2\,\mathrm{MJ\,m^{-2}\,h^{-1}}$$

4.5 (a) needed (b) (i) 12.7 h, 11.3 h (ii) 18.5 h, 5.5 h.

4.6 (b) Summer $H_{0h} = 41.5\,\mathrm{MJ\,m^{-2}}$, winter $H_{0h} = 8.5\,\mathrm{MJ\,m^{-2}}$; $K_T = 07$.

## Chapter 5

5.1 b

$$h_{v,\,pg} = 2.4\,\mathrm{W\,m^{-2}\,K^{-1}}$$
$$h_{r,\,pg} = 6.2\,\mathrm{W\,m^{-2}\,K^{-1}}$$

hence

$$r_{pg} = 0.12\,\mathrm{m^2\,K\,W^{-1}}$$
$$h_{v,\,ga} = 24.7\,\mathrm{W\,m^{-2}\,K^{-1}}$$
$$h_{r,\,ga} = 7.9\,\mathrm{W\,m^{-2}\,K^{-1}}$$

hence

$$r_{ga} = 0.030\,m^2\,\mathrm{K\,W^{-1}}$$
$$r_{pa} = 0.15\,\mathrm{m^2\,K\,W^{-1}}(\mathrm{cf.}0.13\,\mathrm{for}\,T_g = 45°\,\mathrm{C}).$$

c 30 °C. No.

5.2 (a) 4 mm (b) 44 mm.

5.3 In effect, the plate to glass circuit of Figure 5.4(b) occurs twice. $r_{pa} = 0.22\,\mathrm{m^2\,K\,W^{-1}}$.

5.4 Same circuit as Figure 5.4(b) but with $\varepsilon = 0.1$, $r_{pa} = 0.40\,\mathrm{m^2\,K\,W^{-1}}$.

5.5 Circuit as for Problem 5.3 but with $\varepsilon = 0.1$, $r_{pa} = 0.45\,\mathrm{m^2\,K\,W^{-1}}$.

5.6 Use (5.2) in the form

$$mc\Delta T/\Delta t = A\tau_{cov}\alpha G - (T_p - T_a)/r_{pa}$$

with $m = 65\,000$ bottles × 50 kg/bottle, $\Delta T = 30\,°\mathrm{C}$, $\Delta t = 8\,\mathrm{h}$, $G = (20\,\mathrm{MJ\,m^2})/(8\,\mathrm{h}) = 700\,\mathrm{W\,m^{-2}}$ (average), to give

| | $\tau_{cov}$ | $\alpha$ | $r_{pa}$ | $A$ | *Collector price* | *Collector price* |
|---|---|---|---|---|---|---|
| | | | $m^2 K W^{-1}$ | $10^4 m^2$ | $\$ m^{-2}$ | M$ |
| (a) | 1.0 | 1.0 | $\infty$ | 2.0 | – | – |
| (b) | 0.9 | 0.9 | 0.13 | 4.3 | 100 | 4.3 |
| (c) (i) | $(0.9)^2$ | 0.9 | 0.22 | 3.8 | 200 | 7.6 |
| (c) (ii) | $(0.9)^2$ | 0.9 | 0.40 | 3.3 | 250 | 8.2 |

5.7 a Note that $\tau_\alpha + \tau_\alpha(1-\alpha)\rho_d + \tau_\alpha(1-\alpha)^2\rho_d^2 + \cdots = \tau_a/[1-(1-\alpha)\rho_d]$
b 1.08

5.8 a Conduction term is $k\delta dy[(dT/dx)_x - (dT/dx)_{x+dx}]$
c 0.98.

## Chapter 6

6.1 a $2.7\,N\,m^{-2}$ (b) $650\,N\,m^{-2}$.

6.2 a $\varepsilon = n\pi a^2/A_0$ (c) 0.13 m, $10\,m^2$.

6.3 a $r_n = 1.7\,m^2\,K\,W^{-1} > 0.4\,m^2\,K\,W^{-1}$ (for good flat plate)
b $r_v = 0.0018\,m^2\,K\,W^{-1} \ll r_n$ of (a).
c $c \sim 30$ g NaCl per kg $H_2O$) $\ll$ saturated concentration.
d $RC \approx 1.3 \times 10^6\,s = 15$ day (neglected bottom loss) Sunrise temp $= T_a + (T_{store} - T_a)_0 \exp[-t/(RC)] = 78.3\,°C$.
e Diffusion time $\approx X^2/D \approx 5$ y.
f Salt gradient easily established and maintained against diffusion. Heat well retained, therefore theoretically viable. Practicalities: see text.

6.5 a Larger $\eta$
b Use $P = -Ak\partial T/\partial r$
c Use $P_u = \alpha GXA - \varepsilon\sigma\left(T_p^4 - T_a^4\right)A$; $X = 10$, easy (NB $P_u \neq 0$)
d Use (C.4), $\mathcal{A} = 10^5$, $P = 6\,kW$
e $0.2\,km^2$ (for $\eta = 0.25$).

6.6 a $50 \times 10^3\,m^2$ (assuming $\eta = \eta_{Carnot}$, and average power = 1/3 peak) $\approx$40% of area for cells.
b Night-time; beam radiation; feedback from sun sensor at focus.
c 3 kW
d $560\,kW\,m^{-2}$

6.7 a 4 mm, 2 kJ
b large $T \rightarrow$ large $X \rightarrow$ large (energy) density.

## Chapter 7

7.1 $h\nu = E_g$; $\lambda = 0.88\,\mu\text{m}$

7.2 Graph of $I = (10^{-8}\,\text{A m}^2)[\exp(eV/kT) - 1]$

7.3 a From Figure 7.12(b), the photons with average photon number flux density have energy ~1 eV. Thus

$$N \approx \frac{(0.8\,\text{kW m}^{-2})(10^3\,\text{W kW}^{-1})(1\,\text{photon/eV})}{(1.6 \times 10^{-19}\,\text{J eV}^{-1})}$$
$$\approx 0.5 \times 10^{21}\,\text{photon m}^{-2}\,\text{s}^{-1}$$

b Assume 10% of the photons each produce one electron and one hole as charge carriers. Then

$$I = (0.5 \times 10^{21}\,\text{photon m}^{-2}\,\text{s}^{-1})(2\,\text{carrier/photon})(10^{-2}\,\text{m}^{-2})$$
$$\times (1.6 \times 10^{-19}\,\text{C/carrier})$$
$$= 1.6\,\text{A}$$

(NB These are approximate 'ballpark' answers. Mathematical rigor would produce accurate answers.)

7.4 Cells should produce about 10 V to charge an 8 V storage battery. Each cell on peak load has EMF ~0.5 V, so we need parallel arrangements of 20 cells in series. Each night 12 A h is discharged, so each day 12 A h/0.8 = 15 A h is needed from the cells. Assume 3 h of direct sunlight each day so 5 A charging current is needed. In series connection this could be obtained from $5\,\text{A}/(200 \times 10^{-4}\,\text{A cm}^{-2}) = 250\,\text{cm}^2$. With a series arrangement of 20 cells, each cell needs an area of $250\,\text{cm}^2$ (radius ~5.0 cm). Thus a series arrangement of 20 cells each of radius 5.0 cm is reasonable for charging the batteries with the *assumed* solar radiation flux.

Test for short circuit current, open circuit voltage, in direct sunshine normal to the cells.

7.5 a Total energy input $2700\,\text{kWh kWp}^{-1}$, from 4700 kWh of primary energy. At $1\,\text{kW m}^{-2}$ standard insolation for peak power, area of modules $6.7\,\text{m}^2$, so PV electricity produced is $(6.7\,\text{m}^2 \times 1450\,\text{kWh}\,(\text{m}^2\text{y})^{-1} \times 0.15) = 1450\,\text{kWh y}^{-1}$. Primary energy payback $4700\,\text{y}/1450 = 3.2\,\text{y}$

b Per kWh unit, electricity has more 'value' than heat, so the two should not be compared directly (a more meaningful method is to consider exergy ratios). Nevertheless increased production reduces the proportion of 'overhead energy', and using hydroelectricity would not involve the thermal inefficiencies of thermal power stations. Note that 'simple energy payback' is certainly not a simple concept! Other information in Knapp and Jester (2001).

7.6 Facing downward, so avoiding snow cover and allowing reflection onto the module from the snow!

7.8 See [almost] any textbook on 'modern physics' for a description of the photoelectric effect.

7.9 29 W (use inverse square law, allowing for little atmosphere on Mars, so relevant irradiance at Earth is that above the atmosphere, i.e. $G_0{}^* = 1367\,\mathrm{W\,m^{-2}}$, as in Section 4.1).

## Chapter 8

8.1 Viti Levu (Fiji). The catchment area taken has in fact been developed as Fiji's first major hydroelectric station at Monosavu.

a $P_1 = (1000\,\mathrm{kg\,m^{-3}})(4\,\mathrm{m\,y^{-1}})(2000\,\mathrm{km^2})(9.8\,\mathrm{m\,s^{-2}})(300\,\mathrm{m}) = 800$ MW.
b Guesses by A.D.W. : (i) ~50% (ii) ~50% (iii) ~50%. Hence $P_2 = 100\,\mathrm{MW}$.
C Monasavu stage 1 is 40 MW.

8.2 a Note hints in question.
b $Q = \int_{h=0}^{H} u_\mathrm{h} L\,\mathrm{d}h$ Assume $u_1^2/2g \ll h$.
C Turbulence on entry and exit implies loss of KE (cf. Bernoulli).
d $Q = 0.16\,\mathrm{m^3\,s^{-1}}$, $u_1 = 0.03\,\mathrm{m\,s^{-1}}$.
Hence $[u_1^2/(2g) \ll h]$ for $h > 1\,\mathrm{mm}$.

8.3 Simple algebra. See text.

8.4 a $2.4\,\mathrm{m^3\,s^{-1}}$ (b) $65\,\mathrm{rad\,s^{-1}}$ (c) 2.4, smaller pulley on generator.

8.5 Follow the derivations of Section 2.3. Efficiency reduced by 3%. Laboratory angle $\approx \frac{1}{2}$ cup angle.

8.6 a (i) $20\,\mathrm{m\,s^{-1}}$ (ii) 9.8 kW (iii) 2.0 cm.
b Consider for example $\eta = 0.9$: (iv) 25 cups (v) 52 cm (vi) $33\,\mathrm{rad\,s^{-1}}$.
c (vii) $\mathcal{R} = 4 \times 10^5$, $H_\mathrm{a} = 16\,\mathrm{m}$, P = 7.2 kW (viii) First approx. $H_\mathrm{f} \approx 800\,\mathrm{m}$ (see Problem 2.5). Therefore $P \approx 0$ (nearly all potential energy goes into friction and useful output power is insignificant).

## Chapter 9

9.1 Simple algebra. $\mathrm{d}c_\mathrm{p}/\mathrm{d}a = 0$

9.2 Simple algebraic substitution.

9.3 a By conservation of mass, $\rho A_0 u_0 = \rho A_1 u_1$.
From (9.11) and with $a = 1/3$ for maximum power extraction (9.17), $2u_0 = 3u_1$; so $A_1 = 3A_0/2$.
At maximum power extraction:

$$\frac{\text{output power}}{\text{input power}} = \frac{(16/27)A_1 u_0^3}{A_0 u_0^3} = \left(\frac{16}{27}\right)\left(\frac{3}{2}\right) = \frac{8}{9}$$

b Into equation (9.29), substitute from (9.1), (9.31) and (9.33).

9.4 See equation (2.11). If Reynolds number $\sim uX/\nu < 2000$; then $u < 200 \times 10^{-6}\,\text{m}^2\,\text{s}^{-1}/1.0\,\text{m} = 0.2\,\text{mm}\,\text{s}^{-1}$. Flow therefore turbulent always, and laminar flow unrealistic.

9.5 a See (9.69):

$$\Phi(u) = au\exp\left[-bu^2\right], \text{ where } b = \pi/(4\bar{u}^2)$$

$$\frac{d\Phi(u)}{du} = 0 \text{ at } u^2 = 2\bar{u}^2/\pi, u = 0.80\bar{u}$$

b

$$\Phi(u)u^3 = au^4\exp[-bu^2]$$

$$\frac{d\left[\Phi(u)\,u^3\right]}{du} = 0 \text{ at } u^2 = 8\bar{u}^2/\pi, \quad u = 1.60\bar{u}$$

c See (9.71)

$$\overline{u^3} = \frac{\pi}{2\bar{u}^2}\int_0^\infty u^4\exp\left[-\frac{\pi}{4}\left(\frac{u}{\bar{u}}\right)^2\right]du$$

Let $bu^2 = v$, where $b = \pi/4\bar{u}^2$ and so $v = \pi u^2/(2\bar{u})^2$. Hence $u = (v/b)^{1/2}$
and

$$du = \frac{1}{2}\left(\frac{1}{bv}\right)^{1/2} dv$$

Hence

$$\overline{u^3} = \frac{\pi}{2\bar{u}^2}\int_0^\infty \frac{v^2}{\alpha^2}\exp[-v]\frac{1}{2(bv)^{1/2}}dv$$

$$= \frac{\pi}{4\bar{u}^2 b^{5/2}}\int_0^\infty v^{3/2}e^{-v}\,dv$$

$$= \frac{\pi}{4\bar{u}^2}\left[\frac{4\bar{u}^2}{\pi}\right]^{5/2}\left[\frac{3}{2}\right]$$

$$= \left[\frac{4\bar{u}^2}{\pi}\right]^{3/2} \frac{3}{2}\left(\frac{1}{2}!\right)$$

$$= \frac{8(\bar{u})^3}{\pi^{3/2}} \frac{3}{2} \frac{\sqrt{\pi}}{2}$$

$$= \frac{6}{\pi}(\bar{u})^3 = 1.91(\bar{u})^3$$

Hence

$$\left[\overline{u^3}\right]^{1/3} = 1.24\bar{u}$$

9.6

| | $\dfrac{u[\Phi(u)_{max}]}{\bar{u}}$ | $\dfrac{u[\Phi(u)u^3]_{max}}{\bar{u}}$ | $\dfrac{[\overline{u^3}]^{1/3}}{\bar{u}}$ | $\dfrac{u[\Phi(u)u^3]_{max}}{u[\Phi(u)]_{max}}$ |
|---|---|---|---|---|
| Rayleigh distribution | 0.80 | 1.60 | 1.24 | 2.00 |
| North Ronaldsay data | 0.76 | 1.53 | 1.24 | 2.00 |

9.7 Using prime notation for the first set of blades, and double prime for the second, the overall power extraction is $P = \frac{1}{2}\rho A u_0^3 C_P$
But

$$P = P' + P'' = \frac{1}{2}\rho A\left[u_0^3 C'_P + (u'_2)C''_P\right]$$

$$= \frac{1}{2}\rho A u_0^3\left[C'_P + \left(\frac{u'_2}{u_0}\right)^3 C''_P\right]$$

so

$$C_P = C'_P + \left(\frac{u'_2}{u_0}\right)^3 C''_P$$

But

$$a' = \frac{u_0 - u_2}{2u_0}$$

So

$$\frac{u'_2}{u_0} = 1 - 2a'$$

and

$$C'_P = 4a'(1-a')^2$$
$$C_P = 4a'(1-a')^2 + (1-2a')^3 C''_P$$

$C''_p$ is independent of a′, so $C'_p$ is maximum when $C''_p$ is maximum at 16/27.
Thus

$$C_P = 4a'(1-a')^2 + (1-2a')^3(16/27)$$

$C_P$ is a maximum when $a' = 0.2$:

$$C_P = 0.8^3 + (0.6^3)\left(\frac{16}{27}\right) = 0.640.$$

9.8 a $v_{tip} = \lambda u_0$ If $v_{tip} = 330\,\mathrm{m\,s^{-1}}$, then $u_0 = \frac{330\,\mathrm{m\,s^{-1}}}{8} = 41\,\mathrm{m\,s^{-1}}$
b $\Omega = v_{tip}/R = (330/50)\mathrm{s^{-1}} = 6.6\,\mathrm{rad^{-1}}$
$f = \Omega/2\pi = 1.1\,\mathrm{Hz}.$

9.9 Use equation 9.24, with $C_F = 1$ for the 'maximum possible'.

9.10 From equation (9.12) $u_1 = (1-a)u_0$ For maximum power, $a = 1/3$, so $u_1 = 2u_0/3$. Cot $\phi = R\Omega/u_1 = (3/2)R\Omega/u_0 = 1.5\lambda$.

9.11 (a) 293 kW, 49% (b) 200 kW, 33%. (*Hint*: Use 9.54 to estimate the speed at hub height, and then construct a table similar to Table 9.3 to calculate the power obtained in each speed range.)

9.13

$$\bar{P}_T = \frac{\rho C_p}{2}\left(\int_{u=u_{ci}}^{u=u_R} \Phi_u u_0^3 du + \Phi_{u_R<u_0<u_{co}} P_R\right)$$

$$\bar{P}_T = \frac{\rho C_p}{2}\int_{u=u_{ci}}^{u=u_R} \frac{\pi u^4}{2\bar{u}^2}\exp\left[-\frac{\pi}{4}\left(\frac{u_0}{\bar{u}}\right)^2\right] du + P_R \exp\left[-\frac{\pi}{4}\left(\frac{u_R}{\bar{u}}\right)^2\right]$$
$$= \frac{\rho C_p}{2}\frac{6}{\pi}(\bar{u})^3 + P_R \exp\left[-\frac{\pi}{4}\left(\frac{u_R}{\bar{u}}\right)^2\right]$$

## Chapter 10

10.2 4.8 eV/atomC

## Chapter 11

11.1 a Gas yield $200\,\mathrm{MJ\,day^{-1}}$. Car requires 4 litre petrol per day $=$ $160\,\mathrm{MJ\,day^{-1}}$. Compressor work $\bar{p}V \sim 30\,\mathrm{MJ}$.
b See text.

11.2 a $mc\Delta T \approx 0.6\,\mathrm{MJ}$ (heat losses from pot imply actual requirement is higher). $\eta \approx 3\%$.
b 70 tonnes; 7 ha.

11.3 a (i) $3\,\mathrm{m^3}$ gas (ii) 63 MJ (iii) from 1.7 litre kerosene.
b (i) Smaller tank, smaller cost. (ii) Heat required $6\,\mathrm{MJ\,day^{-1}}$. (iii) Heat evolved: 0.3 MJ/(mole sucrose) $= 3.6\,\mathrm{MJ\,day^{-1}}$.

11.4 a 680 litre at 100% yield.
b About 90% if suitable machinery available.

11.5 Note that the oven dry mass remains the same throughout. (o) 1000 kg, 6.3 GJ, 6.3 MJ/(wet kg) (i) 800 kg, 6.8 GJ, 8.5 MJ/(wet kg) (ii) 500 kg, 7.45 GJ, 14.9 MJ/(wet kg).

## Chapter 12

12.1 The wave surface is $h = a \sin kx$. Consider unit width of wave front, and a wavelength $\lambda$. Elements of water of mass $\rho\mathrm{d}x\mathrm{d}z$ are lifted a height of $2z$. The potential energy per wavelength is

$$E_{\mathrm{p},\,\lambda} = \int_{x=0}^{x=\lambda/2} \int_{z=0}^{z=h} (\rho\mathrm{d}x\mathrm{d}z)g(2z)$$

$$= \rho g \int_{x=0}^{x=\lambda/2} h^2\mathrm{d}x$$

$$= \rho g a^2 \int_{x=0}^{x=\lambda/2} (1/2)(1-\cos 2kx)\mathrm{d}x$$

$$= \rho g a^2 \lambda/4$$

The potential energy per unit length is therefore

$$E_{\mathrm{p}} = \rho g a^2/4$$

which equals the kinetic energy (12.21).

12.2 (b) Possibilities: Connect many ducks along a common axis (but it will still need some flexibility); Non-return valves in the water "bearing".

12.3 Forward push if $F_{\mathrm{L}} \sin\phi - F_{\mathrm{L}} \cos\phi > 0$ in the notation of Figure 9.14.

## Chapter 13

13.1 a The centrifugal force about O is $F_{OZ} = m(L'^2 + r^2)^{1/2}\omega^2$. Resolve along EZ, the radial (vertical) component

$$F_{EZ} = F_{OZ} \cos < ZOE = mr\omega^2$$

b From (13.6)

$$F_t = F_X - F_{EZ} = mr\omega^2 2L'/D$$

With (13.4)

$$F_t = (GMm/D^2)(2r/D) = 2MmGr/D^3$$

c The difference in gravitational attraction from the Earth between $r$ and $(r+R)$ is $2M'mR/r^3$. Equating this to $F_t$,

$$R = (M/M')(r^4/D^3)$$

Substituting the data from Figure 13.1, $R = 0.36\,\text{m}$.

13.2 In figure 13.3(a) point A can be taken to represent the Moon revolving around the Earth E. In (13.10) we replace $t^*$ by $T^*$, $t_s$ by $T_m$, and hence

$$T_m = \frac{T^*}{1-(T^*/T_s)}$$
$$= \frac{27.32t_s}{1-(27.32t_s)/(365.256t)_s)} = 29.53t_s$$

13.3 a
$$u = \surd(gh) = \surd(9.8\,\text{m}\,\text{s}^{-2})(4400\,\text{m})$$
$$= 210\,\text{m}\,\text{s}^{-1} = 760\,\text{km}\,\text{h}^{-1}$$

b
$$v = 2\pi r/t_s = (2\pi)(6.38\times10^6\,\text{m})/(86\,400\,\text{s}) = 464\,\text{m}\,\text{s}^{-1}$$
$$= 1670\,\text{km}\,\text{h}^{-1}$$

c The tidal wave cannot keep up with the tidal forcing function and so the equilibrium tide never materialises.

13.4 a $v = \surd(gh) = [(10\,\text{m}\,\text{s}^{-2})(4400\,\text{m})]^{1/2} = 210\,\text{m}\,\text{s}^{-1}$

b $t = 2\pi r/v = 2\pi(64\times10^6\,\text{m})/(210\,\text{m}\,\text{s}^{-1}) = 53\,\text{h}$ (NB $53\,\text{h} \gg 24\,\text{h}$)

c The freely travelling tidal wave propagates at less than half the speed necessary for continual reinforcement by the Moon's tidal influence.

Thus each ocean basin tends to have independent tidal properties from tides that dissipate their energy in shallow water and do not couple with neighbouring oceans.

13.5 Consider a mass $M$ of pumped water. Input energy to pump 1.0 m at 50% efficiency is

$$2 \times [Mg(0.5\,\text{m})] = (1.0\,\text{m})Mg$$

Output energy at low tide is

$$1/2[Mg(5.5\,\text{m})] = 2.7\,\text{m}\ Mg$$

Energy gain/energy input is $(2.7 - 1)/1 = 1.7$.

## Chapter 14

14.1 $A_{\text{wf}} = 19 \times 10^3\,\text{m}^2$, $Q = 5\,\text{m}^3\,\text{s}^{-1}$, $\mathcal{R} = 1.1 \times 10^5$, $u = 2.2\,\text{m}\,\text{s}^{-1}$, $n = 1200$, $l = 100\,\text{m}$

14.2 $f = 0.007$ (when clean), $P_{\text{f}} = 35\,\text{kW}$

14.3 $\eta_{\text{Carnot}}/4$

14.4 $P \propto (T_1 - T_2)^2/T_1$. Hence $\partial P/\partial T_2 = -(T_1 - T_2)/T_1$
$\Delta P/(1\,°\text{C}) \approx 20/300 \approx 7\%$.

## Chapter 15

15.1 a Balancing heat flow through the surface material $kAT/d$ against heat lost by granite mass $\rho Vc\text{d}T/\text{d}t$ implies $\text{d}T/\text{d}t = [kA/pVc]T/d$, so that $T = T_{\text{o}}\text{e}^{-t/\tau}$ where $\tau = phcd/k$

b $\tau \approx 5 \times 10^{11}\,\text{s} \approx 10\,000\,\text{y}$

15.2 a Mass of granite $= (5 \times 10^9\,\text{m}^3)(2700\,\text{kg}\,\text{m}^{-3}) = 13.5 \times 10^{12}\,\text{kg}$
Mass of $^{238}\text{U} = (4 \times 10^{-5})(13.5 \times 10^{12}\,\text{kg}) = 54.0 \times 10^7\,\text{kg}$

$$\text{Heat produced} = \frac{(54 \times 10^7\,\text{kg})(3000\,\text{J}\,\text{kg}^{-1})}{(3.1 \times 10^7\,\text{s})} = 50\,\text{kW}$$

b Total thermal power produced $= 50/0.4\,\text{kW} = 120\,\text{kW}$. This is insignificant compared with geothermal heat power extraction of $\sim$100 MW (thermal) from such a mass of dry rock.

## Chapter 16

16.1 3800 K (assuming $\Delta H$ etc. independent of $T$)

6.2 a 20 MJ
b 16 min.

16.3 0.5 MJ kg$^{-1}$; $1.8 \times 10^3$ rad s$^{-1}$ = 17 000 rpm

16.4 a 2 TOE y$^{-1}$ × 50 M people = 150 GW
b Even, 60 GW; manufacturing etc.
c 14 kW per household. No. household really ~ 1 kW.
d Use Figure 4.16. 25 m$^2$ per house in summer 360 m$^2$ in winter. Possible, yet unlikely without the added use of passive solar design in substantial quantities of new housing stock.
e 90 W m$^2$. If blade diameter = 100 m, need about 30 000 turbines to yield 30 GW. Area for each = $2 \times (2 \times 10^{11}$ m$^2$)/30 000. Average spacing about 3 km.
f 400 km at 70 kW m$^{-1}$. Fits.

*Note*: These 'ballpark' answers show that renewable energy supplies are of significant potential, even for a country with independent supplies of oil and coal.

16.5 0.4 MJ m$^{-3}$

16.6 See Table 16.1.

16.7 a 11 kg s$^{-1}$
b 540 MW
c $1.3 \times 10^6$ m$^3$ day$^{-1}$

16.8 For copper wire, total cross-sectional area ~ 1500 mm$^2$ (e.g. four wires each 22 mm diameter).

16.9 $s = 0.1 = (50\,\text{Hz} - 6f_s)/50\,\text{Hz}$, so $f_s = 55/6\,\text{Hz} = 9.2\,\text{Hz}$; $f_2 = f_1 - nf_s = 5\,\text{Hz}$.

## chapter 17

17.1 13 years. (*Hint*: Complete a spreadsheet as Table 17.2)

17.2 (a) 9.0¢ kWh$^{-1}$ (b) 14.4 (c) 16.2¢ kWh$^{-1}$

17.4 About 1.0 kg $CO_2$/[kW$_e$h(electricity)].

# Index